2004

Numerical Analysis

Numerical Analysis

A Mathematical Introduction

MICHELLE SCHATZMAN

Directeur de Recherche au CNRS
Université Claude Bernard
Lyon

Translated by
JOHN TAYLOR
Edinburgh Petroleum Services

CLARENDON PRESS · OXFORD
2002

OXFORD
UNIVERSITY PRESS

Great Clarendon Street, Oxford OX2 6DP

Oxford University Press is a department of the University of Oxford.
It furthers the University's objective of excellence in research, scholarship,
and education by publishing worldwide in

Oxford New York

Auckland Bangkok Buenos Aires Cape Town Chennai
Dar es Salaam Delhi Hong Kong Istanbul Karachi Kolkata
Kuala Lumpur Madrid Melbourne Mexico City Mumbai Nairobi
São Paulo Shanghai Taipei Tokyo Toronto

with an associated company in Berlin

Published in the United States
by Oxford University Press Inc., New York

Translation of Analyse Numérique by Michelle Schatzman
originally published in French by InterEditions, Paris 1991

Aidé par le ministère français chargé de la culture

First published in English 2002

A catalogue record for this book is available from the British Library

Library of Congress Cataloging in Publication Data

ISBN 0 19 850852 2 (Paperback)

ISBN 0 19 850279 6 (Hardback)

Typeset using the translator's files by Julie Harris

Printed in Great Britain
on acid-free paper by
Biddles Ltd,
Guildford & King's Lynn

Preface

Ne sois pas lascif et peureux
Comme le lièvre et l'amoureux.
Mais que toujours ton cerveau soit
la hase pleine qui conçoit.[1]

Guillaume Apollinaire, *Bestiaire* (1920)

Fairy tales

Fair reader, the preface of a mathematical book is the right place for telling how it came about, and maybe spawn a few legends.

Once upon a time, in a far away country[2], there was a young research mathematician who specialized in nonlinear partial differential equations. In 1984, this young mathematician was promoted to professor of numerical analysis, a subject that she mostly ignored. So, in order to teach, she learnt, a fact already known from the Talmud: 'Rav Hanina said, "Much have I learnt from my masters, more from my colleagues, but the most from my own students"' (*Talmud of Babylon*, Tractate Taanit, 6)[3].

You know how a fairy tale is composed: the heroine has to fall in love: I fell in love with numerical analysis[4]. And here is the result, or rather the second iteration of the result, since a first edition appeared in French in 1991.

Sönke Adlung, who is an editor with Oxford University Press, and to whom I had been introduced by John Ball, thought that it would be nice to have this book translated and also revised.

The heroine must go through hard times and meet a few fire-spitting dragons on her way, which would change her into stone just for the fun of it. I was moving

---

[1]Be not lewd and fearful/ as the hare and the love-fool./ But let your brain ever be/ the hare-doe that conceives.

[2]We have to stick to the traditional format of fairy tales.

[3]אמר רב חנינא: הרבה למדתי מרבותי, ומחבירי יותר מרבותי, ומתלמידי יותר מכולן. (תלמוד בבלי, תענית ו.)

[4]Some would consider that bad taste; theirs is the loss.

so slowly at times that Sönke may well have thought that I had been turned into a piece of rock.

Since I had agreed to update the 1991 edition, I had to rewrite parts of it, and add some material. This did not turn out to be easy. One reason is that my extra-scientific activities graduated from raising two children[5] to chairing a research group of about 50 people, a task which includes dealing with five different administrations. Another reason is that I am still a research mathematician, writing papers, advising students, and doing whatever the form of life called mathematics demands. My natural tendency is to consider that the most important mathematics are tomorrow's, not yesterday's or yesteryear's.

Though I should not admit it, I may have also fallen in love with my research group, MAPLY (Laboratoire de Mathématiques Appliquées de Lyon), and with its future; when I was born, there were trees; so now, I feel obligated to plant some. I did meet quite a few dragons in the forest[6]. This being a legend, fair reader, you have to remember that dragons are in the mind of the beholder.

Acknowledgements

The heroine needs to receive help from many quarters, and so did I.

Sönke found a translator, John Taylor. John, you did a pretty good job, keeping as much as you could of the colloquialism of the original French style.

Thank you, Sönke, for the idea which enabled me to conclude the task: that I should find myself someone who would not be so busy, who would be able to spot typos and mixed up indices[7], and who would be knowledgeable enough to criticize or praise wherever applicable.

And this is how Jean-François Coulombel came in: in February 2000, he was sitting in one of my graduate courses, and one month later, he was bravely starting to push me forward, so that the slow motion would not be so slow. Thank you, Jean-François, you did well, and that must not have been easy for you.

And thank you again, Sönke, for never relenting before my procrastination.

In the final runs, I shared an office with Stéphane Descombes in a faraway

---

[5] Claude and René, you have built me.

[6]
   Nel mezzo del cammin di nostra vita
   mi ritrovai per une selva oscura
   che la diritta via era smarrita.

(In the middle of our life's path/ I found myself in dark woods,/ the clear direct way being lost.)
Dante, *La Divina Commedia*

[7] I guess that I became a mathematician because I have so much trouble with $+$ and $-$, and with i, j, k, l, and m. And n too; so if there are still some errors, I am the one who put them in, and I apologize.

country[8], and he was kind enough to provide a stern examination of the spline chapter. I also benefited from a rereading of the multistep chapter by Magali Ribot, who had a first-hand opportunity to learn about the fallibility of some thesis advisors.

I have had many sources and I have been influenced in person or in print by the following authors: K. E. Atkinson [5], J. C. Butcher [13], P. G. Ciarlet [16], M. Crouzeix and A. L. Mignot [19], C. de Boor [20], G. H. Golub and C. F. van Loan [35], E. Hairer, S. P. Nørsett, and G. Wanner [43,42], P. Henrici [45], E. Isaacson and H. B. Keller [51], A. Iserles [52], D. E. Knuth [38,53,54], H.-O. Kreiss [56], Y. Meyer [61], A. Ralston and P. Rabinowitz [68], R. D. Richtmyer and K. W. Morton [70], L. L. Schumaker [71], J. Stoer and R. Bulirsch [73], and H. S. Wilf [78]. To these, I am deeply indebted.

I have now two institutions to thank: CNRS gave me the initial help in my research career; since 1995, it gave me the charge of planting mathematical trees; it told me, in very direct language, that one has to go forward and take responsibility, or get out of the way; it gave me a temporary research position which enabled me to remain a mathematician instead of turning into a full-time administrator. Finally, it gave me a permanent research position.

The other institution is the Technion in Haifa, where I have been a frequent visitor since 1994. Some of the funding came from the binational Keshet/Arc-en-ciel binational agreement; I also got funding from the CNRS–MOSA. So, twice a year, I have some continuous time to myself, with only a few e-mails and faxes for French business. There is this wonderful library, from which I take out books on the account of my friend and colleague Koby Rubinstein, who from time to time gets a message telling him to return some. And the friends that I have there kindly let me be as bearish as I want; their hospitality is a blessing; I owe much to the kindness of Iris and Yehuda Pinchover.

Finally, in a fairy tale, the heroine has the use of some life-saving tricks, spells or magic formulae. They are called TEX, $\mathcal{A}_{\mathcal{M}}\mathcal{S}$-LATEX, SCILAB, GNUPLOT, and XFIG; they are all free software, and they are wonderful tools without which this book would simply not exist.

Contents: a subjective approach

Since this preface is so romantic, let me mention that love for a scientific subject is exactly the same as love for a human being: you always hate quite a few things in the loved one, but the balance looks good enough to keep the attachment going.

So, let me talk subjectively about my subject.

Looks are so important, and, alas, numerical analysis with all its burden of notations looks more like a heavy matron than a gracious ballerina. So what? Maybe a heavy matron can take care of a large brood of children. Better a heavy

---

[8]Israel, according to the age-old joke: this is far? far from where?

theory with many applications than an elegant one without offspring.

Numerical analysis has to do with real life computation: an understanding of the floating number system and the machine arithmetic is essential. Chapter 1 gives an exercises guided tour to this world. It is highly recommended that you do the exercises.

I observed that numerical analysis requires more maturity from mathematics students than other subjects at the same level of difficulty. A naïve vision puts numerical analysis between two stools: the physicist or the engineer wants methods, and is satisfied with experimental numerics; the mathematician wants beautiful problems, and is not much interested in constructive solutions using a fallible and limited machine. However, I believe that numerical analysis sits on both stools, and has the best of both worlds: motivated problems where getting a solution and getting it fast can make a difference, and tools which can be, at the same time, elementary and powerful: not necessarily a contradiction according to the words of S. S. Abhyankar [1]. Chapter 2 gives a flavour of numerical analysis, constructing the logarithm and the exponential from scratch, using methods which are the daily bread of numerical analysis, and which moreover generalize to situations where power series do not work.

The daily life of numerical analysts includes much linear algebra; Chapter 3 summarizes some of the required knowledge, and adds to it the theory of block matrices.

This makes up Part I. Part II describes polynomial approximation and piecewise polynomial approximation, in algebraic or trigonometric versions: interpolation and divided differences in Chapter 4, least-squares approximation in Chapter 5, and splines in Chapter 6. The recent surge of the use of splines in computer-aided geometrical design and image processing is one of the motivations for the spline chapter. It also turns out that splines are a nice generalization of Bernstein polynomials, and that they fit very well with two approaches: divided differences and convex algorithms. I used to hate splines, a baseless and despicable prejudice. I hate them no longer.

Chapter 7 is on Fourier series in one space dimension; it does not tackle any of the hard questions of which Fourier analysis is so replete. However, it treats easy and essential questions, including convolution and regularization, and it makes room for the Gibbs phenomenon, so important in applications.

Chapter 8 is about quadrature: approximation by algebraic polynomials leads to the classical formulae for numerical integration; trigonometric approximation leads to the Euler–MacLaurin formula and to the fast Fourier transform (FFT), which may be the most important algorithm in scientific computation. The FFT is the prototype of recursive algorithms; it is the ancestor of multigrid and wavelet algorithms; it is the epitome of easy and powerful tools.

Part III relates to numerical linear algebra. This part is important because operation counts are the limiting factor for any serious computation. Any scientific computation program spends most of its time solving linear systems or approximating the solution of linear systems, even when trying to solve nonlinear systems.

Chapter 9 provides the direct methods for the resolution of linear systems of equations, with an emphasis on operation counts. Operation counts justify all of the acrobatics of iterative methods, treated in Chapter 11 after extra information on linear algebra is produced in Chapter 10. Chapter 12 relates orthogonality methods to the resolution of linear systems and introduces the QR decomposition.

Part IV treats a selection of nonlinear or complex problems: resolution of linear equations and systems in Chapter 14, ordinary differential equations in Chapter 15, single-step schemes in Chapter 16 and multistep schemes in Chapter 17, and introduction to partial differential equations in Section 18.4. It would have been my natural tendency to put in much more of these, but, since I decided that this book would avoid any functional spaces beyond spaces of continuous functions or Lebesgue spaces and, in particular, Sobolev spaces, there was little possibility to include more than a tiny introduction to partial differential equations. I tried to select a few important things which are accessible and attractive on an elementary level. I did the same for the ordinary differential equations part: I skipped the detailed analysis of the Runge–Kutta methods by trees, because it is long and difficult; however, I have given a full theory of the analysis and convergence of multistep schemes, because the use of appropriate norms makes it possible without tears.

This book started as an elementary book; the revision put in some more advanced layers, but the layered structure remains; the less elementary parts are Sections 4.4 and 4.5, Chapter 6, Section 7.2, Section 8.6, and Chapters 17 and 18.

A number of problems describe some classical algorithms together with some newer ones. Since mathematics is not a spectator sport, the more advanced parts are put into problems: the most exciting things can be found there.

The prerequisites are linear algebra, calculus, and a tiny bit of Lebesgue theory, which is used only in Chapter 5 on polynomial least-squares approximation, Chapter 7 on Fourier analysis, and Chapter 18 which introduces partial differential equations. I do not use the theory of distributions, though I disguise some of its ideas in the spline Chapter 6.

Just a short word about notation: I decided not to use bold face for matrices or vectors, with very few exceptions. The reason is that I very often use block decomposition of matrices; if I decompose an $n \times n$ matrix into an $(n-1) \times (n-1)$, an $(n-1) \times 1$, a $1 \times (n-1)$, and a 1×1 block, what notation would make sense? I could not imagine an efficient answer, and so I dropped the bold faces.

Computation and numerical analysis

There is always a question about the rôle of computations and software in a numerical analysis book.

I did not include any numerical software, or even quasi-programs which enable one to write one's own software.

The first reason is that good numerical software requires quite a bit of thought to be really efficient. Some really good scientific software, such as MATLAB can be bought; but there are also free products of quality, such as SCILAB, distributed by INRIA, http//www-rocq.inria.fr/scilab, for use under several different operating systems.

The second is that one must be sceptical of scientific computations: numerical analysis is concerned with the essence of the scientific method.

But what is the foundation of such an attitude? It is conceivable if we know that the results are inaccurate, or plainly wrong. But how do we know that? Well, if we know that our mathematical equations are a good approximation of a natural phenomenon, and if the computations do not agree with the observations, then the scientific computation software must be guilty; or must it?

Finding the guilty party may be a very difficult endeavour, because many factors may be involved: maybe the equations were wrong; maybe the numerical method was inappropriate; maybe the parameters of the method were badly chosen; maybe the software was incorrect.

One of the purposes of numerical analysis is to find the specifically mathematical factors which govern the success or failure of numerical computations.

The third reason is that the successful numerical analyst destroys their own job by finding algorithms which are so clear and efficient that they can be safely implemented into software.

Therefore, in order to stay in employment, he or she must keep finding new areas where existing software does not do the job.

Mathematics provides the light which enables us to explore new territory. Mathematics is also a very cultural subject: for mathematicians, a hundred years old result can be as good as a new one, even if we do not use it for the purpose for which it had been initially crafted.

In this book, I try to explain how to make your own light, and how to find your own way with it. This is obviously much harder—for you and for me—than inviting you to contemplate nice pictures without telling you how I might have come across them and how they fit together.

Nevertheless, fair reader, do not believe that I have shyed away from computing: I have only hidden it, since showing it would have led to a completely different book.

So, in order to gain more understanding, try, dear reader, to program the most algorithms that you can think of. It may not be of the same high quality as the commercially or freely available software, but it will teach you much about the behaviour of computational methods, and about the difficulty of putting mathematics into code.

Lyon   M. S.
September 2000

Contents

Part I

The entrance fee

Most of numerical analysis uses recursive procedures: do the same thing again and again, until a reasonable degree of accuracy is reached. Even apparently algebraic problems admitting a solution in finite terms can be efficiently treated by iterative techniques, as we shall see in Chapter 11. Therefore, since numerical analysis involves a large number of machine computations, the error must be analysed.

Numerical analysts are scared stiff of a phenomenon called instability: uncontrolled amplification of error. But where does error come from? It comes from the fact that we represent real numbers in a finite system, called floating numbers; therefore, any arithmetic operation leads to loss of precision. The idiosyncrasies of floating-point operations are examined in Chapter 1.

Of course, instability is quite visible: usually one gets a computer message which says 'overflow', and at this point the computation stops. What is important is to find a cure for instability, and this requires a comprehensive understanding of the mathematical methods which have been used for the computation.

As presented here, numerical analysis is a part of mathematics, but it works on questions which are strongly related to the use of computers and to applications from other sciences. Therefore, numerical analysts create in their minds visions of mathematical objects which may be slightly different from the visions of other mathematicians. In particular, numerical analysis is about constructing or approximating effective solutions. Of course, numerical analysts do use existence theorems, and they combine constructive and non-constructive information.

I have tried to give a flavour of numerical analysis in Chapter 2 through concrete examples, which are probably well known to the reader.

The fastest way of constructing the natural logarithm is to say that it is the integral of the function $x \mapsto 1/x$ which vanishes at 1. This information can be made constructive: in order to compute the integral of $x \mapsto 1/x$, we use approximate formulae, and these formulae enable us to construct from scratch the natural logarithm. In the same fashion, we can define the exponential as the reciprocal function of the logarithm, or using its famous entire series expansion. Unfortunately, neither of these methods leads to an efficient numerical construction. A good construction method for the exponential is the original method of Euler, based on a product formula; the exponential can be constructed from scratch with this method. It is of even more interest that such product formulae are the core of numerical integration for ordinary and partial differential equations.

Chapter 3 provides a review of standard results from linear algebra. Linear algebra is pervasive in this book. We cannot do anything without linear algebra, and we need a bit more than is usually taught at the elementary level. The only not completely standard feature is block matrices.

1

Floating numbers

The beginning of a course on numerical analysis naturally includes some fairly abstract considerations of how real numbers are represented in a computer, known as floating-point representation, and incidentally on systems of counting.

To convince the reader that there is material here which is both surprising and thought-provoking, the following exercises are more effective than a long discourse. It is strongly recommended that the reader sharpens his or her mind by trying them. The level of mathematics and programming required is entirely elementary, though this does not spoil the fun of doing them.

This chapter owes an enormous debt to the paper by G.E. Forsythe [30], which presents some striking examples of the calculation difficulties linked to floating-point numbers.

1.1. Counting in base β

Exercise 1.1.1. Let β be an integer greater than 1. Show that, for every integer n greater than or equal to 1, there exists a unique integer p and integers d_i, $0 \leqslant i \leqslant p$, between 0 and $\beta - 1$ inclusive, with $d_p \neq 0$, such that

(1.1.1)
$$n = \sum_{i=0}^{p} \beta^i d_i.$$

The right-hand side of eqn (1.1.1) gives the representation of n in base β, also denoted by

$$n = \overline{d_p d_{p-1} \cdots d_1 d_0}.$$

Normally, we represent numbers in base $\beta = 10$ using the figures 0, 1, 2, 3, 4, 5, 6, 7, 8, 9, and from an early age we use the result of Exercise 1.1.1 without question, at least in base $\beta = 10$.

The choice of base 10 is linked to an anatomical peculiarity of the human species. We could also have counted in base 20, like the Mayans.

Nevertheless, in the history of humanity, counting with positional numerals is a relatively recent development which we owe to the Hindus and the Arabs. Arithmetic operations in roman numerals are very awkward. It is only thanks to positional numerals that efficient arithmetic algorithms could be developed, and these spread across Europe only from the twelfth century, finally triumphing in the eighteenth century. The reader who is interested in the history of numbers and systems of counting should consult, for example, [48] and [49], and their translation [50].

Other systems of counting have been devised by mathematicians, see, in particular, in [38, pp. 115–23].

Computers, which do not have ten fingers, count in base 2, with the figures 0 and 1 (binary or dyadic), in base 8 (octal), with the figures ˊ0 to ˊ7, and in base 16 (hexadecimal), with the figures ˝0 to ˝9, to which are added the letters ˝A to ˝F.

The list of 128 ASCII characters is a list of standard characters corresponding to codes understood by all computers.

Exercise 1.1.2. The ASCII characters are numbered from 0 to 127; the letter b has octal number ˊ142. Give its decimal and hexadecimal numbers.

What do we do with the fractional part of a number? By analogy with the representation of Exercise 1.1.1, consider expressions of the form

$$x = \sum_{i=-q}^{p} \beta^i d_i,$$

where p and q are positive integers or zero and the d_i are integers from 0 to $\beta - 1$ inclusive. To fix p, we insist that $d_p \neq 0$. We write

$$x = \overline{d_p d_{p-1} \cdots d_1 d_0 . d_{-1} \cdots d_{-q}} .$$

We can also take $q = \infty$. What does this infinite sum mean? If we know the properties of real numbers, we claim that the sequence of rationals

$$r_n = \sum_{i=-n}^{p} \beta^i d_i$$

is increasing and bounded by

$$1 + \sum_{i=0}^{p} \beta^i d_i.$$

It is therefore a convergent sequence whose limit is the sum of the series with general term $\left(\beta^i d_i\right)_{i \leqslant p}$.

Warning: this course assumes that the reader is familiar with the properties of the real numbers. It is wise to revise them before continuing.

A standard difficulty is the occurrence of a real number which can have two distinct representations in base β, as we see in the following exercise:

Exercise 1.1.3. Let $b = \beta - 1$. Show that, in every base β,

$$1 = \overline{0.bbbb\cdots}.$$

How many reals are there which have two distinct representations in base β? Few, as we will see in the exercise below:

Exercise 1.1.4. Give the general form of the reals possessing two distinct representations in base β. Show that the set of all the reals for which there exists a base β in which the real number has two distinct representations is exactly the set of the rationals.

1.2. Expansion of the rational numbers in base β

In the decimal base, the quotient of two integers 'falls exactly' or 'does not fall exactly', but after a certain point it is always periodic. The following exercises allow us to verify this periodicity result in any base β. This section forms a short problem on elementary arithmetic. Before working on the general case, we will solve the following particular case:

Exercise 1.2.1. Calculate the decimal expansion of $1/7$.

Exercise 1.2.2. Let m and n be two relatively prime integers such that $m < n$. Let $r_0 = m$, and define d_{-j} and r_{-j} iteratively as being the quotient and the remainder, respectively, of the Euclidean division of βr_{-j+1} by n:

$$\beta r_{-j+1} = nd_{-j} + r_{-j}, \quad 0 \leqslant r_{-j} < n.$$

Show that, for every $j \geqslant 1$, $0 \leqslant d_{-j} < \beta$.

Exercise 1.2.3. Show that $\overline{0.d_{-1}d_{-2}d_{-3}\cdots}$ is the expansion of m/n in base β.

Exercise 1.2.4. Show that there exist two integers k and ℓ such that r_k and r_ℓ are equal. (Argue by contradiction.)

Exercise 1.2.5. Deduce from this that the expansion of m/n in base β is periodic from a certain point. Generalize this to the case $m/n \geqslant 1$.

Exercise 1.2.6. When do the divisions m/n 'fall exactly' in base β?

Exercise 1.2.7. Give the expansion of $1/5$ in bases 2, 8, and 16. (Observe carefully the result in base 16.)

1.3. The machine representation of numbers

Several types of numbers can be represented in a computer. The machine integers are nothing special, the only point to note is that the set of integers which can be represented is finite, for example from -32768 to $+32767$. Recall that a bit of information is a binary digit, i.e., 0 or 1, and that a byte is a group of 8 bits.

Since in this example there are exactly 2^{16} different numbers, it is sufficient to have 16 bits or 2 bytes to represent them.

The floating-point numbers are more interesting. The set of these numbers is described by a base β, a number of significant figures r, and two integers m and M. Every floating-point number is of the form

$$s \, \overline{0.d_{-1}d_{-2}\cdots d_{-r}} \, \beta^j$$

with j between m and M, $1 \leqslant d_{-1} < \beta$, and, for $k > 1$, $0 \leqslant d_{-k} < \beta$. The letter s designates the sign of the number. We add the zero to the set of floating-point numbers. We thus obtain a subset $F(\beta, r, m, M)$ of $\mathbb{R}$ which is formed from numbers whose expansion in base β is finite. The expansion $\overline{0.d_{-1}d_{-2}\cdots d_{-r}}$ is the mantissa of the number and j is its exponent.

The normalization $1 \leqslant d_{-1} < \beta$ is very important. It ensures that all the figures of all the nonzero floating-point numbers are significant.

If the result of an operation is greater in absolute value than the largest floating-point number then the machine generally returns 'overflow'. If the result of a nonzero operation is rounded to zero then the machine returns 'underflow'.

Exercise 1.3.1. Forsythe's toy floating-point system: take $\beta = 2$, $r = 3$, $m = -1$, and $M = 2$. Determine all the floating-point numbers and draw them as a scale on a straight line segment centred on zero. Does $\overline{0.111}\, 2^{-1}$ belong to this system?

As we see from Exercise 1.3.1, there are many gaps between the floating-point numbers. We therefore need a rounding function A which has the following properties:

- A is defined for all $\mathbb{R}$;

- A leaves $F(\beta, r, m, M)$ invariant;

- Let $x \in \mathbb{R}$. Let $[f, f']$ be the smallest interval containing x, and whose extremities are floating-point numbers. Then $A(x)$ is equal to whichever of the numbers f and f' is closest to x;

- If x is equidistant from f and f', then $A(x)$ is determined in a variety of ways, which can be dependent on the machine.

Exercise 1.3.2. Calculate $A(1/3)$ in the toy system $F(2, 3, -1, 2)$.

Our machine can only recognize the numbers belonging to its floating-point system. We define arithmetic operations on the floating-point numbers by letting

$$f \oplus f' = A\left(f + f'\right),$$
$$f \ominus f' = A\left(f - f'\right),$$
$$f \otimes f' = A\left(ff'\right),$$
$$f \oslash f' = A\left(f/f'\right).$$

Exercise 1.3.3. What are the results of the following operations in the toy floating-point system:

$$\frac{5}{2} \oplus \frac{5}{2}, \quad \frac{5}{16} \ominus \frac{7}{16}, \quad \frac{5}{2} \otimes 3, \quad \frac{1}{4} \otimes \frac{3}{8}?$$

The program must be able to recognize the overflow and underflow situations.

Exercise 1.3.4. Still in the toy system, calculate

$$\left(3 \oplus \frac{5}{16}\right) \oplus \left(-\frac{3}{2}\right) \quad \text{and} \quad 3 \oplus \left(\frac{5}{16} \oplus \left(-\frac{3}{2}\right)\right),$$

$$\left(\frac{5}{16} \otimes 3\right) \otimes \frac{7}{8} \quad \text{and} \quad \frac{5}{16} \otimes \left(3 \otimes \frac{7}{8}\right).$$

What can we say about the algebraic properties of the operations $\oplus$ and $\otimes$?

Exercise 1.3.5. Simulate a small floating-point system in base $\beta = 10$ by using a programming language such as PASCAL or FORTRAN and making a rounding procedure.

Begin by fixing $r = 3$, $m = -3$, and $M = 5$. From this rounding procedure, program the floating-point arithmetic operations in $F(10, 3, -3, 5)$.

Return overflow and underflow when appropriate. A more elaborate version should allow the simulation of floating-point systems for various values of r, m, and M, with $\beta = 10$. Passing to any base β is more delicate, but represents an interesting exercise for those who can program.

Remark 1.3.6. The unit of speed of a computer, used principally for scientific calculations, is the *flops*, or FLoating-point OPeration per Second. We would therefore talk about a machine calculating at 100 megaflops, that is 10^8 floating-point operations per second. Traditionally, when we evaluate the efficiency of a scientific calculation algorithm, we only count the number of multiplications and divisions that it demands. It is more reasonable, with the current state of technology, to also count the additions and subtractions, since the relative time for multiplications and divisions has decreased.

In the era of parallel machines, we cannot be content with evaluating speed in flops. We must also take account of the number of processors, and note that certain algorithms use the structure of the machine more efficiently than others.

1.4. Summation of series in floating-point numbers

When we have a computer at our disposal we are tempted to calculate things that we did not know how to, or did not want to, calculate by hand. The preceding exercises have shown that operations on floating-point numbers suffer from significant arithmetic faults. We will see others which are more analytic.

Let $(u_n)_{n \geqslant 0}$ be a sequence of reals. We let

$$\Sigma_0 = A(u_0) \quad \text{and} \quad \Sigma_n = \Sigma_{n-1} \oplus A(u_n).$$

This is therefore the sum of the numeric series whose general term is u_n.

Exercise 1.4.1. Show that if u_n tends to 0 as n tends to infinity, and if the sums Σ_n stay below the level of overflow, the sequence of Σ_n is stationary from a certain point.

Exercise 1.4.2. We let $u_n = 1/n$ (harmonic series). Is this numeric series convergent? Show that the partial sum of this series is equivalent to $\ln n$.

Exercise 1.4.3. Show that, for $n > \beta^r$, the sequence of floating-point partial sums of the harmonic series is stationary. Find an upper bound for the partial sum thus obtained.

Exercise 1.4.4. Calculate the sum of the harmonic series working in the sets $F(10, 3, -3, 5)$ and in $F(10, 5, -5, 5)$. This is the moment when the sequence of Σ_n is stationary.

Although all the series with terms tending to zero are convergent in a machine, their machine sum depends considerably on the floating-point system used. We would expect that a series with terms tending to zero is divergent if its sum depends on the machine.

Exercise 1.4.5. The partial sums of a divergent series with positive terms exceeds, from a certain point, every given positive number. What must the minimum size of the mantissa be in base 10 so that the partial sums of the harmonic series exceed 100? On a computer doing 10^9 flops how much time, in years, is required to do this? (Take 1 year $= 3 \times 10^7$ seconds and make suitable approximations.)

Another peculiarity of the floating-point numbers is calculation instability. Recall that the series

$$e^x = \sum_{j=0}^{\infty} \frac{x^j}{j!}$$

is convergent for all $x \in \mathbb{R}$.

Exercise 1.4.6. Program: calculate e^5 and e^{-5}, first using the exponential function of the chosen scientific programming language, and then in the floating-point numbers $F(10, 3, -3, 5)$. Explain the difference between the relative error for e^5 and the relative error for e^{-5}. What happens if we reverse the summation order of the partial sums for e^{-5}? What happens if we sum the positive and negative terms separately?

1.5. Even the obvious problems are rotten

Consider the recurrence relation

(1.5.1) $u_{n+1} = (q + 1)\, u_n - p.$

Exercise 1.5.1. Verify that the sequence, whose general term is the constant $u_n = p/q$, is a solution of eqn (1.5.1). Take $q = 3$ and $p = 1$ or $p = 2$, and program the above recurrence. What do you observe?

Exercise 1.5.2. Take $q = 4$ and $p = 1$, 2, or 3, and do the same calculation. Are the phenomena the same?

Exercise 1.5.3. Give an interpretation of the different behaviours observed.

1.6. Even the easy problems are hard

The case of solving the second degree equation

$$(1.6.1) \qquad ax^2 + bx + c = 0$$

allows us to see the difficulties linked to the orders of magnitude of the numbers with which we are working, and how the simplest nonlinear formulae can tie us up in knots.

The roots of eqn (1.6.1) are given by the formulae

$$(1.6.2) \qquad x_1 = \frac{-b + \sqrt{b^2 - 4ac}}{2a} \quad \text{and} \quad x_2 = \frac{-b - \sqrt{b^2 - 4ac}}{2a}.$$

Exercise 1.6.1. Write a program giving the roots of eqn (1.6.1), including the complex case and a test to ensure that a is not zero.

Exercise 1.6.2. Let r be the maximum number of decimals representable in floating-point, and q the integer part of $1 + (r/2)$. We take

$$a = 1, \quad b = -10^q, \quad c = 1.$$

Calculate the roots of eqn (1.6.1). Also calculate them by machine and compare the result.

Exercise 1.6.3. To avoid the above difficulty, we can write the formulae (1.6.2) a little differently by noting that

$$\sqrt{y} - \sqrt{z} = \frac{y - z}{\sqrt{y} + \sqrt{z}}.$$

Write the corresponding new formulae, which the program will apply if it detects the need after having made a test.

Apply the new program to the following choices of coefficients, denoting the exponent of the largest power of 10 which can be represented by the machine by M, and denoting the integer part of $1 + (M/2)$ by n:

	$a = 6,$	$b = 5,$	$c = -4,$
(1.6.3)	$a = 6 \times 10^n,$	$b = 5 \times 10^n,$	$c = -4 \times 10^n,$
(1.6.4)	$a = 10^{-n},$	$b = -10^n,$	$c = 10^n,$
(1.6.5)	$a = 1,$	$b = -2,$	$c = 1 - 10^{-r-1}.$

Note that the case (1.6.3) can be solved by using a suitable scale. For the case (1.6.4) it is necessary to use more drastic means, for example, the change of variable $y = 1/x$. Provide tests and modifications of the program to allow the solution of these problems.

Denote the smallest root of the equation

(1.6.6) $x^2 - 2x + 1 = \epsilon$

by $x_1(\epsilon)$ and the largest root by $x_2(\epsilon)$. Calculate the derivatives of x_1 and x_2 with respect to ϵ, for $\epsilon > 0$. Explain the phenomena observed and describe the remedies that can be employed.

1.7. A floating conclusion

If the floating-point numbers have so many faults, why not calculate with other representations of the numbers? In certain cases, it is wise to use exact representations, such as rational numbers, or, more generally, a representation in some suitable field or ring. For applications arising directly from engineering science or from nature, such representations are often not convenient since they assume precise knowledge of the numerical data for the problem. In general, this knowledge is not accessible. There are, however, less obvious applications: if we seek the coefficients of a series that we know are rational *a priori*, it is natural to use an exact representation, that is, the rationals. More generally, formal calculation tools allow us to calculate the derivatives, and sometimes the integrals, of functions for which we possess an explicit expression, to find the explicit solutions of differential equations, and to give a large palette of tools which are later available for use in the heart of a scientific calculation program (it is possible for some formal calculation software to produce a procedure in FORTRAN, C, or some other language). It is particularly interesting if the program relies on complex formulae which are difficult to check. In this case, the symbolic calculation tool can be remarkably effective, when it is reliable.

But there is always the other side of the coin: the use of symbolic manipulation software leads to extremely complex calculations more rapidly than the use of scientific calculation software, and is occasionally totally infeasible. Furthermore, this software is still new, and clearly less reliable than scientific calculation software. If, for example, we ask a formal calculation program to take the integral of a function which has distinct forms according to the interval considered, the program is frequently susceptible to giving the wrong answer. Similarly, formal calculation programs are often bad at simplifying complicated expressions. Finally, everyone who leaves the beaten track requires tailor-made code and there exist far fewer libraries of coherent programs than for scientific programming.

For a numerical analyst, floating-point representation is the devil that we know. Formal calculation software is the devil that we know less well. Whatever happens, with these two devils, it is necessary to proceed with caution.

2

A flavour of numerical analysis

Teaching numerical analysis, or, more generally, the analysis of calculation procedures, often has the reputation of being pointless.

Chapter 1 will, I hope, have convinced the reader that it is useful to think about numerical methods. The present chapter introduces some mathematical techniques which are extremely common in numerical analysis.

Since mathematics is certainly not a spectator sport, but more an activity where one only acquires skill and strength by doing it, studying proofs, and solving exercises, the format of Chapter 2 is similar to that of Chapter 1. It therefore consists of a self-guided visit in the garden of approximations of the continuous by the discrete.

We place the discrete in the care of Don Knuth [38], and the continuous in that of the great Euler [25].

There are no surprising results in this chapter, just the comparison of powers and exponentials, and the construction of the logarithm and exponential functions. It allows us to sum up results which are generally known to the reader, but by using an approach which is independent of the classical results of elementary real analysis. This approach rests on simple arithmetic identities and inequalities, and has a large place in familiar numerical analysis procedures, but is little used elsewhere.

We must not be afraid of throwing ourselves 'in at the deep end' of indices and limits of sequences of functions. At the beginning, we will, perhaps, not feel completely at ease, but as we go along and use these tools, it will become beautiful and natural.

2.1. Comparison of exponentials and powers

There are numerous methods for proving the comparison theorems between exponentials and powers. The first exercises of this project will allow the reader to obtain these comparisons in a totally elementary manner, using only algebraic identities and simple inequalities. The results are eminently classical; the proofs guided by these exercises are perhaps not.

We begin with some elementary identities.

Exercise 2.1.1. Let m be a strictly positive integer. Let (using notation from [38])

$$(2.1.1) \qquad y^{\overline{m}} = y\,(y+1)\cdots(y+m-1)\,, \quad y^{\overline{0}} = 1.$$

Verify the following identities:

$$(2.1.2) \qquad y^{\overline{m}} - (y-1)^{\overline{m}} = m y^{\overline{m-1}},$$

$$(2.1.3) \qquad (y+z)^{\overline{m}} = \sum_{j=0}^{m} C_m^j\, y^{\overline{m-j}}\, z^{\overline{j}}.$$

Hint: for identity (2.1.3), let

$$F(y,z,m) = (y+z)^{\overline{m}} - \sum_{j=0}^{m} C_m^j\, y^{\overline{m-j}}\, z^{\overline{j}}.$$

Calculate the difference $F(y,z,m) - F(y,z,m-1)$ with the aid of identity (2.1.2) and conclude by means of a recurrence on m.

Exercise 2.1.2. Let x be a number (rational, real, or complex) which is different from 1. Calculate the sum

$$S(x,m,n) = \sum_{k=1}^{n} x^k\, k^{\overline{m}},$$

using the fact that

$$k^{\overline{m}} = m \sum_{j=0}^{k} j^{\overline{m-1}}$$

and changing the order of the summations. Deduce from this calculation that, if $0 \leqslant x < 1$, then

$$0 \leqslant S(x,m,n) \leqslant \frac{m!}{(1-x)^{m+1}}$$

and

$$(2.1.4) \qquad 0 \leqslant n^{\overline{m}}\, x^{n+1} \leqslant \frac{m!}{(1-x)^m}.$$

Exercise 2.1.3. Show that for every $y \in]x, 1[$ there exists a number C such that, for every $n \geqslant 0$, we have the estimate

$$n^{\overline{m}} x^n \leqslant C y^n.$$

Show, by using eqn (2.1.3), that there exists a C' such that, for every $n \geqslant 0$ and for every $N > n$,

$$(2.1.5) \qquad \sum_{k=n+1}^{N} x^k k^{\overline{m}} \leqslant C' (n+1)^{\overline{m}} x^{n+1}.$$

Use the formula found in Exercise 2.1.2 and inequality (2.1.5) to calculate the limit of $S(x, m, n)$ as n tends to infinity. What can we say if x is a complex number of modulus strictly less that 1?

The preceding exercises thus show, in an elementary way, that increasing exponentials dominate all power functions.

2.2. Convergence and divergence of classic series

The divergence of the harmonic series is a well-known fact, but generally proved by comparison with the logarithm function. There follows a proof of this divergence which is entirely independent of all knowledge of logarithms:

Let

$$(2.2.1) \qquad H_n = 1 + \frac{1}{2} + \ldots + \frac{1}{n}.$$

Exercise 2.2.1. Show that, for every $n \geqslant 1$,

$$H_{2^n} - H_{2^{n-1}} \geqslant \frac{1}{2}.$$

Deduce the divergence of the harmonic series.

The same technique can be used, suitably modified, to show convergence:

Exercise 2.2.2. Show that, for every $\alpha > 1$, the series

$$\sum_{n=1}^{\infty} n^{-\alpha}$$

converges.

Hint: do not delimit the packets exactly as in the preceding exercise.

2.3. Discrete approximation of the logarithm

This section gives a construction of the natural logarithm from scratch; we do not suppose that the reader has never seen a natural logarithm before, but we wish to lead him or her to explore known ground with new eyes.

We call the largest integer less than or equal to the real x the 'floor' of x and denote it by $\lfloor x \rfloor$. We call the smallest integer greater than or equal to x the 'ceiling' of x and denote it by $\lceil x \rceil$. These ideas differ from the widely known idea of the integer part, but are often rather easier to use. Recommendation: draw graphs of the floor, ceiling and integer part functions to see how they are different.

We define the following functions for $x \geqslant 1$:

(2.3.1) $$\underline{L}(x,n) = H_{\lceil nx \rceil} - H_n,$$

(2.3.2) $$\overline{L}(x,n) = H_{\lfloor nx \rfloor} - H_{n-1}.$$

Exercise 2.3.1. Show that $\underline{L}(x,n)$ is strictly positive if $x > 1$ and $n \geqslant n_0(x) = \lceil (x-1)^{-1} \rceil$, and that $\overline{L}(x,n)$ is strictly positive if $x \geqslant 1$ and $n \geqslant 2$.

Exercise 2.3.2. Show that, for every real $x > 1$ and every sufficiently large integer n, we have the inequalities

$$\underline{L}(x,n) \leqslant \underline{L}(x,2n) \leqslant \overline{L}(x,2n) \leqslant \overline{L}(x,n).$$

Exercise 2.3.3. Show that the limits

$$\lim_{n \to \infty} \underline{L}(x,2^n) \quad \text{and} \quad \lim_{n \to \infty} \overline{L}(x,2^n)$$

exist and are equal to a function of x, which we will denote by $L(x)$.

Remark 2.3.4. Some simple numerical experiments with $x = \sqrt{2}$ will convince the reader that the sequence of numbers $(\underline{L}(x,n))_{n \geqslant n_0(x)}$ is not monotonic and that, consequently, its convergence is not obvious. On the other hand, we have just shown that the sequence of numbers $(\underline{L}(x,2^n))_{2^n \geqslant n_0(x)}$ is monotonic. We have thus extracted a subsequence whose convergence is easy to show. There are a lot of procedures for extracting subsequences. Here we have used an arithmetic argument but we often call on a compactness argument. It would be a shame to be happy with only demonstrating the convergence of a subsequence: we will therefore show the convergence of the two sequences $(\underline{L}(x,n))_{n \geqslant 1}$ and $(\overline{L}(x,n))_{n \geqslant 1}$, beginning with the convergence of $(\underline{L}(x,n))_{n \geqslant 1}$ when x is an integer.

Exercise 2.3.5. Show that, if p is an integer which is greater than or equal to 2, then for every $n \geqslant 1$

$$\underline{L}(p,n) \leqslant \underline{L}(p,n+1).$$

Deduce from this that, for every integer p which is greater than or equal to 2, the sequence $(\underline{L}(p,n))_n$ converges to $L(p)$.

What is the value of $L(1)$?

The function L should be the Napierian logarithm. It is therefore necessary to verify that it has the additive property

(2.3.3) $$L(xy) = L(x) + L(y).$$

We will first verify this on the integers:

Exercise 2.3.6. Show that, for all integers p, q, and n greater than or equal to 2,

$$\underline{L}(pq, n) = \underline{L}(p, n) + \underline{L}(q, np).$$

Deduce from this that $L(pq) = L(p) + L(q)$, for all integers p and q greater than or equal to 1.

This property is immediately generalized as follows:

Exercise 2.3.7. Let x be a dyadic number with finite expansion: $x = 2^{-l}r > 1$. Show that, for every integer $p \geq 1$ and for every integer $n \geq l$, we have

$$\underline{L}(xp, 2^n) = \underline{L}(x, 2^n) + \underline{L}(p, 2^{n-l}r),$$

and deduce from Exercise 2.3.5 that $L(px) = L(p) + L(x)$.

Then show that $L(x) = L(r) - lL(2)$ and, therefore, for all dyadic numbers x and y which are greater than or equal to 1 and have a finite expansion, L has the additive property (2.3.3).

The function L has the additive property when it acts on dyadic numbers of finite expansion. How does it behave on reals greater than or equal to 1? A density and continuity argument gives the answer:

Exercise 2.3.8. Show that, for all x and y such that $1 \leq x < y$, and every integer $n \geq n_0(x)$,

$$\frac{\lfloor ny \rfloor - \lfloor nx \rfloor}{\lfloor ny \rfloor} \leq \underline{L}(y, n) - \underline{L}(x, n) \leq \frac{\lfloor ny \rfloor - \lfloor nx \rfloor}{\lfloor nx \rfloor}.$$

From this deduce the inequalities

$$(2.3.4) \qquad \frac{y - x}{y} \leq L(y) - L(x) \leq \frac{y - x}{x}.$$

Show that eqn (2.3.3) holds for all reals x and y greater than or equal to 1, by combining the continuity relation (2.3.4) and the additive property (2.3.3) already proved for dyadic numbers.

Show that the function L is differentiable on $[1, +\infty)$. What is the value of its derivative?

Now we are ready to show that every sequence $\underline{L}(x, n)$ converges to its limit:

Exercise 2.3.9. Deduce from the inequalities (2.3.4) that

$$\frac{1}{k + 1} \leq L(k + 1) - L(k) \leq \frac{1}{k}$$

and that, consequently,

$$L(\lfloor xn \rfloor + 1) - L(n + 1) \leq H_{\lfloor xn \rfloor} - H_n \leq L(\lfloor xn \rfloor) - L(n).$$

Use the properties (2.3.3) and (2.3.4), and the definition of the floor function to show that

$$L(x) - \frac{1}{n} \leqslant \underline{L}(x,n) \leqslant L(x).$$

What can we say about $\overline{L}(x,n)$?

We have momentarily forgotten the logarithms of numbers less than 1, but they are treated in an analogous way:

Exercise 2.3.10. Show that, for every $x < 1$ and for every sufficiently large n, we have the inequalities

$$-\frac{1}{n-1} - \overline{L}(x,n) \leqslant -\frac{1}{2n-1} - \overline{L}(x,2n)$$

$$\leqslant \frac{1}{\lfloor 2xn \rfloor} + \frac{1}{2n-1} - \underline{L}(x,2n) \leqslant \frac{1}{\lfloor xn \rfloor} + \frac{1}{n-1} - \underline{L}(x,n).$$

Exercise 2.3.11. Show that, if p and q are integers such that $q < p$, then

$$\underline{L}\left(\frac{p}{q}, nq\right) = -\underline{L}\left(\frac{q}{p}, np\right).$$

Conclude, by using the toolkit developed above, that $L(x) = -L(1/x)$ for any strictly positive real x.

Calculate the derivative of L at every positive real number.

The function L thus obtained is the Napierian logarithm. It was introduced without calling on its definition as the integral of $1/x$ which vanishes at $x = 1$. Certainly, the integral definition has been hidden: without saying it, we approximated an integral with a finite sum by means of a rectangle method, allowing us to even drop the small pieces at the ends of the interval (draw a picture).

It is recommended that the reader draw the areas defined by $\overline{L}(x,n)$ and $\underline{L}(x,n)$ for $x > 1$, and compare them with the area defined by $\int_1^x dy/y$. It will be useful to write, for example,

$$\underline{L}(x,n) = \frac{1}{n}\left[\left(1 + \frac{1}{n}\right)^{-1} + \left(1 + \frac{2}{n}\right)^{-1} + \ldots + \left(1 + \frac{\lfloor nx \rfloor}{n}\right)^{-1}\right].$$

2.4. Comparison of means

We are also going to construct the exponential function by elementary methods. To do this we will need classical inequalities between different means.

The proof of these inequalities is very simple, and curiously dyadic—as the reader will appreciate.

We recall some terminology: the arithmetic mean of a and b is equal to half of the sum of a and b; the geometric mean of a and b is the square root of the product of a and b; and the harmonic mean of a and b is the inverse of the arithmetic mean of $1/a$ and $1/b$.

Exercise 2.4.1. Show that, for all positive real a and b, we have the inequalities

$$2\left(\frac{1}{a} + \frac{1}{b}\right)^{-1} \leqslant \sqrt{ab} \leqslant \frac{1}{2}(a+b).$$

We are going to show that, for every integer n, and for every choice of strictly positive reals $a_1, a_2, \ldots, a_n$, we also have the following inequality:

$$(2.4.1) \qquad n\left(\sum_{j=1}^{n} \frac{1}{a_j}\right)^{-1} \leqslant \sqrt[n]{a_1 a_2 \cdots a_n} \leqslant \frac{1}{n}\sum_{j=1}^{n} a_j.$$

Exercise 2.4.2. Show that inequality (2.4.1) is true when n is a positive integer power of 2.

Exercise 2.4.3. Let $n < 2^m$. Show that we can deduce the second inequality in expression (2.4.1) for n from the corresponding inequality for 2^m, on the condition that we choose

$$b_j = \begin{cases} a_j & \text{if } 1 \leqslant j \leqslant n; \\ g & \text{if } n+1 \leqslant j \leqslant 2^m \end{cases}$$

and we take g to be the geometric mean of the a_j.

This elementary proof of the inequality between the geometric and the arithmetic mean of n numbers can be found in [44, Chapter II, Section 5].

Exercise 2.4.4. Use a similar procedure to show the first inequality in expression (2.4.1).

2.5. Elementary construction of the exponential

The construction described below is entirely independent of the construction of the logarithm which was presented above. We will see in the differential equations chapter that this construction is a precursor to a lot of the ideas used for the numerical integration methods for ordinary differential equations.

Just as the logarithm was obtained by an additive approximation of integration, the exponential will be obtained by a multiplicative approximation of integration.

I owe the relation between the inequalities on the means and the approximation of the number e to the small book by P. P. Korovkin [55], and I have generalized this idea to the approximation of the exponential function. On the other hand, his proof of the inequalities on the means is more complicated than that given in the preceding section.

Suppose that x is strictly positive, until indicated otherwise.

We define the following functions:

$$\underline{E}(x,m) = \left(1 + \frac{x}{m}\right)^m \quad \text{and} \quad \overline{E}(x,m) = \left(1 - \frac{x}{m}\right)^{-m}.$$

It is proved in all calculus courses that $\overline{E}(\cdot, m)$ and $\underline{E}(\cdot, m)$ converge to the exponential as m tends to infinity. Here, we shall see that these functions give a means to construct the exponential from scratch; moreover, we show that the exponential is the inverse function of the logarithm.

Exercise 2.5.1. Let $x > 0$. Show that, for every integer $m \geqslant 1$, we have

$$\underline{E}(x, m) \leqslant \underline{E}(x, m + 1).$$

Hint: apply the second of the inequalities (2.4.1) with $a_1 = 1$, $a_2 = \cdots = a_{m+1} = 1 + x/m$, and $n = m + 1$.

Exercise 2.5.2. In the same way, show that, if $m > x$,

$$\overline{E}(x, m) \geqslant \overline{E}(x, m + 1),$$

using the first of the inequalities (2.4.1).

Exercise 2.5.3. Show that, for every $p > x$ which is sufficiently large and for every m greater than or equal to 1,

$$\overline{E}(x, p) \geqslant \underline{E}(x, m).$$

Exercise 2.5.4. Show that, for every $m > x$, we have

$$0 \leqslant \overline{E}(x, m) - \underline{E}(x, m) \leqslant \frac{x^2}{m} \overline{E}(x, m).$$

Exercise 2.5.5. Finally, show that, for every $x > 0$, the sequences $\left(\overline{E}(x, m)\right)_m$ and $\left(\underline{E}(x, m)\right)_m$ each have a limit as m tends to infinity and that these limits coincide.

We will denote by $E(x)$ the common limit of $\overline{E}(x, m)$ and $\underline{E}(x, m)$ as m tends to infinity.

Exercise 2.5.6. Calculate $E(0)$.

Exercise 2.5.7. Show that, if p is greater than or equal to 1, then

$$E(px) = E(x)^p.$$

Exercise 2.5.8. Deduce from the preceding result that, for every positive or zero rational p/q, we have

$$E\left(\frac{p}{q}\right) = E(1)^{p/q}.$$

Exercise 2.5.9. Show that, if x is positive or zero and if y is strictly greater than x, then

$$(y - x)\,\underline{E}(x, m)\left(1 + \frac{x}{m}\right)^{-1} \leqslant \underline{E}(y, m) - \underline{E}(x, m)$$

$$\leqslant (y - x)\,\underline{E}(y, m)\left(1 + \frac{y}{m}\right)^{-1}.$$

From this, deduce that in the limit as m tends to infinity,

$$(y - x) E(x) \leqslant E(y) - E(x) \leqslant (y - x) E(y).$$

Show that E is differentiable for every $x > 0$ and right differentiable at 0. Calculate $E'(x)$.

Exercise 2.5.10. Deduce from Exercises 2.5.8 and 2.5.9 that, for every x and y greater than or equal to 0, E satisfies the multiplicative property

(2.5.1) $$E(x + y) = E(x) E(y).$$

From now on we are interested in the behaviour of E for $x < 0$.

Exercise 2.5.11. If x is strictly negative, show that the two sequences

$$\left(\underline{E}(x, m)\right)_{m > -x} \quad \text{and} \quad \left(\overline{E}(x, m)\right)_{m \geqslant 1}$$

are well defined and converge monotonically to $E(-x)^{-1}$. Then, show that the property (2.5.1) holds for all real x and y.

The function E is therefore the exponential function. It remains to verify that it is the inverse of the logarithm.

Exercise 2.5.12. Deduce from expression (2.3.4) that, for every $x \geqslant 0$,

$$\frac{x}{1 + x} \leqslant L(1 + x) \leqslant x.$$

Deduce from this inequality that

$$\frac{x}{1 + x/m} \leqslant L\left(\underline{E}(x, m)\right) \leqslant x$$

and that consequently, for every $x \geqslant 0$,

$$L(E(x)) = x.$$

Show that this relation is still true for $x < 0$, and that we also have, for $y > 0$,

$$E(L(y)) = y.$$

Thus, we have constructed the exponential and logarithm functions by entirely elementary methods and we have shown that one is the inverse of the other.

Equally simple methods can be useful elsewhere—and this is what we are going to see in the next subsection.

2.6. Exponentials of matrices

Let $\mathcal{M}_n$ be the set of real or complex square matrices with n rows and n columns. We will verify that $\mathcal{M}_n$ can be equipped with a norm in the following way: let $A = (A_{ij})_{1 \leqslant i,j \leqslant n}$ and

$$\|A\|_1 = \max_{1 \leqslant j \leqslant n} \sum_{i=1}^{n} |A_{ij}|.$$

Exercise 2.6.1. Show that the expression thus defined really is a norm, that is to say that, for all matrices A and B of $\mathcal{M}_n$ and every scalar λ, we have

$$\|A\|_1 \geqslant 0, \qquad\qquad \text{positivity,}$$
$$\|A\|_1 = 0 \text{ if and only if } A = 0, \qquad \text{the norm is positive definite,}$$
$$\|A + B\|_1 \leqslant \|A\|_1 + \|B\|_1, \qquad \text{triangle inequality,}$$
$$\|\lambda A\|_1 = |\lambda|\,\|A\|_1, \qquad\qquad \text{homogeneity of order 1.}$$

The matrix norm that we have just defined is linked in a simple way to a certain n-component vector norm as follows:

Exercise 2.6.2. If $x_1, x_2, \ldots, x_n$ are the coordinates of some vector x, we let

$$|x|_1 = \sum_{j=1}^{n} |x_j|\,.$$

Show that we have the inequality

(2.6.1) $$|Ax|_1 \leqslant \|A\|_1 |x|_1.$$

Furthermore, show that, for every matrix A, we can always find a vector $x \neq 0$ such that

$$|Ax|_1 = \|A\|_1 |x|_1.$$

Then, show that

(2.6.2) $$\|A\|_1 = \max_{x \neq 0} \frac{|Ax|_1}{|x|_1}.$$

Exercise 2.6.3. Show that, if A and B are in $\mathcal{M}_n$, then

$$\|AB\|_1 \leqslant \|A\|_1\,\|B\|_1.$$

Denote by $\mathbf{1}$ the identity matrix. Calculate $\|\mathbf{1}\|_1$.

Exercise 2.6.4. Let $(A(m))_{m \geqslant 1}$ be a sequence of matrices belonging to $\mathcal{M}_n$. Show that the sequence of norms $\|A(m)\|_1$ tends to 0 if and only if each of the sequences $(A_{ij}(m))_{m \geqslant 1}$ tends to 0 for any i and j between 1 and n.

Exercise 2.6.5. Show that the sequence of matrices $A(m)$ converges if and only if it is a Cauchy sequence; in other words, if and only if, for every given number $\varepsilon > 0$, we can find an integer M such that, for every m and p greater than or equal to M, $\|A(m) - A(p)\|_1$ is less than or equal to ε.

The result of the previous exercise states that $\mathcal{M}_n$ is a complete space, i.e., all Cauchy sequences converge.

The preceding ideas are sufficient to construct the exponential of a matrix A without using power series. We begin by showing an elementary criterion of invertibility:

Exercise 2.6.6. Let $A \in \mathcal{M}_n$ be a matrix with corresponding norm strictly less than 1. Show, with the aid of inequality (2.6.1), that $\mathbf{1} - A$ is injective and, therefore, invertible. Also show, using eqn (2.6.2), that $\|(\mathbf{1} - A)^{-1}\|_1 \leqslant (1 - \|A\|_1)^{-1}$.

We will now show an elementary identity:

Let C and D be matrices belonging to $\mathcal{M}_n$. Show that the following identity is valid for every integer $p \geqslant 1$:

$$(2.6.3) \qquad C^p - D^p = \sum_{j=0}^{p-1} C^j (C - D) D^{p-1-j}.$$

We now use the following approximations to the exponential:

$$(2.6.4) \qquad \underline{E}(A, m) = \left(1 + \frac{1}{m} A\right)^m \quad \text{and} \quad \overline{E}(A, m) = \left(1 - \frac{1}{m} A\right)^{-m}.$$

Exercise 2.6.7. Verify that the approximations (2.6.4) are well defined for every sufficiently large m.

Compare $\|\underline{E}(A, m)\|_1$, $\|\overline{E}(A, m)\|_1$, $\underline{E}(\|A\|_1, m)$, and $\overline{E}(\|A\|_1, m)$.

Exercise 2.6.8. Show that, for every m greater than or equal to 1,

$$(2.6.5) \qquad \|\underline{E}(A, 2m) - \underline{E}(A, m)\|_1 \leqslant \frac{\|A\|_1^2 \exp(\|A\|_1)}{m}.$$

Hint: apply eqn (2.6.3) with $C = (1 + A/2m)^2$ and $D = 1 + A/m$.

This inequality allows us to show the convergence of a subsequence:

Exercise 2.6.9. Verify that the sequence $\underline{E}(A, 2^m)$ has a limit, which will be denoted by $E(A)$, as m tends to infinity.

What can we say about the sequence $\overline{E}(A, 2^m)$?

Exercise 2.6.10. Show that, even if A and B do not commute, we have

$$(2.6.6) \qquad \|E(A) - E(B)\|_1 \leqslant \|A - B\|_1 \exp(\max(\|A\|_1, \|B\|_1)).$$

Hint: this is another application of identity (2.6.3).

It remains to illustrate something of the multiplicative property of the exponential:

Exercise 2.6.11. Show that, if A and B commute, then

$$(2.6.7) \qquad E(A+B) = E(A)E(B) = E(B)E(A).$$

Show that, if p is an integer greater than or equal to 1, then

$$E(A)^p = E(pA).$$

Exercise 2.6.12. Prove that

$$(2.6.8) \qquad \|E(A) - \mathbf{1} - A\|_1 \leqslant 2\|A\|_1^2 \exp(\|A\|_1).$$

Hint: estimate $\|\underline{E}(A, 2^j) - \underline{E}(A, 2^{j-1})\|_1$ using expression (2.6.5) and sum with respect to j.

The inequality (2.6.8) allows us to solve the remaining questions, and, in particular, the convergence of the sequence $\underline{E}(A, p)$ to $E(A)$:

Exercise 2.6.13. Show that, if p is an integer greater than or equal to 1,

$$\|\underline{E}(A, p) - E(A)\|_1 \leqslant \frac{2\exp(2\|A\|_1)\|A\|_1^2}{p}.$$

This allows us to verify the convergence of $\overline{E}(A, n)$ to $E(A)$:

Exercise 2.6.14. Show that there exists a constant C such that

$$\|\overline{E}(A, p) - E(A)\|_1 \leqslant \frac{C}{p}$$

and estimate its value.

This allows us to consider the real variable function which associates t with $E(tA)$:

Exercise 2.6.15. Show that the function $t \mapsto E(tA)$ is infinitely differentiable and calculate all of its derivatives.

We can also make some calculations of exponentials in practice:

Exercise 2.6.16. Let A be a diagonal matrix. Calculate $E(A)$.

This was easy. Let us try a more complicated case:

Exercise 2.6.17. Let A be a nilpotent matrix, that is, there exists p (less than or equal to n) such that $A(p)$ is zero. Calculate $\exp(A)$. Show that, for any m, $\underline{E}(A, m)$ is a polynomial in A, whose degree is bounded independently of m.

The case of nilpotent matrices is particularly curious, because we have an analogous behaviour for $\overline{E}(A, m)$:

Exercise 2.6.18. Under the conditions of the preceding exercise, show that $\overline{E}(A, m)$ is also a polynomial in A whose degree is bounded independently of m, and find its limit as m tends to infinity.

There are many ways of approximating a matrix exponential. Assume that the matrix A and the function $B(t)$ are given such that

$$\|B(t) - E(tA)\|_1 = t\varepsilon(t),$$

where $\varepsilon(t)$ is a function of t which tends to 0 as t tends to 0.

Exercise 2.6.19. Show that there exists a constant C such that, for all m,

$$\left\| B\left(\frac{1}{m}\right)^m - E(A) \right\|_1 \leqslant C\varepsilon\left(\frac{1}{m}\right).$$

Therefore, if we assume that $\varepsilon(t)$ is small compared with t, the approximation of $E(A)$ by $B(1/m)^m$ is more precise than the approximation of $E(A)$ by $\underline{E}(A, m)$ or by $\overline{E}(A, m)$.

Exercise 2.6.20. For sufficiently small t, let

$$(2.6.9) \qquad B(t) = \left(1 + \frac{1}{2}tA\right)\left(1 - \frac{1}{2}tA\right)^{-1}.$$

Show that

$$\|B(t) - E(tA)\|_1 = t^3\eta(t),$$

where η is a function of t which is bounded for sufficiently small t. What is the relationship between $B(t/m)$, $\overline{E}(tA, m)$, and $\underline{E}(tA, m)$?

Not only is this type of approximation more precise, but it allows us to obtain interesting information. Denote the Euclidean scalar product in $\mathbb{R}^n$ (or the Hermitian scalar product in $\mathbb{C}^n$) by (x, y). A matrix A is self-adjoint if, for any x and y, $(Ax, y) = (x, Ay)$. It is skew-adjoint if, for any x and y, $(Ax, y) = -(x, Ay)$. Finally, it is unitary if, for any x and y, $(Ax, Ay) = (x, y)$.

Exercise 2.6.21. Let A be a self-adjoint matrix. Show that $E(A)$ is also self-adjoint. Let A be a skew-adjoint matrix. Show that the matrix B, defined by eqn (2.6.9), is unitary and, consequently, that $E(A)$ is unitary.

This is not all; we can estimate matrix exponentials when we have little information on the matrix itself:

Exercise 2.6.22. Suppose that A is a self-adjoint matrix which is positive in the sense of quadratic forms, that is

$$(2.6.10) \qquad (Ax, x) \geqslant 0,$$

for any x in $\mathbb{R}^n$ (or $\mathbb{C}^n$). Show that, for every x, we have

$$0 \leqslant (E(-A)x, x) \leqslant (x, x).$$

Hint: use $\overline{E}(A, n)$ rather than $\underline{E}(A, n)$ to obtain the answer.

What use is such an example? It happens (and we will illustrate this in a particular case in the last chapter of this book) that a number of practical numerical analysis operators have the property (2.6.10), but their norm grows rapidly with the number of degrees of freedom, n, of the discretization; they arise from differential operators which have no chance of being bounded in a reasonable vector space. This does not prevent us from analysing them and doing the calculations which require them, even in practice.

One property of the exponential is lost when we consider matrices: we still do not have $E(A + B) = E(A)E(B)$. Below is an elementary example of this phenomenon. Let

$$A = \begin{pmatrix} 0 & 1 \\ 0 & 0 \end{pmatrix} \quad \text{and} \quad B = \begin{pmatrix} 0 & 0 \\ -1 & 0 \end{pmatrix}.$$

Exercise 2.6.23. Calculate $E(A)$ and $E(B)$. Is $E(A)E(B)$ unitary?

All of this section can be generalized without difficulty to the case of a unitary normed algebra, that is, a set $\mathcal{A}$ which has a vector space structure and, furthermore, which possesses one multiplication which is distributive with respect to addition, and a norm for which this space is complete.

3

Algebraic preliminaries

This chapter contains some elementary and some very elementary information. Normally, at level n one has not entirely assimilated the lessons of level $n - 1$. Therefore, the consultation of first year courses and books is strongly recommended in case of difficulty.

3.1. Linear algebra refresher

We assume the reader to be familiar with the ideas of a vector space over the field $\mathbb{K}$, which could be the real field $\mathbb{R}$ or the complex field $\mathbb{C}$, and linear mappings. We also assume the reader to be familiar with the ideas of linearly independent sets, spanning sets, dimension, and basis.

3.1.1. The matrix of a linear mapping

Let V be a vector space of dimension n, and W a vector space of dimension m, over the field $\mathbb{K}$. We choose a basis $(v_1, \ldots, v_n)$ of V and a basis $(w_1, \ldots, w_m)$ of W. Recall how we determine the matrix of the mapping f between these two bases: the images $f(v_j)$ of the vectors of the basis of V have the following decomposition on the basis w_j:

$$f(v_j) = \sum_{i=1}^{m} w_i A_{ij}, \quad j = 1, \ldots, n.$$

Every element x of V has the following unique decomposition:

$$x = \sum_{j=1}^{n} x_j v_j.$$

By the linearity of f,

$$f\left(\sum_{j=1}^{n} x_j v_j\right) = \sum_{j=1}^{n} x_j f(v_j) = \sum_{j=1}^{n} x_j \sum_{i=1}^{m} w_i A_{ij},$$

25

and, therefore,

$$(3.1.1) \qquad f(x) = \sum_{i=1}^{m} \left(\sum_{j=1}^{n} A_{ij} x_j \right) w_i.$$

The matrix of the mapping f between these two bases is therefore the following table of numbers belonging to $\mathbb{K}$:

$$A = (A_{ij})_{1 \leqslant i \leqslant m, 1 \leqslant j \leqslant n} = \begin{pmatrix} A_{11} & A_{12} & \cdots & A_{1n} \\ A_{21} & A_{22} & \cdots & A_{2n} \\ \vdots & \vdots & & \vdots \\ A_{m1} & A_{m2} & \cdots & A_{mn} \end{pmatrix}.$$

We frequently denote the element of the matrix A which is situated at the intersection of the i-th row with the j-th column by A_{ij} or $(A)_{ij}$. The notational convention consists of first giving the row index and then the column index. The set of matrices with m rows, n columns and coefficients in $\mathbb{K}$, equipped with matrix addition and the multiplication by a scalar, forms a vector space denoted by $\mathcal{M}_{m,n}(\mathbb{K})$. If $m = n$, this space is formed from square matrices and we denote it $\mathcal{M}_n(\mathbb{K})$. If the choice of the field $\mathbb{K}$ is not important then we will denote it simply by $\mathcal{M}_{m,n}$ or $\mathcal{M}_n$. The space $\mathcal{M}_{m,n}(\mathbb{K})$ therefore corresponds to the space of linear mappings from $\mathbb{K}^n$ to $\mathbb{K}^m$, once these two spaces are equipped with bases. Matrix multiplication corresponds to the composition of linear mappings. If B belongs to $\mathcal{M}_{m,n}(\mathbb{K})$ and A to $\mathcal{M}_{n,p}(\mathbb{K})$, we define the product AB by

$$(AB)_{ik} = \sum_{j=1}^{n} A_{ij} B_{jk}.$$

In $\mathcal{M}_n(\mathbb{K})$, multiplication is an internal law. The identity matrix $\mathbf{1}$ is the identity in $\mathcal{M}_n(\mathbb{K})$; it will also be denoted by I, or by I_n, if we wish to specify the dimension. A synonym of invertible is regular. A matrix which is not invertible is said to be singular.

Remark 3.1.1 (Geometric Remark). If we *identify* V and $\mathbb{R}^n$ (or $\mathbb{C}^n$), and similarly W and $\mathbb{R}^m$ (or $\mathbb{C}^m$), then we can say that *the column vectors of the matrix A are the images of the vectors of the starting basis.*

We can perceive the matrix of the linear mapping as the assembly (in the right order) of the images of the basis vectors. Thus, the identity matrix is formed from the assembly of the canonical basis vectors of $\mathbb{K}^n$.

The above identification will be automatic in finite dimensions. Unless indicated otherwise, we will identify linear form and row vector, linear mapping from the basis field $\mathbb{K}$ to V and vector of V, and linear mapping from V to W and the matrix of this linear mapping.

3.1.2. The determinant

We recall the definition of the determinant as well as its principal properties. To do this, we identify families of n vectors of $\mathbb{K}^n = V$ and matrices formed from the assembly of these n vectors. The set Λ^n of alternate multilinear forms from V^n to $\mathbb{K}$ is formed of mappings f, from V^n to $\mathbb{K}$, which have the following properties:

$$f(v_1 + \lambda v_1', v_2, \ldots, v_n) = f(v_1, v_2, \ldots, v_n) + \lambda f(v_1', v_2, \ldots, v_n),$$
$$\forall v_1, v_1', v_2, \ldots, v_n \in \mathbb{K}^n, \ \forall \lambda \in \mathbb{K},$$

$$f(v_{\sigma(1)}, v_{\sigma(2)}, \ldots, v_{\sigma(n)}) = \epsilon(\sigma) f(v_1, v_2, \ldots, v_n),$$

where σ is a permutation of n objects and $\epsilon(\sigma)$ is its signature.

It can be proved that Λ^n is a vector space of dimension 1. The determinant is the one of these alternate multilinear forms on $\mathbb{K}^n$ which has the value 1 on the canonical basis of V. The determinant of a matrix is the determinant of the family of its n column vectors. The determinant of a linear mapping from W to itself is the determinant of its matrix, provided that we choose the same basis before and after the mapping. It can be proved that this determinant does not depend on the chosen basis.

The explicit expression for the determinant of a matrix is given by the formula

$$(3.1.2) \qquad \det A = \sum_\sigma \epsilon(\sigma) \prod_{i=1}^n A_{i,\sigma(i)},$$

where the sum extends to all permutations σ of n objects.

We know that a square matrix A is invertible if and only if its determinant $\det A$ is not zero. Similarly, the determinant of a family of vectors is nonzero if and only if this family is a basis. We also know that the determinant of a product of matrices is the product of their determinants.

However, this description is of little practical interest, since determinants are difficult to calculate. We happily calculate the determinant of a 2×2 numeric matrix, and we frequently solve a 2×2 system by employing the Cramer formulae. For $n \geqslant 4$, it is already a bad idea to calculate a determinant with formula (3.1.2). We will see in Chapter 9 how the tools used for solving linear systems give efficient methods for the numerical calculation of determinants.

The determinant of a family of n vectors remains an interesting theoretical tool, if only because of its geometric significance. It is the volume of a parallelepiped constructed on n vectors. Note that we allow this volume to be negative if the basis made out of these n vectors has the opposite orientation to the reference basis of the vector space. Consequently, the determinant of a matrix is the volume of the parallelepiped constructed from the column vectors of the matrix. The determinant therefore serves a crucial purpose when we make a change of variable in a multiple integration and, consequently, in the study of differential operators in geometry.

For higher dimensions, the calculation of the determinant can be simplified by considering structure or symmetry.

As an example of the exploitation of symmetry, we prove the following classical result, which concerns the determinant of Vandermonde:

Lemma 3.1.2. The following identity holds:

(3.1.3)
$$\begin{vmatrix} 1 & 1 & 1 & \cdots & 1 \\ x_0 & x_1 & x_2 & \cdots & x_n \\ x_0^2 & x_1^2 & x_2^2 & \cdots & x_n^2 \\ \vdots & \vdots & \vdots & & \vdots \\ x_0^n & x_1^n & x_2^n & \cdots & x_n^n \end{vmatrix} = \prod_{0 \leqslant i < j \leqslant n} (x_j - x_i).$$

The determinant thus calculated is called the Vandermonde determinant.

Proof. To verify this identity let

$$P(x_0, x_1, \ldots, x_n)$$

be the Vandermonde determinant. This is a polynomial in the $n+1$ variables $x_0, x_1, \ldots, x_n$. If two of the numbers x_i are identical, the determinant has two identical columns and, therefore, it vanishes. Consequently, P must be of the form

$$Q(x_0, x_1, \ldots, x_n) \prod_{0 \leqslant i < j \leqslant n} (x_j - x_i),$$

with Q being another polynomial. Moreover, examining the formula (3.1.2) indicates that each term of the determinant, given by Cramer's formulae, is of global degree

$$1 + 2 + \ldots + n = \frac{1}{2}n(n+1).$$

Consequently, P is a homogeneous polynomial of degree $n(n+1)/2$ with real coefficients and, therefore, Q is a real constant. It remains to calculate this constant and, to do this, we note that, in the formula which gives the Vandermonde determinant, the monomial

$$x_0^0 x_1^1 \cdots x_n^n$$

can only be obtained by taking σ to be the identity permutation. The coefficient of this monomial in the determinant is 1. It remains to find the coefficient of this monomial in

$$\prod_{0 \leqslant i < j \leqslant n} (x_j - x_i).$$

To obtain the term in x_n^n we must make x_n appear n times and, therefore, we must take

$$\prod_{i=0}^{n-1} (x_n - x_i).$$

The coefficient of our monomial is therefore the coefficient of

$$x_0^0 \, x_1^1 \cdots x_{n-1}^{n-1}$$

in

$$\prod_{0 \leqslant i < j \leqslant n-1} (x_j - x_i).$$

By induction, this is 1. □

We will come across this determinant again later when investigating the theory of polynomial interpolation.

To conclude this section, we recall the following two facts:

- A nonzero matrix can have a zero determinant. For example,

$$A = \begin{pmatrix} 1 & 0 \\ 0 & 0 \end{pmatrix}.$$

- Except in 1 dimension, the determinant of the sum of two matrices is not related to the sum of their determinants. The reader who is in doubt would do well to compare $\det(A + B)$ and $\det A + \det B$, where B is given by

$$B = \begin{pmatrix} 0 & 0 \\ 0 & 1 \end{pmatrix}.$$

3.1.3. The fundamental theorem of linear algebra and its consequences

The following theorem will be required frequently:

Theorem 3.1.3. Let f be a linear mapping from a finite-dimensional space V to a finite-dimensional space W. Then, the dimension of the domain of f is equal to the sum of the dimension of the kernel of f, and the dimension of the image of f:

$$\dim \operatorname{Im} f + \dim \ker f = \dim V. \qquad \diamond$$

Here is an interesting exercise:

Exercise 3.1.4. Prove, without using the determinant, that it is equivalent for a square matrix to be invertible, to have a left inverse, or to have a right inverse. *Hint: if A has a left inverse B, then the kernel of A is reduced to 0 and the image of B is the whole space; use then the fundamental theorem of linear algebra, Theorem 3.1.3.*

We can immediately deduce from this the following corollary for the linear system

(3.1.4) $$Ax = b,$$

whose matrix A is square:

Corollary 3.1.5. The following properties are equivalent:

(i) For every vector b, eqn (3.1.4) has at least one solution;

(ii) There exists a vector b for which eqn (3.1.4) has at most one solution.

The linearity of the problem implies that if we have uniqueness for a vector b then we have uniqueness for all vectors b.

There is a simple criterion which guarantees that a square matrix is invertible: we say that an n by n matrix is strictly diagonally dominant if, for all $j = 1, \ldots, n$, the following inequality holds:

$$|a_{jj}| > \sum_{\{k : k \neq j\}} |a_{jk}| .$$

Lemma 3.1.6. A strictly diagonally dominant matrix is invertible.

Proof. We will show that the kernel of a strictly diagonally dominant matrix A is reduced to 0. We assume that there exists a vector $x \neq 0$ such that Ax vanishes, and let i be the index of the component of x with maximum absolute value. The i-th equation can be written

$$A_{ii} x_i = - \sum_{\{k : k \neq i\}} A_{ik} x_k .$$

Using the triangle inequality and the definition of i, this relation implies the inequality

$$|A_{ii}| \, |x_i| \leqslant \sum_{\{k : k \neq i\}} |A_{ik}| \, |x_i| ,$$

which, after division by $|x_i|$, requires that

$$|A_{ii}| \leqslant \sum_{\{k : k \neq i\}} |A_{ik}| ,$$

contradicting the assumption on A. Therefore, due to the fundamental theorem of algebra, A is invertible.   □

3.1.4. Eigenvalues and eigenvectors

Let f be a linear mapping from a vector space V of finite dimension n to itself. The spectrum of f is the complement of the set of complex numbers λ for which $f - \lambda I$ is invertible. As we are in finite dimensions $f - \lambda I$ is invertible if and only if its kernel is reduced to zero. Therefore, if λ is in the spectrum of f, there exists a vector x (clearly nonzero) for which

$$f(x) = \lambda x .$$

In this case, λ is an eigenvalue of f, and x an eigenvector of f. If we fix the basis of the space V, we can identify f with its matrix A in this basis, and thus we will refer to the eigenvalues of A and its eigenvectors. The characteristic polynomial of f, or of A, is

$$P(X) = \det(XI - A).$$

The eigenvalues of A are the roots of its characteristic polynomial. It can be proved that every matrix is similar in the complex field to a triangular matrix of the form $D + N$, where D is a diagonal matrix and N is a nilpotent matrix (that is $N^n = 0$) which commutes with D.

3.1.5. Scalar products, adjoints, and company

Recall that a scalar product on a real (respectively, complex) vector space V is a positive definite bilinear (respectively, sesquilinear) form on $V \times V$.

On $V = \mathbb{R}^n$ the canonical bilinear scalar product is given by

(3.1.5)
$$(x, x')_V = \sum_{j=1}^{n} x_j x'_j,$$

and on $W = \mathbb{C}^m$ the canonical sesquilinear scalar product is given by

$$(y, y')_W = \sum_{i=1}^{m} y_i \bar{y}'_i,$$

where $\bar{y}_i$ is the complex number conjugate to y_i. The Euclidean length of a vector $x \in V$ is

$$\|x\|_V = \sqrt{(x, x)_V},$$

with an analogous definition in W. Given two finite-dimensional vector spaces V and W, each equipped with a scalar product denoted by $(\cdot, \cdot)_V$ and $(\cdot, \cdot)_W$, respectively, the adjoint of a linear mapping f from V to W is the unique linear mapping f^* from W to V such that

$$(f(x), y)_W = (x, f^*(y))_V, \quad \forall x \in V, \ \forall y \in W.$$

If V (respectively, W) is identified with $\mathbb{K}^n$ (respectively, $\mathbb{K}^m$) equipped with its canonical basis, if the two spaces are equipped with their canonical scalar products, and if f has $A \in \mathcal{M}_{m,n}$ as its matrix, then the matrix $A^* \in \mathcal{M}_{n,m}$ of f^* is given by

$$(A^*)_{ij} = \bar{A}_{ji}.$$

The transpose of a matrix A of m rows and n columns is the matrix $A^\top$ of n rows and m columns whose coefficients are defined by

$$\left(A^\top\right)_{ij} = A_{ji}.$$

The transpose is the same as the adjoint in the real field; in the complex field, the adjoint is the conjugate transposed matrix. It is obvious that

$$(A^*)^* = A,$$

and that

(3.1.6) $(Ax, y)_W = (x, A^*y)_V, \quad \forall x, \forall y.$

Sometimes, the transpose of A is denoted by $^t A$, which is somewhat inconsistent with the notation for the adjoint.

Definition 3.1.7. A matrix A is said to be Hermitian if it is square and $A^* = A$. We also say that it is self-adjoint. A matrix A is said to be skew-Hermitian if it is square and $A^* = -A$.

To be totally rigorous, it is not the same thing for a matrix to be Hermitian and to be self-adjoint: if we use a non-canonical scalar product, a matrix can be self-adjoint with respect to this scalar product without being Hermitian. As we do not treat mappings using these non-canonical scalar products in $\mathbb{K}^n$ in this book, we will use these two terms interchangeably without causing confusion.

The spectral properties of Hermitian matrices are summarized by the following assertion:

Theorem 3.1.8. A Hermitian matrix is diagonalizable in an orthonormal basis and its eigenvalues are real.   ◇

If the matrix A has real coefficients and is Hermitian, it is said to be symmetric. In this case, the diagonalization basis can be taken to be real.

A matrix A is said to be unitary if it is square and its inverse is equal to its adjoint:

(3.1.7) $A^*A = AA^* = I.$

The eigenvalues of A are complex with modulus 1 and A is diagonalizable in an orthonormal basis. If A is a unitary matrix with all real coefficients, we say that it is orthogonal. However, it should be noted that for an orthogonal matrix the eigenvectors are, in general, not real. To construct an orthonormal basis of eigenvectors of this matrix it is, therefore, necessary to use complex numbers. This idea will rarely be used in this course.

We say that a linear mapping f from V to W conserves the Euclidean length if

$$\|f(x)\|_W = \|x\|_V, \quad \forall x \in V.$$

Lemma 3.1.9. Let f be a linear mapping with corresponding matrix A, which is assumed to be square. Then, A conserves Euclidean lengths if and only if it is unitary.

Proof. If A is unitary, we have, on using eqn (3.1.6), that

$$(Ax, Ax) = (A^*Ax, x) = (x, x).$$

Therefore, A conserves lengths.

Conversely, let A be a linear mapping matrix which conserves lengths. We then have the relation:

$$(3.1.8) \qquad (Ax, Ax) = (x, x), \quad \forall x.$$

To extract information from this relation we are going to pass to the *polar form* of the quadratic form $x \mapsto (Ax, Ax) - (x, x)$. We first of all remark that we have, on using the sesquilinearity of the scalar product,

$$(Ax + Ay, Ax + Ay) - (Ax - Ay, Ax - Ay) = 4\, \Re\, (Ax, Ay), \quad \forall x, \, \forall y.$$

Similarly,

$$(x + y, x + y) - (x - y, x - y) = 4\, \Re\, (x, y), \quad \forall x, \, \forall y.$$

It follows, by using eqn (3.1.8), that

$$(3.1.9) \qquad \Re\, (Ax, Ay) = \Re\, (x, y), \quad \forall x, \, \forall y.$$

If we replace y by iy, the preceding relation becomes

$$(3.1.10) \qquad \Im\, (Ax, Ay) = \Im\, (x, y), \quad \forall x, \, \forall y.$$

We deduce from eqns (3.1.9) and (3.1.10) that

$$(3.1.11) \qquad (Ax, Ay) = (x, y), \quad \forall x, \, \forall y.$$

Using the definition of the adjoint we obtain

$$((A^*A - I)\, x, y) = 0, \quad \forall x, \, \forall y.$$

which clearly implies that $A^*A = I$. Since A is square, A^* is also a right inverse of A. $\qquad \square$

The Schur lemma states that, for every matrix A, there exists a unitary matrix U such that $U^{-1}AU$ is upper triangular. We call such an upper triangular matrix, which is unitarily equivalent to A, a Schur form.

More generally, a square matrix is said to be normal if it commutes with its adjoint. Normal matrices are diagonalizable in an orthonormal basis.

3.1.6. Triangular matrices

Numerical analysts love triangular matrices because systems (3.1.4), where A is triangular, are very easy to solve. Suppose, for instance, that A is upper triangular. Then, the linear system in which we are interested is written

$$a_{11}x_1 + a_{12}x_2 + \ldots + a_{1,n-1}x_{n-1} + a_{1n}x_n = b_1,$$
$$a_{22}x_2 + \ldots + a_{2,n-1}x_{n-1} + a_{2n}x_n = b_2,$$
$$\vdots$$
$$a_{n-1,n-1}x_{n-1} + a_{n-1,n}x_n = b_{n-1},$$
$$a_{nn}x_n = b_n.$$

Assuming that none of the a_{ij} are zero, we see that

$$x_n = \frac{b_n}{a_{nn}},$$
$$x_{n-1} = \frac{b_{n-1} - a_{n-1,n}x_n}{a_{n-1,n-1}},$$
$$\vdots$$
$$x_1 = \frac{b_1 - a_{12}x_2 - \ldots - a_{1n}x_n}{a_{11}}.$$

Solving this is completely elementary and requires only a few operations, see Operation Counts 9.2.2 and 9.3.5.

3.2. Block matrices

Block matrices generalize the concept of decomposition in coordinates. They come up almost systematically when we discretize differential equations in more than one variable. They are also very useful.

3.2.1. Block decomposition of a linear mapping or matrix

Suppose that V, the domain space of the mapping f which we have already considered, decomposes into a direct sum of subspaces V_j:

$$V = \bigoplus_{j=1}^{N} V_j.$$

Similarly, suppose that W, the image space of f, decomposes into a direct sum of subspaces W_i:

$$W = \bigoplus_{i=1}^{M} W_i.$$

Therefore, every element x of V decomposes on V_j in a unique way in the form

$$x = \sum_{j=1}^{N} x_j.$$

The mapping $x \to x_j$ is the canonical projection of V to V_j and is denoted by $\mathcal{P}_j$. The canonical injection $\mathcal{J}_j$ is the mapping

$$\mathcal{J}_j : V_j \mapsto V,$$
$$x_j \to x_j.$$

In the same way, $\mathcal{I}_i$ is the canonical injection from W_i to W, and $\mathcal{Q}_i$ is the canonical projection from W to W_i.

We can then write the linear mapping f by decomposing it on V_j and W_i as follows:

$$f(x) = \sum_{j=1}^{N} f(\mathcal{J}_j x_j),$$

and therefore the component of $f(x)$ on W_i is given by

$$\sum_{j=1}^{N} \mathcal{Q}_i f(\mathcal{J}_j x_j).$$

Let

$$f_{ij} = \mathcal{Q}_i \circ f \circ \mathcal{J}_j.$$

Then, since $x_j = \mathcal{P}_j x$,

(3.2.1)
$$f(x) = \sum_{i=1}^{M} \mathcal{I}_i \sum_{j=1}^{N} f_{ij}(\mathcal{P}_j x).$$

Note the resemblance between eqns (3.2.1) and (3.1.1). Relation (3.2.1) is a generalization of eqn (3.1.1), and eqn (3.2.1) leads to eqn (3.1.1) provided that $V_i = \mathbb{K} v_i$, and similarly for the W_j.

By analogy with matrix notation, we can write the block decomposition of f in the form

(3.2.2)
$$\begin{pmatrix} f_{11} & f_{12} & \cdots & f_{1N} \\ f_{21} & f_{22} & \cdots & f_{2N} \\ \vdots & \vdots & & \vdots \\ f_{M1} & f_{M2} & \cdots & f_{MN} \end{pmatrix}.$$

We can obviously replace the f_{ij} by their matrices A_{ij}, by fixing the bases of each of V_j and W_i and obtaining the block decomposition of the matrix A of f:

(3.2.3)
$$A = \begin{pmatrix} A_{11} & A_{12} & \cdots & A_{1N} \\ A_{21} & A_{22} & \cdots & A_{2N} \\ \vdots & \vdots & & \vdots \\ A_{M1} & A_{M2} & \cdots & A_{MN} \end{pmatrix}.$$

We say that the decompositions $V = \oplus_j V_j$ and $W = \oplus_j W_j$ equip f (or A) with a block structure. It so happens that certain problems impose a block structure. We are going to show that block decomposition makes matrix multiplication easier.

3.2.2. Block multiplication

Suppose that f is a mapping from V to W which has a block structure given by eqn (3.2.2). In the same way, g is a mapping from W to X. If we suppose that

$$X = \bigoplus_{h=1}^{L} X_h,$$

with the canonical projections $\mathcal{R}_h$ and the canonical injections $\mathcal{H}_h$, then g has the block structure

(3.2.4)
$$\begin{pmatrix} g_{11} & g_{12} & \cdots & g_{1M} \\ g_{21} & g_{22} & \cdots & g_{2M} \\ \vdots & \vdots & & \vdots \\ g_{L1} & g_{L2} & \cdots & g_{LM} \end{pmatrix}.$$

Clearly,
$$g_{hi} = \mathcal{R}_h \circ g \circ \mathcal{I}_i,$$

and, as in eqn (3.2.1),

$$g(y) = \sum_{h=1}^{L} \mathcal{H}_h \sum_{i=1}^{M} g_{hi}(\mathcal{Q}_i y).$$

Consequently, the composition $g \circ f$ becomes

$$(g \circ f)(y) = \sum_{h=1}^{L} \mathcal{H}_h \sum_{i=1}^{M} g_{hi}(\mathcal{Q}_i \circ f(x))$$

$$= \sum_{h=1}^{L} \mathcal{H}_h \sum_{i=1}^{M} g_{hi}\left(\mathcal{Q}_i \sum_{i'=1}^{M} \mathcal{I}_{i'} \sum_{j=1}^{N} f_{i'j}(\mathcal{P}_j x) \right)$$

$$= \sum_{h=1}^{L} \mathcal{H}_h \sum_{i=1}^{M} g_{hi}\left(\sum_{j=1}^{N} f_{ij}(\mathcal{P}_j x) \right),$$

since $Q_i \mathcal{I}_{i'} = 0$ except when $i' = i$. In this case $Q_i \mathcal{I}_i$ is the identity of W_i. It follows that $g \circ f$ has the block decomposition $(g \circ f)_{hj}$ with

$$(g \circ f)_{hj} = \sum_{i=1}^{M} g_{hi} \circ f_{ij}, \quad h = 1, \ldots, L, \quad j = 1, \ldots, N.$$

This formula is obviously the generalization of the matrix product. If the matrix B of g is decomposed into matrix blocks B_{hi}, we obtain an analogous formula. To give an example, let

$$A = \begin{pmatrix} A_{11} & A_{12} & A_{13} \\ A_{21} & A_{22} & A_{23} \end{pmatrix} \quad \text{and} \quad B = \begin{pmatrix} B_{11} & B_{12} \end{pmatrix}.$$

Then,

$$BA = \begin{pmatrix} B_{11}A_{11} + B_{12}A_{21} & B_{11}A_{12} + B_{12}A_{22} & B_{11}A_{13} + B_{12}A_{23} \end{pmatrix}.$$

It must be remembered from all of this that block multiplication is identical to normal matrix multiplication, provided that the dimensions of the blocks are compatible: block B_{hi} must have as many columns as block A_{ij} has rows.

The Jordan decomposition of a square matrix A of order n makes use of the idea of block decomposition. Such a matrix has one Jordan form, that is, there exists one invertible matrix P such that $P^{-1}AP = J$, where

$$J = \begin{pmatrix} J(\lambda_1, n_1) & 0 & & \cdots & & 0 \\ 0 & J(\lambda_2, n_2) & 0 & & & 0 \\ \vdots & & \ddots & & & \vdots \\ & & 0 & J(\lambda_{r-1}, n_{r-1}) & & 0 \\ 0 & & \cdots & & 0 & J(\lambda_r, n_r) \end{pmatrix}.$$

The Jordan blocks $J(\lambda, m)$ are $m \times m$ matrices of the form

$$J(\lambda, m) = \begin{pmatrix} \lambda & 1 & 0 & \cdots & 0 \\ 0 & \lambda & 1 & & 0 \\ \vdots & & \ddots & & \\ & & & \lambda & 1 \\ 0 & \cdots & & 0 & \lambda \end{pmatrix}.$$

We remark here that the Jordan form is numerically less stable than the Schur form, see [15].

3.3. Exercises from Chapter 3

3.3.1. Elementary algebra

Exercise 3.3.1. Let P be a square matrix of order n with generic element p_{ij} given by the formula

$$p_{ij} = \delta_{ij} + a\delta_{mi}\delta_{rj},$$

where a is a real number, δ_{ij} is the Kronecker delta, and m and r are integers between 1 and n.

(i) Is the matrix P invertible?

(ii) Calculate the inverse of P if it exists;

(iii) Show that an endomorphism whose matrix is independent of basis is a scaling. To do this, use a change of basis whose transformation matrix is the form P.

Exercise 3.3.2. Let A be a rectangular matrix of m rows and n columns. Show that, by applying a change of basis in the domain and image spaces, we can put A in the form

$$\left(\begin{array}{cc} \begin{pmatrix} 1 & 0 & \cdots & 0 \\ 0 & 1 & \cdots & 0 \\ \vdots & & \ddots & 0 \\ 0 & 0 & \cdots & 1 \end{pmatrix} & (0) \\ (0) & (0) \end{array}\right).$$

Reason in the following manner:

• If the matrix A is identically zero there is nothing to do;

• Suppose now that j_1 is the index of the first column vector of A which is not identically zero. With the first change of basis we can move column j_1 to the position of column 1. Then, with a second change of basis, we replace the first column by a column containing only a 1 in its first row and zeros underneath;

• The general case follows by induction.

3.3.2. Block decomposition

Exercise 3.3.3. Let A be a square matrix of order n. Suppose that it has a block decomposition A_{ij} of dimensions $n_i \times n_j$, where $1 \leqslant i, j \leqslant m$. We suppose, furthermore, that A is Hermitian and, therefore, diagonalizable in an orthonormal basis. Determine a sufficient condition such that the transformation matrix to this orthogonal basis has the same block structure.

Exercise 3.3.4. Let A be a square matrix of order n. Suppose that it has a block decomposition A_{ij} of dimensions $n_i \times n_j$, where $1 \leqslant i, j \leqslant m$. We say that A is block triangular if $A_{ij} = 0$ for every index i and j such that $i > j$.

Calculate the determinant of A as a function of the determinants of the blocks.

Hint: begin with a matrix A decomposed into 2×2 blocks. There are at least two possible proofs of the result:

(i) *By induction on the dimension of the diagonal blocks;*

(ii) *Show that A is similar to a triangular matrix by a similarity which does not destroy the block structure.*

To complete the proof, argue by induction on the number of blocks.
Show by means of a counterexample that, in general,

$$\det \begin{pmatrix} A_{11} & A_{12} \\ A_{21} & A_{22} \end{pmatrix} \neq \det(A_{11}) \det(A_{22}) - \det(A_{12}) \det(A_{21}).$$

3.3.3. Graphs and matrices

A simple graph is defined as follows: we begin with a finite set X and a finite subset U of $X \times X$. Suppose that U contains no element of the form (x, x). The graph is the pair $G = (X, U)$. If, for every $(x, y) \in U$, we also have $(y, x) \in U$, we say that the graph is symmetric. We will only consider symmetric graphs, which is the same as considering U as a set of subsets of two distinct elements of X. If $u = (x, y)$ we say that x and y are the end-points of u.

We say that (x, y) is an edge of the graph if (x, y) belongs to U. We say that two vertices x and y are adjacent if x and y are linked by an edge. Finally, we say that an edge u is incident to a vertex x towards the interior if $u = (x, y)$ or if $u = (y, x)$. The degree $d_G(x)$ of a vertex x is the number of edges incident to x.

We define the associated matrix of a graph to be the matrix $A(G)$ such that

$$\left(A(G) \right)_{xy} = \begin{cases} 1 & \text{if } (x, y) \in U; \\ 0 & \text{otherwise.} \end{cases}$$

Exercise 3.3.5. We say that a graph is properly coloured if we can colour each vertex in such a way that two adjacent vertices are always of different colours. It is k-colourable if we can properly colour it with k colours.

Show that if a graph is k-colourable, its associated matrix has a block structure indexed by the colours.

Exercise 3.3.6. The vertex-edge incidence matrix of a graph is the matrix $R(G)$ defined by

$$\left(R(G) \right)_{xu} = \begin{cases} 1 & \text{if } u \text{ has } x \text{ as an end-point;} \\ 0 & \text{otherwise.} \end{cases}$$

The degree matrix of G is given by

$$\left(D(G) \right)_{xy} = \begin{cases} d_G(x) & \text{if } x = y; \\ 0 & \text{otherwise.} \end{cases}$$

Show that we have

$$A(G) = R(G) R(G)^{\top} - D(G).$$

Exercise 3.3.7. If G is a simple graph, we define a new graph $S(G)$ by adding a new vertex $\xi(u)$ for each edge $u = (x, y)$ and defining the new set of edges as the set of the $(x, \xi(u))$ and the $(\xi(u), y)$. Show that the matrix associated to $S(G)$ is given by

$$A\big(S\,(G)\big) = \begin{pmatrix} 0 & R(G)^{\top} \\ R(G) & 0 \end{pmatrix}.$$

Exercise 3.3.8. We define a matrix B in the following way: for every $u \in U$, consider a strictly positive number $a(u)$ and let

$$B_{xy} = \begin{cases} -\sum_{\{u:x \text{ is an end-point of } u\}} a(u) & \text{if } x = y; \\ a(u) & \text{if } u = (x, y); \\ 0 & \text{otherwise.} \end{cases}$$

Show that the matrix thus defined is symmetric and either positive or zero. What is the dimension of its kernel?

3.3.4. Functions of matrices

We begin by recalling some essential facts about square matrices. Every square matrix A of order n has a decomposition of the form

$$A = T + N,$$

where T and N commute together and with A and, furthermore, T is diagonalizable and N is nilpotent, that is, $N^n = 0$. This fact is clear in the complex field; in the real field one has to be careful, and to proceed as follows: complexify the problem and combine the projections on the generalized eigenspace corresponding to two complex conjugate eigenvalues.

Exercise 3.3.9. Calculate the powers of A as functions of T and N.

Exercise 3.3.10. Let Q be a polynomial of the variable x. Show that we have

$$(3.3.1) \quad Q\,(A) = Q\,(T) + Q'\,(T)\,N + \frac{Q''\,(T)\,N^2}{2} + \ldots + \frac{Q^{(n-1)}\,(T)\,N^{n-1}}{(n-1)!}.$$

Exercise 3.3.11. Let f be a function of the variable x. Suppose that the domain of definition of f contains the spectrum of the operator A. If this spectrum is purely real, we will suppose that f is C^{n-1} on an open subset of $\mathbb{R}$ containing this spectrum. If the spectrum of A contains points which are not real, we will suppose that f is $n-1$ times continuously differentiable with respect to the complex variable x in a neighbourhood of the spectrum of A. The theory of analytic functions allows us to confirm that this condition is fulfilled if f is once continuously differentiable, and at every point in the neighbourhood of the spectrum of A it has a convergent series expansion. Given a finite sequence $a_1, \ldots, a_d$, we denote by $\text{diag}(a_i)$ the diagonal matrix whose elements are $a_1, \ldots, a_d$, in this order. We extend the definition (3.3.1) by defining, for a diagonal matrix B, a diagonal matrix

(3.3.2) $$f(B) = \operatorname{diag}(f(b_i)),$$

for $C = P^{-1}BP$ a diagonalizable matrix,

(3.3.3) $$f(C) = P^{-1}f(B)P,$$

and, for any matrix A,

(3.3.4) $$f(A) = f(T) + f'(T)N + \frac{f''(T)N^2}{2} + \ldots + \frac{f^{(n-1)}(T)N^{n-1}}{(n-1)!}.$$

Show that the definition (3.3.3) does not depend on the matrix P which is employed for the diagonalization.

Exercise 3.3.12. Show that if g is the reciprocal function of f, with g regular in a suitable neighbourhood of the spectrum of $f(A)$, we have

$$g\big(f(A)\big) = A.$$

3.3.5. Square roots, cosines, and sines of matrices

Exercise 3.3.13. Let A be a real matrix of order n which is symmetric and positive definite. By using the diagonalization of A, show that there exists a symmetric positive definite matrix B such that

$$B^2 = A.$$

Exercise 3.3.14. Let B' be a real symmetric positive definite matrix of order n such that

$$A = B'^2.$$

Show that A and B' commute. Deduce from this that B' is equal to matrix B of the preceding question.

Exercise 3.3.15. We want to solve the differential equation system

$$\frac{d^2u}{dt^2} + Au = 0,$$

where the unknown u is a function from $\mathbb{R}$ to $\mathbb{R}^n$. To do this, let

$$\frac{du}{dt} = Bv.$$

What is the first-order differential equation system that is satisfied by the vector $(u, v)^\top$?

Exercise 3.3.16. Let

$$\cos tB = \frac{e^{itB} + e^{-itB}}{2} \quad \text{and} \quad \sin tB = \frac{e^{itB} - e^{-itB}}{2i}.$$

Show that, if $(u, v)^\top$ satisfies the system described in Exercise 3.3.15, then

$$\begin{pmatrix} u\,(t) \\ v\,(t) \end{pmatrix} = \begin{pmatrix} \cos\,(tB) & B^{-1}\sin\,(tB) \\ -B\sin\,(tB) & \cos\,(tB) \end{pmatrix} \begin{pmatrix} u\,(0) \\ v\,(0) \end{pmatrix}.$$

3.3.6. Companion matrices and bounds of matrix powers

Exercise 3.3.17. Given $q + 1$ real numbers $\alpha_0, \alpha_1, \ldots, \alpha_q$, with $\alpha_0 \neq 0$, $\alpha_q = 1$, consider the recurrence

$$(3.3.5) \qquad\qquad \sum_{j=0}^{q} \alpha_j U_{j+n} = 0, \quad n \geqslant 0.$$

Let

$$P\,(x) = x^q + \alpha_{q-1}x^{q-1} + \ldots + \alpha_0$$

be the characteristic polynomial of the recurrence. Write

$$V_n = \begin{pmatrix} U_n \\ U_{n+1} \\ \vdots \\ U_{n+q-1} \end{pmatrix}$$

and determine the square matrix A of order q such that eqn (3.3.5) can be written in the equivalent form

$$(3.3.6) \qquad\qquad V_{n+1} = AV_n.$$

Exercise 3.3.18. Let $\mathcal{P}$ be the set of polynomials of degree q of the form

$$P\,(x) = x^q + \beta_{q-1}x^{q-1} + \ldots + \beta_1 x + \beta_0.$$

We identify $\mathcal{P}$ with a subset of $\mathbb{C}^q$ equipped with the Hermitian distance. Show that the set of elements of $\mathcal{P}$ whose roots are all simple is a dense open subset of $\mathcal{P}$.

Exercise 3.3.19. Let v_k be an eigenvector of A associated to the eigenvalue λ_k. What relation is satisfied by the λ_k and the α_i? From this, deduce the identity

$$(3.3.7) \qquad\qquad P\,(x) = \det\,(xI - A).$$

This question can be answered without calculating any determinant. First solve the case where all of the eigenvalues of A are distinct, then argue by continuity using Exercise 3.3.18.

Exercise 3.3.20. Let

$$V(x) = \begin{pmatrix} 1 \\ x \\ x^2 \\ \vdots \\ x^{q-1} \end{pmatrix}.$$

Calculate for all x the following vector with polynomial coefficients:

$$W(x) = (A - xI)V(x).$$

Exercise 3.3.21. If P has n distinct roots, express the eigenvectors of A as functions of the roots of P and the vector function V.

Exercise 3.3.22. Show that we have the following identity for every integer j:

$$(A - xI)V^{(j)}(x) = j V^{(j-1)}(x) + \begin{pmatrix} 0 \\ \vdots \\ 0 \\ P^{(j)}(x) \end{pmatrix}.$$

Exercise 3.3.23. Deduce from the preceding exercise that A is diagonalizable if and only if all the roots of P are distinct. Give the dimension of the Jordan blocks of A as a function of the multiplicity of the roots of P.

Exercise 3.3.24. Let J be a Jordan block of order n:

(3.3.8)
$$J(\lambda, n) = \begin{pmatrix} \lambda & 1 & 0 & \cdots & 0 \\ 0 & \lambda & 1 & & 0 \\ \vdots & & \ddots & & \\ & & & \lambda & 1 \\ 0 & \cdots & & 0 & \lambda \end{pmatrix}.$$

Calculate the powers J^m of J for every integer $m \geqslant 1$. Show that $\|J^m\|_1$ is bounded independently of m if and only if

(i) $|\lambda| \leqslant 1$; and

(ii) If $|\lambda| = 1$ then $n = 1$.

Exercise 3.3.25. Recall that every square matrix of order n is similar to its Jordan decomposition

$$J = \begin{pmatrix} J(\lambda_1, n_1) & 0 & & \cdots & & 0 \\ 0 & J(\lambda_2, n_2) & 0 & & & 0 \\ \vdots & & \ddots & & & \vdots \\ & & & 0 & J(\lambda_{r-1}, n_{r-1}) & 0 \\ 0 & & \cdots & & 0 & J(\lambda_r, n_r) \end{pmatrix},$$

where the $J(\lambda_s, n_s)$ are the Jordan blocks of the form (3.3.8). What is the Jordan decomposition of a diagonal matrix?

Exercise 3.3.26. Show that the matrix A, defined in eqn (3.3.6), satisfies

$$\|A^m\|_1 \leqslant C, \quad \forall m \geqslant 1$$

if and only if

(i) All of the roots of P are of modulus at most equal to 1; and

(ii) If λ is a root of P of modulus 1, it is simple.

3.3.7. The Kantorovich inequality

Exercise 3.3.27. Let a and $A > a$ be positive numbers. Prove the following inequality:

$$\frac{x}{y} + \frac{y}{x} \leqslant \frac{a}{A} + \frac{A}{a}, \quad \forall x, y \in [a, A].$$

Exercise 3.3.28. Let $\lambda_1 < \lambda_2 < \cdots$ be a strictly increasing sequence of strictly positive numbers, and denote by $\mu_i = 1/\lambda_i$ the sequence of its reciprocals. To each finite subset J of $\mathbb{N}$, we associate a $|J|$-dimensional simplex $\Sigma(J)$ defined as

$$\Sigma(J) = \{(x_j)_{j \in J} : x_j \geqslant 0, \sum_{j \in J} x_j = 1\}.$$

We define a function on $\Sigma(J)$ by

$$f(x, J) = \left(\sum_{j \in J} \lambda_j x_j\right)\left(\sum_{j \in J} \mu_j x_j\right)$$

and we let

$$\rho(J) = \max\{f(x, J) : x \in \Sigma(J)\}.$$

Find $\rho(J)$ when J has exactly two elements.

Exercise 3.3.29. Define the following vectors of $\mathbb{R}^J$:

$$\lambda(J) = (\lambda_j)_{j \in J} \quad \text{and} \quad \mu(J) = (\mu_j)_{j \in J},$$

and let $\omega(J)$ be the vector of $\mathbb{R}^J$ whose components are all equal to 1.

Assume that J has at least three elements. Check, then, that $\lambda(J)$, $\mu(J)$, and $\omega(J)$ are linearly independent.

We shall show by contradiction that f cannot attain its maximum value in the interior of $\Sigma(J)$. Suppose, indeed, that there exists x in the interior of $\Sigma(J)$ such that

$$f(y, J) \leqslant f(x, J), \quad \forall y \in \Sigma(J).$$

Prove that, for all z orthogonal to $\omega(J)$, we must have

$$\left(\mu\left(J\right)^\top x\right)\left(\lambda\left(J\right)^\top z\right) + \left(\mu\left(J\right)^\top z\right)\left(\lambda\left(J\right)^\top x\right) = 0,$$

and, therefore, there exists a scalar β such that

$$\left(\mu\left(J\right)^\top x\right)\lambda\left(J\right) + \left(\lambda\left(J\right)^\top x\right)\mu\left(J\right) = \beta\omega\left(J\right).$$

Hence, infer a contradiction.

Exercise 3.3.30. Let $J = \{1, \ldots, n\}$. Show that $\rho(J)$ is given by

$$(3.3.9) \qquad\qquad \rho\left(J\right) = \frac{\lambda_1}{\lambda_n} + \frac{\lambda_n}{\lambda_1}.$$

Exercise 3.3.31. Show that eqn (3.3.9) still holds if the λ_j are not all distinct.

Exercise 3.3.32. Let A be a symmetric positive definite $n \times n$ real matrix and let $\lambda_1 \leqslant \lambda_2 \leqslant \cdots \leqslant \lambda_n$ be its eigenvalues. Prove the inequality

$$\frac{x^\top A x \, x^\top A^{-1} x}{|x|^4} \leqslant \frac{(\lambda_1 + \lambda_n)^2}{4\lambda_1\lambda_n}, \qquad \forall x \in \mathbb{R}^n \setminus \{0\}.$$

Exercise 3.3.33. Let A and B be symmetric positive definite $n \times n$ real matrices and define a scalar product on $\mathbb{R}^n$ by

$$(x, y)_B = x^\top B y.$$

Show that $B^{-1}A$ is self-adjoint relative to the scalar product $(x, y)_B$. Hence, show that the eigenvalues of $B^{-1}A$ are real and strictly positive. They will be denoted by $\lambda_1 \leqslant \lambda_2 \leqslant \cdots \leqslant \lambda_n$. Derive the following inequality for all $x \in \mathbb{R}^n \setminus \{0\}$ (the Kantorovich inequality):

$$(3.3.10) \qquad\qquad \frac{\left(x^\top A x\right)\left(x^\top B^{-1}AB^{-1}x\right)}{\left(x^\top B x\right)^2} \leqslant \frac{(\lambda_1 + \lambda_n)^2}{4\lambda_1\lambda_n}.$$

Part II

Polynomial and trigonometric approximation of functions

Polynomials are the easiest functions to calculate. Not everything is polynomial, but everything can be approximated, in some sense or another, by polynomials. We are studying three great classes of approximation. The first two use approximation by polynomials:

- Interpolation constrains the values of the approximating polynomial to coincide with those of the function at a finite number of points;

- Least-squares approximation, which constrains the average of the square of the difference between the function and the polynomial to be small.

These two types of approximation have different properties, but are both analysable by linear methods. The last class of approximation methods is spline approximation. A spline is a function which coincides with a polynomial on intervals between knots, and which satisfies some continuity requirements at the knots.

Splines generalize nicely many ideas used both for interpolation and for least-squares approximation, and they have also recently proved extremely useful in image analysis and computer-aided design.

It must also be mentioned that certain classes of wavelets are constructed from splines with uniformly spaced knots. An introduction to wavelets is given in Subsection 8.8.3.

Much of the lore of polynomial and piecewise polynomial approximation is no longer used for what it was intended for originally. For instance, divided differences, which generalize derivatives, were extensively used to construct numerical tables, and to analyse the noisiness of data.

Numerical tables are now generated by programs, often without the user being aware of what is going on. For instance, computer scientists can use efficient algorithms to calculate the usual transcendentals: exp, sin, cos, log, $\sqrt{}$.

At times, we may have to calculate a very complicated function. Suppose, for instance, that each value of this function is obtained by solving a nonlinear partial differential system which requires two hours of machine time. Then it makes sense to construct a table for a finite number of values and a means to fill up the gaps between them. For that purpose, we need interpolation and approximation.

Similarly, we may have obtained experimental values, which are usually noisy, and we would like to draw a curve which is at the same time smooth and close enough to our data. This is what smoothing splines are used for.

However, we do not necessarily have to write software. The existing software (including freeware and shareware) can perform smoothing and approximation very efficiently. Some of these packages are user-friendly and it is a good idea to start with the available libraries.

Curve and surface fitting lead to many mathematically interesting problems, which are treated, for example, in [23].

Nowadays, divided differences are most often used to create numerical schemes for ordinary and partial differential equations, and they have non-commutative versions, which have been studied by the Russian school of V. P. Maslov, in order to obtain the asymptotics of some partial differential equations with rapidly oscillating coefficients. Even when their initial motivation is lost, mathematical methods may prove useful for entirely different purposes.

In numerical analysis, polynomial approximation is used in a very systematic fashion to create finite element methods and pseudo-spectral methods—two of the work-horses of the numerical approximation of solutions to partial differential equations.

Least-squares approximation is used in many areas. Statistics and numerical resolution of partial differential equations immediately come to mind. However, more generally, least-squares approximation works because nature seems quite often to use least-energy principles, termed variational principles. If we are able to expand an energy functional up to quadratic terms, and we minimize the resulting expression, we find that the minimizer solves a linear problem.

Fourier analysis is quite close, in many respects, to least-squares polynomial approximation. For this reason, Fourier series are studied immediately after polynomial approximation. The elementary theory of the convergence of Fourier series is proved to be local. At discontinuities, the convergence is not uniform. This is the Gibbs phenomenon, which is a serious problem when one uses Fourier series in order to approximate discontinuous functions, and particularly in the presence of nonlinear phenomena.

Finally, this part concludes on numerical integration, or quadrature, which use tools from the theory of polynomial approximation. Numerical integration formulae are created with the help of the theory of interpolation and also, at times, using orthogonal polynomials. However, integration formulae are refined by dividing the interval of integration into small intervals, and this meets the idea of piecewise polynomial approximation.

The chapter on quadrature concludes with the fast Fourier transform (FFT), which may be the most widely used of all numerical algorithms, and ranks as a major discovery in numerical simulations, since it enables us to use $O(N \log_2 N)$ operations for a discrete Fourier transform on $N = 2^n$ points instead of the naïve N^2. The FFT uses, most efficiently, the idea of decimation—which should instead be halving in the present environment, i.e., treating differently the odd and even indices in one direction, and the first $N/2$ versus the last $N/2$ in the other direction, and doing that again and again in a recursive fashion parameterized by the size of the vectors being manipulated.

The importance of decimation and recursive algorithms cannot be over-emphasized. In one way or another, all efficient numerical algorithms for large-sized problems rely on some version of these ideas. This is the case for multigrid methods, and also the case for wavelets. On top of that, the FFT is easy to understand and easy to program.

Lest the reader think that all problems are linear, I shall conclude this introduction with the following famous story: a man walking in the street at night meets another one, who seems to be searching for something under the street light. 'May I help you, sir?' asks the passer-by. 'Well sure, I can't find my car keys.' 'Do you know where you might have lost them?' asks the helpful stranger. 'Not really', says the motorist. 'So why do you look for them *here*?' 'Here, there *is* light.'

4

Interpolation and divided differences

4.1. Lagrange interpolation

The problem that we consider in this chapter is the following: let f be a function which we assume to be continuous on the interval $[a, b]$ and let $x_0, x_1, \ldots, x_n$ be $n + 1$ pairwise distinct points given in the interval $[a, b]$. We denote by $\mathbb{P}_n$ the vector space of polynomials of degree at most n. It is well known that $\mathbb{P}_n$ is of dimension $n + 1$.

We ask ourselves the following questions:

(i) Can we find an integer m and a polynomial $P \in \mathbb{P}_m$ which coincides with f at the knots $(x_j)_{0 \leqslant j \leqslant n}$? P is called an interpolating polynomial.

(ii) How do we choose $m \in \mathbb{N}$ to have a solution for every given f?

(iii) How do we choose m so that this solution is unique?

(iv) What error do we commit if we replace $f(x)$ by $P(x)$ when x is not a knot?

Interpolating means that we replace a function by a polynomial which takes the same values as the function at a set of given knots. It is, however, necessary to know, and this is a common sense remark, that replacing a function by its interpolant is a step which supposes a minimum of information about the function. Refer to Figures 4.1 and 4.2 to understand graphically the phenomena which can appear, where the original function and its interpolant are indicated by the solid and broken lines, respectively.

4.1.1. The Lagrange interpolation problem

We begin by answering the first three questions posed above.

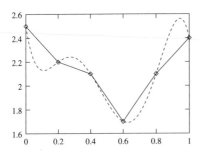

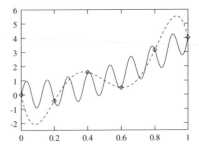

Figure 4.1: The function to be interpolated is not smooth.

Figure 4.2: The function to be interpolated is smooth, but there are not enough interpolation knots.

Theorem 4.1.1. For all choices of $n+1$ pairwise distinct knots $x_0, \ldots, x_n$ and for all given data $f(x_0), \ldots, f(x_n)$, there exists a unique interpolating polynomial P, of degree at most n, which satisfies

$$(4.1.1) \qquad P(x_j) = f(x_j), \quad \forall j = 0, 1, \ldots, n. \qquad \diamond$$

Proof. If we look for a solution in $\mathbb{P}_m$, then we must determine the $m+1$ unknowns which are the coefficients of P. Let

$$P(x) = \sum_{k=0}^{m} a_k x^k,$$

then the equations which must be satisfied are

$$\sum_{k=0}^{m} a_k x_j^k = f(x_j), \quad 0 \leqslant j \leqslant n.$$

We thus have $n+1$ equations for $m+1$ unknowns. It is natural to choose $m = n$ and solve the interpolation problem in $\mathbb{P}_n$. We show that we have uniqueness as follows: let P and $\hat{P}$ be two polynomials of degree at most n, which both interpolate the function f. Consequently, $P - \hat{P}$ is a polynomial of degree at most n, which vanishes at $n+1$ points $x_0, \ldots, x_n$. By Euclidean division, $P - \hat{P}$ is identically zero. Corollary 3.1.5 implies that we have existence for all data. The matrix of the system (4.1.1) is a Vandermonde matrix, which is invertible. $\square$

Note that the proof above is purely algebraic and does not require any hypothesis on f, even on the values of f at the points x which are not knots.

It remains to find a practical interpolation formula. To do this, we will determine the polynomials ϕ_j such that

$$(4.1.2) \qquad \phi_j(x_k) = \delta_{jk}.$$

Theorem 4.1.1 assures us that, for every choice of knots and for all $j = 0, \ldots, n$, there exists a unique polynomial of degree less than or equal to n satisfying eqn (4.1.2). We now explicitly calculate the ϕ_j. For $k \neq j$, we see that ϕ_j vanishes. Therefore, it is of the form

$$\phi_j(x) = a(x) \prod_{k:k\neq j} (x - x_k).$$

This product consists of n factors and, as ϕ_j is of degree at most n, the polynomial $a(x)$ is of degree 0. It is a constant which we calculate using the relation

$$1 = \phi_j(x_j) = a \prod_{k:k\neq j} (x_j - x_k).$$

We therefore deduce that

$$\phi_j(x) = \prod_{k:k\neq j} \left(\frac{x - x_k}{x_j - x_k} \right).$$

The polynomial

$$\sum_{j=0}^{n} f(x_j) \phi_j(x)$$

agrees with the interpolation polynomial of f at the knots x_j. As it is of degree at most n, Theorem 4.1.1 requires that it is therefore equal to the interpolation polynomial. We can now write the interpolation polynomial in the form

$$P(x) = \sum_{j=0}^{n} f(x_j) \phi_j(x).$$

We also see that the ϕ_j form a basis of $\mathbb{P}_n$, the Lagrange basis.

Unfortunately, from a practical point of view, the ϕ_j are not very suitable. The calculation of $P(x)$ is not very difficult, on the condition that we rewrite it in the form

(4.1.3) $$\omega(x) = \prod_{j=0}^{n} (x - x_j),$$

(4.1.4) $$P(x) = \omega(x) \sum_{k=0}^{n} \frac{f(x_k)}{\omega'(x_k)(x - x_k)}.$$

We then need $n + 1$ multiplications, $n + 1$ divisions, and $2n + 1$ additions or subtractions to find $P(x)$. What is particularly annoying, is the simple fact that adding a knot leads to completely changing the basis ϕ_j, without the possibility of reusing the ϕ_j calculated previously. We are therefore going to look for another approach, with a better behaviour when we add extra points.

4.2. Newton's form of interpolation and divided differences

4.2.1. Newton's basis is better than Lagrange's basis

An astronomical application of interpolation is to the problem of plotting the path of celestial objects. One kept adding more and more observations and these could not be made at equidistant time intervals, at least before the age of satellite telescopes, since during the day, or on cloudy nights, we cannot see much in the sky. Thus, Newton came up with an idea for a basis of polynomials, which enabled him to add data without recalculating everything.

With a single interpolation point x_0, the interpolant of f is

$$P^0(x) = f(x_0).$$

When we have two interpolation points x_0 and x_1, the interpolant of f is a polynomial of degree 1 which is chosen to have the form

$$P^1(x) = P^0(x) + R^1(x),$$

since we want to be able to easily add some extra points. Since $P^1(x_0) = f(x_0)$ and $P^1(x_1) = f(x_1)$, we must have

$$R^1(x) = a_1(x - x_0),$$
$$R^1(x_1) = f(x_1) - f(x_0),$$

which implies that

$$a_1 = \frac{f(x_1) - f(x_0)}{x_1 - x_0}.$$

If we pass to three interpolation points x_0, x_1, and x_2, we would like to write the interpolation polynomial P_2 of f at these three points in the form

$$P^2(x) = P^1(x) + R^2(x).$$

As R^2 is of degree at most 2 and vanishes at x_0 and x_1, since $P^2(x_0) = P^1(x_0)$ and $P^2(x_1) = P^1(x_1)$, we must have

$$R^2(x) = a_2(x - x_0)(x - x_1).$$

Newton's form of interpolation therefore consists of writing the interpolation polynomial P of f at the points $x_0, \ldots, x_n$ in the form

$$(4.2.1) \quad P(x) = a_0 + a_1(x - x_0) + a_2(x - x_0)(x - x_1) + \ldots + a_n \prod_{k=0}^{n-1}(x - x_k).$$

This is possible as the sequence of polynomials

$$1, \quad x - x_0, \quad (x - x_0)(x - x_1), \quad \ldots, \quad \prod_{k=0}^{n-1}(x - x_k)$$

forms a basis of $\mathbb{P}_n$, Newton's basis, since the first is exactly of degree 0, the second exactly of degree 1, and so on, to the $(n+1)$-th, which is exactly of degree n. It remains now to calculate the a_j as functions of P, or more precisely, as functions of $P(x_j) = f(x_j)$, since we are in an interpolation setting.

We see immediately that

$$a_0 = f(x_0) \quad \text{and} \quad a_1 = \frac{f(x_1) - f(x_0)}{x_1 - x_0}.$$

We note that a_0 depends only on x_0 and that a_1 depends only on x_0 and x_1. More generally, when

$$P(x) = a_0 + a_1(x - x_0) + a_2(x - x_0)(x - x_1) + \ldots + a_n \prod_{k=0}^{n-1}(x - x_k),$$

we choose a $j \leqslant n$ and we write

$$Q(x) = a_0 + a_1(x - x_0) + a_2(x - x_0)(x - x_1) + \ldots + a_j \prod_{k=0}^{j-1}(x - x_k).$$

This polynomial is in $\mathbb{P}_j$ and it agrees with P at the points $x_0, \ldots, x_j$, since the remaining terms of P contain the product of factors $\prod_{k=0}^{j}(x - x_k)$. This proves that Q is the unique interpolation polynomial of f at the points $x_0, \ldots, x_j$ and, therefore, the coefficients a_j depend only on $x_0, \ldots, x_j$. We therefore introduce the general notation

$$a_j = f[x_0, x_1, \ldots, x_j],$$

as a result of which eqn (4.2.1) is rewritten as

$$(4.2.2) \quad P(x) = f[x_0] + f[x_0, x_1](x - x_0) + \ldots + f[x_0, x_1, \ldots, x_n]\prod_{k=0}^{n-1}(x - x_j).$$

In view of the explicit expressions already found for a_0 and a_1, and of Lemma 4.2.2, which we are going to prove below, $f[x_0, x_1, \ldots, x_j]$ is called the j-th divided difference.

Lemma 4.2.1. For any n, $n+1$ distinct points $x_0, \ldots, x_n$ and, for any permutation σ on $0, \ldots, n$, we have

$$(4.2.3) \quad f[x_0, \ldots, x_n] = f\left[x_{\sigma(0)}, \ldots, x_{\sigma(n)}\right].$$

Proof. To simplify the notation we let

$$y_j = x_{\sigma(j)}.$$

We can consider the decompositions of P, the interpolation polynomial of f at the points $(x_k)_{0 \leqslant k \leqslant n}$, on the two bases

$$1, \quad (x - x_0), \quad \ldots, \quad \prod_{k=0}^{n-1} (x - x_k)$$

and

$$1, \quad (x - y_0), \quad \ldots, \quad \prod_{k=0}^{n-1} (x - y_k).$$

We therefore have

$$P(x) = a_0 + a_1 (x - x_0) + \ldots + a_n \prod_{k=0}^{n-1} (x - x_k)$$

$$= b_0 + b_1 (x - y_0) + \ldots + b_n \prod_{k=0}^{n-1} (x - y_k).$$

The coefficient of the term x^n in P is a_n in the first decomposition, since the only polynomial of the basis $1, (x - x_0), \ldots, \prod_{k=0}^{n-1}(x - x_k)$ containing terms of degree n is the last, and the term of degree n appears with a coefficient of 1. In the same way, the coefficient of the term of degree n in the second decomposition is b_n. We see, therefore, that $b_n = a_n$, which proves our lemma. □

Lemma 4.2.2. We have the following recurrence relation:

$$(4.2.4) \qquad f[x_0, \ldots, x_n] = \frac{f[x_0, \ldots, x_{n-1}] - f[x_1, \ldots, x_n]}{x_0 - x_n}.$$

Proof. Take $y_j = x_{n-j}$. Therefore, we have

$$P(x) = a_0 + a_1 (x - x_0) + \ldots + a_n \prod_{k=0}^{n-1} (x - x_k)$$

$$= b_0 + b_1 (x - x_n) + \ldots + b_n \prod_{k=1}^{n} (x - x_k).$$

We already know that $b_n = a_n$. We equate the terms of degree $n - 1$ in each of the above two expressions for P to give

$$a_{n-1} - a_n \sum_{k=0}^{n-1} x_k = b_{n-1} - a_n \sum_{k=1}^{n} x_k,$$

from which we obtain

$$a_{n-1} - b_{n-1} = a_n (x_0 - x_n).$$

Now

$$a_{n-1} = f [x_0, \ldots, x_{n-1}] \quad \text{and} \quad b_{n-1} = f [x_1, \ldots, x_n],$$

which proves the recurrence relation (4.2.4). □

We have therefore completely justified the name divided differences.

The practical calculation of divided differences is founded on the recurrence relation (4.2.4). By hand, we can construct the following table, from which we will easily deduce an automatic algorithm:

$$x_0 \quad f(x_0)$$
$$\frac{f(x_0) - f(x_1)}{x_0 - x_1} = f[x_0, x_1]$$
$$x_1 \quad f(x_1) \qquad\qquad f[x_0, x_1, x_2]$$
$$\frac{f(x_1) - f(x_2)}{x_1 - x_2} = f[x_1, x_2]$$
$$x_2 \quad f(x_2)$$

We calculate $P(x)$ by a Horner-type algorithm, so that, for example,

$$P(x) = \Big(\big(f[x_0, x_1, x_2, x_3] (x - x_2) + f[x_0, x_1, x_2] \big) (x - x_1)$$
$$+ f[x_0, x_1] \Big) (x - x_0) + f[x_0]$$

when four points are chosen.

To pass from n points to $n + 1$ points demands n calculations of divided differences. For example, in the above table, the addition of the point x_3 demands the calculation of $f(x_3)$, $f[x_2, x_3]$, $f[x_1, x_2, x_3]$, and $f[x_0, x_1, x_2, x_3]$, or 4 calculations of divided differences to pass from 3 to 4 points.

4.2.2. Integral representation of divided differences

Let f be a C^1 function on an interval containing x_0 and x_1. We then have

$$f(x_1) - f(x_0) = \int_0^1 f'\big(x_0 + t(x_1 - x_0)\big)(x_1 - x_0) \, dt,$$

as verified by an elementary calculation. From this, we deduce that

$$f[x_0, x_1] = \int_0^1 f'\big(x_0 + t(x_1 - x_0)\big) \, dt.$$

So far, the first divided difference has appeared as an approximation to the first derivative. Now, we see it as the average of the derivative on the interval with end-points x_0 and x_1.

We are going to generalize this result:

Theorem 4.2.3. Suppose f to be C^n on the interval

$$\left[\min_{0 \leqslant j \leqslant n} x_j, \max_{0 \leqslant j \leqslant n} x_j \right].$$

Then

$$
(4.2.5) \qquad f\left[x_0, \ldots, x_n\right] = \int_0^1 \int_0^{t_1} \cdots \int_0^{t_{n-1}} f^{(n)}\big(x_0 + t_1\left(x_1 - x_0\right)
$$
$$
+ t_2\left(x_2 - x_1\right) + \ldots + t_n\left(x_n - x_{n-1}\right)\big) \, dt_n \cdots dt_1. \quad \diamond
$$

Proof. First of all, examine the domain of integration of this multiple integral. If $n = 1$ it is the segment $[0, 1]$. If $n = 2$ it is the triangle with vertices $(0, 0)$, $(1, 0)$, and $(1, 1)$. If $n = 3$ it is the tetrahedron with vertices $(0, 0, 0)$, $(1, 0, 0)$, $(1, 1, 0)$, and $(1, 1, 1)$. In dimension n it is the n-simplex

$$\Sigma_n = \{t \in \mathbb{R}^n \, : \, 0 \leqslant t_n \leqslant t_{n-1} \leqslant \cdots \leqslant t_1 \leqslant 1\}$$

of vertices $(0, 0, \ldots, 0)$, $(1, 0, \ldots, 0)$, and so on, successively replacing the 0 by 1 each time to give a new vertex, up to $(1, 1, \ldots, 1)$.

We are going to argue by induction: for $n = 1$ we have already seen that formula (4.2.5) is true. Suppose that it is true up to $n - 1$. We note that the inner integral is

$$
\int_0^{t_{n-1}} f^{(n)}\big(x_0 + t_1\left(x_1 - x_0\right) + \ldots + t_n\left(x_n - x_{n-1}\right)\big) \, dt_n
$$
$$
= \left[\frac{f^{(n-1)}\big(x_0 + t_1\left(x_1 - x_0\right) + \ldots + t_n\left(x_n - x_{n-1}\right)\big)}{x_n - x_{n-1}} \right]_{t_n=0}^{t_n=t_{n-1}}
$$
$$
= \bigg(f^{(n-1)}\big(x_0 + t_1\left(x_1 - x_0\right) + \ldots + t_{n-1}\left(x_n - x_{n-2}\right)\big)
$$
$$
- f^{(n-1)}\big(x_0 + t_1\left(x_1 - x_0\right) + \ldots + t_{n-1}(x_{n-1} - x_{n-2})\big) \bigg)
$$
$$
\times \left(x_n - x_{n-1}\right)^{-1}.
$$

We therefore have, due to the induction hypothesis,

$$
\int_0^1 \cdots \int_0^{t_{n-1}} f^{(n)}\big(x_0 + t_1\left(x_1 - x_0\right) + t_2\left(x_2 - x_1\right) + \ldots
$$
$$
+ t_n\left(x_n - x_{n-1}\right)\big) \, dt_1 \cdots dt_n
$$
$$
= \int_0^1 \cdots \int_0^{t_{n-2}} \Big\{ f^{(n-1)}\big(x_0 + t_1\left(x - x_0\right) + \ldots + t_{n-1}\left(x_n - x_{n-2}\right)\big)
$$
$$
- f^{(n-1)}\big(x_0 + t_1\left(x - x_0\right) + \ldots + t_{n-1}\left(x_{n-1} - x_{n-2}\right)\big) \Big\} \frac{dt_1 \cdots dt_{n-1}}{x_n - x_{n-1}}.
$$

This is equal to

$$\frac{f\left[x_0,\ldots,x_{n-2},x_n\right]-f\left[x_0,\ldots,x_{n-2},x_{n-1}\right]}{x_n-x_{n-1}}=f\left[x_0,\ldots,x_{n-2},x_{n-1},x_n\right],$$

due to the induction hypothesis. This proves that the formula (4.2.5) is true for n. □

It follows, from Theorem 4.2.3, that we can define divided differences of f on $n+1$ distinct or repeated points, provided that f is C^n, by means of the integral representation (4.2.5). A particularly important case is that in which $x_0=x_1=\cdots=x_n$. In this case

$$f\underbrace{\left[x_0,\ldots,x_0\right]}_{n+1\text{ arguments}}=\int_0^1\int_0^{t_1}\cdots\int_0^{t_{n-1}}f^{(n)}\left(x_0\right)\mathrm{d}t_n\,\mathrm{d}t_{n-1}\cdots\mathrm{d}t_1$$

$$=f^{(n)}\left(x_0\right)v_n,$$

where v_n is the volume of the n-simplex. We have

$$v_n=\int_0^1\cdots\int_0^{t_{n-1}}\mathrm{d}t_n\cdots\mathrm{d}t_1$$

$$=\int_0^1\cdots\int_0^{t_{n-2}}t_{n-1}\,\mathrm{d}t_{n-1}\cdots\mathrm{d}t_1$$

$$=\int_0^1\cdots\int_0^{t_{n-3}}\frac{1}{2}t_{n-2}^2\,\mathrm{d}t_{n-2}\cdots\mathrm{d}t_1$$

$$=\cdots=\frac{1}{n!}.$$

We have therefore obtained the relation

(4.2.6)
$$f\underbrace{\left[x_0,\ldots,x_0\right]}_{n+1\text{ arguments}}=\frac{f^{(n)}\left(x_0\right)}{n!}.$$

4.3. Interpolation error

With the aid of divided differences, we can evaluate the interpolation error, that is, the difference between the function f and its interpolation polynomial at the points $x_0,\ldots,x_n$.

Theorem 4.3.1. Let f be a C^{n+1} function on an interval $[a,b]$ and let P be its interpolation polynomial on the $n+1$ points $x_0,\ldots,x_n$ belonging to $[a,b]$. Then, for every $x\in[a,b]$, there exists a number ξ_x in the interval

$$\left[\min\left(x,\min_k x_k\right),\ \max\left(x,\max_k x_k\right)\right]$$

such that

$$f(x) - P(x) = \prod_{j=0}^{n} (x - x_j) \frac{f^{(n+1)}(\xi_x)}{(n+1)!}. \qquad \diamond$$

Proof. If $x = x_j$ the conclusion is obvious. If not, let Q be the interpolation polynomial of f at the knots $x_0, x_1, \ldots, x_n, x$. We can then write Q, using its Newton form, as

$$Q(y) = P(y) + f[x_0, \ldots, x_n, x] \prod_{k=0}^{n} (y - x_k).$$

In particular,

$$Q(x) - P(x) = f(x) - P(x) = f[x_0, \ldots, x_n, x] \prod_{k=0}^{n} (x - x_k).$$

But, we know that

$$f[x_0, \ldots, x_n, x] = \int_0^1 \cdots \int_0^{t_n} f^{(n+1)}(x_0 + t_1(x_1 - x_0) + \ldots$$
$$+ t_{n+1}(x - x_n)) \, dt_{n+1} \cdots dt_1.$$

The $(n+1)$-simplex Σ_{n+1} of $\mathbb{R}^{n+1}$ is a connected set. The image of this connected set under the continuous real-valued mapping

$$F : t \mapsto f^{(n+1)}(x_0 + t_1(x_1 - x_0) + \ldots + t_{n+1}(x - x_n))$$

is connected. We can apply the mean value theorem for multiple integrals, to prove that there exists a t_x in Σ_{n+1} such that

$$\int_{\Sigma_{n+1}} F(t) \, dt = F(t_x) \int_{\Sigma_{n+1}} 1 \, dt.$$

Now, we have previously calculated the volume of the n-dimensional simplex for all n. We therefore find that

$$\int_{\Sigma_{n+1}} F(t) \, dt = \frac{F(t_x)}{(n+1)!}.$$

We have $F(t_x) = f^{(n+1)}(\xi_x)$, with

$$\xi_x \in \left[\min\left(x, \min_k x_k\right), \ \max\left(x, \max_k x_k\right) \right],$$

which proves the theorem. $\square$

The convergence of a sequence of interpolation polynomials of increasing degree to the function that we are interpolating is not true in general. We can find an analytic function f such that the sequence of interpolation polynomials P_k of degree at most k, with knots in the interval $[a, b]$ given by

$$x_j = a + \frac{(b-a)\, j}{k}, \quad j = 0, \ldots, k,$$

diverges catastrophically. The classic example consists of taking $f(x) = 1/(1 + x^2)$ and $[a, b] = [-5, 5]$. This example is known as the Runge phenomenon. The sequence $(P_k)_k$ diverges at many points of the interval. It is represented in Figures 4.3, 4.4, and 4.5 for different values of k. On these figures, the divergence phenomenon is quite striking. However, it is also a classical fact that changing the location of the knots may improve the situation. For instance, if we choose to interpolate the same function at the Chebyshev points, i.e., at

$$x_j = 5 \cos\left(\frac{j\pi}{n}\right), \quad j = 0, \ldots, n,$$

then the result is much better, as can be seen in Figures 4.6 and 4.7.

We can even prove that, for every sequence of families of interpolation knots, there exists a function f for which the sequence of interpolation polynomials does not uniformly converge towards the function we want to approximate. Supplementary information on this remark is given in [19, Chapter 1].

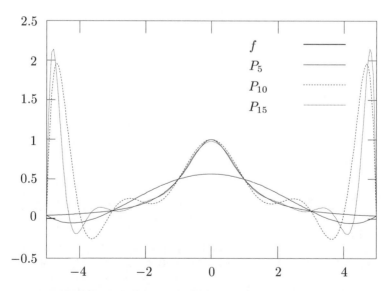

Figure 4.3: Approximation of $f(x) = 1/(1 + x^2)$ by interpolation polynomials of degrees 5, 10, and 15, using equidistant knots.

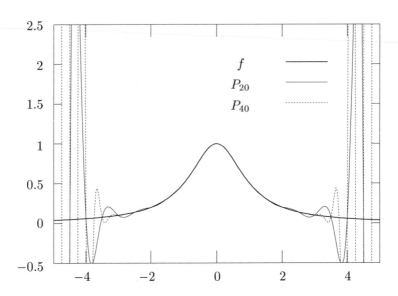

Figure 4.4: Approximation of $f(x) = 1/(1 + x^2)$ by interpolation polynomials of degrees 20 and 40, using equidistant knots.

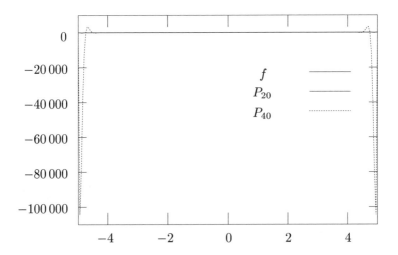

Figure 4.5: The same functions as in Figure 4.4 using an unclipped vertical axis.

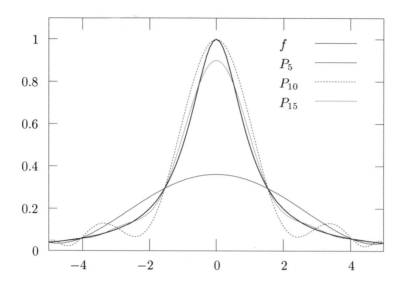

Figure 4.6: Approximation of $f(x) = 1/(1+x^2)$ by interpolation polynomials of degrees 5, 10, and 15, using knots at the Chebyshev points.

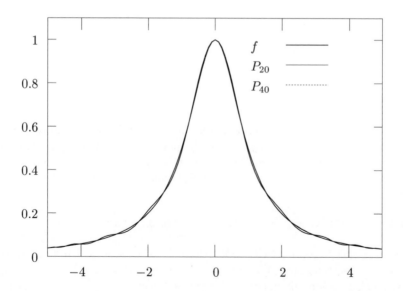

Figure 4.7: Approximation of $f(x) = 1/(1+x^2)$ by interpolation polynomials of degrees 20 and 40, using knots at the Chebyshev points.

4.4. Hermite and osculating interpolation

The section on the integral representation of divided differences shows that it is natural to approximate a sufficiently smooth function f by a polynomial P which coincides with f, together with a number of its derivatives, at a finite number of points. Thus, given $x_0, \ldots, x_n$ and $n + 1$ positive integers $r_0, \ldots, r_n$, we seek a polynomial P of the smallest degree, such that

$$(4.4.1) \qquad f^{(j)}(x_k) = P^{(j)}(x_k), \quad \forall k = 0, \ldots, n, \quad \forall j = 0, \ldots, r_k - 1.$$

Let

$$(4.4.2) \qquad r = r_0 + \ldots + r_n.$$

The existence and the uniqueness of the solution of eqn (4.4.1) in $\mathbb{P}_{r-1}$ is proved in the following lemma:

Lemma 4.4.1. The mapping L from $\mathbb{P}_{r-1}$ to $\mathbb{R}^r$ defined by

$$LP = \left(P(x_0), \ldots, P^{(r_0-1)}(x_0), \ldots, P(x_n), \ldots, P^{(r_n-1)}(x_n) \right)$$

is invertible.

Proof. If LP vanishes for some $P \in \mathbb{P}_{r-1}$, then P is divisible by the polynomials $(x-x_0)^{r_0}, \ldots, (x-x_n)^{r_n}$. These polynomials are relatively prime and, therefore, P is divisible by their product, which is of degree r. Since P is of degree at most $r - 1$, P vanishes. As $\mathbb{P}_{r-1}$ is of dimension r, the fundamental theorem of linear algebra enables us to conclude the result of the lemma.   □

The polynomial $P \in \mathbb{P}_{r-1}$ which satisfies eqn (4.4.1) is called the osculating polynomial to f at the points $x_0, \ldots, x_n$ which agrees at order $r_j - 1$ with f at each x_j.

The following particular case is very important for applications. Suppose that we choose $r_j = 2$ for all j. Then, the polynomial of degree at most $2n + 1$ which satisfies eqn (4.4.1) is called the Hermite interpolation polynomial of f at the points $x_0, \ldots, x_n$. It is the unique polynomial of degree at most $2n + 1$ such that

$$(4.4.3) \qquad P(x_j) = f(x_j), \quad P'(x_j) = f'(x_j), \quad \forall j = 0, \ldots, n.$$

It can be constructed explicitly with the help of a basis which is analogous to the basis of Lagrange interpolation polynomials.

We seek polynomials $h_k(x)$ and $\hat{h}_k(x)$ such that

$$\left. \begin{array}{ll} h_k(x_j) = \delta_{jk}, & h'_k(x_j) = 0, \\ \hat{h}_k(x_j) = 0, & \hat{h}'_k(x_j) = \delta_{jk} \end{array} \right\} \quad \forall j, k.$$

We may take h_k and $\hat{h}_k$ to have the form

$$h_k(x) = \prod_{\substack{0 \leqslant j \leqslant n \\ j \neq k}} \left(\frac{x - x_j}{x_k - x_j} \right)^2 (a_k(x - x_k) + b_k),$$

$$\hat{h}_k(x) = \prod_{\substack{0 \leqslant j \leqslant n \\ j \neq k}} \left(\frac{x - x_j}{x_k - x_j} \right)^2 (\hat{a}_k(x - x_k) + \hat{b}_k).$$

With the aid of the notation (4.1.2) for the basis of Lagrange polynomials, h_k and $\hat{h}_k$ can be rewritten as

$$h_k(x) = \phi_k(x)^2 \left(a_k(x - x_k) + b_k \right) \quad \text{and} \quad \hat{h}_k(x) = \phi_k(x)^2 \left(\hat{a}_k(x - x_k) + \hat{b}_k \right).$$

The condition $h_k(x_k) = 1$ implies that $b_k = 1$. The condition $h'_k(x_k) = 0$ further implies that

$$2\phi_k(x_k)\phi'_k(x_k) + \phi_k(x_k)^2 a_k = 0$$

and hence

$$a_k = -2\phi'_k(x_k),$$

so that

(4.4.4) $$h_k(x) = \left(1 - 2\phi'_k(x_k)(x - x_k) \right) \phi_k(x)^2.$$

A similar argument gives

(4.4.5) $$\hat{h}_k(x) = \phi_k(x)^2 (x - x_k).$$

In this basis, the Hermite interpolation polynomial of f is given by

$$P(x) = \sum_{j=0}^{n} \left(h_j(x) f(x_j) + \hat{h}_j(x) f'(x_j) \right).$$

The interpolation error is

$$f(x) - P(x) = \frac{(\omega(x))^2}{(2n + 2)!} f^{(2n+2)}(\xi),$$

where ξ belongs to the interval $\left[\min_{0 \leqslant j \leqslant n} x_j, \max_{0 \leqslant j \leqslant n} x_j \right]$ and ω has been defined in eqn (4.1.3). The proof of this estimate is completely analogous to the proof of Theorem 4.3.1; the details of this proof depend upon the relationship between divided differences with coincident arguments and osculating polynomials, as described in the following lemma.

Lemma 4.4.2. Let $x_0, \ldots, x_n$ be $n+1$ distinct points in the interval $[a,b]$, and let $r_0, \ldots, r_n$ be $n+1$ integers which are at least equal to 1. Let f be a function of class $C^{r_0 + \cdots + r_n - 1}$ over the interval $[a,b]$. Then, the osculating polynomial of f at the points x_j, which agrees at order $r_j - 1$ with f at each x_j, is given by

$$P(x) = f(x_0) + f[x_0, x_0](x - x_0) + \ldots + f[\underbrace{x_0, \ldots, x_0}_{r_0 \text{ arguments}}](x - x_0)^{r_0 - 1}$$

$$+ f[\underbrace{x_0, \ldots, x_0}_{r_0 \text{arguments}}, x_1](x - x_0)^{r_0} + \ldots$$

(4.4.6)
$$+ f[\underbrace{x_0, \ldots, x_0}_{r_0 \text{arguments}}, \ldots, \underbrace{x_n, \ldots, x_n}_{r_n \text{arguments}}](x - x_0)^{r_0}(x - x_1)^{r_1} \cdots (x - x_n)^{r_n - 1}.$$

Proof. Let r be as in eqn (4.4.2) and let $y_1, \ldots, y_r$ be r distinct points belonging to (a, b). Then, we know from eqn (4.2.2) that the interpolation polynomial of f at the points y_j is given by

$$P(x; y) = f(y_1) + f[y_1, y_2](x - y_1) + \ldots + f[y_1, y_2, \ldots, y_r] \prod_{j=1}^{r-1} (x - y_j).$$

The integral representation (4.2.5) of divided differences shows that $f[y_1, \ldots, y_j]$ is a continuous function of its arguments. It is clear that the polynomial

$$\prod_{j=1}^{r-1} (x - y_j)$$

depends continuously on the y_js. Denote by ∂_1 the derivative of a function of two vector arguments with respect to the first one. Therefore, if y tends to any element $\bar{y}$ of $(a, b)^r$, then, for all $k = 1, \ldots, r-1$, $\partial_1^k P(\cdot, y)$ converges to $\partial_1^k P(\cdot, \bar{y})$ uniformly on compact sets of the real line and, in particular, uniformly on $[a, b]$. The reader should be aware that this conclusion is still true if some of the points $\bar{y}_j$ coincide. Thus, let us assume that

$$\bar{y}_1 = \cdots = \bar{y}_{r_0} = x_0,$$
$$\bar{y}_{r_0 + 1} = \cdots = \bar{y}_{r_0 + r_1} = x_1,$$

$$\vdots$$

$$\bar{y}_{r_0 + \ldots + r_{n-1} + 1} = \cdots = \bar{y}_{r_0 + \ldots + r_n} = x_n.$$

We will now show that $P(\cdot, \bar{y})$ is the osculating polynomial to f at the points x_j, which agrees at order $r_j - 1$ at each x_j. It is clear that, for all $j = 0, \ldots, n$,

$$P(x_j, \bar{y}) = f(x_j).$$

Choose j such that $r_j > 1$. Therefore, we have r_j distinct points $y_{r_0+\ldots+r_{j-1}+1}$, $\ldots, y_{r_0+\ldots+r_j}$ at which $f - P(\cdot, y)$ vanishes and, by Rolle's theorem, we have $r_j - 1$ points in the interval

$$\min\{y_s : r_0 + \ldots + r_{j-1} + 1 \leqslant s \leqslant r_0 + \ldots + r_j\} = I_j(y)$$

at which $f' - \partial_1 P(\cdot, y)$ vanishes. By an obvious recursive argument, we see that there exist $r_j - 2$ points at which $f'' - \partial_1^2 P(\cdot, y)$ vanishes in $I_j(y)$ and, finally, one point in $I_j(y)$ where $f^{(r_j-1)} - \partial_1^{r_j-1} P(\cdot, y)$ vanishes. As y tends to $\bar{y}$, all of these points where the successive derivatives of $f - P(\cdot, y)$ vanish tend to x_j and, therefore, in the limit

$$\partial_1^k P(x_j, \bar{y}) = f^{(k)}(x_j), \quad \forall j = 0, \ldots, n, \quad \forall k = 0, \ldots, r_j - 1,$$

which proves the lemma. $\qquad \square$

The relation (4.4.6) shows very clearly that the osculating polynomial at x_0 which agrees with f up to order $r_0 - 1$ is the Taylor expansion of f truncated after order $r_0 - 1$. In general, the osculating polynomial is a combination of Taylor and interpolation polynomials. This idea is made more precise in Exercise 4.6.1.

Now that Lemma 4.4.2 is proved, it is a simple matter to show that, if f is of class C^r and P is the osculating polynomial which agrees with f at degree $r_j - 1$ at each point x_j, then there exists ξ_x in the smallest convex interval containing x and the x_js such that

$$f(x) - P(x) = \frac{f^{(r)}(\xi_x)}{r!} \prod_{j=0}^{n} (x - x_j)^{r_j}.$$

The proof of this assertion is a complete repetition of the proof of Theorem 4.3.1, and it is left to the reader.

4.5. Divided differences as operators

The divided difference $f[x_0, \ldots, x_n]$ can be considered as a function of $n + 1$ variables. Thus, we may study the mapping

$$f \mapsto f \underbrace{[\cdot, \ldots \ldots, \cdot]}_{n+1 \text{ arguments}}$$

as a mapping δ^n which transforms functions of one variable into functions of $n + 1$ variables. If we want to make explicit the $n + 1$ arguments of the divided difference, we shall write

$$\delta^n (x_0, \ldots, x_n) f = f[x_0, \ldots, x_n].$$

If all of the x_js coincide and f is n times continuously differentiable, we have already seen that $\delta^n(x, \ldots, x)$ coincides with $f^{(n)}(x)/n!$. Therefore, the operator δ^n generalizes differentiation. The notation has been chosen to emphasize the similarity between divided differences and differentiation: we need $n + 1$ parameters in divided differences to make something which resembles an n-th derivative.

It is clear that δ^n is a linear mapping: for all scalars α and all functions f and g

$$\delta^n \left(f + \alpha g\right) = \delta^n f + \alpha \delta^n g.$$

What is even more interesting is that Leibniz' formula can be generalized to divided differences:

Lemma 4.5.1. For all f and g and all $n + 1$ distinct parameters $x_0, x_1, \ldots, x_n$, we have the identity

$$(4.5.1) \qquad \delta^n \left(x_0, \ldots, x_n\right) (fg) = \sum_{j=0}^{n} \left(\delta^j \left(x_0, \ldots, x_j\right) f\right) \left(\delta^{n-j} \left(x_j, \ldots, x_n\right) g\right).$$

If f and g are of class C^n, the above relation also holds even when some of the knots x_j coincide.

Proof. For $n = 0$, the left-hand side of eqn (4.5.1) is equal to $\delta^0(x_0)(fg) = f(x_0)g(x_0)$ and the right-hand side of eqn (4.5.1) is equal to $\delta^0 f(x_0)\delta^0 g(x_0)$. Therefore, the identity (4.5.1) is verified in this case. Before embarking on the general case, let us consider the case of $n = 1$. Then, the left-hand side is

$$\frac{f\left(x_0\right) g\left(x_0\right) - f\left(x_1\right) g\left(x_1\right)}{x_0 - x_1}$$

and the right-hand side is

$$f\left(x_0\right) \frac{g\left(x_0\right) - g\left(x_1\right)}{x_0 - x_1} + \frac{f\left(x_0\right) - f\left(x_1\right)}{x_0 - x_1} g\left(x_1\right),$$

so that the identity (4.5.1) is also clear in this case. Assume now that identity (4.5.1) holds up to some index n. Then, by the definition of divided differences,

$$\delta^{n+1} \left(x_0, \ldots, x_{n+1}\right) (fg) = \frac{\delta^n \left(x_0, \ldots, x_n\right) (fg) - \delta^n \left(x_1, \ldots, x_{n+1}\right) (fg)}{x_0 - x_{n+1}}.$$

We now use the induction hypothesis to show that

$$\delta^n \left(x_0, \ldots, x_n\right) (fg) - \delta^n \left(x_1, \ldots, x_{n+1}\right) (fg)$$
$$= \sum_{j=0}^{n} \Big[\left(\delta^j \left(x_0, \ldots, x_j\right) f\right) \left(\delta^{n-j} \left(x_j, \ldots, x_n\right) g\right)$$
$$- \left(\delta^j \left(x_1, \ldots, x_{j+1}\right) f\right) \left(\delta^{n-j} \left(x_{j+1}, \ldots, x_{n+1}\right) g\right) \Big].$$

However, for each $j = 0, \ldots, n$, the term in brackets in the above sum can be rewritten as

$$\left(\delta^j \left(x_0, \ldots, x_j\right) f\right) \left(\delta^{n-j} \left(x_j, \ldots, x_n\right) g - \delta^{n-j} \left(x_{j+1}, \ldots, x_{n+1}\right) g\right)$$
$$+ \left(\delta^j \left(x_0, \ldots, x_j\right) f - \delta^j \left(x_1, \ldots, x_{j+1}\right) f\right) \left(\delta^{n-j} \left(x_{j+1}, \ldots, x_{n+1}\right) g\right)$$

which, by the definition of divided differences, is equal to

$$\left(x_j - x_{n+1}\right) \left(\delta^j \left(x_0, \ldots, x_j\right) f\right) \left(\delta^{n+1-j} \left(x_j, \ldots, x_{n+1}\right) g\right)$$
$$+ \left(x_0 - x_{j+1}\right) \left(\delta^{j+1} \left(x_0, \ldots, x_{j+1}\right) f\right) \left(\delta^{n-j} \left(x_{j+1}, \ldots, x_{n+1}\right) g\right).$$

We sum these expressions with respect to j and divide by $x_0 - x_{n+1}$ to give

$$\delta^{n+1} \left(x_0, \ldots, x_{n+1}\right) \left(fg\right)$$
$$= \sum_{j=0}^{n} \frac{x_j - x_{n+1}}{x_0 - x_{n+1}} \left(\delta^j \left(x_0, \ldots, x_j\right) f\right) \left(\delta^{n+1-j} \left(x_j, \ldots, x_{n+1}\right) g\right)$$
$$+ \sum_{j=0}^{n} \frac{x_0 - x_{j+1}}{x_0 - x_{n+1}} \left(\delta^{j+1} \left(x_0, \ldots, x_{j+1}\right) f\right) \left(\delta^{n-j} \left(x_{j+1}, \ldots, x_{n+1}\right) g\right).$$

We replace the index j by $j + 1$, changing accordingly the summation range, in the second sum of the above expression to obtain

$$\delta^{n+1} \left(x_0, \ldots, x_{n+1}\right) \left(fg\right)$$
$$= \sum_{j=0}^{n} \frac{x_j - x_{n+1}}{x_0 - x_{n+1}} \left(\delta^j \left(x_0, \ldots, x_j\right) f\right) \left(\delta^{n+1-j} \left(x_j, \ldots, x_{n+1}\right) g\right)$$
$$+ \sum_{j=1}^{n+1} \frac{x_0 - x_j}{x_0 - x_{n+1}} \left(\delta^j \left(x_0, \ldots, x_j\right) f\right) \left(\delta^{n+1-j} \left(x_j, \ldots, x_{n+1}\right) g\right),$$

and it is now clear that the identity (4.5.1) holds. □

There is a rather obvious corollary to Lemma 4.5.1:

Corollary 4.5.2 (Leibniz' formula). If f and g are of class C^m, then, for all integers $k \leqslant m$, the following formula holds:

$$\frac{d^k}{dx^k} \left(fg\right) = \sum_{j=0}^{k} C_k^j \frac{d^j f}{dx^j} \frac{d^{k-j} g}{dx^{k-j}}.$$

Proof. Use the relation (4.2.6) and the previous lemma. □

Divided differences can also be used in several variables. We will need later only the case of two variables, the case of N variables being analogous.

Let f be a function of two variables $x \in [a, b]$ and $y \in [c, d]$. Given two sequences of distinct knots $x_0 < x_1 < \cdots < x_n$ in $[a, b]$ and $y_0 < y_1 < \cdots < y_p$ in $[c, d]$, we may apply $\delta^n(x_0, \ldots, x_n)$ to $f(\cdot, y)$. We write the result as

$$(4.5.2) \qquad \delta_x^n (x_0, \ldots, x_n) f (\cdot, y),$$

where the subscript x emphasizes that the finite differences are applied to the first variable, x, and this parallels the notations ∂_x or $\partial/\partial x$ for partial derivatives. Then, we apply $\delta^p(y_0, \ldots, y_p)$ to eqn (4.5.2), and the result will be denoted by

$$\delta_y^p (y_0, \ldots, y_p) \, \delta_x^n (x_0, \ldots, x_n) f.$$

The identity

$$\delta_y^p (y_0, \ldots, y_p) \, \delta_x^n (x_0, \ldots, x_n) f = \delta_x^n (x_0, \ldots, x_n) \, \delta_y^p (y_0, \ldots, y_p) f$$

is an algebraic fact, which is an immediate consequence of the linearity of δ_x^n and δ_y^p. The reader may want to set up a recursive proof, if they feel so inclined.

If we assume, for instance, that f and its partial derivatives with respect to y up to order p are continuous, then the same formula holds with repeated y knots. Hence, we obtain the following identity for $0 \leqslant l \leqslant p$:

$$(4.5.3) \qquad \frac{\partial^l}{\partial y^l} \delta_x^n (x_0, \ldots, x_n) f (\cdot, y) = \delta_x^n (x_0, \ldots, x_n) \frac{\partial^l f (\cdot, y)}{\partial y^l}.$$

This identity will prove very useful in the following sections.

4.5.1. Finite differences on uniform grids

The forward and backward finite difference operators are defined by

$$(4.5.4) \qquad (\Delta_h f) (x) = f (x + h) - f (x),$$
$$(4.5.5) \qquad (\nabla_h f) (x) = f (x) - f (x - h).$$

If Δ or ∇ are written without index, it means that h is taken to be equal to 1.

Divided differences on uniformly spaced points can be expressed in terms of forward and backward differences:

Lemma 4.5.3. The following identities hold, for all $n \geqslant 1$:

$$(4.5.6) \qquad \delta^n (x_0, \ldots, x_0 + nh) f = \frac{1}{h^n n!} (\Delta_h^n f) (x_0),$$

$$(4.5.7) \qquad \delta^n (x_0 - nh, \ldots, x_0) f = \frac{1}{h^n n!} (\nabla_h^n f) (x_0).$$

Proof. For $n = 1$, the identities are clear. Assume that they hold up to some integer $n \geqslant 1$. Now, by the definition of divided differences,

$$\delta^{n+1} (x_0, \ldots, x_0 + nh, x_0 + (n + 1) h) f$$
$$= \frac{\delta^n (x_0 + h, \ldots, x_0 + (n + 1) h) f - \delta^n (x_0, \ldots, x_0 + nh) f}{(n + 1) h}.$$

We use the induction hypothesis to show that the right-hand side of the above expression is equal to

$$\frac{1}{(n+1)\,h}\frac{1}{h^n n!}\left(\left(\Delta_h^n f\right)(x_0+h)-\left(\Delta_h^n f\right)(x_0)\right),$$

which is clearly equal to

$$\frac{1}{h^{n+1}\,(n+1)!}\left(\Delta_h^{n+1} f\right)(x_0).$$

The proof of the second identity is completely analogous.   □

Due to eqn (4.5.6), Newton's form of the interpolation polynomial of a function f at the points $x_0, x_0 + h, \ldots, x_0 + nh$ is given by

(4.5.8)
$$P(x_0+hs)=f(x_0)+s\left(\Delta_h f\right)(x_0)+\frac{s(s-1)\left(\Delta_h^2 f\right)(x_0)}{2!}$$
$$+\ldots+\frac{s(s-1)\cdots(s-n+1)\left(\Delta_h^n f\right)(x_0)}{n!}.$$

If we generalize the definition of the binomial coefficients to all real or complex values of s by letting

(4.5.9)
$$\binom{s}{k}=\frac{s(s-1)\cdots(s-k+1)}{k!},$$

then the formula (4.5.8) can be rewritten in the following more convenient form:

(4.5.10)
$$P(x_0+hs)=\sum_{k=0}^{n}\binom{s}{k}\left(\Delta_h^k f\right)(x_0).$$

There is an analogous formula when the knots make up a decreasing arithmetic progression:

$$x_i = x - ih.$$

The argument which gave us eqn (4.5.10) now gives

(4.5.11)
$$P(x_0+hs)=\sum_{i=0}^{n}\binom{-s}{i}\left(\nabla_h^i f\right)(x_0).$$

Of significant practical importance are the following central differences: the central difference approximation to the first derivative of a function is

$$\frac{\left(\left(\nabla_h+\Delta_h\right)f\right)(x)}{2h}=\frac{f(x+h)-f(x-h)}{2h}$$

and the central difference to the second derivative of a function is

$$\frac{(\nabla_h \Delta_h f)(x)}{h^2} = \frac{f(x+h) - 2f(x) + f(x-h)}{h^2}.$$

Finite differences on regular meshes were used extensively in the era of numerical tables; this motivation for using them is now quite slight. However, they remain an important device to generate numerical methods for the solution of partial differential equations, being mainly applied to equations where convection is dominant and time-dependent equations. As we shall see in Chapter 17, they are an essential ingredient for the construction of some numerical integration schemes for ordinary differential equations.

4.6. Exercises from Chapter 4

4.6.1. More on divided differences

An osculating polynomial is a combination of Taylor and interpolation polynomials. Beyond that, the maximum regularity needed in order to have the general integral representation of finite differences is not required: a divided difference with coincident knots can be defined, provided that the local regularity of the function at these knots is good enough. The purpose of this problem is to prove the above statements in a straightforward and elementary fashion.

Exercise 4.6.1. Let f be a function defined on an interval $]a, b[$ and let $x_0, \ldots, x_n$ be $n + 1$ distinct points from $[a, b]$. Prove the identity

$$f[x_0, \ldots, x_n] = \sum_{j=0}^{n} f(x_j) \prod_{\substack{0 \leqslant k \leqslant n \\ k \neq j}} (x_j - x_k)^{-1}.$$

Exercise 4.6.2. Assume that f is a function of class C^n in (a, b). Show that, for all $k \leqslant n$,

$$\lim_{h \to 0} f[x, x+h, \ldots, x+kh] = \frac{1}{k!} f^{(k)}(x).$$

Exercise 4.6.3. Assume in the remainder of this problem that $y_0, \ldots, y_n$ are given distinct points in (a, b), that the intervals $(y_j - \varepsilon, y_j + \varepsilon)$ are disjoint for some $\varepsilon > 0$, and that, for each $j = 0, \ldots, n$, f is of class $C^{r(j)}$ on the interval $(y_j - \varepsilon, y_j + \varepsilon)$. Write

$$U = \prod_{j=0}^{n} (y_j - \varepsilon, y_j + \varepsilon).$$

Show that, for each j and for each $k = 0, \ldots, r(j)$, the mapping

$$x \mapsto \frac{\partial^{r(j)}}{\partial x_j^{r(j)}} f[x_0, x_1, \ldots, x_k]$$

from U to $\mathbb{R}$ is continuous.

Exercise 4.6.4. Introducing the notation

$$\phi(x_0) = f[x_0, x_1, \ldots, x_n],$$

show that, for all positive integers k and for all sufficiently small h,

$$\phi[x_0, x_0 + h, \ldots, x_0 + kh] = f[x_0, x_0 + h, \ldots, x_0 + kh, x_1, \ldots, x_n].$$

Exercise 4.6.5. Deduce from the previous exercise that, for all $x \in U$,

$$f[\underbrace{x_0, \ldots, x_0}_{r(0)+1 \text{ times}}, x_1, \ldots, x_n] = \frac{1}{r(0)!} \frac{\partial^{r(0)}}{\partial x_0^{r(0)}} f[x_0, x_1, \ldots, x_n].$$

Exercise 4.6.6. Let

$$\psi(x_1) = \frac{1}{r(0)!} \frac{\partial^{r(0)}}{\partial x_0^{r(0)}} f[x_0, x_1, \ldots, x_n].$$

Show that, for all positive integers k and for all sufficiently small h,

$$\psi[x_1, x_1 + h, \ldots, x_1 + kh]$$
$$= \frac{1}{r(0)!} \frac{\partial^{r(0)}}{\partial x_0^{r(0)}} f[x_0, x_1, x_1 + h, \ldots, x_1 + kh, x_2 \ldots, x_n],$$

and deduce that, for all $x \in U$,

$$f[\underbrace{x_0, \ldots, x_0}_{r(0)+1 \text{ times}}, \underbrace{x_1, \ldots, x_1}_{r(1)+1 \text{ times}}, x_2, \ldots, x_n]$$
$$= \frac{1}{r(0)! r(1)!} \frac{\partial^{r(0)+r(1)}}{\partial x_0^{r(0)} \partial x_1^{r(1)}} f[x_0, x_1, \ldots, x_n].$$

Exercise 4.6.7. Show, in general, that, for all $x \in U$ and all sequences of integers $k(j)$, $j = 0, \ldots, n$, such that $k(j) \leqslant r(j)$, we have the identity

$$f[\underbrace{x_0, \ldots, x_0}_{k(0)+1 \text{ times}}, \underbrace{x_1, \ldots, x_1}_{k(1)+1 \text{ times}}, \ldots, \underbrace{x_n, \ldots, x_n}_{k(n)+1 \text{ times}}]$$
$$= \frac{1}{k(0)! k(1)! \cdots k(n)!} \frac{\partial^{k(0)+k(1)+\ldots+k(n)}}{\partial x_0^{k(0)} \partial x_1^{k(1)} \cdots \partial x_n^{k(n)}} f[x_0, x_1, \ldots, x_n].$$

Exercise 4.6.8. Let f be a function of class C^2. Assume that x_0 and x_1 are distinct points, and set $y_0 = y_2 = y_4 = x_0$ and $y_1 = y_3 = y_5 = x_1$. Determine the expressions $f[y_0, \ldots, y_j]$ for $0 \leqslant j \leqslant 5$. Letting $f(x) = \sin x$, $x_0 = 0$, and $x_1 = \pi/6$, calculate an approximation to $\sin(\pi/12)$. It may be useful to take advantage of a program for symbolic manipulation, such as MAPLE.

4.6.2. Numerical approximation to the solution of a boundary value problem for a differential equation by a finite difference method

Exercise 4.6.9. Let f be a function of class C^4 on the interval $[0, 1]$. Derive the following estimate:

$$h^{-2} \nabla_h \Delta_h f(x) - f''(x) = O(h^4).$$

Show that, if f is assumed to be of class C^2, then we have

$$\lim_{h \to 0} \max \left\{ |h^{-2} \nabla_h \Delta_h f(x) - f''(x)| : h \leqslant x \leqslant 1 - h \right\} = 0.$$

Exercise 4.6.10. We want to approximate the solution of

$$(4.6.1) \qquad\qquad -u''(x) + c(x) u = f(x), \quad x \in [0, 1]$$

by a finite difference method. For this purpose, we assume that c is non-negative and continuous, and we complement eqn (4.6.1) by the Dirichlet boundary conditions

$$(4.6.2) \qquad\qquad u(0) = u(1) = 0.$$

At this point, we do not know that there exists a solution of eqns (4.6.1) and (4.6.2), but we will assume that such a solution exists and that it is unique.

Exercise 4.6.11. Let $n \geqslant 1$ be given and define $h = 1/(n+1)$. Consider the bilinear form α on $\mathbb{R}^n$ given by

$$\alpha(U, V) = \sum_{j=0}^{n} h \frac{U_{j+1} - U_j}{h} \frac{V_{j+1} - V_j}{h} + \sum_{j=1}^{n} c(jh) U_j V_j,$$

where we define, by convention, $U_0 = V_0 = 0$ and $U_{n+1} = V_{n+1} = 0$. Show that α is symmetric and positive definite.
Hint: $\alpha(U, U) = 0$ then $U_{j+1} = U_j$, $\forall j$, and conclude.

Exercise 4.6.12. Let l be the linear form on $\mathbb{R}^n$ defined by

$$lU = \sum_{j=1}^{n} h f(jh) U_j.$$

Show that there exists a unique minimizer of the functional

$$\phi(U) = \frac{1}{2} \alpha(U, U) - lU.$$

Show that the minimizer U solves a linear system of equations, and give this system explicitly. The matrix of this system will be denoted by A.

Exercise 4.6.13. Solve explicitly the equation

(4.6.3) $$-\frac{V_{j+1} - 2V_j + V_{j-1}}{h^2} = 1$$

under the boundary conditions

(4.6.4) $$V_0 = V_{n+1} = 0.$$

Hint: seek the solution as a polynomial in j of degree 2.

Exercise 4.6.14. Assume that $F = (F_j)_{1 \leqslant j \leqslant n}$ is a vector in $\mathbb{R}^n$ with non-negative components. Let W be the solution of

(4.6.5) $$AW = F.$$

Show that the coordinates of W are non-negative.

Hint: argue by contradiction and consider the i-th equation in the system (4.6.5), assuming that i is an index at which $j \mapsto W_j$ attains a strictly negative minimum.

Exercise 4.6.15. Denote by $|F|_\infty = \max_{1 \leqslant j \leqslant n} |F_j|$ the maximum norm of $F \in \mathbb{R}^n$. Infer from the inequalities

$$|F|_\infty - F_j \geqslant 0 \quad \text{and} \quad |F|_\infty + F_j \geqslant 0, \quad \forall j = 1, \ldots, n$$

that, for any F, the solution of eqn (4.6.5) satisfies the inequalities

$$-|F|_\infty V_j \leqslant W_j \leqslant |F|_\infty V_j, \quad \forall j = 1, \ldots, n,$$

where V is the function defined by eqns (4.6.3) and (4.6.4).

Exercise 4.6.16. Let

$$F_j = f(jh) - c(jh) u(jh) + \frac{u((j+1)h) - 2u(jh) + u((j-1)h)}{h^2}.$$

Show that

$$\lim_{h \to 0} |F|_\infty = 0.$$

Exercise 4.6.17. Deduce from the previous question that

$$\lim_{h \to 0} \max \{|U_j - u(jh)| : 1 \leqslant j \leqslant n\} = 0.$$

4.6.3. Extrapolation to the limit

The process to be described here is also called Richardson's extrapolation.

Exercise 4.6.18. Suppose that a function f has the limited expansion near $x = 0$ given by

$$f(x) = f_0 + x f_1 + O(x^2),$$

and assume that we know neither f_0 nor f_1, but that we have a reliable process for calculating $f(x)$ for arbitrarily small values of $x > 0$. Find a combination of $f(x)$ and $f(2x)$ which gives an approximation of order 2 to f_0.

Exercise 4.6.19. Suppose that a function f has a limited expansion near $x = 0$ given by

$$f(x) = P(x) + O(x^{n+1}), \quad P \in \mathbb{P}_n.$$

The polynomial P is unknown, but we have, as in Exercise 4.6.18, a reliable process for calculating f. Show that, for every choice of $n + 1$ positive distinct numbers r_j, there is a linear combination with coefficients μ_j^0 such that

$$\sum_{j=0}^{n} \mu_j^0 P(r_j x) = f(0) + O(x^{n+1}).$$

More generally, show that, for each $k = 1, \dots, n$, there exists a linear combination such that

$$\sum_{j=0}^{n} \mu_j^k P(r_j x) = k! f^k(0) x^k + O(x^{n+1}).$$

Exercise 4.6.20. What kind of advantage is there in performing the above procedure? What difficulties do you foresee?

5

Least-squares approximation for polynomials

Section 4.1 was dedicated to polynomial interpolation: we replace a function by a polynomial of degree at most n which coincides with the function at $n+1$ points. We have seen that the result is not always satisfactory in terms of convergence.

5.1. Posing the problem

In the present chapter, given *a priori* a distance, we seek a polynomial of degree at most n for which the distance to the given function f is minimized. The term least-squares approximation describes the distance under consideration. It is the distance given by the quadratic mean, whose square is

$$\int_a^b |P(x) - f(x)|^2 \, w(x) \, \mathrm{d}x,$$

where $[a, b]$ is a compact interval of $\mathbb{R}$ (with $a < b$) and w is a weight, integrable on $[a, b]$, that we assume to be strictly positive almost everywhere.

5.1.1. Least-squares is Pythagoras in many dimensions

We have, first of all, the following theorem on existence and uniqueness:

Theorem 5.1.1. Let w be an integrable function which is strictly positive almost everywhere on the compact interval $[a, b]$. There exists a unique polynomial $P \in \mathbb{P}_n$, such that

$$(5.1.1) \quad \int_a^b |P(x) - f(x)|^2 \, w(x) \, \mathrm{d}x \leqslant \int_a^b |Q(x) - f(x)|^2 \, w(x) \, \mathrm{d}x, \quad \forall Q \in \mathbb{P}_n.$$

This polynomial is called the least-squares approximation to f and is of degree at most n.   ◇

The proof of this theorem depends on a very neat trick, which will be used several times later on, and which is described in the following lemma:

Lemma 5.1.2. Let V be a real or complex finite-dimensional vector space and let α be a bilinear symmetric or sesquilinear Hermitian form. Assume that α is non-negative and let l be a semilinear form on V. Then, x minimizes the function

$$\phi(x) = \frac{1}{2}\alpha(x, x) - \Re(lx)$$

if and only if the following identity holds:

(5.1.2) $\alpha(x, z) - lz = 0, \quad \forall z \in V.$

Moreover, α is positive definite if and only if for all $z \in V$, there exists a unique minimizer of ϕ over V.

Proof. We treat the case when V is a complex vector space, the real case being somewhat simpler. If x is a minimizer of ϕ over V, then, for all $t \in \mathbb{R}$ and all $z \in V$, we must have

$$\phi(x + tz) - \phi(x) \geqslant 0.$$

Expanding the above inequality, we obtain

(5.1.3) $t\alpha(x, z) + t\alpha(z, x) + t^2\alpha(z, z) \geqslant 2\Re(tlz).$

First, suppose that t is strictly positive and divide eqn (5.1.3) by t. Letting t tend to zero, we obtain

$$\alpha(x, z) + \alpha(z, x) \geqslant 2\Re(lz).$$

If, instead of t being positive, we assume t to be negative, the analogous operation implies that

$$\alpha(x, z) + \alpha(z, x) \leqslant 2\Re(lz),$$

i.e.,

(5.1.4) $\Re\alpha(x, z) = \Re(lz), \quad \forall z \in V.$

We now replace z in eqn (5.1.4) by iz to obtain

$$\Im\alpha(x, z) = \Im(lz), \quad \forall z \in V,$$

which indeed proves that any minimizer verifies the identity (5.1.2).

Conversely, let x verify eqn (5.1.2). Then, for any $w \in V$,

$$\phi(w) - \phi(x) = \Re\alpha(x, w - x) + \frac{1}{2}\alpha(w - x, w - x) - \Re l(w - x).$$

However, due to the assumption (5.1.2), the term

$$\Re\alpha\,(x, w - x) - \Re\big(l\,(w - x)\big)$$

vanishes and there remains

(5.1.5) $$\phi\,(w) - \phi\,(x) = \frac{1}{2}\alpha\,(w - x, w - x)\,,$$

which is non-negative, owing to the assumption on α.

Assume now that α is positive definite. Relation (5.1.5) immediately implies the uniqueness of minimizers of ϕ. Conversely, assume that the minimizers of ϕ are unique. Then, if A is the matrix of the form α in an arbitrary basis of V, the relation (5.1.2) is equivalent to

(5.1.6) $$A\xi = \eta,$$

where ξ is the vector of coordinates of x, η is the vector of coordinates of y, and y is the unique vector in V such that

$$lz = z^*y, \quad \forall z \in V.$$

The system (5.1.6) is linear and, therefore, if for all data it admits at most one solution, then it also possesses a solution for all data η. Conversely, the uniqueness assumption implies that A is positive definite: if its kernel were not reduced to 0 then, for any w in the kernel of α, we would have

$$\phi\,(x + w) - \phi\,(x) = \frac{1}{2}\alpha\,(w, w)\,,$$

which contradicts the uniqueness assumption. $\qquad\square$

Proof of Theorem 5.1.1. For all continuous f and g on $[a, b]$, we define

$$\alpha\,(f, g) = \int_a^b w\,(x)\,f\,(x)\,\overline{g\,(x)}\,\mathrm{d}x,$$

which is a sesquilinear form. We let $V = \mathbb{P}_n$ and also define a semilinear form L on $\mathbb{P}_n$ by

$$LP = \int_a^b w\,(x)\,\overline{P\,(x)}f\,(x)\,\mathrm{d}x.$$

The restriction of α to $V = \mathbb{P}_n$ is Hermitian and positive definite. Theorem 5.1.1 then implies that there exists a unique P in $\mathbb{P}_n$ which minimizes

$$\phi\,(P) = \frac{1}{2}\int_a^b w\,(x)\,|P\,(x)|^2\,\mathrm{d}x - \Re\int_a^b w\,(x)\,\overline{P\,(x)}f\,(x)\,\mathrm{d}x.$$

P also minimizes

$$\frac{1}{2}\int_a^b w\,(x)\,|P\,(x) - f\,(x)|^2\,\mathrm{d}x,$$

which differs from the previous expression by the constant

$$\frac{1}{2} \int_a^b w\left(x\right) |f\left(x\right)|^2 \, \mathrm{d}x.$$

In particular, relation (5.1.2) can be rewritten as

(5.1.7) $$\qquad\qquad\qquad a\left(f - P, R\right) = 0, \quad \forall R \in \mathbb{P}_n.$$

Moreover, if we choose the basis of monomials $1, x, \ldots, x^n$ for $\mathbb{P}_n$, and if P is of the form

$$P\left(x\right) = \sum_{j=0}^n a_j x^j,$$

then the relation (5.1.2) is equivalent to the following system of $n + 1$ equations with $n + 1$ unknowns $a_0, \ldots, a_n$:

(5.1.8) $$\qquad \sum_{j=0}^n a_j \int_a^b x^{j+k} w\left(x\right) \, \mathrm{d}x = \int_a^b x^k f\left(x\right) w\left(x\right) \, \mathrm{d}x, \quad 0 \leqslant k \leqslant n.$$

Then, Lemma 5.1.2 tells us that eqn (5.1.8) possesses a unique solution, which determines the least-squares approximation of f in $\mathbb{P}_n$, with respect to the weight w. It is useful to rewrite eqn (5.1.5) and, in our case, it becomes

(5.1.9) $\quad a\left(f - Q, f - Q\right) = a\left(f - P, f - P\right) + a\left(P - Q, P - Q\right), \quad \forall Q \in \mathbb{P}_n,$

which is simply Pythagoras' theorem.   □

Remark 5.1.3. This proof also works on an infinite interval I, with a weight w, such that every power of x is integrable on I with respect to this weight. Important examples of weights are: $x \mapsto \exp(-x)$ (Laguerre weight) on $I = \mathbb{R}^+$ and $x \mapsto \exp(-x^2/2)$ on $I = \mathbb{R}$ (Hermite weight).

5.1.2. Is it really calculable?

We immediately note that the matrix of the system of eqns (5.1.8) is symmetric. Furthermore, it is positive definite since

$$\sum_{j,k=0}^n a_j \overline{a_k} \int_a^b w\left(x\right) x^{j+k} \, \mathrm{d}x = \int_a^b w\left(x\right) |P\left(x\right)|^2 \, \mathrm{d}x = \left(P, P\right).$$

We may believe that the system (5.1.8) is very easy to solve numerically, but it is nothing of the sort, as we are going to see in a simple case.

When the weights w are equal to 1 and the interval $[a, b]$ is equal to $[0, 1]$, the matrix of the system (5.1.8) is given by

$$H_{n+1} = \begin{pmatrix} 1 & 1/2 & 1/3 & \cdots & 1/(n+1) \\ 1/2 & 1/3 & 1/4 & \cdots & 1/(n+2) \\ 1/3 & 1/4 & 1/5 & \cdots & 1/(n+3) \\ \vdots & \vdots & \vdots & & \vdots \\ 1/(n+1) & 1/(n+2) & 1/(n+3) & \cdots & 1/(2n+1) \end{pmatrix}.$$

This matrix is called the Hilbert matrix. We can explicitly calculate the inverse of this matrix. An explicit formula can be found in [40]. I will be content with reproducing the inverse of H_6, displaying only the elements in the lower triangle, since H_6 is symmetric. The inverse of H_6 is given by

$$\begin{pmatrix} 36 & & & & & \\ -630 & 14700 & & & & \\ 3360 & -88200 & 564480 & & & \\ -7560 & 211680 & -1411200 & 3628800 & & \\ 7560 & -220500 & 1512000 & -3969000 & 4410000 & \\ -2772 & 83160 & -582120 & 1552320 & -1746360 & 698544 \end{pmatrix}.$$

The examination of the components of this matrix shows that its largest element is of the order of 4×10^6 and, therefore, it is necessary to know the second term of eqn (5.1.8) with an error which is small relative to 10^{-6}, to obtain acceptable results.

In other words, the matrices H_n are very poorly conditioned. Let $|\cdot|$ be some vector norm and let $\|\cdot\|$ be its subordinate operator norm. The *condition number* of a matrix A is the number $\|A\| \|A^{-1}\|$ which allows us to write the sensitivity of the system

$$Ax = b$$

to errors δA in A and δb in b. We shall prove in Subsection 9.5.3 that the relative error $|\delta x|/|x|$ can be estimated by means of the formula

$$\frac{|\delta x|}{|x|} \leqslant \frac{\kappa(A)}{1 - \kappa(A) \|\delta A\|/\|A\|} \left(\frac{|\delta b|}{|b|} + \frac{\|\delta A\|}{\|A\|} \right).$$

The conditioning of H_n, based on the Euclidean norm, is of order

$$\kappa(A_n) \sim e^{7n/2}.$$

We therefore have a problem which is untreatable numerically.

5.2. Orthogonal polynomials

The solution of the system (5.1.8) is very difficult because the column vectors of the matrix of this system are nearly collinear. We notice this particularly in the case of the Hilbert matrix H_n.

Nevertheless, the least-squares approximation is commonly used; we remove the difficulty by choosing an appropriate basis. We return to the system in the form of eqn (5.1.7) and choose a basis of $\mathbb{P}_n$ leading to a system which behaves as well as possible. The best possible behaviour, corresponding to the technical notion of conditioning, as developed in Subsection 9.5.3 (measured with Euclidean norms), is that of unitary matrices. This, therefore, leads us back to finding an orthonormal basis of $\cup_n \mathbb{P}_n$.

5.2.1. Definition and construction of orthogonal polynomials

Definition 5.2.1. We call the sequence of polynomials $P_0, P_1, \ldots, P_n, \ldots$ *orthogonal* relative to a weight w, which is strictly positive almost everywhere and integrable on an interval $[a, b]$, if it has the following properties:

(i) For any n, P_n is of degree n and the coefficient of its term of degree n is 1;

(ii) For any n, P_n is orthogonal to $\mathbb{P}_{n-1}$, that is, all the polynomials of degree strictly less than n. The orthogonal polynomials are ordered from number zero and the n-th orthogonal polynomial is always of degree n.

We call the normalized polynomials

$$\tilde{P}_n = \frac{P_n}{\|P_n\|}$$

orthonormal to a weight w.

We show, first of all, that such sequences exist and give their most elementary properties.

Lemma 5.2.2. For any weight w, which is integrable on the closed bounded interval $[a, b]$, there exists a sequence of orthogonal polynomials satisfying Definition 5.2.1. If

$$(5.2.1) \qquad P_n = x^n - \sum_{i=0}^{n-1} c_{in} P_i$$

then

$$(5.2.2) \qquad c_{in} = \frac{(x^n, P_i)}{(P_i, P_i)}.$$

Proof. By the process of Gram–Schmidt orthonormalization (Theorem 12.1.1) applied to the monomials 1, x, x^2, $\ldots$, x^j in this order, we obtain the sequence of orthonormal polynomials. If we divide each of them by the coefficient of their highest term, we obtain the sequence of orthogonal polynomials. The uniqueness is immediate. □

Using relations (5.2.1) and (5.2.2), we can determine the coefficients of orthogonal polynomials relative to a weight w on an interval $[a, b]$ by quadrature. Generally, it would be necessary to integrate numerically, which is not necessarily very economical. But the situation can be a lot better, in various important cases.

5.2.2. Examples of orthogonal polynomials

For a certain number of simple weights, we know explicit analytic expressions for the orthogonal polynomials.

Theorem 5.2.3. If $[a, b] = [-1, 1]$ and if $w = 1$, then the orthonormal polynomials are given by the following formula:

$$(5.2.3) \qquad P_n\left(x; -1, 1\right) = \sqrt{n + \frac{1}{2}} \frac{1}{n! \, 2^n} \frac{\mathrm{d}^n}{\mathrm{d}x^n}\left[\left(x^2 - 1\right)^n\right]. \qquad \diamond$$

Proof. We show that eqn (5.2.3) defines orthonormal polynomials. To lighten the notation we will let

$$R_n\left(x\right) = \frac{\mathrm{d}^n}{\mathrm{d}x^n}\left[\left(x^2 - 1\right)^n\right].$$

It is clear that R_n is a polynomial of degree n. We will verify that if $p < n$ then there exists a polynomial $r_p(x)$ such that

$$\frac{\mathrm{d}^p}{\mathrm{d}x^p}\left[\left(x^2 - 1\right)^n\right] = r_p\left(x\right)\left(x^2 - 1\right)^{n-p}.$$

This relation is clearly true when $p = 0$. Suppose that it is true for $p-1$. Then,

$$\frac{\mathrm{d}^p}{\mathrm{d}x^p}\left[\left(x^2 - 1\right)^n\right] = \frac{\mathrm{d}}{\mathrm{d}x}\left[r_{p-1}\left(x\right)\left(x^2 - 1\right)^{n-p+1}\right]$$
$$= \left[r'_{p-1}\left(x\right)\left(x^2 - 1\right) + 2x\left(n - p + 1\right)r_{p-1}\left(x\right)\right]\left(x^2 - 1\right)^{n-p}.$$

Suppose that $m \leqslant n$. We calculate the scalar products (R_n, R_m) using integration by parts as follows:

$$\int_{-1}^1 R_n\left(x\right) R_m\left(x\right)\mathrm{d}x = \frac{\mathrm{d}^{n-1}}{\mathrm{d}x^{n-1}}\left[\left(x^2 - 1\right)^n\right] \frac{\mathrm{d}^m}{\mathrm{d}x^m}\left[\left(x^2 - 1\right)^m\right]\Bigg|_{x=-1}^{x=1}$$
$$- \int_{-1}^1 \frac{\mathrm{d}^{n-1}}{\mathrm{d}x^{n-1}}\left[\left(x^2 - 1\right)^n\right] \frac{\mathrm{d}^{m+1}}{\mathrm{d}x^{m+1}}\left[\left(x^2 - 1\right)^m\right]\mathrm{d}x.$$

The integrated term vanishes at $x = \pm 1$. By an elementary induction, we see that

$$\int_{-1}^1 R_n\left(x\right) R_m\left(x\right)\mathrm{d}x = (-1)^p \int_{-1}^1 \frac{\mathrm{d}^{n-p}}{\mathrm{d}x^{n-p}}\left[\left(x^2 - 1\right)^n\right] \frac{\mathrm{d}^{m+p}}{\mathrm{d}x^{m+p}}\left[\left(x^2 - 1\right)^m\right]\mathrm{d}x,$$

for all $p \leqslant n$.

If $m < n$, we can take $p = m + 1$. In this case, the derivative of order $m + p = 2m + 1$ of the term $\left[(x^2 - 1)^m \right]$ is zero, and we see that $(R_n, R_m) = 0$.

If $m = n$, we take $p = n$. The derivative of order $2n$ of the term $(x^2 - 1)^n$ is $(2n)!$ and we have that

$$(R_n, R_n) = (2n)! \int_{-1}^{1} \left(1 - x^2 \right)^n \, dx.$$

Let

$$I_n = \int_{-1}^{1} \left(1 - x^2 \right)^n \, dx$$

and integrate by parts to obtain

$$\int_{-1}^{1} \left(1 - x^2 \right)^n \, dx = x \left(1 - x^2 \right)^n \Big|_{x=-1}^{x=1} - \int_{-1}^{1} x \left(-2nx \right) \left(1 - x^2 \right)^{n-1} \, dx.$$

We therefore have the recurrence

$$I_n = 2n \left(I_{n-1} - I_n \right),$$

so that

$$I_n = \frac{2n}{2n+1} I_{n-1},$$

and, since $I_0 = 2$, we see that

$$I_n = \frac{2n \left(2n - 2 \right) \cdots 2}{\left(2n+1 \right) \left(2n-1 \right) \cdots 3} \, 2 = \frac{2^{n+1} n!}{\left(2n+1 \right) \left(2n \right)!/2^n n!} = \frac{2^{2n+1} \left(n! \right)^2}{\left(2n+1 \right) \left(2n \right)!}.$$

Finally,

$$(R_n, R_n) = \frac{2^{2n+1} \left(n! \right)^2}{\left(2n+1 \right)}.$$

This shows that the $P_n \left(\cdot ; -1, 1 \right)$ form an orthonormal family of polynomials. The degree of $P_n \left(\cdot ; -1, 1 \right)$ is n and the coefficient of the highest term of $P_n \left(\cdot ; -1, 1 \right)$ is positive. They are, therefore, orthonormal polynomials relative to the weight 1 on the interval $[-1, 1]$. □

We call the polynomials Q_n given by

$$Q_n \left(x \right) = \frac{1}{n! \, 2^n} \frac{d^n}{dx^n} \left[\left(x^2 - 1 \right)^n \right]$$

Legendre polynomials.

We can deduce from eqn (5.2.3) the orthonormal polynomials relative to the weight 1 on any interval $[a, b]$ in the following way: we seek them in the form

$$P_n \left(x; a, b \right) = \alpha P_n \left(\frac{2x - a - b}{b - a} ; -1, 1 \right),$$

since the affine mapping $x \mapsto (2x - a - b)/(b - a)$ transforms $[a, b]$ to $[-1, 1]$. We have orthogonality since

$$\int_a^b P_n \left(\frac{2x - a - b}{b - a}; -1, 1 \right) P_m \left(\frac{2x - a - b}{b - a}; -1, 1 \right) dx$$

$$= \frac{1}{2} (b - a) \int_{-1}^1 P_n (x; -1, 1) P_m (x; -1, 1) \, dx = \frac{1}{2} (b - a) \, \delta_{mn}.$$

By choosing

$$\alpha = \sqrt{\frac{2}{b - a}}$$

we ensure the normalization.

Consider now the weight

$$w(x) = \left(1 - x^2 \right)^{-1/2}$$

on the interval $[-1, 1]$. The weight w is singular on $[-1, 1]$, but it is integrable. We are going to explicitly determine the orthogonal polynomials with respect to this weight:

Theorem 5.2.4. Let Arc cos be the inverse function of cos defined by

$$\theta = \text{Arc} \cos x \quad \Longleftrightarrow \quad \theta \in [0, \pi] \quad \text{and} \quad x = \cos \theta.$$

The functions

$$Q_n(x) = \cos(n \, \text{Arc} \cos x),$$

defined on the interval $[-1, 1]$, are relatively orthogonal to the weight

$$w(x) = \left(1 - x^2 \right)^{-1/2}.$$

Furthermore, Q_n is a polynomial of degree n and

$$(Q_n, Q_n) = \begin{cases} \pi/2 & \text{if } n \geqslant 1; \\ \pi & \text{if } n = 0. \end{cases} \qquad \diamond$$

Proof. With the change of variable $x = \cos \theta$, we calculate the following scalar product:

$$(Q_n, Q_m) = \int_{-1}^1 Q_n(x) Q_m(x) w(x) \, dx = \int_0^\pi \cos(n\theta) \cos(m\theta) \, d\theta.$$

If $m \neq n$, we then have

$$(Q_n, Q_m) = \frac{1}{2} \left[\frac{\sin(n + m)\theta}{n + m} + \frac{\sin(n - m)\theta}{n - m} \right]_0^\pi = 0.$$

Therefore, the Q_n are pairwise orthogonal. Furthermore, if $m = n \neq 0$,

$$(Q_n, Q_n) = \frac{1}{2} \left[\frac{\sin 2n\theta}{2n} + \theta \right]_0^{\pi} = \frac{\pi}{2}.$$

If $n = m = 0$,

$$(Q_0, Q_0) = \pi.$$

We show that Q_n are polynomials. Indeed, for $n = 0$, we have

$$Q_0 = 1,$$

which is obviously a polynomial. For $n = 1$,

$$Q_1 = x,$$

which is again a polynomial. We calculate $Q_2(x)$:

$$Q_2(x) = \cos(2 \operatorname{Arc\,cos} x) = 2 \cos^2(\operatorname{Arc\,cos} x) - 1 = 2x^2 - 1.$$

We are going to establish a recurrence relation on the Q_n. Still letting $\theta = \operatorname{Arc\,cos} x$, we can write

$$Q_{n-1}(x) + Q_{n+1}(x) = \cos(n-1)\theta + \cos(n+1)\theta$$
$$= 2 \cos\theta \cos n\theta = 2x\, Q_n(x).$$

Consequently,

$$(5.2.4) \qquad\qquad Q_{n+1}(x) = 2x\, Q_n(x) - Q_{n-1}(x).$$

Since the degree of Q_0 is 0 and the degree of Q_1 is 1, it suffices to refer to formula (5.2.4) to see that the degree of Q_n is n. Furthermore, if a_n is the coefficient of the term of degree n in Q_n, we deduce from the recurrence relation (5.2.4) that $a_n = 2^{n-1}$, for $n \geqslant 1$. $\qquad\square$

The functions $\tilde{P}_n$ defined by

$$\tilde{P}_n = \begin{cases} Q_n \sqrt{2/\pi} & \text{if } n \neq 0; \\ Q_0 \sqrt{1/\pi} & \text{if } n = 0 \end{cases}$$

form an orthonormal family of polynomials. These are the polynomials which are orthonormal relative to the weight $(1-x^2)^{-1/2}$. The polynomials $P_n = 2^{-n+1} Q_n$ are the orthogonal polynomials relative to the weight w in the sense of Definition 5.2.1. These are called Chebyshev polynomials.

5.2.3. Revival of special functions

The study of orthogonal polynomials, a great subject of the nineteenth century, has known a recent resurge of interest. Indeed, it is a chapter in the theory of special functions to which mathematicians have paid little attention in the twentieth century, but which theoretical physicists, or, at least, certain specialists in quantum mechanics, have found to be of great interest, as we can obtain explicit expressions of coherent states and their energies by means of special functions. In a more mathematical language, special functions play a part in the spectra of certain infinite-dimensional self-adjoint operators. The extreme weakness of communication between physicists and mathematicians on the subject is cuttingly described in the preface of [62] by Richard Askey. It tells an impressive story of the duplication of effort and of mutual ignorance, compounded by the difficulty of getting access to certain slightly old works.

A very readable work on special functions from a classical point of view, which has several applications in physics, is given in Nikiforov and Uvarov [64]. Equally good is Miller [62] at a higher level.

Special functions and, in particular, the celebrated hypergeometric function, appear in all sorts of counting problems. They play as much a part in combinatorics and computing as in probability and the theory of numbers. To see some combinatorial applications consult the marvellous Graham *et al.* [38], and the more difficult Fine [29]. Some applications in probability are presented in [38], but, above all, it is recommendable to read Feller [27], which is a masterpiece on discrete probability.

From a strictly numerical point of view, orthogonal polynomials are useful outside of polynomial approximation theory. They appear in the convergence acceleration theory of Padé approximations. In this theory, we approximate functions by rational fractions, whilst requiring that the order of approximation at a point is maximal amongst all rational functions whose numerator and denominator are of a given maximum degree. We refer to, for example, the work of Brezinski [10], or the older book by G. A. Baker [6].

We also find orthogonal polynomials when we solve partial differential equations by spectral and pseudo-spectral methods, which are very powerful in simple geometries. These are used, for better or for worse, in weather forecasting and the study of global climatic models. The greenhouse effect, due to carbon dioxide and other gases, seems to lead to global warming. This deduction depends on a large number of calculations using orthogonal polynomials, and their use seems to be growing a lot quicker than the level of the oceans—fortunately! Introductions can be found in [36, 37].

Special functions have never ceased to be part of physics culture, although they have almost disappeared from the training of mathematicians contemporary with the author of these lines. Weather calculations are the calculations of physicists, who have never asked the permission of mathematicians to do them, luckily. It is therefore very difficult to declare a theory dead, as some seem, like the phoenix, to be reborn from their ashes when the time is right.

5.2.4. Orthogonal polynomials and least-squares

We can now solve the system (5.1.7). Since the $\tilde{P}_j$ form an orthonormal basis of $\mathbb{P}_n$, it is equivalent to have the relation (5.1.7) and

$$(5.2.5) \qquad (P, \tilde{P}_j) = (f, \tilde{P}_j), \quad \forall j \in \{0, \dots, n\}.$$

From this, we immediately deduce that

$$P = \sum_{j=0}^{n} (f, \tilde{P}_j) \tilde{P}_j.$$

Relation (5.1.7), or equivalently eqn (5.2.5), expresses that P is the orthogonal projection of f on the space $\mathbb{P}_n$. Furthermore, relation (5.1.9) for $Q = 0$ implies that

$$(f - P, f - P) + (P, P) = (f, f),$$

that is

$$(5.2.6) \qquad \sum_{j=0}^{n} |(f, \tilde{P}_j)|^2 \leqslant (f, f).$$

The right-hand term of eqn (5.2.6) is independent of n. Consequently, we can bound from above the supremum of the left-hand term by (f, f), that is

$$(5.2.7) \qquad \sum_{j=0}^{\infty} |(f, \tilde{P}_j)|^2 \leqslant (f, f).$$

This is *Bessel's inequality*.

5.3. Polynomial density: Bernstein polynomials

Amongst other things, the Stone–Weierstrass approximation theorem permits us to confirm that the polynomials on the compact interval $[a, b]$ are dense in $C^0([a, b])$. The original proof of this result is not very constructive. We will present here the proof by Bernstein.

Recall that Lagrangian interpolation has poor convergence properties. In fact, the error estimate (Theorem 4.3.1) requires a lot of regularity on the functions that we are interpolating. The idea is to use polynomials in greater number than in the case of interpolation, without demanding that the values coincide at the knots.

5.3.1. Modulus of continuity

Recall the definition of the modulus of continuity of a function f which is continuous on a compact set K. It is the increasing function ω from $\mathbb{R}^+$ to itself, vanishing at zero, and such that

$$|f(x) - f(y)| \leqslant \omega(|x - y|), \quad \forall x, y \in K.$$

The function ω can be obtained from the formula

$$\omega(h) = \sup_{\substack{x,y \in K \\ |x-y| \leqslant h}} |f(x) - f(y)|.$$

We deduce immediately that such a function is increasing and vanishes at zero.

Lemma 5.3.1. Every function f which is continuous on a compact convex set K of $\mathbb{R}^n$ possesses a modulus of continuity which is continuous at 0.

Proof. We are going to verify that the function ω defined above is continuous. The continuity of ω at 0 is equivalent to the uniform continuity of f, which is true since f is continuous on a compact set K.

Furthermore, ω has the property of sub-additivity. In other words, for any h_1 and h_2, we have

$$\omega(h_1 + h_2) \leqslant \omega(h_1) + \omega(h_2).$$

Indeed, if x and y are two elements of K such that $|x - y| \leqslant h_1 + h_2$, we can find a point z on the segment $[x, y]$ joining x and y such that $|x - z| \leqslant h_1$ and $|z - y| \leqslant h_2$. We then have

$$|f(x) - f(y)| \leqslant |f(x) - f(z)| + |f(z) - f(y)| \leqslant \omega(h_1) + \omega(h_2).$$

Then, let $h > 0$ be fixed and let h' tend to zero from above. We have

$$\omega(h + h') \leqslant \omega(h) + \omega(h'),$$

from which we deduce that

$$\limsup_{h' \to 0} \omega(h + h') \leqslant \omega(h).$$

However, as ω is increasing, that is, for h and h' positive or zero, $\omega(h+h') \geqslant \omega(h)$, we can pass to the limit and

$$\liminf_{h' \to 0} \omega(h + h') \geqslant \omega(h).$$

The combination of this relation with the preceding one shows that ω is right-continuous. In the same way,

$$\omega(h) \leqslant \omega(h - h') + \omega(h'),$$

which implies that

$$\liminf_{h' \to 0} \omega(h - h') \geqslant \omega(h).$$

Combining this last property with the fact that ω is increasing, we see that ω is left-continuous. □

5.3.2. Bernstein polynomials and Bernstein approximation

We naturally restrict ourselves to the interval $[0, 1]$ and let

$$\beta_{n,j}(x) = C_n^j \, x^j \, (1 - x)^{n-j} .$$

If f is a continuous function on $[0, 1]$, the Bernstein polynomial approximation of f is defined by

$$(5.3.1) \qquad B_n(f, x) = \sum_{j=0}^{n} \beta_{n,j}(x) \, f\left(\frac{j}{n}\right).$$

Bernstein's theorem is stated as follows:

Theorem 5.3.2. Let f be a continuous function of $[0, 1]$, let ω be its modulus of continuity, and let $B_n(f, x)$ be the Bernstein approximation polynomial of degree n. We have the following estimate:

$$(5.3.2) \qquad \max_{x \in [0,1]} |f(x) - B_n(f, x)| \leqslant \frac{9}{4} \omega\left(n^{-1/2}\right). \qquad\qquad \diamond$$

Proof. From the binomial theorem

$$(5.3.3) \qquad (a + b)^n = \sum_{j=0}^{n} C_n^j \, a^j \, b^{n-j},$$

we will deduce that the $\beta_{n,j}$ satisfy the following relations:

$$(5.3.4) \qquad \sum_{j=0}^{n} \beta_{n,j}(x) = 1,$$

$$(5.3.5) \qquad \sum_{j=0}^{n} \frac{j}{n} \beta_{n,j}(x) = x,$$

$$(5.3.6) \qquad \sum_{j=0}^{n} \frac{j^2}{n^2} \beta_{n,j}(x) = \left(1 - \frac{1}{n}\right) x^2 + \frac{1}{n} x.$$

First of all, we choose $a = x$ and $b = 1 - x$ in eqn (5.3.3), which gives us eqn (5.3.4). If we differentiate eqn (5.3.3) with respect to a once, we obtain

$$n(a + b)^{n-1} = \sum_{j=1}^{n} j \, C_n^j \, a^{j-1} \, b^{n-j},$$

from which we deduce, on multiplying by a and dividing by n, that

$$a(a + b)^{n-1} = \sum_{j=1}^{n} \frac{j}{n} C_n^j \, a^j \, b^{n-j}.$$

On substituting $a = x$ and $b = 1 - x$, we obtain eqn (5.3.5).

We differentiate eqn (5.3.3) once more with respect to a, to get

$$n\,(n-1)\,(a+b)^{n-2} = \sum_{j=2}^{n} C_n^j\, j\,(j-1)\, a^{j-2}\, b^{n-j}.$$

On multiplying by a^2 and dividing by n^2, we have

$$\left(1 - \frac{1}{n}\right) a^2\,(a+b)^{n-2} = \sum_{j=0}^{n} \left(\frac{j^2}{n^2} - \frac{j}{n^2}\right) C_n^j\, a^j\, b^{n-j}.$$

Employing the same substitution as the preceding one, we obtain

$$\sum_{j=0}^{n} \left(\frac{j^2}{n^2} - \frac{j}{n^2}\right) \beta_{n,j}\,(x) = \left(1 - \frac{1}{n}\right) x^2.$$

From this, and using eqn (5.3.5), we deduce that

$$\sum_{j=0}^{n} \frac{j^2}{n^2} \beta_{n,j}\,(x) = \sum_{j=0}^{n} \left(\frac{j^2}{n^2} - \frac{j}{n^2} + \frac{j}{n^2}\right) \beta_{n,j}\,(x) = \left(1 - \frac{1}{n}\right) x^2 + \frac{1}{n} x,$$

which is eqn (5.3.6).

The error between f and $B_n\,(f, \cdot)$ is defined by

$$e_n\,(f, x) = f\,(x) - \sum_{j=0}^{n} \beta_{n,j}\,(x)\, f\!\left(\frac{j}{n}\right).$$

Identity (5.3.4) allows us to transform this error expression into

$$e_n\,(f, x) = \sum_{j=0}^{n} \beta_{n,j}\,(x) \left(f\,(x) - f\!\left(\frac{j}{n}\right)\right),$$

which we bound from above, using the triangle inequality, by

$$(5.3.7) \qquad |e_n\,(x)| \leqslant \sum_{j=0}^{n} \beta_{n,j}\,(x) \left| f\,(x) - f\!\left(\frac{j}{n}\right) \right|.$$

Fixing x, we are going to bound from above the terms of the sum appearing in eqn (5.3.7), differently according to whether j/n is close to x or not. To do this, we use a positive parameter δ, which we will fix later, and note that, if $|x - (j/n)| \leqslant \delta$, then

$$\left| f\,(x) - f\!\left(\frac{j}{n}\right) \right| \leqslant \omega\,(\delta).$$

Consequently,

$$(5.3.8) \qquad \sum_{j:|x-(j/n)|\leqslant\delta} \beta_{n,j}(x) \left| f(x) - f\left(\frac{j}{n}\right) \right| \leqslant \omega(\delta).$$

Conversely, if $|x - (j/n)| > \delta$, let p be the integer part of $|x - (j/n)|/\delta$, that is, the unique integer such that

$$(5.3.9) \qquad p\delta \leqslant \left| x - \frac{j}{n} \right| < (p+1)\delta.$$

Let $y_0, y_1, \ldots, y_{p+1}$ be the points

$$y_0 = x, \quad \ldots, \quad y_k = x + \frac{k}{p+1}\left(\frac{j}{n} - x\right), \quad \ldots, \quad y_{p+1} = \frac{j}{n}.$$

As the points y_k are pairwise separated by a distance which is, at most, equal to δ, we see that

$$\left| f(x) - f\left(\frac{j}{n}\right) \right| \leqslant |f(x) - f(y_1)| + \ldots + |f(y_k) - f(y_{k+1})| + \ldots$$

$$+ \left| f(y_p) - f\left(\frac{j}{n}\right) \right| \leqslant (p+1)\omega(\delta).$$

By virtue of eqn (5.3.9), we see that

$$\left| f(x) - f\left(\frac{j}{n}\right) \right| \leqslant \omega(\delta)\left(1 + \frac{1}{\delta}\left| x - \frac{j}{n} \right|\right).$$

However, since $|x - (j/n)| > \delta$, we bound $|x - (j/n)|/\delta$ by its square and, therefore,

$$\left| f(x) - f\left(\frac{j}{n}\right) \right| \leqslant \omega(\delta)\left(1 + \frac{1}{\delta^2}\left(x - \frac{j}{n}\right)^2\right).$$

Hence, we have

$$\sum_{j:|x-(j/n)|>\delta} \beta_{n,j}(x) \left| f(x) - f\left(\frac{j}{n}\right) \right|$$

$$\leqslant \omega(\delta)\left[\sum_{j=0}^{n} \beta_{n,j}(x) + \frac{1}{\delta^2}\sum_{j=0}^{n}\left(x - \frac{j}{n}\right)^2 \beta_{n,j}(x)\right].$$

The first sum within the bracket is equal to 1 and the second is calculated by means of the formulae (5.3.4)–(5.3.6), and as follows:

$$\sum_{j=0}^{n}\left(x - \frac{j}{n}\right)^2 \beta_{n,j}(x) = x^2 - 2x^2 + \left(1 - \frac{1}{n}\right)x^2 + \frac{1}{n}x = \frac{x(1-x)}{n}.$$

The maximum of the function $x \mapsto x(1-x)$ on the interval $[0,1]$ is achieved at $x = 1/2$ and has the value $1/4$. We obtain

$$\sum_{j:|x-(j/n)|>\delta} \beta_{n,j}(x) \left| f(x) - f\left(\frac{j}{n}\right) \right| \leqslant \omega(\delta) \left(1 + \frac{1}{4n\delta^2} \right).$$

Combining this last expression with eqn (5.3.8), we have the error bound

$$(5.3.10) \qquad\qquad |e_n| \leqslant \omega(\delta) \left(2 + \frac{1}{4n\delta^2} \right).$$

If we choose $\delta = 1/\sqrt{n}$, we obtain eqn (5.3.2). $\qquad\square$

In Figs 5.1 and 5.2 we present the graphs of the four Bernstein polynomials of degree 3, and the Bernstein approximation of degree 3 of $x \mapsto \sin(\pi x/2)$. Notice that this Bernstein approximation is not very accurate. It is precisely for this reason that it will be stable. An essential trait of Bernstein polynomials is that they oscillate very little, and it is this that allows the proof to work.

A consequence of Theorem 5.3.2 is the following density result, the proof of which is left to the reader:

Corollary 5.3.3. Let $[a,b]$ be a compact interval. Polynomials are dense in $C^0([a,b])$.

If the function f is Lipschitz with respect to its coefficients, its modulus of continuity ω satisfies

$$(5.3.11) \qquad\qquad \omega(\delta) \leqslant L\delta,$$

for some finite positive number L. Consequently, the estimate (5.3.10) becomes

$$|e_n| \leqslant L\delta \left(2 + \frac{1}{4n\delta^2} \right),$$

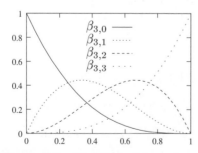

Figure 5.1: The Bernstein polynomials of degree 3.

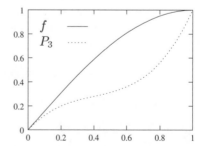

Figure 5.2: Approximation of $f(x) = \sin(\pi x/2)$ over $[0,1]$ by $P_3(x) = B_3(f,x)$.

and the choice $\delta = 1/(2\sqrt{2n})$ leads to the estimate

$$|e_n| \leqslant L\sqrt{\frac{2}{n}}.$$

Even if f is very differentiable, the Bernstein approximation polynomials cannot converge quicker than n^{-2}. Indeed, we can show that, if f is C^2, then

$$\lim_{n \to \infty} n^2 \left[B_n\left(f, x\right) - f\left(x\right) \right] = \frac{1}{2} f''\left(x\right) x \left(1 - x\right).$$

5.3.3. Application of Bernstein polynomials to graphics software: the Bézier curves

Bernstein polynomials have known a new vogue since Bézier [9] and Casteljau [22] proposed numerical methods for the approximation of surfaces by bi-dimensional generalizations of them. First used in the context of the automobile industry, these generalizations have appeared extensively in graphics software of recent years. Indeed, Bézier curves and surfaces have great numerical stability, conveniently allowing the calculation of certain partial derivatives, and are obtained by economic algorithms. The interest in graphics software is creating a new field of research, at the crossroads of algebraic geometry, differential geometry, and computing. Here is an example of an open problem: there are formal calculation programs which can find the intersections of two algebraic surfaces which have equations with rational coefficients. However, in an industrial context, the data is rarely known with very great precision, and we do not know of a good result on the stability of the intersections with respect to the coefficients.

Given $n + 1$ points $\boldsymbol{x}_i$, $0 \leqslant i \leqslant n$, in the space $\mathbb{R}^d$, a Bézier curve is parameterized by

$$\boldsymbol{X}\left(t; \boldsymbol{x}_0, \ldots, \boldsymbol{x}_n\right) = \sum_{j=0}^{n} \beta_{n,j}\left(t\right) \boldsymbol{x}_j.$$

The nice feature of a Bézier curve is the geometric insight given by this parameterization, in contrast to a representation in another basis of polynomials, such as the basis of monomials, the Lagrange basis, or the Newton basis. Indeed, we see immediately that

$$\boldsymbol{X}\left(0; \boldsymbol{x}_0, \ldots, \boldsymbol{x}_n\right) = \boldsymbol{x}_0 \quad \text{and} \quad \boldsymbol{X}\left(1; \boldsymbol{x}_0, \ldots, \boldsymbol{x}_n\right) = \boldsymbol{x}_n.$$

Moreover, if $n \geqslant 2$, the tangents at $\boldsymbol{X}(0)$ and $\boldsymbol{X}(1)$ have directions given by

$$\boldsymbol{X}'\left(0; \boldsymbol{x}_0, \ldots, \boldsymbol{x}_n\right) = n\left(\boldsymbol{x}_1 - \boldsymbol{x}_0\right) \quad \text{and} \quad \boldsymbol{X}'\left(1; \boldsymbol{x}_0, \ldots, \boldsymbol{x}_n\right) = n\left(\boldsymbol{x}_n - \boldsymbol{x}_{n-1}\right).$$

More generally, the k-th derivative at 0, $\boldsymbol{X}^{(k)}(0)$, can be expressed as a linear combination of the vectors $\boldsymbol{x}_0, \ldots, \boldsymbol{x}_k$, and a similar statement holds for the k-th derivative at 1.

Furthermore, there is a very geometric construction of the points of a Bézier curve, obtained by a triangular algorithm which is reminiscent of the Pascal triangle. The Bernstein basis functions satisfy the following recursive definition:

$$\beta_{n,j}(t) = t\beta_{n-1,j-1}(t) + (1-t)\beta_{n-1,j}(t).$$

Therefore, it is clear that the curves $X(\cdot\,;x_0,\ldots,x_n)$ also satisfy the recursive identity:

$$X(t;x_0,\ldots,x_n) = tX(t;x_0,\ldots,x_{n-1}) + (1-t)X(t;x_1,\ldots,x_n).$$

Geometrically, this means that $X(t;x_0,\ldots,x_n)$ is obtained as the barycentre of $X(t;x_0,\ldots,x_{n-1})$ with weight t and $X(t;x_1,\ldots,x_n)$ with weight $1-t$. Hence, it can be obtained by taking the barycentres between x_j and x_{j+1}, $0 \leqslant j \leqslant n-1$, and then the $n-1$ barycentres between the previously constructed barycentres, and so on, in $n(n+1)/2$ operations. If n is not very large, this is a very efficent algorithm, which is depicted in Fig. 5.3. This construction is called the de Casteljau algorithm.

The points x_j are called control points. Graphics users learn very quickly that when pulling out a control point, the Bézier curve follows it. However, Bézier curves suffer from the limitations of all polynomial approximations: rigidity, and hence lack of stability. Indeed, it is obvious that changing one of the control

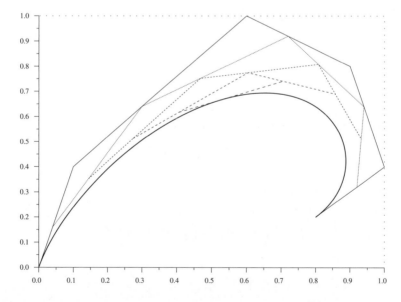

Figure 5.3: The thin solid line is the control polygon, the thicker solid curve is the Bézier curve, and the successive dotted lines are the barycentres' lines, the weight being $t = 0.6$.

points in a Bézier curve changes the whole of the curve. This is the reason why it is more advantageous to use piecewise Bézier curves; but how does one control the continuity at the break points? Obtaining this control is the reason for the use of B-splines in CAGD, see Subsection 6.3.5.

5.4. Least-squares convergence of a polynomial approximation

The density of the polynomials in the space of continuous functions on a compact interval allows us to show that the least-squares approximation to a function f converges to this function in a sense that we will now define precisely:

Theorem 5.4.1. Let w be a weight which is strictly positive almost everywhere and integrable on the compact interval $[a, b]$. Let f be a continuous function on $[a, b]$ and let Q_n be the polynomial of degree at most n which is its least-squares approximation relative to w. Then, as n tends to infinity, Q_n converges to f in the quadratic mean (with the weight w), that is

$$\lim_{n \to \infty} \int_a^b |f(x) - Q_n(x)|^2 \, w(x) \, \mathrm{d}x = 0.$$

Furthermore, we have Parseval's relation:

$$(5.4.1) \qquad \sum_{j=0}^{\infty} |(f, \tilde{P}_j)|^2 = \int_a^b |f(x)|^2 \, w(x) \, \mathrm{d}x. \qquad \diamond$$

Proof. By the definition of least-squares approximation, the following inequality holds for every polynomial $R \in \mathbb{P}_n$:

$$(f - Q_n, f - Q_n) \leqslant (f - R, f - R).$$

If we take the polynomial R to be the Bernstein polynomial $B_n(f, x)$, then, since $|f(x) - B_n(f, x)| \leqslant 9\,\omega(1/\sqrt{n})/4$, we see that

$$(f - Q_n, f - Q_n)^{1/2} \leqslant \frac{9\omega(1/\sqrt{n})}{4} \left(\int_a^b w(x) \, \mathrm{d}x \right)^{1/2}.$$

This proves the first assertion.

Relation (5.1.9) with $P = Q_n$ and $Q = 0$ can be written as

$$(f - Q_n, f - Q_n) + (Q_n, Q_n) = (f, f).$$

From this, we get

$$\sum_{j=0}^{n} |(f, \tilde{P}_j)|^2 \geqslant (f, f) - (f - Q_n, f - Q_n).$$

We can pass to the limit due to the first assertion and then conclude with the aid of the Bessel inequality (5.2.7). □

Convergence in the quadratic mean is not very precise. In particular, it in no way implies uniform convergence. We give an example of this phenomenon. On the line $\mathbb{R}$, consider a continuous function f which is positive, not identically zero, and which has compact support. We define a sequence of functions f by writing

$$f_n(x) = \frac{1}{\alpha_n} f\left(\frac{x}{\beta_n}\right).$$

If we choose a sequence α_n which decreases to zero, the maximum of f_n tends to infinity. Now we choose β_n in such a way that f_n tends to 0 in its quadratic mean. For this, it is necessary that

$$\int_{\mathbb{R}} f_n(x)^2 \, dx = \frac{1}{\alpha_n^2} \int f\left(\frac{x}{\beta_n}\right)^2 dx = \frac{\beta_n}{\alpha_n^2} \int f(y)^2 \, dy$$

tends to zero when n tends to infinity. We therefore choose β_n such that

$$\frac{\beta_n}{\alpha_n^2} \to 0.$$

Nevertheless, we can obtain a uniformly convergent result in the case where the weight is 1 by imposing regularity conditions on f. For example, we have the following result:

Theorem 5.4.2. Let f be C^2 on the interval $[0,1]$ and let Q_n be its least-squares approximation relative to the weight 1. Then, for every $\epsilon > 0$, there exists an N such that, for all $n \geqslant N$,

$$\max_{x\in[0,1]} |f(x) - Q_n(x)| \leqslant \frac{\epsilon}{\sqrt{n}}. \qquad \diamond$$

The proof of this result may be found in [51]. From the analytical point of view, the situation is much better than in the interpolation case since, for C^2 functions, we always have uniform convergence of least-squares approximations on any interval when the weight is 1.

5.5. Qualitative properties of orthogonal polynomials

We now present some general properties of orthogonal polynomials, which will be of use later.

Theorem 5.5.1. Let w be an integrable weight which is strictly positive almost everywhere on the compact interval $[a,b]$. Then, for any n, all the roots of the n-th orthogonal polynomial P_n are real and simple. Moreover, these roots belong to the interval $]a,b[$. $\qquad \diamond$

Proof. Let $x_1, x_2, \ldots, x_j$ be the roots of P_n in the interval $]a,b[$, listed with their multiplicities. The number j is at most equal to n and it could be zero.

Suppose that it is strictly less than n. Since the coefficients of P_n are real, P_n will change sign at every root with odd multiplicity. If $j > 0$, we let

$$Q(x) = \prod_{k=1}^{j} (x - x_k)^{\varepsilon(k)},$$

where $\varepsilon(k)$ is 1 if the multiplicity of x_j is odd and 0 otherwise. If $j = 0$, we take $Q = 1$. The product $P_n Q$ does not change sign in $]a, b[$. On the other hand, Q is of degree at most $n - 1$. Therefore, we have

$$\int_a^b P_n(x) Q(x) w(x) \, dx = 0.$$

As the expression to be integrated does not change sign, it must vanish almost everywhere, which is a contradiction.

The roots of P_n are all in $]a, b[$. It remains to see that they are simple. Suppose that there is a multiple root, denoted by x_1. Then, $P_n(x) = p(x)(x - x_1)^2$, and p and P_n have the same sign. Since p is of degree at most $n - 2$, we see that

$$\int_a^b P_n(x) p(x) w(x) \, dx = 0.$$

As before, we have a contradiction. □

Orthogonal polynomials satisfy a remarkable recurrence relation given by the theorem which follows:

Theorem 5.5.2. Let w be a weight which is integrable and strictly positive almost everywhere on the compact interval $[a, b]$. Then, for all $n \geqslant 1$, the orthonormal polynomials P_{n+1}, P_n, and P_{n-1} are linked by the following recurrence relation:

$$P_{n+1} = (A_n x + B_n) P_n - C_n P_{n-1},$$

where the constants A_n, B_n, and C_n depend only on the polynomials P_{n+1}, P_n, and P_{n-1}. ◇

Proof. Denote by a_k and b_k the coefficients of the terms of P_k of degree k and $k - 1$, respectively. The polynomial $P_{n+1} - Ax P_n$ is, in general, of degree $n + 1$. It will be of degree at most n if its term of highest degree vanishes, that is, if $a_{n+1} - A a_n = 0$. We therefore let

$$A_n = \frac{a_{n+1}}{a_n}.$$

Let $Q_n = P_{n+1} - A_n x P_n$. We expand this polynomial over the basis P_j, for $0 \leqslant j \leqslant n$, as follows:

$$Q_n = \sum_{j=0}^{n} \alpha_j P_j.$$

We have

$$\alpha_j = (Q_n, P_j) = (P_{n+1}, P_j) - A_n (xP_n, P_j)$$
$$= (P_{n+1}, P_j) - A_n (P_n, xP_j).$$

If xP_j is of degree less than or equal to $n - 1$, then P_n is orthogonal to xP_j and, therefore, α_j is zero for $j \leqslant n - 2$. Consequently,

$$Q_n = \alpha_n P_n + \alpha_{n-1} P_{n-1}.$$

As xP_{n-1} is a polynomial of degree n, we can write it in the form

$$xP_{n-1} = \frac{a_{n-1}}{a_n} P_n + q_{n-1},$$

where the degree of q_{n-1} is at most equal to $n - 1$. We can therefore calculate

$$\alpha_{n-1} = (P_{n+1} - A_n xP_n, P_{n-1}) = -A_n \left(\frac{a_{n-1}}{a_n} P_n + q_{n-1}, P_n \right)$$
$$= -A_n \frac{a_{n-1}}{a_n}.$$

Hence, we take

$$C_n = -\alpha_{n-1} = \frac{a_{n-1} a_{n+1}}{a_n^2}.$$

It remains to calculate B_n. We write

$$xP_n = \frac{a_n}{a_{n+1}} P_{n+1} + \left(b_n - \frac{b_{n+1} a_n}{a_{n+1}} \right) P_n + r_{n-1},$$

where r_{n-1} is a polynomial of degree at most $n - 1$. We therefore have

$$\alpha_n = (P_{n+1} - A_n xP_n, P_n) = -A_n \left(b_n - \frac{b_{n+1} a_n}{a_{n+1}} \right),$$

from which we get

$$B_n = \frac{b_{n+1} a_n - b_n a_{n+1}}{a_n}.$$

We have thus calculated the three coefficients A_n, B_n, and C_n. $\qquad \square$

5.6. Exercises from Chapter 5

5.6.1. Laguerre polynomials

Exercise 5.6.1. Let P be a polynomial of degree d and let p be any positive integer or zero. Show that the function

$$Q(x) = e^x \frac{d^p}{dx^p} \left(P(x) e^{-x} \right)$$

is a polynomial and calculate its degree. Derive the coefficient of the highest term of Q as a function of p and the highest degree term of P.

Exercise 5.6.2. We define the Laguerre polynomials by

$$L_n(x) = \frac{e^x}{n!} \frac{d^n}{dx^n} \left(x^n e^{-x} \right).$$

We denote by E the set of continuous functions f on $[0, +\infty)$ such that

$$\int |f(x)|^2 e^{-x} \, dx < \infty.$$

We equip E with a pre-Hilbertian scalar product

$$(f \mid g) = \int_0^\infty f(x) g(x) e^{-x} \, dx.$$

Show that the L_n are orthogonal relative to this scalar product. It is sufficient to show that L_n is orthogonal to x^p for $p < n$, by integrating by parts a sufficient number of times.

Exercise 5.6.3. Calculate $(L_n \mid x^n)$.

Exercise 5.6.4. Calculate the coefficient of the term of L_n which has degree n. From this, deduce $(L_n \mid L_n)$.

Exercise 5.6.5. Let $L'_n = L_n + M_n$. Calculate the highest degree term of M_n and deduce the value of $(L_n \mid M_n)$.

Exercise 5.6.6. Calculate $(M_n \mid M_n)$ and show that

$$(L'_n \mid L'_n) = n.$$

Exercise 5.6.7. Calculate the decomposition of $\phi(x) = e^{-\alpha x}$, for $\alpha > 0$, over the basis L_n. Show that the partial sums

$$\sum_{n=0}^m (L_n, \phi) L_n \quad \text{and} \quad \sum_{n=0}^m (L_n, \phi) L'_n$$

converge in E when m tends to infinity. What are their respective limits?

Exercise 5.6.8. Calculate $L_n(0)$. Show that the partial sums

$$\sum_{n=0}^m (L_n, \phi) L_n(0)$$

converge in $\mathbb{R}$ when m tends to infinity.

Exercise 5.6.9. Deduce from the two preceding questions that the partial sums

$$\phi_m = \sum_{n=0}^m (L_n, \phi) L_n$$

uniformly converge to ϕ on compact subsets of $\mathbb{R}^+$. We can estimate the term $\phi_m(x) - \phi(x)$ by noting that

$$\left|\phi_m(x)^2 - \phi(x)^2\right| \leqslant \left|\phi_m(0)^2 - \phi(0)^2\right|$$
$$+ \int_0^x \left|2\phi_m(x)\,\phi'_m(x) - 2\phi(x)\,\phi'(x)\right| dx.$$

5.6.2. Padé type and Padé approximations

Introduction: formal series and orthogonality with respect to indefinite quadratic forms

The vector space of formal series $\mathbb{R}[[x]]$ is formed from the sequences $(c_j)_{j\in\mathbb{N}}$ with coefficients in $\mathbb{R}$. We do not impose any restriction on the growth of the $|c_j|$ at infinity. We associate with $(c_j)_{j\in\mathbb{N}}$ the following expression, for which the radius of convergence could be zero:

$$(5.6.1) \qquad\qquad f(x) = \sum_{j\geqslant 0} c_j x^j.$$

For formal series we define some operations analogous to those on polynomials. The sum of two formal series $(c_j)_j$ and $(c'_j)_j$ is $(c_j + c'_j)_j$. The product of a formal series with a scalar λ is $(\lambda c_j)_j$. Finally, by analogy with the product of two polynomials, the product of two formal series $(c_j)_j$ and $(c'_j)_j$ is the formal series defined by

$$c''_j = \sum_{k=0}^{j} c_k c'_{j-k}.$$

It goes without saying that if f and f', associated with $(c_j)_j$ and $(c'_j)_j$, respectively, have a strictly positive radius of convergence, then the same applies for f'' associated with $(c''_j)_j$ and, as expected,

$$f''(x) = f(x)\,f'(x).$$

The order of a formal series is the largest index j such that, for every $k < j$, $c_k = 0$.

If $W(t)/V(t)$ is a rational function such that $V(t)f(t) - W(t)$ is of order k, we will use the simpler notation

$$f(t) - \frac{W(t)}{V(t)} = O\left(t^k\right).$$

In the case where f converges in a neighbourhood of 0, this notation agrees with the usual notation.

In this problem, we focus on a formal series $(c_j)_j$ and the corresponding expression f.

We define a linear function on the space of real polynomials $\mathbb{P}$ by giving its value on each element of the basis of monomes $1, x, x^2, \ldots, x^j, \ldots$ as follows:

$$(5.6.2) \qquad\qquad c\left(x^j\right) = c_j.$$

We will think of eqn (5.6.2) as a generalization of the formula

$$c_j = \int_a^b x^j w\left(x\right) \mathrm{d}x,$$

where w is a weight, which is positive almost everywhere and integrable on $[a, b]$.

If $g(x, t)$ is a formal series in two variables of the form

$$(5.6.3) \qquad\qquad g\left(x, t\right) = \sum_{j,k \geqslant 0} \gamma_{jk} x^j t^k,$$

which satisfies the condition

$$(5.6.4) \qquad\qquad \{j : \gamma_{jk} \neq 0\} \text{ is finite}, \quad \forall k \geqslant 0,$$

we naturally define a new formal series by

$$c\big(x \mapsto g\left(x, t\right)\big) = \sum_{k \geqslant 0} t^k \sum_{j \geqslant 0} \gamma_{jk} c_j.$$

In summary, and with the exception of the verification of condition (5.6.4), we shall work with formal series in the same way as with polynomials.

First part: Padé type approximations

Exercise 5.6.10. Show that

$$f\left(t\right) = c\left(x \mapsto \frac{1}{1 - xt}\right).$$

Use the formal series expansion

$$\frac{1}{1 - xt} = 1 + xt + x^2 t^2 + \ldots$$

and show that it satisfies the condition (5.6.4).

Exercise 5.6.11. Let $\mathbb{P}_k$ be the space of polynomials of degree at most k, and let $v \in \mathbb{P}_k$ be given by

$$v\left(x\right) = \sum_{j=0}^k b_j x^j.$$

We write $v[x,t] = (v(x) - v(t))/(x - t)$. This is a divided difference. Let

$$(5.6.5) \qquad w(t) = c\left(x \mapsto v\left[x, t\right]\right).$$

Show that w belongs to $\mathbb{P}_{k-1}$. Calculate the coefficient a_j of its term of degree j, for j between 0 and $k - 1$.

Exercise 5.6.12. Let $\tilde{v}(t) = t^k v(1/t)$ and $\tilde{w}(t) = t^{k-1} w(1/t)$. Show that $\tilde{v}$ and $\tilde{w}$ are polynomials. Give their degree. Show, using Exercise 5.6.10, that

$$\tilde{v}(t) f(t) - \tilde{w}(t) = c\left(x \mapsto \frac{v(x)}{1 - xt}\right) t^k.$$

From this, deduce that $f(t) - \tilde{w}(t)/\tilde{v}(t) = O(t^k)$.

From now on we denote $\tilde{w}/\tilde{v} = (k - (1/k))_f$, and we will say that $(k - (1/k))_f$ is a Padé type approximation of f having v as its polynomial generator.

Exercise 5.6.13. Let $v(t) = t^k$. Calculate $(k - (1/k))_f$.

Exercise 5.6.14. Let

$$(5.6.6) \qquad c_j^\ell = \begin{cases} c_{j+\ell} & \text{if } j \geqslant 0 \text{ and } j + \ell \geqslant 0; \\ 0 & \text{otherwise.} \end{cases}$$

We associate with $(c_j^\ell)_{j \geqslant 0}$ the formal integration rule defined by

$$c^\ell\left(x^j\right) = c_j^\ell$$

and the formal series

$$f^\ell(x) = \sum_{j \geqslant 0} c_j^\ell x^j.$$

For $k \geqslant 0$ and $\ell \geqslant 1 - k$ let
$$(5.6.7)$$
$$(k + \ell - (1/k))_f(t) = \begin{cases} c_0 + \ldots + c_{\ell-1} t^{\ell-1} + t^\ell (k - (1/k))_{f^\ell} & \text{if } \ell > 0; \\ t^\ell (k - (1/k))_{f^\ell} & \text{if } \ell < 0. \end{cases}$$

Show that, for all $k \geqslant 0$ and $\ell \geqslant 1 - k$, $f(t) - (k + \ell - (1/k))_f(t) = O(t^{k+\ell})$. Note that the case $\ell = 0$ has been treated previously and distinguish between the cases $\ell > 0$ and $\ell < 0$. In each of these cases, verify that the numerator of the rational fraction is of degree at most $k + \ell - 1$ and the denominator of degree at most k.

Second part: higher-order Padé type approximations

Exercise 5.6.15. Using Exercise 5.6.10, show that, if $c(x^j v(x)) = 0$ for all $0 \leqslant j \leqslant m - 1$, then

$$(5.6.8) \qquad f(t) - (k - (1/k))_f(t) = \frac{t^{k+m}}{\tilde{v}(t)} c\left(x \mapsto \frac{v(x) x^m}{1 - xt}\right).$$

Establish and use the identity

$$(1 - xt)^{-1} = 1 + xt + \ldots + x^{m-1}t^{m-1} + x^m t^m (1 - xt)^{-1}.$$

Third part: Padé approximations

Exercise 5.6.16. Suppose that $k = m$. Show that we obtain the coefficients of a polynomial v of degree exactly k such that $c(x^j v(x)) = 0$, for $j = 0, \ldots, k - 1$, as the solution to a linear system, which should be given explictly. We can establish a parallel between this question and the following, and the results from the chapters on orthogonal polynomials and Gaussian quadrature formulae.

Let H_k denote the determinant of this system. From now on, we suppose that

$$(5.6.9) \qquad\qquad H_k \neq 0, \quad \forall k \geqslant 0.$$

Exercise 5.6.17. Let Q be a polynomial given by the determinant

$$Q(x) = \begin{vmatrix} A_{00} & A_{01} & \cdots & A_{0r} \\ A_{10} & A_{11} & \cdots & A_{1r} \\ \vdots & \vdots & & \vdots \\ A_{r-1,0} & A_{r-1,1} & \cdots & A_{r-1,r} \\ Q_0(x) & Q_1(x) & \cdots & Q_r(x) \end{vmatrix},$$

where the $Q_j(x)$ are polynomials of degree exactly j and $(A_{ij})_{0 \leqslant i,j \leqslant r-1}$ is a regular matrix. Show that $Q(x)$ is a polynomial of degree exactly r and that

$$c(Q) = \begin{vmatrix} A_{00} & A_{01} & \cdots & A_{0r} \\ A_{10} & A_{11} & \cdots & A_{1r} \\ \vdots & \vdots & & \vdots \\ A_{r-1,0} & A_{r-1,1} & \cdots & A_{r-1,r} \\ c(Q_0) & c(Q_1) & \cdots & c(Q_r) \end{vmatrix}.$$

Use the fact that the determinant is a multilinear function with respect to its columns and rows and, in particular, with respect to the last row.

Exercise 5.6.18. Let

$$P_k(x) = D_k \begin{vmatrix} c_0 & c_1 & \cdots & c_k \\ c_1 & c_2 & \cdots & c_{k+1} \\ \vdots & \vdots & & \vdots \\ c_{k-1} & c_k & \cdots & c_{2k-1} \\ 1 & x & \cdots & x^k \end{vmatrix},$$

where D_k is real and nonzero.

Show that, for every $j = 0, \ldots, k - 1$, $c(P_k(x)x^j) = 0$. Deduce from condition (5.6.9) that, for every k, $c(P_k^2) \neq 0$.

Exercise 5.6.19. The Padé type approximation $(k/k - 1)_f$ of the polynomial generator P_k is called a Padé approximation. We denote it by $[k - (1/k)]_f$. Show that

$$f(t) - [k - (1/k)]_f = \frac{t^{2k}}{\tilde{P}_k(t)} c \left(x \mapsto \frac{x^k P_k(x)}{1 - xt} \right).$$

As in Exercise 5.6.12, we construct the Padé approximations $[p/q]_f$ for every $p \geqslant 0$ and $q \geqslant 0$, with the aid of the f^ℓ, on which we make a hypothesis analogous to eqn (5.6.8). Show that

$$[p/q]_f - f = O\left(t^{p+q+1}\right).$$

Exercise 5.6.20. Calculate $[p/q]_f$ for $f(x) = e^x$ and $0 \leqslant p, q \leqslant 2$. Place them in a square table using the convention that $p = $ row index and $q = $ column index. Use the orthogonality relations previously shown. Calculate $[2, 2]_f(1)$ and compare it with the number e.

6

Splines

Until now, we have studied two different approaches to polynomial approximation: interpolation and mean square approximation. Suppose that we seek a
smooth function u from $[a, b]$ to $\mathbb{R}$, which is required to take given values y_j at
points x_j, $1 \leqslant j \leqslant n$. Then it makes sense to minimize a quantity which measures the 'wiggliness' of u, a good candidate for which is the following energy,
provided that u is m times continuously differentiable over $[a, b]$:

$$(6.0.1) \qquad E_m(u) = \int_a^b \left| u^{(m)}(x) \right|^2 dx,$$

under the constraints

$$(6.0.2) \qquad u(x_j) = y_j, \quad \forall j = 1, \ldots, n.$$

It is convenient to define

$$x_0 = a, \quad x_{n+1} = b.$$

This minimization problem is still somewhat vague; its solutions, if they exist,
are called interpolating splines. Originally, a spline was a draughtsman's tool.
A spline is a thin flexible beam, which draughtsmen of the pre-CAD (computer
aided design) age would shape by moving weights (called ducks or rats) with
attached arms designed to fit inside a groove of the beam. This device was used
to draw free-form curves. The elastic energy of the spline deformed into the curve
parameterized by $(t, u(t))$ is given by the integral of the square of the curvature,
namely

$$(6.0.3) \qquad \int_a^b \frac{|u''(t)|^2}{(1 + |u'(t)|^2)^{5/2}} \, dt.$$

Therefore, the lowest order approximation to eqn (6.0.3) is the energy $E_2(u)$. It
is valid only when the gradient of u is small.

The constraints (6.0.2) are very rigid and maybe we should not put too much trust in them. For instance, the y_j may have been obtained by chopping digits off some calculations or some measurements, or, perhaps, our measurements are not very precise, but we have some evidence that the phenomenon described by these measurements is reasonably smooth. Therefore, we do not want to enforce the constraints (6.0.2) strictly and we should instead estimate how well these constraints are satisfied. Thus, given strictly positive numbers ρ_j, $1 \leqslant j \leqslant n$, we define

$$F_\rho(u, y) = \sum_{j=1}^{n} \frac{(u(x_j) - y_j)^2}{\rho_j},$$

and we would like to minimize the energy

(6.0.4) $$E_m(u) + F_\rho(u, y).$$

A solution of this problem is called a smoothing spline. If the ρ_j tend to 0, we expect minimizers of the energy (6.0.4) to converge to minimizers of expression (6.0.1) under the conditions (6.0.2). There is much freedom for choosing the value of the coefficients ρ_j and there are algorithms which play on the values of these coefficients to obtain an answer with desirable properties, such as monotonicity, convexity, concavity, and more.

Both kinds of splines, and much more general ones, are currently used in areas of contemporary high interest, such as image analysis and manipulation, robotics, and data smoothing.

We start with natural splines, for which we discuss two different types of questions: how to ascertain the existence and the uniqueness of a solution (Section 6.1) of the above two minimization problems, and how to calculate numerically these solutions (Section 6.2). The numerical calculation of interpolating or smoothing splines for $m = 2$ (cubic splines) is not difficult. For higher degree splines, it is a good idea to consider a more general situation, and to work with the so-called B-splines, which give a very useful basis of spline space. They are also quite useful for constructing generalizations of Bézier curves for computational geometry (Section 6.3).

6.1. Natural splines: the functional approach

In order to find a minimizer of E_m, or of $E_m + F_\rho(\cdot, y)$, we need a functional lemma, which will enable us to take 'weak derivatives', i.e., derivatives defined via integration by parts. The reader who is knowledgeable in distributions will recognize a derivative in the sense of distributions. However, we do not assume any non-elementary knowledge.

6.1.1. Weak equality of functions

The first result gives a weak version of the equality of functions. In all of this section, we will denote by C_0^k the space of k times continuously differentiable functions on $]a, b[$ which vanish outside of a compact subset of $]a, b[$. We say that such functions have compact support.

Lemma 6.1.1. Let z be a continuous function from $[a, b]$ to $\mathbb{R}$. If, for all ζ in C_0^0, the following equality holds:

$$\int_a^b z\zeta \, \mathrm{d}t = 0,$$

then z is the zero function over $[a, b]$.

Proof. Let ϕ be an arbitrary function in C_0^0. Then $\phi^2 z$ also belongs to C_0^0 and we must have that

$$\int_a^b z^2\phi^2 \, \mathrm{d}t = 0,$$

and so ϕz vanishes. However, we can construct fairly arbitrary functions ϕ. If a' is an arbitrary point of $]a, b[$ and b' an arbitrary point of $]a', b[$, we let, for instance,

$$\phi(t) = \begin{cases} 0 & \text{if } a \leqslant t \leqslant a'; \\ (t - a')\,(b' - t) & \text{if } a' \leqslant t \leqslant b'; \\ 0 & \text{if } b' \leqslant t \leqslant b. \end{cases}$$

The function ϕ is continuous and has compact support, and, therefore, z vanishes on $]a', b'[$. However, a' and b' are arbitrary and, therefore, z vanishes on the open set $]a, b[$ and, by continuity, on the closed set $[a, b]$.   □

This was really easy; let us graduate to something more interesting:

6.1.2. Weak integrals of functions

Lemma 6.1.2. Let g be a continuous function over $[a, b]$. Define its successive integrals by

$$g_0 = g, \quad g_k\,(t) = \int_a^t g_{k-1}\,(s)\,\mathrm{d}s.$$

For all integers m, let z be a continuous function over $[a, b]$ such that the following relation holds for all $\zeta \in C_0^m$:

$$(-1)^m \int_a^b z\zeta^{(m)} \, \mathrm{d}t = \int_a^b g\zeta \, \mathrm{d}t.$$

Then $z - g_m$ is a polynomial of degree at most $m - 1$.

Proof. For $m = 0$, the result has been proved in Lemma 6.1.1. If we could perform integration by parts m times, the result would be obvious; the point is that we do not know (yet!) that z is m times continuously differentiable. Thus, we have to substitute something else for integration by parts. We introduce the notation

$$\langle f \rangle = \int_a^b f(t)\, dt,$$

where f is a continuous function over $[a, b]$. An integral of a function $\eta \in C_0^k$ can belong to C_0^{k+1} only if $\langle \eta \rangle$ vanishes. Thus, we introduce a function ξ which is $m + 1$ times continuously differentiable over $[a, b]$, and is equal to 0 in a neighbourhood of a and to 1 in a neighbourhood of b. We could take, for instance,

$$\xi_1(t) = \begin{cases} 0 & \text{if } a \leqslant t \leqslant (2a+b)/3; \\ \left(t - \dfrac{2a+b}{3}\right)^{m+1} \left(\dfrac{a+2b}{3} - t\right)^{m+1} & \text{if } (2a+b)/3 \leqslant t \leqslant (a+2b)/3; \\ 0 & \text{if } (a+2b)/3 \leqslant t \leqslant b, \end{cases}$$

and

$$\xi(t) = \frac{1}{\langle \xi_1 \rangle} \int_a^t \xi_1(s)\, ds.$$

Observe that the integral of ξ_1 is strictly positive, so that the division is legitimate. Then, we define

$$(L\eta)(t) = \int_a^t \eta(s)\, ds - \langle \eta \rangle \, \xi(t).$$

It is immediate that, for $k \leqslant m+1$, L maps C_0^{k-1} to C_0^k. Moreover, the following identity holds, for all $p \leqslant m + 1$:

(6.1.1)
$$(L\eta)^{(p)} = \eta^{(p-1)} - \langle \eta \rangle \, \xi^{(p)}.$$

Assume, therefore, that the conclusion of the lemma holds for all integers k up to m. If η is an arbitrary function in C_0^m, then

$$(-1)^{m+1} \int_a^b z\,(L\eta)^{(m+1)}\, dt = \int_a^b gL\eta\, dt,$$

so that, with the help of the identity (6.1.1), we may write

$$(-1)^{m+1} \left\{ \int_a^b z\eta^{(m)}\, dt - \langle \eta \rangle \int_a^b z\xi^{(m+1)}\, dt \right\}$$

$$= \int_a^b g(t) \left(\int_a^t \eta(s)\, ds - \langle \eta \rangle \, \xi(t) \right) dt.$$

Observe now that

$$\int_a^b g(t) \int_a^t \eta(s)\, ds\, dt = \int_a^b \eta(t) \int_t^b g(s)\, ds\, dt = \int_a^b \eta(t)\, (g_1(b) - g_1(t))\, dt.$$

Thus, we can now see that, for all $\eta \in C_0^m$, the following identity holds:

$$(-1)^{m+1} \int_a^b z\eta^{(m)} \, dt = \int_a^b \eta(t) \left(g_1(b) - g_1(t) - \langle g\xi \rangle + (-1)^{m+1} \left\langle z\xi^{(m+1)} \right\rangle \right) dt.$$

We observe that the m-th integrals of

$$t \mapsto g_1(b) - g_1(t) - \langle g\xi \rangle + (-1)^{m+1} \left\langle z\xi^{(m+1)} \right\rangle$$

are equal to the sum of $-g_{m+1}$ and a polynomial of degree at most m. The induction hypothesis enables us to conclude the required result.   □

We denote by $V^m(x, y)$ the set of functions u of class C^m over $[a, b]$ which satisfy

$$u(x_j) = y_j, \quad \forall j = 1, \ldots, n.$$

This set is an affine space, obtained by translation from $V^m(x, 0)$.

6.1.3. The space of natural splines

We draw the following important consequence from Lemma 6.1.2:

Theorem 6.1.3.   (i) Let u be a minimizer of E_m over $V^m(x, y)$. Then, u is a
function of class C^{2m-2} over $[a, b]$ which coincides with a polynomial of
degree at most $2m - 1$ on each interval $]x_j, x_{j+1}[$, for $1 \leqslant j \leqslant n - 1$, and
with a polynomial of degree at most $m - 1$ on the end intervals $]x_0, x_1[$ and
$]x_n, x_{n+1}[$.

(ii) Let ρ_j, $1 \leqslant j \leqslant n$, be strictly positive numbers and let u minimize $E_m +$
$F_\rho(\cdot, y)$ over $C^m([a, b])$. Then, u is a function of class C^{2m-2} over $[a, b]$
which coincides with a polynomial of degree at most $2m - 1$ on each interval
$]x_j, x_{j+1}[$, for $1 \leqslant j \leqslant n - 1$, and with a polynomial of degree at most
$m - 1$ on the end intervals $]x_0, x_1[$ and $]x_n, x_{n+1}[$. Moreover, at the knots,
u satisfies the relation

$$(6.1.2) \qquad (-1)^m \left(u^{(2m-1)}(x_j + 0) - u^{(2m-1)}(x_j - 0) \right) + \frac{u(x_j) - y_j}{\rho_j} = 0,$$

$$\forall j = 1, \ldots, n. \;\; \diamond$$

Proof. In the interpolating spline case, we observe that, if u belongs to
$V^m(x, y)$ and v to $V^m(x, 0)$, then $u + v$ belongs to $V^m(x, y)$. If u is a minimizer of E_m over $V^m(x, y)$, then, arguing as for the proof of Lemma 5.1.2, we
get

$$\int_a^b u^{(m)} v^{(m)} \, dt = 0, \quad \forall v \in V^m(x, 0).$$

Let $[a', b']$ be a compact sub-interval of $]x_j, x_{j+1}[$, $1 \leqslant j \leqslant n - 1$, with non-
empty interior. Then, v can be chosen arbitrarily in $C_0^m([a', b'])$ and extended

by 0 to the whole interval $[a, b]$. This v belongs to $V^m(x, 0)$. Lemma 6.1.2 now immediately implies that $u^{(m)}$ coincides on $[a', b']$ with a polynomial of degree at most $m - 1$. This proves, indeed, that the restriction of u to each interval $]x_j, x_{j+1}[$ is a polynomial of degree at most $2m - 1$, for $1 \leqslant j \leqslant n - 1$. On the end intervals, there are less restrictions on v. Let w be any continuous function which vanishes for $t \geqslant x_1$. Then, the function v defined by

$$v(x) = (-1)^m \int_x^{x_1} \frac{(y - x)^{m-1}}{(m - 1)!} w(y)\, dy$$

belongs to $V^m(x, 0)$ and, therefore, Lemma 6.1.1 shows that $u^{(m)}$ must vanish on $[x_0, x_1]$, so that u coincides with a polynomial of degree at most $m - 1$ on this interval.

In the smoothing spline case, we argue as in the proof of Lemma 5.1.2 and we see that, for all $v \in C^m([a, b])$,

$$(6.1.3) \qquad \int_a^b u^{(m)} v^{(m)}\, dt + \sum_{j=1}^n \frac{(u(x_j) - y_j)\, v(x_j)}{\rho_j} = 0.$$

If v belongs to $V^m(x, 0)$, the sum over the knots vanishes and we are left with the condition

$$\int_a^b u^{(m)} v^{(m)}\, dt = 0, \quad \forall v \in V^m(x, 0).$$

Then, the argument made for the interpolating case implies that the conclusion also holds for the smoothing case.

In the case of interpolating splines, let us show now the continuity of u and its first $2m - 2$ derivatives across the knots x_j. By construction, u is m times continuously differentiable, so that we must only look at the derivatives of order greater than or equal to $m + 1$. It is also enough to observe what is happening around a single knot. Indeed, let v belong to $V^m(x, 0)$ and assume that v has compact support in $]a', b'[$, where this interval contains exactly one knot x_j, for some $1 \leqslant j \leqslant n$. With $m - 1$ integrations by parts, we can write

$$0 = \int_{a'}^{b'} u^{(m)} v^{(m)}\, dt$$

$$= \left(u^{(m+1)}(x_j + 0) - u^{(m+1)}(x_j - 0) \right) v^{(m-2)}(x_j)$$

$$- \left(u^{(m+2)}(x_j + 0) - u^{(m+2)}(x_j - 0) \right) v^{(m-3)}(x_j) + \ldots$$

$$+ (-1)^{m-3} \left(u^{(2m-2)}(x_j + 0) - u^{(2m-2)}(x_j - 0) \right) v'(x_j).$$

However, the derivatives of v of order 1 to $m - 2$ are arbitrary. This proves the desired result.

We can now establish the case of smoothing splines. In eqn (6.1.3) we take v in C^m, not in $V^m(x,0)$, so that when integrating by parts the product $u^{(m)}v^{(m)}$ we get one more term. We also assume that the support of v contains only the knot x_j. We now obtain the relation

$$\sum_{k=0}^{m-1} (-1)^{k+1} \left(u^{(m+k)}(x_j+0) - u^{(m+k)}(x_j-0) \right) v^{(m-k-1)}(x_j)$$

$$+ \frac{(u(x_j) - y_j)v(x_j)}{\rho_j} = 0.$$

Since $v(x_j)$ and all of its derivatives up to order m are arbitrary, the continuity of the derivatives of u of order at most $2m-2$ across knots is proved, together with the relation (6.1.2). □

The space $S_N^{2m-1}(x)$ of natural splines with knots at $x_1, \ldots, x_n$ is the space of functions of class C^{2m-2} which coincide with a polynomial of degree at most $2m-1$ on each of the intervals $[x_j, x_{j+1}]$, $1 \leqslant j \leqslant n-1$, and with a polynomial of degree at most $m-1$ on each of the end intervals $[x_0, x_1]$ and $[x_n, x_{n+1}]$. The space $S_N^{2m-1}(x)$ is clearly a space of finite dimension. Later on, we shall compute its dimension.

The interesting fact is that there is a converse to Theorem 6.1.3 which shows that the necessary conditions for minimization are also sufficient.

Lemma 6.1.4.   (i) Let u belong to $S_N^{2m-1}(x)$ and define $y_j = u(x_j)$. Then, u minimizes E_m over $V^m(x,y)$.

(ii) Let u belong to $S_N^{2m-1}(x)$ and define y_j by relation (6.1.2). Then, u minimizes $E_m + F_\rho(\cdot, y)$ over $C^m([a,b])$.

Proof. Let u belong to C^{2m-2} and assume that v belongs to $V^m(x,y)$. We have the following identity:

$$E_m(v) - E_m(u) = 2\int_a^b u^{(m)} \left(v^{(m)} - u^{(m)} \right) dt + E_m(v-u).$$

Using integration by parts,

$$\int_a^b u^{(m)} \left(v^{(m)} - u^{(m)} \right) dt$$

$$= \sum_{j=0}^{n} \sum_{k=0}^{m-1} (-1)^k u^{(m+k)} \left(v^{(m-k-1)} - u^{(m-k-1)} \right) \Big|_{x_j+0}^{x_{j+1}-0} - \int_a^b u^{(2m)}(v-u)\, dt.$$

If u is a natural spline and v is of class C^m, then many terms in the above expression vanish, and we are left with

$$\int_a^b u^{(m)} \left(v^{(m)} - u^{(m)} \right) dt$$

$$= (-1)^m \sum_{j=0}^n \left(v\left(x_j\right) - u\left(x_j\right) \right) \left(u^{(2m-1)}\left(x_j + 0\right) - u^{(2m-1)}\left(x_j - 0\right) \right).$$

Finally, we obtain the identity

$$E_m\left(v\right) - E_m\left(u\right) = E_m\left(u - v\right) + 2\left(-1\right)^m \sum_{j=0}^n \Big\{ \left(v\left(x_j\right) - u\left(x_j\right)\right)$$

(6.1.4)
$$\times \left(u^{(2m-1)}\left(x_j + 0\right) - u^{(2m-1)}\left(x_j - 0\right) \right) \Big\}.$$

In the case of the interpolating spline, take $v \in V^m(x,y)$, so that $v - u$ belongs to $V^m(x,0)$. Then, we have the identity

$$(6.1.5) \qquad E_m\left(v\right) - E_m\left(u\right) = E_m\left(v - u\right)$$

and the conclusion is immediate.

In the case of the smoothing spline, we remark that

$$F_\rho\left(v,y\right) - F_\rho\left(u,y\right) = 2\sum_{j=1}^n \frac{\left(v\left(x_j\right) - u\left(x_j\right)\right)\left(u\left(x_j\right) - y_j\right)}{\rho_j} + \sum_{j=1}^n \frac{\left(v\left(x_j\right) - u\left(x_j\right)\right)^2}{\rho_j}.$$

Using again the identity (6.1.4), we find that

$$E_m\left(v\right) - E_m\left(u\right) + F_\rho\left(v,y\right) - F_\rho\left(u,y\right) = E_m\left(v - u\right) + \sum_{j=1}^n \frac{\left(v\left(x_j\right) - u\left(x_j\right)\right)^2}{\rho_j}.$$

(6.1.6)

This concludes the proof of the lemma. □

In order to prove the uniqueness of the smoothing or interpolating splines, we need to assume that $n \geqslant m$. Then, we have the following lemma:

Lemma 6.1.5. Assume $n \geqslant m$. If u and v are splines (belonging to $S_N^{2m-1}(x)$) which coincide at the knots, then they are identical.

Proof. Define $y_j = u(x_j)$. We know from Lemma 6.1.4 that u and v must minimize E_m over $V^m(x,y)$. Then, due to the identity (6.1.5), $E_m(v - u)$ must vanish, which implies that $v - u$ coincides over $[a, b]$ with a polynomial of degree at most $m - 1$. But $u - v$ vanishes at $n \geqslant m$ distinct points, which means that $u - v$ must vanish identically. □

The corollary to this is the following uniqueness result:

Theorem 6.1.6. Assume $n \geqslant m$. There exists at most one interpolation spline through the points (x_j, y_j), $1 \leqslant j \leqslant n$, and one smoothing spline relative to the points (x_j, y_j), $1 \leqslant j \leqslant n$, with weights $\rho_j > 0$.   ◇

Proof. In the case of the interpolating spline, the uniqueness is immediate. In the case of the smoothing spline, we infer from eqn (6.1.6) that any two minimizers of $E_m + F_\rho(\cdot, y)$ coincide at the knots, and Lemma 6.1.5 then gives the required conclusion.   □

We now obtain our final theoretical result on splines:

Theorem 6.1.7. Assume $n \geqslant m \geqslant 1$. For any finite sequence of knots $x_1 < x_2 < \cdots < x_n$ in the interval $]a, b[$ and any sequence of numbers $y_1, \ldots, y_n$, there exists a unique interpolating spline $u \in S_N^{2m-1}(x)$ satisfying the constraint (6.0.2). Given positive weights ρ_j, there also exists a unique smoothing spline relative to the points (x_j, y_j) and these weights.   ◇

Proof. In order to prove the existence of interpolating and smoothing splines, we just have to count dimensions. Thus, we represent splines as polynomials on each of the intervals between knots as follows:

$$u(t) = \sum_{k=1}^{2m} c_{j,k} \frac{t^{k-1}}{(k-1)!}, \quad 0 \leqslant j \leqslant n, \quad x_j \leqslant t \leqslant x_{j+1}.$$

Hence, we describe the space $S_N^{2m-1}(x)$ using $2m(n+1)$ parameters, but they are not free. First, the coefficients $c_{0,k}$ and $c_{n,k}$ vanish for $m+1 \leqslant k \leqslant 2m$. Next, we must write the transmission conditions at the knots, that is

$$\sum_{k=l}^{2m} c_{j-1,k} \frac{x_j^{k-l}}{(k-l)!} = \sum_{k=l}^{2m} c_{j,k} \frac{x_j^{k-l}}{(k-l)!}, \quad \forall j = 1, \ldots, n, \quad \forall l = 1, \ldots, 2m-1.$$

Define $2m - 1$ by $2m$ matrices A_j, for $1 \leqslant j \leqslant n$, by

$$A_j = \begin{pmatrix} 1 & x_j & x_j^2/2! & \cdots & x_j^{2m-1}/(2m-1)! \\ & 1 & x_j & \cdots & x_j^{2m-2}/(2m-2)! \\ & & \ddots & \ddots & \vdots \\ & & & 1 & x_j \end{pmatrix}$$

and m by $2m$ matrices A_0 and A_{n+1} by

$$A_0 = A_{n+1} = \begin{pmatrix} 0 & \cdots & 0 & 1 & 0 & \cdots & 0 \\ 0 & \cdots & 0 & 0 & 1 & \cdots & 0 \\ \vdots & & \vdots & \vdots & & \ddots & \vdots \\ 0 & \cdots & 0 & 0 & 0 & \cdots & 1 \end{pmatrix}.$$

Let A denote the matrix

$$A = \begin{pmatrix} A_0 & & & & \\ A_1 & -A_1 & & & \\ & A_2 & -A_2 & & \\ & & \ddots & \ddots & \\ & & & A_n & -A_n \\ & & & & A_{n+1} \end{pmatrix}.$$

Then, the compatibility condition may be written as

$$Ac = 0,$$

with c the transpose of

$$\begin{pmatrix} c_{0,1} & \cdots & c_{0,2m} & c_{1,1} & \cdots & c_{1,2m} & \cdots & c_{n,1} & \cdots & c_{n,2m} \end{pmatrix}.$$

We claim that the rank of the matrix A is equal to the number of its rows, i.e., $n(2m - 1) + 2m = 2m(n + 1) - n$. Let l be a row vector with $2m(n + 1) - n$ columns. If lA vanishes, then l_j vanishes for $j \geqslant 2m$. We just have to perform an induction starting from the last column. It remains to find the rank of the matrix

$$\begin{pmatrix} A_0 & 0 \\ A_1 & -A_1 \end{pmatrix}.$$

By column combinations, the rank of this matrix is equal to the rank of the matrix

(6.1.7)
$$\begin{pmatrix} A_0 & 0 \\ B_1 & -A_1 \end{pmatrix},$$

where B_1 is the matrix whose first m columns are the same as the first m columns of A_1 and whose last m columns vanish. It is now clear that the matrix (6.1.7) is of rank $3m - 1$. Thus, we have shown that the space $S_N^{2m-1}(x)$ is of dimension n. Solving for an interpolating spline is a linear problem in c, since it can be written as

$$Ac = 0, \qquad \sum_{k=1}^{2m} \frac{x_j^{k-1}}{(k - 1)!} c_{j,k} = y_j, \quad \forall j = 1, \ldots, n.$$

Therefore, if $n \geqslant m$, the uniqueness result of Lemma 6.1.5 implies the existence of an interpolating spline, due to the fundamental theorem of linear algebra. Solving for a smoothing spline is also a linear problem, which can be written as

$$Ac = 0,$$

$$(-1)^m \left(c_{j,2m} - c_{j-1,2m} \right) + \frac{1}{\rho_j} \left(\sum_{k=1}^{2m} \frac{c_{j,k} x_j^{k-1}}{(k - 1)!} - y_j \right) = 0, \quad \forall j = 1, \ldots, n.$$

Then, the previous argument also works and the uniqueness for $n \geqslant m$ implies the existence of smoothing splines. $\qquad \square$

6.2. Numerics for cubic natural splines

The construction of the previous section can be made numerical, but it is extremely awkward, since it requires the solution of a system of $2m(n+1)$ equations in $2m(n+1)$ unknowns, where only n of these equations are not homogeneous. Therefore, in order to be efficient, one has to be smart in choosing the representation of spline functions.

In the case of cubic natural splines, i.e., $m = 2$, the numerical calculation of the interpolating or of the smoothing spline is a very simple problem, as will be shown now. Of course, the choice of coordinates is an essential question. When $m = 2$, we know that natural splines are of class C^2. Thus, we choose as our unknowns the values of the second derivative of u at the knots, namely

$$z_j = u''(x_j), \quad 1 \leqslant j \leqslant n.$$

We know that

$$z_1 = z_n = 0$$

since u is of degree at most 1 on the end intervals. Therefore, we shall solve a system of $n-2$ equations with $n-2$ unknowns, assuming $n > 2$. The case $n = 2$ is quite boring (why?). Let us write

$$\Delta x_j = x_{j+1} - x_j, \quad \Delta y_j = y_{j+1} - y_j, \quad \Delta z_j = z_{j+1} - z_j.$$

On each interval $[x_j, x_{j+1}]$, $1 \leqslant j \leqslant n-1$, u'' is of degree at most 1 and is given by

$$u''(x) = \frac{x_{j+1} - x}{\Delta x_j} z_j + \frac{x - x_j}{\Delta x_j} z_{j+1}.$$

Therefore, on the interval $[x_j, x_{j+1}]$, u is given by the expression

$$(6.2.1) \qquad u(x) = A_j + B_j(x - x_j) + \frac{(x_{j+1} - x)^3}{6\Delta x_j} z_j + \frac{(x - x_j)^3}{6\Delta x_j} z_{j+1}.$$

We shall express the values of A_j and B_j in terms of the other parameters of the problem by solving the pair of linear equations

$$u(x_j) = y_j, \quad u(x_{j+1}) = y_{j+1}.$$

The first equation gives

$$(6.2.2) \qquad\qquad A_j = y_j - \frac{z_j \Delta x_j^2}{6},$$

and the second gives

$$(6.2.3) \qquad\qquad B_j = \frac{\Delta y_j}{\Delta x_j} - \frac{\Delta x_j \Delta z_j}{6}.$$

We now maintain the continuity of u' at knots. If $2 \leqslant j \leqslant n-1$, then we have

$$u'(x_j + 0) = B_j - \frac{\Delta x_j z_j}{2} \quad \text{and} \quad u'(x_j - 0) = B_{j-1} + \frac{\Delta x_{j-1} z_j}{2}.$$

We replace Δz_j and Δz_{j-1} by their values, put all the terms including one of the z_i on the left-hand side, and we then obtain the following system of equations for $2 \leqslant j \leqslant n-1$:

$$(6.2.4) \qquad \left(\frac{\Delta x_j}{3} + \frac{\Delta x_{j-1}}{3} \right) z_j + \frac{\Delta x_{j-1}}{6} z_{j-1} + \frac{\Delta x_j}{6} z_{j+1} = \frac{\Delta y_j}{\Delta x_j} - \frac{\Delta y_{j-1}}{\Delta x_{j-1}}.$$

Recall that z_1 and z_n vanish, so that the system (6.2.4) is an $n-2$ by $n-2$ system. Its matrix is tridiagonal and strictly diagonally dominant. Hence, it is very clear now that there exists a cubic interpolating natural spline provided that $n \geqslant 2$. If we define

$$\alpha_j = \frac{\Delta x_j}{6},$$

then the matrix of the system is given by:

$$(6.2.5) \qquad \begin{pmatrix} 2(\alpha_1 + \alpha_2) & \alpha_2 & & & \\ \alpha_2 & 2(\alpha_2 + \alpha_3) & \alpha_3 & & \\ & \ddots & \ddots & \ddots & \\ & & & & \alpha_{n-2} \\ & & & \alpha_{n-2} & 2(\alpha_{n-2} + \alpha_{n-1}) \end{pmatrix}.$$

The z solution gives the A_js, due to eqn (6.2.2), the B_j, due to eqn (6.2.3), and u on the intervals $[x_j, x_{j+1}]$, $1 \leqslant j \leqslant n-1$, due to eqn (6.2.1). On the end intervals, we use the formulae

$$u(x) = y_1 + u'(x_1)(x - x_1), \quad u(x) = y_n + u'(x_n)(x - x_n).$$

Let us now consider the case of the smoothing spline. We have to enforce the following transmission condition at the knots:

$$(6.2.6) \qquad u'''(x_j + 0) - u'''(x_j - 0) + \frac{u(x_j) - y_j}{\rho_j} = 0.$$

If we define $u(x_j) = u_j$, then the above argument implies that, for $2 \leqslant j \leqslant n-1$,

$$(6.2.7) \qquad \left(\frac{\Delta x_j}{3} + \frac{\Delta x_{j-1}}{3} \right) z_j + \frac{\Delta x_{j-1}}{6} z_{j-1} + \frac{\Delta x_j}{6} z_{j+1} = \frac{\Delta u_j}{\Delta x_j} - \frac{\Delta u_{j-1}}{\Delta x_{j-1}},$$

but we now have to get the value of u_j from the following form of condition (6.2.6):

$$u_j = y_j - \rho_j \left(u'''(x_j + 0) - u'''(x_j - 0) \right).$$

The values of the third derivatives can be readily obtained from the values of z as:

$$u'''(x_j + 0) = \frac{\Delta z_j}{\Delta x_j}, \quad u'''(x_j - 0) = \frac{\Delta z_{j-1}}{\Delta x_{j-1}}.$$

Therefore,

$$(6.2.8) \quad \frac{\Delta u_j}{\Delta x_j} = \frac{\Delta y_j}{\Delta x_j} - \frac{1}{\Delta x_j}\left(\rho_{j+1}\left(\frac{\Delta z_{j+1}}{\Delta x_{j+1}} - \frac{\Delta z_j}{\Delta x_j}\right) - \rho_j\left(\frac{\Delta z_j}{\Delta x_j} - \frac{\Delta z_{j-1}}{\Delta x_{j-1}}\right)\right).$$

Substituting the expression (6.2.8) into eqn (6.2.7), we obtain, after the same kinds of manipulations as those used to obtain (6.2.4),

$$\left(\frac{\Delta x_j}{3} + \frac{\Delta x_{j-1}}{3}\right)z_j + \frac{\Delta x_{j-1}}{6}z_{j-1} + \frac{\Delta x_j}{6}z_{j+1}$$

$$+ \frac{\rho_{j+1}}{\Delta x_{j+1}\Delta x_j}z_{j+2} - \left(\frac{\rho_{j+1}}{\Delta x_j\Delta x_{j+1}} + \frac{\rho_{j+1} + \rho_j}{\Delta x_j^2} + \frac{\rho_j}{\Delta x_j\Delta x_{j-1}}\right)z_{j+1}$$

$$+ \left(\frac{\rho_{j+1} + \rho_j}{\Delta x_j^2} + \frac{2\rho_j}{\Delta x_j\Delta x_{j-1}} + \frac{\rho_{j-1} + \rho_j}{\Delta x_{j-1}^2}\right)z_j$$

$$- \left(\frac{\rho_j}{\Delta x_j\Delta x_{j-1}} + \frac{\rho_{j-1} + \rho_j}{\Delta x_{j-1}^2} + \frac{\rho_{j-1}}{\Delta x_{j-1}\Delta x_{j-2}}\right)z_{j-1} + \frac{\rho_{j-1}}{\Delta x_{j-1}\Delta x_{j-2}}z_{j-2}$$

$$= \frac{\Delta y_j}{\Delta x_j} - \frac{\Delta y_{j-1}}{\Delta x_{j-1}},$$

which is a pentadiagonal system of $n - 2$ equations with $n - 2$ unknowns. Of course, we have to let $z_1 = z_n = 0$. Introducing the notation

$$\beta_j = \frac{\rho_j}{\Delta x_j\Delta x_{j-1}}, \quad \gamma_j = \frac{\rho_j}{\Delta x_j^2}, \quad \delta_j = \frac{\rho_j}{\Delta x_{j-1}^2},$$

$$a_j = \gamma_{j-1} + \gamma_j + 2\beta_j + \delta_j + \delta_{j+1}, \quad b_j = \beta_j + \gamma_j + \beta_{j+1} + \delta_{j+1},$$

we see that the matrix of this system is the sum of the matrix (6.2.5) and

$$\begin{pmatrix} a_2 & -b_2 & \beta_3 & & & \\ -b_2 & a_3 & -b_3 & \beta_4 & & \\ \beta_3 & -b_3 & a_4 & -b_4 & \beta_5 & \\ & \beta_4 & -b_4 & a_5 & -b_5 & \beta_6 \\ & & \ddots & \ddots & \ddots & \ddots & \ddots \end{pmatrix}.$$

If we are interested in higher degree splines then we have to use more sophisticated methods, which are explained in the next section.

6.3. Spaces of splines, B-splines

6.3.1. Splines with distinct knots

More generally, given n distinct knots $x_1 < \cdots < x_n$ in the open interval $]a, b[$ and an integer $k \geqslant 1$, the space $S^k(x)$ is the space of functions of class C^{k-1} over $[a, b]$ which coincide with polynomials of degree at most k on each interval $[x_j, x_{j+1}]$, for $0 \leqslant j \leqslant n$. Here, we keep the convention $x_0 = a$, $x_{n+1} = b$. The space $S^k(x)$ is called the space of splines of degree k. For historical reasons, much of the literature on splines prefers to call this space the space of splines of order $k + 1$. I feel uncomfortable with this terminology, because order is used in many other circumstances, and this creates confusion.

The first main fact is that the derivative of a spline of degree k is a spline of degree $k - 1$.

Let us calculate the dimension of $S^k(x)$, arguing as in the proof of Theorem 6.1.7. We represent each function $u \in S^k(x)$ by the coefficients of its polynomial expansion on the interval $[x_j, x_{j+1}]$ as follows:

$$u(t) = \sum_{l=0}^{k} c_{j,l} \frac{t^l}{l!},$$

so that we immerse $S^k(x)$ into a space of dimension $(n+1)(k+1)$. Then, the compatibility conditions can be written

$$\sum_{l=m}^{k} (c_{j-1,l} - c_{j,l}) \frac{x_j^{l-m}}{(l-m)!} = 0, \quad \forall j = 1, \ldots, n, \quad \forall m = 0, \ldots, k-1.$$

These relations can be put into the form

$$Ac = 0,$$

where A is the nk by $(n+1)(k+1)$ matrix

$$\begin{pmatrix} A_1 & -A_1 & & & \\ & A_2 & -A_2 & & \\ & & \ddots & \ddots & \\ & & & A_n & -A_n \end{pmatrix}$$

and A_j is the matrix

$$A_j = \begin{pmatrix} 1 & x_j & x_j^2/2! & \cdots & & x_j^k/k! \\ & 1 & x_j & \cdots & & x_j^{k-1}/(k-1)! \\ & & \ddots & & & \vdots \\ & & & 1 & & x_j \end{pmatrix}.$$

The matrix A is of maximal rank, as can be immediately checked, so that $S^k(x)$ is of dimension $k + n + 1$.

In order to work easily with splines, we produce good basis elements for the space $S^k(x)$.

Denote by r_+ the positive part of the real number r, i.e.,

$$r_+ = \max(r, 0).$$

By convention, r_+^0 is the characteristic function of $\mathbb{R}^+$.

We see immediately that the function $t \mapsto (t - x_j)_+^k$ is defined for all integers $j \in \{1, \ldots, n\}$ and that it is of class C^{k-1} over $[a, b]$. It agrees with a polynomial in each of the intervals $[x_l, x_{l+1}]$, so that it belongs to $S^k(x)$.

The functions $t \mapsto (t - x_j)_+^k$ are linearly independent. However, the support of these functions is large, which can be a serious numerical inconvenience, and there are not enough such functions to make a basis. Of course, this can be cured by adding to the basis the functions $(x - a)^l$, $0 \leqslant l \leqslant k$, and then we have the required number of independent elements to make a basis (check that!).

6.3.2. The beautiful properties of B-splines

However, there is a much better choice of basis. It is possible to define the so-called B-splines. They are elements of $S^k(x)$ with the smallest possible support, there is a very stable numerical algorithm to construct them, they are positive on their support, and they add up to 1 on $[a, b]$.

However, in order to define B-splines, we have to increase the number of knots. In this section we limit ourselves to distinct knots, but in the problem section (see Subsection 6.4.3) the generalization to the case of coincident knots is taken up.

Assume that we are given $2k$ additional points satisfying the inequalities

$$x_{-k} < \cdots < x_{-1} < x_0 \quad \text{and} \quad x_{n+1} < \cdots < x_{n+k+1}.$$

We define a B-spline by the following divided differences formula, where the index s on δ means that the finite difference operator works on the s variable:

$$(6.3.1) \qquad N_{i,k}(t) = (x_{i+k+1} - x_i)\, \delta_s^{k+1}(x_i, \ldots, x_{i+k+1})\, (s - t)_+^k.$$

Let us first show that the family $N_{i,k}$, for $-k \leqslant i \leqslant n$, can be constructed by a recursive formula in k, due simultaneously to de Boor [20] and Cox [18].

Lemma 6.3.1. The $N_{i,j}$ satisfy the following recursion:

$$(6.3.2) \qquad N_{i,0} = 1_{[x_i, x_{i+1}[}, \quad -k \leqslant i \leqslant n + k,$$

and, for $1 \leqslant j \leqslant k$ and $-k \leqslant i \leqslant n + k - j$,

$$(6.3.3) \qquad N_{i,j}(t) = \frac{t - x_i}{x_{i+j} - x_i} N_{i,j-1}(t) + \frac{x_{i+j+1} - t}{x_{i+j+1} - x_{i+1}} N_{i+1,j-1}(t).$$

Proof. Relation (6.3.2) is clear. Assume now that j is at least equal to 1 and observe that

$$(s-t)_+^j = (s-t)_+^{j-1}(s-t).$$

We apply the generalization of Leibniz' formula given in Lemma 4.5.1 to the finite difference $\delta_s^{j+1}(x_i, \ldots, x_{i+j+1})$ appearing in the definition (6.3.1) of the B-spline $N_{i,j}$. For this purpose, we remark that

$$\delta_s^m(x_l, \ldots, x_{l+m})(s-t)$$

vanishes unless $m = 0$ or $m = 1$. Then, we see that

$$N_{i,j}(t) = (x_{i+j+1} - x_i)\left(\delta_s^j(x_i, \ldots, x_{i+j})(s-t)_+^{j-1}\right)$$

(6.3.4)
$$+ (x_{i+j+1} - x_i)\left(\delta_s^{j+1}(x_i, \ldots, x_{i+j+1})(x_{i+j+1} - t)_+^{j-1}\right)(x_{i+j+1} - t).$$

The first term on the right-hand side of eqn (6.3.4) is equal to

$$\frac{x_{i+j+1} - x_i}{x_{i+j} - x_i}N_{i,j-1}(t).$$

In order to identify the second term on the right-hand side of eqn (6.3.4), we use the definition of divided differences as follows:

$$\delta_s^{j+1}(x_i, \ldots, x_{i+j+1})(s-t)_+^{j-1}$$
$$= \frac{\delta_s^j(x_{i+1}, \ldots, x_{i+j+1})(s-t)_+^{j-1} - \delta_s^j(x_i, \ldots, x_{i+j})(s-t)_+^{j-1}}{x_{i+j+1} - x_i}.$$

Therefore, using the definition of the B-splines, we see that

(6.3.5)
$$(x_{i+j+1} - x_i)\left(\delta_s^{j+1}(x_i, \ldots, x_{i+j+1})(s-t)_+^{j-1}\right)(x_{i+j+1} - t)$$
$$= \frac{x_{i+j+1} - t}{x_{i+j+1} - x_{i+1}}N_{i+1,j-1}(t) - \frac{x_{i+j+1} - t}{x_{i+j} - x_i}N_{i,j-1}(t).$$

When we substitute eqn (6.3.5) into the right-hand side of eqn (6.3.4), we obtain the formula (6.3.3). □

Formula (6.3.3) gives a B-spline as a convex combination of splines of lower degree, which is a very stable numerical process.

To be precise, $B_{i,k}(t) = N_{i,k}(t)/(x_{i+k+1} - x_i)$ is a convex combination of $B_{i,k-1}(t)$ and $B_{i+1,k-1}(t)$, as the reader may verify. However, we have not yet shown that a B-spline is a spline. This, and other properties, are consequences of lemma 6.3.1, and they are summarized in next theorem.

Theorem 6.3.2. The B-splines of degree k belong to $S^k(x)$. The support of $N_{i,k}$ is included in $[x_i, x_{i+k+1}]$ and $N_{i,k}$ is non-negative on its support. For $k \geqslant 1$, the derivative of $N_{i,k}$ satisfies the identity

$$(6.3.6) \qquad N_{i,k}'(t) = k \left(\frac{N_{i,k-1}(t)}{x_{i+k} - x_i} - \frac{N_{i+1,k-1}(t)}{x_{i+k+1} - x_{i+1}} \right).$$

B-splines form a partition of unity:

$$(6.3.7) \qquad \sum_{i=-k}^{n} N_{i,k}(t) = 1, \quad \forall t \in \,]a, b[\, . \qquad\qquad \diamond$$

Proof. That $N_{i,k}$ belongs to the space of splines is immediate from the formula (6.3.3) by recurrence, as is the statement on the support and on the positivity of the B-spline.

The formula for derivatives is proved with the help of the identity (4.5.3). We apply this formula to the t derivative of $N_{i,k}$ and we see immediately that

$$N_{i,k}'(t) = (x_{i+k+1} - x_i) \frac{\partial}{\partial t} \delta_s^{k+1} (x_i, \ldots, x_{i+k+1}) (s - t)_+^k$$
$$= -(x_{i+k+1} - x_i) k \delta_s^{k+1} (x_i, \ldots, x_{i+k+1}) (s - t)_+^{k-1}.$$

In order to find the value of this expression, we use the definition of divided differences, and we find

$$N_{i,k}'(t) = -k \left(\delta_s^k (x_{i+1}, \ldots, x_{i+k+1}) (s - t)_+^{k-1} - \delta_s^k (x_i, \ldots, x_{i+k}) (s - t)_+^{k-1} \right),$$

from which the formula (6.3.6) follows.

That the splines of degree 0 form a partition of unity on $]a, b[$ is quite clear. Assume that eqn (6.3.7) holds up to some integer $k - 1$. We use the de Boor–Cox recursion formula to write

$$\sum_{i=-k}^{n} N_{i,k}(t)$$

$$= \frac{t - x_{-k}}{x_0 - x_{-k}} N_{-k,k-1}(t) + \sum_{i=-k+1}^{n} N_{i,k-1}(t) + \frac{x_{n+k+1} - t}{x_{n+k+1} - x_{n+1}} N_{n+1,k-1}(t).$$

However, the support of $N_{-k,k-1}(t)$ is included in $[x_{-k}, x_0]$ and the support of $N_{n+1,k-1}$ is included in $[x_{n+1}, x_{n+k+1}]$, both of which do not intersect $]a, b[$. This enables us to conclude the required result. □

Now, the important fact is that the $N_{i,k}$ form a basis of $S^k(x)$, for $-k \leqslant i \leqslant n$. There are $n + k + 1$ such B-splines $N_{i,k}$. It is already clear that they belong to $S^k(x)$. It remains to prove that they are independent, as is done in following lemma:

Lemma 6.3.3. For all integers k, the B-splines are independent as functions on $]a, b[$.

Proof. For $k = 0$, it is clear that the B-splines $N_{i,0}$ are independent, since they have disjoint supports. Assume then that the B-splines $N_{i,k-1}$ are independent, and consider the linear relation

$$\sum_{i=-k}^{n} \lambda_i N_{i,k}(t) = 0, \quad \forall t \in]a, b[.$$

Differentiating this relation with respect to t, we find that

$$k \sum_{i=-k}^{n} \lambda_i \left(\frac{N_{i,k-1}(t)}{x_{i+k} - x_i} - \frac{N_{i+1,k-1}(t)}{x_{i+k+1} - x_{i+1}} \right) = 0,$$

with the help of the identity (6.3.6). This relation can be rewritten as

$$\sum_{i=-k}^{n+1} \mu_i N_{i,k-1}(t) = 0.$$

However, the first and last terms vanish on $]a, b[$. It suffices to reduce the summation limits to $-k + 1$ and n, where the μ_is are given by

$$\mu_i = \frac{k(\lambda_i - \lambda_{i-1})}{x_{i+k} - x_i}, \quad \forall i = -k + 1, \ldots, n.$$

The induction hypothesis implies that all of the μ_i should vanish. This gives us a linear system on the λ_i whose solution is a constant vector. As the $N_{i,k}$ sum up to 1, we may conclude the required result. $\square$

The following is an interesting formula relating divided differences and B-splines:

Lemma 6.3.4. For all $n \geqslant 1$ and all distinct knots $x_0 < \cdots < x_n$, the following identity holds:

$$(6.3.8) \qquad \delta^n(x_0, \ldots, x_n) f = \int \frac{f^{(n)}(t) N_{0,n-1}(t)}{(n-1)!(x_n - x_0)} \, dt.$$

Proof. For $n = 1$, the left-hand side of eqn (6.3.8) is equal to

$$\frac{f(x_1) - f(x_0)}{x_1 - x_0}$$

and its right-hand side is equal to

$$\int f'(t) \frac{1_{[x_0, x_1[}(t)}{x_1 - x_0} \, dt,$$

which is clearly equal to the left-hand side.

Assume then that the identity (6.3.8) holds up to some index n, and for all choice of knots and functions. Due to the support properties of B-splines, we may rewrite the right-hand side of (6.3.8) as follows:

$$\frac{1}{n!} \int \frac{f^{(n+1)} N_{0,n}}{x_{n+1} - x_0} \, dt = \frac{1}{n!} \int_{x_0}^{x_{n+1}} f^{(n+1)} (t) \, \delta_s^{n+1} (x_0, \ldots, x_{n+1}) (s - t)_+^n \, dt,$$

which we integrate by parts, with the help of the identity (4.5.3), to obtain

$$\frac{n}{n!} \int_{x_0}^{x_{n+1}} f^{(n)} (t) \, \delta_s^{n+1} (x_0, \ldots, x_{n+1}) (s - t)_+^{n-1} \, dt.$$

We now use the definition of divided differences and of the B-splines to obtain the following expression for the right-hand side of the identity (6.3.8):

$$\frac{1}{(n-1)! \, (x_{n+1} - x_0)} \left(\int_{x_1}^{x_{n+1}} f^{(n)} \frac{N_{1,n-1}}{x_{n+1} - x_1} \, dt - \int_{x_0}^{x_n} f^{(n)} \frac{N_{0,n-1}}{x_n - x_0} \, dt \right).$$

The induction hypothesis then yields the desired result.   □

B-splines are interesting in themselves. The reader is invited to try the exercises in Subsection 6.4.3 in order to understand B-splines with coincident

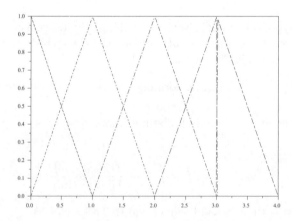

Figure 6.1: The splines of degree 1 are piecewise affine functions. There are 6 such splines, but, with the choice (6.3.9) of knots, the first one vanishes. If, among the three knots appearing in the divided difference, two coincide, then the B-spline is not continuous. This is the case for the second, fifth, and sixth B-splines of this collection.

knots. With the vector of knots

(6.3.9) $(x_0, \ldots, x_7) = (0, 0, 0, 1, 2, 3, 3, 4),$

we have put a 'portrait gallery' of B-splines of degrees 1, 2, 3, and 4, as shown in Figs 6.1, 6.2, 6.3, and 6.4, respectively.

6.3.3. Numerics with B-splines

We can now give the general ideas used for computing with splines and represent any spline in a basis of B-splines. For instance, if we wish to find the interpolating natural spline of degree $2m - 1$, we write the linear system

(6.3.10)
$$\sum_{i=-2m+1}^{n} z_i N_{i,2m-1}(x_j) = y_j, \qquad \forall j = 1, \ldots, n,$$

(6.3.11)
$$\sum_{i=-2m+1}^{n} z_i N_{i,2m-1}^{(l)}(x_1 - 0) = 0, \qquad \forall l = m, \ldots, 2m - 1,$$

(6.3.12)
$$\sum_{i=-2m+1}^{n} z_i N_{i,2m-1}^{(l)}(x_n + 0) = 0, \qquad \forall l = m, \ldots, 2m - 1.$$

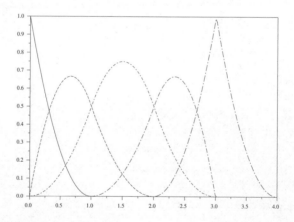

Figure 6.2: The splines of degree 2 are piecewise quadratic functions. For the choice (6.3.9) of knots, there are 5 such splines. If, among the four knots appearing in the divided difference, three coincide, then the B-spline is not continuous. This is the case for the first B-spline of degree 2. If, among the four knots appearing in the finite difference, two coincide, then the B-spline is of class C^0, but not of class C^1. This is the case for the fifth element of the collection.

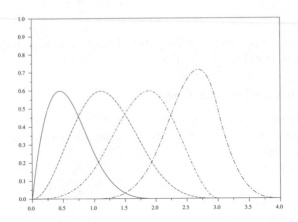

Figure 6.3: The splines of degree 3 are piecewise cubic functions. For the choice (6.3.9) of knots, there are 4 such splines. If, among the five knots appearing in the divided difference, three coincide, then the B-spline is not of class C^1. This is the case for the first B-spline of degree 3.

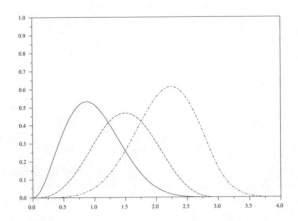

Figure 6.4: The three B-splines of degree 4, for the choice (6.3.9) of knots.

This is a system of $n + 2m$ equations with $n + 2m$ unknowns, and Theorem 6.1.7 guarantees the existence of a solution. Moreover, the matrix has few nonzero coefficients: we know that the support of $N_{i,2m-1}$ is included in $[x_i, x_{i+2m}]$. Therefore, the lines corresponding to eqns (6.3.10) contain at most $2m-1$ nonzero coefficients, and the same holds for the lines corresponding to eqns (6.3.11) and (6.3.12), for $m \leqslant l \leqslant 2m - 2$. Finally, the lines of eqns (6.3.11) and (6.3.12) corresponding to $l = 2m - 1$ contain at most $2m$ nonzero coefficients.

An example shows the structure of the matrix. Let $m = 3$ and $n = 4$. Then, if $\bullet$ denotes the non-vanishing coefficients, the matrix of the linear system has the following structure:

$$
\begin{pmatrix}
\bullet & \bullet & \bullet & \bullet & \bullet & \bullet & & & & \\
 & \bullet & \bullet & \bullet & \bullet & \bullet & & & & \\
 & & \bullet & \bullet & \bullet & \bullet & \bullet & & & \\
 & & & \bullet & \bullet & \bullet & \bullet & & & \\
 & & & & \bullet & \bullet & \bullet & \bullet & \bullet & \\
 & & & & & \bullet & \bullet & \bullet & \bullet & \bullet \\
 & & & & & & \bullet & \bullet & \bullet & \bullet & \bullet \\
 & & & & & & & \bullet & \bullet & \bullet & \bullet & \bullet \\
 & & & & & & & & \bullet & \bullet & \bullet & \bullet \\
 & & & & & & & & & \bullet & \bullet & \bullet & \bullet & \bullet
\end{pmatrix}
$$

We will see in Chapter 9 that the numerical resolution of a linear system whose matrix has the above structure is simple.

6.3.4. Using B-splines to understand natural splines

There is another way to look at natural splines with the help of B-splines:

Lemma 6.3.5. Assume $n \geqslant m$, and let T be the linear span of the B-splines $N_{1,m-1}, \ldots, N_{n-m,m-1}$, seen as functions from $[a, b]$ to $\mathbb{R}$. Let f be a function of class C^m and let $y_j = f(x_j)$. Then, u is the interpolating natural spline through the points (x_j, y_j) if and only if $u^{(m)}$ is the projection on T of $f^{(m)}$ in the mean square sense.

Proof. If u is a natural spline of degree $2m - 1$, its derivative of order m belongs to the space $S^{m-1}(x)$, and it vanishes on the end intervals $[x_0, x_1]$ and $[x_n, x_n + 1]$. Therefore, it is clear that $u^{(m)}$ belongs to T. Let v be the least-squares projection of $f^{(m)}$ on T. The existence of such a projection has been proved in Lemma 5.1.2, and it satisfies the following relations:

$$
\int_a^b \left(f^{(m)} - v \right) N_{i,m-1} \, dt = 0, \quad \forall i = 1, \ldots, n - m
$$

and

$$
\int_a^b v^2 \, dt \leqslant \int_a^b \left(f^{(m)} \right)^2 \, dt.
$$

We choose an m-th integral w of v which satisfies the following m conditions:

$$w\left(x_j\right) = f\left(x_j\right), \quad 1 \leqslant j \leqslant m.$$

If $n = m$, then it is clear that w is an interpolating spline through the points (x_j, y_j). If $n > m$, we observe that

$$\int_a^b \left(f^{(m)} - w^{(m)}\right) N_{1,m-1} \, dt = 0$$

and we use the identity (6.3.8) to infer from this that

$$f\left[x_1, \ldots, x_{m+1}\right] - w\left[x_1, \ldots, x_{m+1}\right] = 0,$$

which immediately implies that $w(x_{m+1})$ is equal to $f(x_{m+1})$. By recurrence, we can see that

$$w\left(x_j\right) = f\left(x_j\right), \quad \forall j = 1, \ldots, n.$$

This proves that w is a natural interpolating spline through the points (x_j, y_j). Let us prove uniqueness without appealing to the results of Section 6.1. If u and w coincide at all points x_j, then $u[x_i, \ldots, x_{i+m}]$ coincides with $w[x_i, \ldots, x_{i+m}]$, for $i = 1, \ldots, m - n$, and, therefore, due again to the identity (6.3.8), we must have

$$\int_a^b \left(w^{(m)} - u^{(m)}\right) N_{i,m-1} \, dt = 0,$$

for $i = 1, \ldots, n - m$. Therefore, $w^{(m)} - u^{(m)}$ is orthogonal to T. However, it also belongs to T. Therefore, it vanishes, and $w - u$ is a polynomial of degree at most $m - 1$. This polynomial vanishes at $n \geqslant m$ points, which means that it vanishes identically. This proves the lemma.   □

For the specifics of the numerical analysis of splines and, in particular, the choice of knots, we refer the interested reader to the literature of the subject, and, in particular, [23] or the older, but highly readable, [21].

6.3.5. B-splines in CAGD

The use of B-splines and, in particular, of fractions of B-splines in CAGD (computer aided geometric design) is very nicely detailed in [67], which contains many pretty figures and a large number of algorithms in C which allow for numerous geometrically motivated operations on curves and surfaces.

The advantage of rational fractions over polynomials comes from a classic observation: while it is impossible to parameterize exactly an arc of a circle by polynomials, the unit circle, without the point $(-1, 0)$, is parameterized by

(6.3.13) $$x\left(t\right) = \frac{1 - t^2}{1 + t^2}, \quad y\left(t\right) = \frac{2t}{1 + t^2}.$$

The choice (6.3.13) is not necessarily the most convenient. It is even better to use homogeneous coordinates, i.e.,

$$X(t) = 1 - t^2, \quad Y(t) = 2t, \quad Z(t) = 1 + t^2.$$

This means that we add a dimension to the space, but we consider two points to be equivalent if they are not zero and if they lie on the same line through 0. The set of lines of $\mathbb{R}^3$ through 0 is called the projective plane. The homogeneous coordinates of a point of the projective plane are the coordinates of any nonzero vector on the line of $\mathbb{R}^3$ associated with that point. The affine plane is identified with a subset of the projective plane: to the point (x_1, x_2) we associate the line through $(x_1, x_2, 1)$.

This notion is easily generalized to three dimensions; the homogeneous coordinates in projective space are nonzero vectors in $\mathbb{R}^4$, with the same equivalence relation as above.

Curves in three-dimensional space are parameterized with the help of three splines of one variable:

$$X(t) = \sum_{j=1}^{n} w_j N_{j,k}(t)\,\xi_j, \qquad Y(t) = \sum_{j=1}^{n} w_j N_{j,k}(t)\,\eta_j,$$

$$Z(t) = \sum_{j=1}^{n} w_j N_{j,k}(t)\,\zeta_j, \qquad T(t) = \sum_{j=1}^{n} w_j N_{j,k}(t).$$

The numbers w_j are non-negative and they do not all vanish; they give more freedom to the user. The vectors with coordinates (ξ_j, η_j, ζ_j) are the vertices of the control polygon. In the same fashion, a surface can be represented by products of B-splines:

$$X(s,t) = \sum_{i=1}^{m} \sum_{j=1}^{n} w_{i,j} N_{i,l}(s)\, N_{j,k}(t)\,\xi_{i,j},$$

$$Y(s,t) = \sum_{i=1}^{m} \sum_{j=1}^{n} w_{i,j} N_{i,l}(s)\, N_{j,k}(t)\,\eta_{i,j},$$

$$Z(s,t) = \sum_{i=1}^{m} \sum_{j=1}^{n} w_{i,j} N_{i,l}(s)\, N_{j,k}(t)\,\zeta_{i,j},$$

$$T(s,t) = \sum_{i=1}^{m} \sum_{j=1}^{n} w_{i,j} N_{i,l}(s)\, N_{j,k}(t).$$

Thus, we have defined NURBS or Non-Uniform Rational B-Splines, which have invaded all industries where shape is important: from its origin in metalwork to the textile and shoe industries, and now to the design of character fonts (through Postscript and its descendents); tools from this section are also used in image manipulation and virtual reality.

Though NURBS use almost nothing of the theory of algebraic curves and surfaces, they lead to many computational problems. In particular, good algorithms must be fast and flexible; they must allow the construction of surfaces bounded by given curves, the deformation of a curve or a surface into another by a finite number of steps, an easy search for intersections, and so on.

In any case, a rather simple program enables us to draw with B-splines; an example is given in Figure 6.5.

6.4. Exercises from Chapter 6

6.4.1. Varied exercises on splines

Exercise 6.4.1. Given a degree m and a sequence of $n \geqslant m$ distinct knots $x_1 < x_2 < \cdots < x_n$ in the open interval $]a, b[=]x_0, x_{n+1}[$, and a sequence of data y_j, let u be an interpolation natural spline of degree $2m - 1$ relative to this data. Let ε be a strictly positive number and let u^ε be the natural smoothing spline relative to this data and the uniform weights $\rho_j = \varepsilon$. Show that u^ε converges to u in $C^{2m-2}([a, b])$.

Exercise 6.4.2. Let the knot x_j be equal to j. Show that the B-splines of degree k can be deduced by translation from the B-spline of degree k with support in $[0, k + 1]$, which will be denoted by N_k. Calculate and plot N_k for $0 \leqslant k \leqslant 3$.

Exercise 6.4.3. With the notation of Exercise 6.4.2, let

$$M_k(x) = N_k \left(x + \frac{k+1}{2} \right).$$

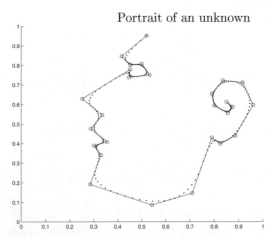

Portrait of an unknown

Figure 6.5: Drawing with B-splines. The mouse captures the control polygon drawn on screen by the user. The user defines the degree of the spline and gives the vector of knots.

Show that, for all integers k and l, the convolution of M_k and M_l is given by

$$(M_k * M_l)(x) = \int_{\mathbb{R}} M_k(x-y) M_l(y) \, dy = M_{k+l+1}(x),$$

and deduce from this identity that the Fourier transform of M_k is

$$\widehat{M_k}(\xi) = \int e^{-ix\xi} M_k(x) \, dx = \left(\frac{\sin(\xi/2)}{\xi/2} \right)^{k+1}.$$

Hint: use the following fact of Fourier analysis: the Fourier transform of a convolution is the product of the Fourier transforms of the factors.

Exercise 6.4.4. Let

$$x_i = \cos\left(\frac{(m-i)\pi}{m} \right), \quad i = 0, \dots, m.$$

Show that the absolute value of the $(m-1)$-th derivative of the B-spline of degree $m-1$ on these knots is independent of x on the support of this B-spline. Show that this B-spline is even and plot it for $m = 1$, 2, and 3.

6.4.2. Approximation by splines

Exercise 6.4.5. Let f be a function of class C^0 on $[a, b]$. Suppose that the x_j are given knots, $-k \leqslant j \leqslant n+k+1$, and denote by ξ the maximum of $\Delta x_i = x_{i+1} - x_i$, $-k \leqslant i \leqslant n + k$. Define the Lagrange-type spline by

$$L_{x,k} f(t) = \sum_{i=-k}^{n} f(x_i) N_{i,k}(t).$$

Let ω be the modulus of continuity of f, i.e.,

$$\omega(f, \tau) = \max \{ |f(x+h) - f(x)| : a \leqslant x \leqslant x+h \leqslant b, \ 0 \leqslant h \leqslant \tau \}.$$

Derive the following estimate:

$$\sup_{a \leqslant t \leqslant b} |f(t) - L_{x,k} f(t)| \leqslant \omega(f, (k+1)\xi).$$

Hint: use the identity (6.3.7) and the fact that the support of the B-splines is small.

Exercise 6.4.6. Assume now that f is of class C^l. Prove that

$$\sup_{a \leqslant t \leqslant b} |f(t) - L_{x,k} f(t)| \leqslant C \omega(f^{(l)}, (k+1)\xi)\xi^l,$$

for all $l \leqslant k$.

Exercise 6.4.7. From Exercise 6.4.6, determine an upper bound on the distances in $L^2(a,b)$ and in $L^\infty(a,b)$ from a function f to $S^k(x)$.

Exercise 6.4.8. Let f be a function of class C^m on $[a,b]$ and let $x_0^n,\ldots,x_n^n$ be a sequence of knots whose diameter $\xi^n = \max(x_{i+1} - x_i)$ tends to 0 as n tends to infinity. Show that the interpolating natural spline through the points $(x_j^n, f(x_j^n))$ tends uniformly to f as n tends to infinity.
Hint: use Lemma 6.3.5 and Exercise 6.4.5.

Exercise 6.4.9. If f is assumed to be of class C^m, show that the convergence of the natural interpolating splines sequence of Exercise 6.4.8 is faster.

6.4.3. Coincident knots

Exercise 6.4.10. Assume that $x = (x_1, x_2, \ldots, x_n)$ is a non-decreasing sequence of not necessarily distinct points. We will assume that no more than $k + 1$ of these points coincide. The B-splines $N_{i,k}$ on these points are still defined by eqn (6.3.1). Show that the recursion formulae (6.3.2) and (6.3.3) still hold, assuming that any indeterminate expression $0/0$ is replaced by 0.

Exercise 6.4.11. Let x_0, x_1, $x_2 = x_1 + \varepsilon$, and x_3 be distinct knots. Show that the B-spline of degree 2 on these knots tends to a limit as ε tends to 0. What is this limit? Now let x_1 be equal to $x_0 + \varepsilon$. What is the limit of the B-spline when ε tends to 0?

Exercise 6.4.12. Suppose that in the set $x_i, \ldots, x_{i+k+1}$ a number $m \leqslant k + 1$ of the knots coincide. Show, then, that at this point $N_{i,k}$ is not of class C^{k-1}, but of class C^{k-m+1}. Show that if $m = k + 1$ of these knots coincide, then $N_{i,k}$ is discontinuous.

Exercise 6.4.13. Let $x_1 = \cdots = x_p = 0$ and $x_{p+1} = \cdots = x_{n+2} = 1$. Give the explicit expression for the B-spline of degree n on these knots.

Exercise 6.4.14. Given a list of integers $0 \leqslant m_i \leqslant k - 1$ and the knots $x_1 \leqslant \cdots \leqslant x_n$, define the spline space $S^k(x,m)$ as the space of functions on $[a,b] = [x_0, x_{n+1}]$ which coincide with polynomials of degree at most k on each of the intervals $[x_i, x_{i+1}]$ and which satisfy the following continuity condition at the x_is: a function of this space is of class C^{m_i} in a neighbourhood of x_i. Find the dimension of $S^k(x,m)$.

Exercise 6.4.15. Replace the sequence of knots x_i by another sequence in which x_i is repeated $k - m_i$ times. Add to this sequence k auxiliary knots belonging to $(-\infty, x_0]$ and k auxiliary knots belonging to $[x_{n+1}, \infty)$. The new sequence will be denoted by x_i^*. Show that the B-splines of degree k on the knots x_i^* form a basis of $S^k(x,m)$.

7

Fourier's world

At the end of the eighteenth century and the beginning of the nineteenth century lived two important men in France. Although they were contemporaries and possessed the same surname, they were not related. One was Charles Fourier (1772–1837), philosopher and Utopian, the inspiration behind phalanstery and of a communism founded on free cooperation in a harmonious climate of human goodness. The other was Joseph Fourier, inventor of the series which bears his name. His series was known to Euler, at least, and played an important rôle in [31]. The phalansters did not work and I leave the reader to analyse the causes since I cannot expound on this subject with all the scientific competence required in a university text. The Fourier (Joseph) transformation and series are alive and well, being the subject of multiple theoretical and applied works. As for human goodness and the free cooperation between individuals, who would not like to see a little more?

By following these ideas, which, once again, come straight from the eighteenth and nineteenth centuries, we are trying to approximate functions, this time by trigonometric polynomials which are, after ordinary polynomials, the easiest to actually calculate. The theory is, in part, parallel to that of least-squares polynomial approximation, but it also has some different characteristics. I have not shirked from some repetition from the preceding chapter, for which I hope the reader will forgive me.

7.1. Trigonometric approximation and Fourier series

In this chapter, we approximate periodic functions by trigonometric polynomials in the least-squares sense. We show, by a convolution technique, that the trigonometric polynomials of period 1 are dense in the space of continuous complex periodic functions of period 1. We link trigonometric approximation and Fourier series and give some elementary results on the convergence of Fourier series.

7.1.1. Trigonometric polynomials

A trigonometric polynomial is an expression of the form

$$(7.1.1) \qquad \sum_{|k| \leqslant N} a_k e^{2i\pi k x},$$

where the numbers a_k are complex, and N is a positive integer or zero. Such a trigonometric polynomial is said to be of degree at most N. It is exactly of degree N if a_N or a_{-N} is not zero. The vector space of trigonometric polynomials of degree at most N is a vector space on $\mathbb{C}$ of dimension $2N + 1$. It will be denoted by $\mathbb{T}_N$. The trigonometric polynomials have period 1.

We denote by $C_\sharp^0$ the space of continuous periodic functions of period 1, from $\mathbb{R}$ to $\mathbb{C}$. This space is equipped with the maximum norm

$$\|u\|_\infty = \max \left\{ |u(x)| : x \in \mathbb{R} \right\}.$$

More generally, $C_\sharp^k$ is the set of k times continuously differentiable functions on $\mathbb{R}$ and of period 1. We are going to approximate, in the least-squares sense, the functions $f \in C_\sharp^0$ by elements of $\mathbb{T}_N$. The proof technique will be the same as for ordinary polynomial approximation since, geometrically, we are making an orthogonal projection in a pre-Hilbertian space on a space of finite dimension.

7.1.2. Integration of periodic functions

To properly express the operations that we are going to make, we need a coherent description of the theory of integration of periodic functions of period 1.

Let $\mathcal{L}_\sharp^1$ be the vector space of measurable functions from $\mathbb{R}$ to $\mathbb{C}$, which are of period 1 (that is $x \mapsto f(x+1) - f(x)$ is a negligible function), and are integrable on every compact subset of $\mathbb{R}$. We note that, if f is in $\mathcal{L}_\sharp^1$ and if a is some real number, then the expression

$$(7.1.2) \qquad \int_a^{a+1} f(x)\, dx$$

does not depend on a, as we can immediately verify. The common value of the expressions (7.1.2) will be denoted by

$$(7.1.3) \qquad \int_\sharp f(x)\, dx = \int_a^{a+1} f(x)\, dx, \quad \forall a \in \mathbb{R}.$$

We equip $\mathcal{L}_\sharp^1$ with the semi-norm

$$\|f\|_1 = \int_\sharp |f(x)|\, dx.$$

The kernel of this semi-norm is formed from functions which are negligible on $\mathbb{R}$. It is a classical fact that the quotient $L^1_\sharp$ of $\mathcal{L}^1_\sharp$ by negligible functions is a Banach space if we equip it with the norm

$$(7.1.4) \qquad \|f\|_1 = \int_\sharp |f(x)|\,\mathrm{d}x.$$

Almost all of the time functions and their equivalence classes modulo negligible functions are denoted identically. In the same way, the vector space $\mathcal{L}^2_\sharp$ is the set of measurable functions from $\mathbb{R}$ to $\mathbb{C}$, which have period 1 and which are square-integrable on every compact subset of $\mathbb{R}$. Its quotient by negligible functions is a Hilbert space denoted by $L^2_\sharp$. It is normed by

$$(7.1.5) \qquad \|f\|_2 = \left(\int_\sharp |f(x)|^2\,\mathrm{d}x \right)^{1/2}.$$

The corresponding scalar product is denoted by

$$(f,g)_\sharp = \int_\sharp f(x)\,\bar{g}(x)\,\mathrm{d}x.$$

Note that it is sesquilinear since we are referring to *complex* Hilbert spaces.

7.1.3. Least-squares approximation for trigonometric polynomials

We begin with an approximation theorem for which the proof is completely parallel to that of Theorem 5.1.1, and which will therefore be given in brief.

Theorem 7.1.1. Let $f \in C^0_\sharp$ be a periodic function of period 1. For any N in $\mathbb{N}$, there exists a unique trigonometric polynomial $P \in \mathbb{T}_N$, such that

$$(7.1.6) \qquad \int_\sharp |f - P|^2\,\mathrm{d}x \leqslant \int_\sharp |f - Q|^2\,\mathrm{d}x, \quad \forall Q \in \mathbb{T}_N.$$

Furthermore, if the k-th Fourier coefficient of f is defined by

$$(7.1.7) \qquad \hat{f}(k) = \int_\sharp f(x)\,\mathrm{e}^{-2\mathrm{i}\pi kx}\,\mathrm{d}x,$$

then P is given explicitly by

$$(7.1.8) \qquad P(x) = \sum_{|k| \leqslant N} \hat{f}(k)\,\mathrm{e}^{2\mathrm{i}\pi kx}.$$

Furthermore, for any $k \in \mathbb{Z}$,

$$(7.1.9) \qquad |\hat{f}(k)| \leqslant \|f\|_1. \qquad \diamond$$

Proof. The proof follows by the application of Lemma 5.1.2.   □

Remark 7.1.2. The expression $(\cdot,\cdot)_\sharp$ defines a pre-Hilbertian complex scalar
product on $C^0_\sharp$, and we have only used the pre-Hilbertian structure in the preceding proof. However, for every function $f \in L^1_\sharp$, the Fourier coefficients of f
are defined by eqn (7.1.7). In particular, if f is in $L^2_\sharp$, it is also in $L^1_\sharp$, has Fourier
coefficients, and we can approximate f in the least-squares sense by a trigono-
metric polynomial of degree at most N. In this case, and without changing a
comma of the preceding proof, the trigonometric polynomial which minimizes
$\|f - Q\|_2$, for Q in $\mathbb{T}_N$ is given by eqns (7.1.7) and (7.1.8). We recognize this
polynomial P as the partial sum of the Fourier series of f. This partial sum is
given by

$$(7.1.10) \qquad S_N f(x) = \sum_{|k| \leqslant N} \hat{f}(k)\, e^{2i\pi kx}.$$

As for ordinary polynomials, we have a Bessel inequality:

Corollary 7.1.3. The Fourier coefficients of a function $f \in L^2_\sharp$ satisfy the Bessel
inequality

$$(7.1.11) \qquad \sum_{|k| \leqslant N} \left|\hat{f}(k)\right|^2 \leqslant \int_\sharp |f|^2 \, dx.$$

Proof. The left-hand side of inequality (7.1.11) is the square of the norm of
the orthogonal projection of f and the right-hand side is the square of the norm
of f. Therefore the inequality is clear.   □

The Fourier coefficients of a periodic function are very often used in physics
and engineering. This is because physicists need expansions in Fourier series to
explain the vibrations of continuous media with simple geometric boundaries,
and therefore, every sort of phenomenon in acoustics, elasticity, and electromag-
netism, as well as non-vibratory phenomena such as the propagation of heat.

It was precisely to explain the heat equation that Joseph Fourier used the
series which has since borne his name. Section 18.4 presents some of his ideas
on heat. The original work of J. Fourier [31] is lacking in what we would today
call 'rigour', in that he believed that his series converged without problem.

7.1.4. Density of trigonometric polynomials in the space of continuous periodic functions

Just as ordinary polynomials are dense in $C^0(K)$ for the maximum norm on a
compact interval K, the trigonometric polynomials are dense in $C^0_\sharp$. The Weierstrass approximation theorem allows us to obtain this result, but we are going
to verify it by using convolution to construct a uniformly convergent sequence
of trigonometric approximations to a continuous periodic function f. We begin
with the lemma:

Lemma 7.1.4. We define the functions Q_n by letting

$$P_n(x) = (1 + \cos(2\pi x))^n \quad \text{and} \quad Q_n(x) = P_n(x) \left(\int_\sharp P_n(x)\, dx \right)^{-1}.$$

Then, the Q_n are trigonometric polynomials of degree n, they are positive or zero, their integral is 1 and, furthermore, for every $\epsilon \in\,]0, 1/2]$,

$$(7.1.12) \qquad \lim_{n\to\infty} \int_\epsilon^{1-\epsilon} Q_n(x)\, dx = 0.$$

Proof. It is clear that P_n and therefore Q_n are positive functions. We can write

$$P_n(x) = \left(1 + \frac{1}{2}\left(e^{2i\pi x} + e^{-2i\pi x} \right) \right)^n,$$

and it is clear that P_n is a trigonometric polynomial of degree at most n, of period 1, and so, therefore, is Q_n. Q_n is also non-negative and has integral 1 by construction. To prove eqn (7.1.12), it is necessary to show that

$$\lim_{n\to\infty} \int_\epsilon^{1-\epsilon} P_n(x)\, dx \Big/ \int_0^1 P_n(x)\, dx = 0.$$

Since P_n is invariant under the transformation $x \mapsto 1 - x$, we note that

$$\int_\epsilon^{1/2} P_n(x)\, dx = \int_{1/2}^{1-\epsilon} P_n(x)\, dx.$$

It is therefore equivalent to show that

$$(7.1.13) \qquad \lim_{n\to\infty} \int_\epsilon^{1/2} P_n(x)\, dx \Big/ \int_0^{1/2} P_n(x)\, dx = 0.$$

To do this, we need to bound the first term of eqn (7.1.13) from above and the second term from below. Note that

$$\max_{\epsilon \leqslant x \leqslant 1/2} P_n(x) \leqslant (1 + \cos(2\pi\epsilon))^n.$$

Moreover,

$$\int_0^{1/2} P_n(x)\, dx \geqslant \int_0^{\epsilon/2} P_n(x)\, dx \geqslant \frac{\epsilon}{2}(1 + \cos(\pi\epsilon))^n.$$

We see that

$$\int_\epsilon^{1/2} P_n(x)\, dx \Big/ \int_0^{1/2} P_n(x)\, dx \leqslant \frac{2}{\epsilon}\left(\frac{1 + \cos(2\pi\epsilon)}{1 + \cos(\pi\epsilon)} \right)^n = \frac{2}{\epsilon}\left(\frac{\cos(\pi\epsilon)}{\cos(\pi\epsilon/2)} \right)^{2n}.$$

Since $0 < \epsilon \leqslant 1/2$, we see that eqn (7.1.13) holds, which proves the lemma. $\qquad\square$

We can now state and prove the density theorem:

Theorem 7.1.5. The periodic trigonometric polynomials of period 1 are dense in the space $C_\sharp^0$ of continuous periodic functions of period 1.   ◇

Proof. Let f be in $C_\sharp^0$. Since $y \mapsto f(x-y)Q_n(y)$ is continuous and of period 1, the function

$$(7.1.14) \qquad f_n(x) = \int_\sharp f(x-y)\, Q_n(y)\, dy$$

is well defined for every x. We are going to show that, for every n, f_n is a trigonometric polynomial (and therefore, in particular, a continuous function). Indeed, Q_n is a trigonometric polynomial and also a linear combination of the monomials $t \mapsto e^{2i\pi kt}$. It is sufficient to verify that

$$x \mapsto \int_\sharp e^{2i\pi ky} f(x-y)\, dy$$

is a trigonometric polynomial. We make the change of variables $y = x - t$ in the integral

$$\int_\sharp e^{2i\pi ky} f(x-y)\, dy = \int_0^1 e^{2i\pi ky} f(x-y)\, dy.$$

We obtain

$$\int_0^1 e^{2i\pi ky} f(x-y)\, dy = \int_{x-1}^x e^{2i\pi k(x-t)} f(t)\, dt$$

$$= \int_\sharp e^{2i\pi k(x-t)} f(t)\, dt = e^{2i\pi kx} \int_\sharp e^{-2i\pi kt} f(t)\, dt.$$

Thus, we see that f_n is in $\mathbb{T}_n$. We can now estimate the difference between f and f_n:

$$|f_n(x) - f(x)| = \left| \int_\sharp f(x-y)\, Q_n(y)\, dy - f(x) \right|$$

$$= \left| \int_{-1/2}^{1/2} f(x-y)\, Q_n(y)\, dy - \int_{-1/2}^{1/2} f(x)\, Q_n(y)\, dy \right|$$

$$= \int_{-1/2}^{1/2} |[f(x-y) - f(x)]\, Q_n(y)\, dy|$$

$$\leqslant \int_{|y|<\epsilon} |f(x-y) - f(x)|\, Q_n(y)\, dy$$

$$+ \int_{\epsilon \leqslant |y| \leqslant 1/2} |f(x-y) - f(x)|\, Q_n(y)\, dy.$$

Let ω be the modulus of continuity of f (see Lemma 5.3.1), and let α be a strictly positive number. If we choose $\epsilon > 0$ such that $\omega(\epsilon) \leqslant \alpha/2$, then

$$\int_{|y|<\epsilon} |f(x-y) - f(x)| Q_n(y)\,\mathrm{d}y \leqslant \int_{-\epsilon}^{\epsilon} \omega(\epsilon) Q_n(y)\,\mathrm{d}y \leqslant \omega(\epsilon) \leqslant \frac{\alpha}{2}.$$

We fix this ϵ, and we note that

$$\int_{\epsilon \leqslant |y| \leqslant 1/2} |f(x-y) - f(x)| Q_n(y)\,\mathrm{d}y \leqslant 2 \max_{[0,1]} |f(x)| \int_{\epsilon \leqslant |y| \leqslant 1/2} Q_n(y)\,\mathrm{d}y.$$

Consequently, Lemma 7.1.4 allows us to choose an n such that this last expression is less than $\alpha/2$. Regrouping the terms, we obtain

$$|f(x) - f_n(x)| \leqslant \alpha,$$

and we have shown the desired density result. $\qquad\square$

7.1.5. Convergence in the mean square of trigonometric approximation to continuous functions

The density result of Theorem 7.1.5 allows us to deduce the convergence in the least-squares sense of the sequence of trigonometric approximations to a continuous function of period 1:

Theorem 7.1.6. For any continuous periodic function f, the partial Fourier sums of f

$$(7.1.15) \qquad\qquad S_N(f) = \sum_{|k| \leqslant N} \hat{f}(k)\, \mathrm{e}^{2\mathrm{i}\pi kx}$$

converge to f in the least-squares sense as follows:

$$\lim_{N \to \infty} \int_\sharp |f - S_N f|^2\,\mathrm{d}x = 0.$$

Furthermore, we have Parseval's relation:

$$(7.1.16) \qquad\qquad \int_\sharp |f|^2\,\mathrm{d}x = \sum_{k \in \mathbb{Z}} |\hat{f}(k)|^2. \qquad\qquad \diamond$$

Proof. From Pythagoras' theorem, eqn (5.1.9), with f_N defined by eqn (7.1.14), we have

$$(f - S_N f, f - S_N f)_\sharp + (S_N f - f_N, S_N f - f_N)_\sharp = (f - f_N, f - f_N)_\sharp.$$

Since $\|f - f_N\|_\infty$ tends to zero as N tends to infinity, we see that $\|f - S_N f\|_2$ tends to zero as N tends to infinity. Moreover, using eqn (5.1.9) again we obtain

$$(f - S_N f, f - S_N f)_\sharp + (S_N f, S_N f)_\sharp = (f, f)_\sharp,$$

and, therefore,

$$\lim_{N \to \infty} \sum_{|k| \leqslant N} |\hat{f}(k)|^2 = \lim_{N \to \infty} (S_N f, S_N f)_\sharp = (f, f)_\sharp .$$

From this we immediately deduce Parseval's relation. $\square$

7.1.6. Asymptotic behaviour of Fourier coefficients

From Parseval's relation it follows that the Fourier coefficients $\hat{f}(k)$ of a periodic
continuous function $f \in C_\sharp^0$ tend to zero as $|k|$ tends to infinity. This phenomenon
can be seen in the following, more general, case:

Lemma 7.1.7 (Riemann–Lebesgue). Let f be in $L_\sharp^1$. Then,

$$\lim_{|k| \to \infty} \hat{f}(k) = 0.$$

Proof. Suppose, first of all, that f is in $C_\sharp^0$. Then, f is also in $L_\sharp^2$ and $\hat{f}(k)$ tends
to zero when k tends to infinity. Suppose now that f is in $L_\sharp^1$. We identify it with
a function in $L^1(0,1)$ and apply the result on the density of continuous functions
on $[0,1]$ in $L^1(0,1)$, see, for example, [28]. Therefore, there exists a sequence of
continuous functions on $[0,1]$ denoted by g_n such that $|g_n - f|_{L^1} \leqslant 1/n$. We can
replace the g_n by the f_n with compact support. If ψ is a function of $\mathbb{R}$ in the
interval $[0,1]$, increasing, continuous, zero if $x \leqslant 1$, and equal to 1 if $x \geqslant 2$, we
note that

$$h_{m,n} = \psi(mx) \, \psi(m(1-x)) \, g_n(x) - g_n(x)$$

has support in $[0, 2/m] \cup [1 - 2/m, 1]$, and is bounded on this interval by $\max |g_n|$.
Consequently, we can choose m large enough so that $\|g_n - h_{m,n}\|_1 \leqslant 1/n$. We
let $f_n = h_{m,n}$ for this choice of m, and we estimate $\hat{f}(k)$ by noting that

$$|\hat{f}(k)| = \left| \int_\sharp f(x) \, e^{-2i\pi kx} \, dx \right|$$

$$\leqslant \left| \int_\sharp (f(x) - f_n(x)) \, e^{-2i\pi kx} \, dx \right| + \left| \int_\sharp f_n(x) \, e^{-2i\pi kx} \, dx \right|$$

$$\leqslant \|f - f_n\|_1 + |\hat{f}_n(k)|,$$

giving us an $\epsilon > 0$, and we fix n so that $\|f - f_n\|_1 \leqslant \epsilon/2$. We see that, for
sufficiently large k, $|\hat{f}_n(k)|$ can be made less than or equal to $\epsilon/2$. $\square$

Remark 7.1.8. The difficulty in the Riemann–Lebesgue lemma is conceptual: indeed, we can construct elements of $L_\sharp^1$ whose Fourier coefficients tend to 0 at
infinity arbitrarily slowly. Besides, we can show that there exist many choices
of Fourier coefficients a_k which decrease to 0 when $|k|$ tends to infinity, and
which are the Fourier coefficients of no integrable function. We refer to Subsections 7.3.3 and 7.3.4 for the construction of these counterexamples.

The technique of the proof which we have just used is completely standard and can be described in the following systematic fashion:

Theorem 7.1.9. Let E and F be Banach spaces on the field $\mathbb{K}$ (that is, normed complete vector spaces), equipped with respective norms $\|\cdot\|_E$ and $\|\cdot\|_F$. Let $(A_n)_{n\in\mathbb{N}}$ be a sequence of linear mappings from E to F whose operator norm is uniformly bounded, that is, there exists a number K such that

$$(7.1.17) \qquad \|A_n(x)\|_F \leqslant K\|x\|_E, \quad \forall x \in E, \forall n \in \mathbb{N}.$$

Suppose that there exists a dense subset D of E such that

$$\lim_{n\to\infty} A_n(x), \quad \forall x \in D,$$

exists. Then, there exists a unique continuous mapping B from E to F which continues the mapping

$$D \to F$$
$$x \mapsto \lim_{n\to\infty} A_n(x).$$

Furthermore, B is linear from E to F and its operator norm is bounded above by K. ◇

Proof. We begin by showing that the sequence $\left(A_n(y)\right)_n$ is a Cauchy sequence for every $y \in E$. For every x in D, we have

$$\|A_n y - A_m y\|_F \leqslant \|A_n y - A_n x\|_F + \|A_n x - A_m x\|_F + \|A_m x - A_m y\|_F$$
$$\leqslant 2K\|x - y\|_E + \|A_n x - A_m x\|_F.$$

Let $\epsilon > 0$. Fix x such that $K\|x - y\|_E \leqslant \epsilon/3$. This is possible since D is dense in E. We can then find an M such that

$$\|A_n x - A_m x\| \leqslant \frac{\epsilon}{3}, \quad \forall n, m \geqslant M.$$

Consequently, the sequence $(A_n y)_{n\in\mathbb{N}}$ is a Cauchy sequence, and it converges to a certain limit which we call By. Clearly, B is linear. Let us show that it is continuous. The operator norm of B is bounded above as follows:

$$\|By\|_F \leqslant \|By - A_n y\|_F + \|A_n y\|_F \leqslant \|By - A_n y\|_F + K\|y\|_E.$$

By passing to the limit when n tends to infinity, we have

$$(7.1.18) \qquad \|B(y)\|_F \leqslant K\|y\|.$$

Let C be another continuous extension of $x \mapsto \lim_n A_n(x)$ to E, and let x_k be a sequence of elements of D converging to $y \in E$. Then,

$$\|Cy - By\|_F \leqslant \|Cy - Cx_k\|_F + \|Bx_k - By\|_F.$$

By passing to the limit in k, we see that C coincides with B. This proves the uniqueness of the continuous extension. □

Remark 7.1.10. Suppose that E and F are not Banach spaces but are complete metric spaces, equipped with distances d_E and d_F, and that the A_n are mappings which are uniformly equi-continuous, in the sense that there exists a modulus of continuity ω (see the definition in Subsection 5.3.1), such that

$$d_F\left(A_n\left(y\right), A_n\left(y'\right)\right) \leqslant \omega\left(d_E\left(y, y'\right)\right), \quad \forall n \in \mathbb{N}, \ \forall y, y' \in E.$$

Then, the convergence of the A_n on the dense subset of E implies the convergence of the A_n on all of E and their limit B satisfies

$$d_F\left(B\left(y\right), B\left(y'\right)\right) \leqslant \omega\left(d_E\left(y, y'\right)\right), \quad \forall y, y' \in E.$$

The proof of this fact is left to the reader.

We now see that Lemma 7.1.7 is a consequence of Theorem 7.1.9 provided that, for the space E we take the space $L^1_\sharp$, for the dense subset D we take the space $C^0_\sharp$, for space F we take the space $\mathbb{C}^2$, and for the sequence A_n we take the operator $f \mapsto (\hat{f}(n), \hat{f}(-n))$.

7.1.7. Convergence of trigonometric approximation to $L^2_\sharp$ functions

Just as we deduced the density of $C^0([a, b])$ in $L^1(a, b)$ and the density of $C^0_\sharp$ in $L^1_\sharp$, we can deduce the density of $C^0([a, b])$ in $L^2(a, b)$ and the density of $C^0_\sharp$ in $L^2_\sharp$. We therefore have the following theorem:

Theorem 7.1.11 (Riesz–Fischer; Parseval). For every function f in $L^2_\sharp$, the partial Fourier sums $S_N(f)$ of f converge to f in $L^2_\sharp$ as N tends to infinity. Furthermore, we have Parseval's relation

$$(7.1.19) \qquad \sum_{k \in \mathbb{Z}} \left|\hat{f}\left(k\right)\right|^2 = \int_\sharp \left|f\left(x\right)\right|^2 \, \mathrm{d}x,$$

with its polarized form

$$\int_\sharp f\left(x\right) \bar{g}\left(x\right) \mathrm{d}x = \sum_{k \in \mathbb{Z}} \hat{f}\left(k\right) \overline{\hat{g}\left(k\right)}, \quad \forall f, g \in L^2_\sharp.$$

Conversely, giving the coefficients $a_k \in \mathbb{C}$ such that

$$\sum_{k \in \mathbb{Z}} \left|a_k\right|^2 < +\infty$$

allows the definition of the function f by the series

$$\sum_{k \in \mathbb{Z}} a_k \mathrm{e}^{2\mathrm{i}\pi k x},$$

which converges in the quadratic mean towards a function $f \in L^2_\sharp$ whose Fourier coefficients satisfy

$$\hat{f}(k) = a_k.$$

In summary, if $\ell^2(\mathbb{Z})$ denotes the vector space of complex sequences indexed by $\mathbb{Z}$ which converge quadratically, the mapping

$$L^2_\sharp \to \ell^2(\mathbb{Z})$$
$$f \mapsto \left(\hat{f}(k)\right)_{k \in \mathbb{Z}}$$

is a bijective isometry. ◇

Proof. We will apply the technique used in the proof of Theorem 7.1.9. Take as spaces E and F the space $L^2_\sharp$, and as operator A_N the mapping S_N of the partial sum. We deduce from Bessel's inequality (7.1.11) that

$$\|S_N f\|_2 \leqslant \|f\|_2.$$

If the dense set D is $C^0_\sharp$, Theorem 7.1.6 shows that we are in the area of applicability of Theorem 7.1.9, with

$$\lim_{n \to \infty} \|S_N f - f\|_2 = 0, \quad \forall f \in C^0_\sharp.$$

We therefore have the convergence of the partial Fourier sums S_N to f in $L^2_\sharp$. Furthermore, as N tends to infinity,

$$\|S_N f\|_2^2 = \sum_{|k| \leqslant N} \left|\hat{f}(k)\right|^2 \to \|f\|_2^2.$$

We have Parseval's relation and we pass to the polarized form of this by noting that

$$4(f,g)_\sharp = (f+g, f+g)_\sharp + (f-g, f-g)_\sharp + (f+ig, f+ig)_\sharp + (f-ig, f-ig)_\sharp.$$

Conversely, if we let

$$f_N(x) = \sum_{|k| \leqslant N} a_k e^{2i\pi kx},$$

we note that, if $M > N$,

$$\|f_N - f_M\|_2^2 = \left\|\sum_{M \geqslant |k| > N} a_k e^{2i\pi kx}\right\|_2^2 = \sum_{M \geqslant |k| > N} |a_k|^2.$$

The sequence of the f_N is therefore a Cauchy sequence for the norm $L^2_\sharp$. Its limit is a certain function $f \in L^2_\sharp$. If we fix $k \in \mathbb{Z}$, we note that for $N \geqslant |k|$,

$$\hat{f}_N(k) = a_k.$$

As the mapping $g \mapsto \hat{g}(k)$ is continuous from $L_\sharp^2$ to $\mathbb{C}$, we see, by passing to the limit, that

$$\hat{f}(k) = a_k.$$

This gives us the converse, and the conclusion of the theorem is immediate. $\square$

Now we show that a regularity hypothesis on f implies an estimate of the decrease of the Fourier coefficients of f at infinity:

Lemma 7.1.12. Let f be a C^p function on $\mathbb{R}$ of period 1. Then, for every $m \leqslant p$, there exists a constant C_m such that

$$\left|\hat{f}(k)\right| \leqslant \frac{C_m}{|k|^m}, \quad \forall k \neq 0.$$

Proof. We note that, if $p \geqslant 1$ and $k \neq 0$,

$$\hat{f}(k) = \int_\sharp f(x)\, e^{-2i\pi kx}\, dx = \int_0^1 f(x)\, e^{-2i\pi kx}\, dx$$

$$= \frac{f(x)\, e^{-2i\pi kx}}{-2i\pi k}\bigg|_0^1 - \int_0^1 \frac{f'(x)\, e^{-2i\pi kx}}{-2i\pi k}\, dx.$$

Consequently, since f is continuous and of period 1, the integrated term vanishes and

$$\hat{f}(k) = \frac{\widehat{f'}(k)}{2i\pi k}.$$

An immediate recurrence gives

$$\hat{f}(k) = \frac{\widehat{f^{(m)}}(k)}{(2i\pi k)^m},$$

and the constant C_m in the theorem can be taken equal to $\|f^{(m)}\|_1 (2\pi)^{-m}$. $\square$

7.1.8. Uniform convergence of Fourier series

In the preceding section, we have seen that the partial Fourier sums of a square-integrable function f converge in $L_\sharp^2$ to f. In this section we give sufficient conditions for the uniform convergence of a Fourier series, which is a lot more precise than convergence in the quadratic mean.

Lemma 7.1.13. Let f be an element of $L_\sharp^1$ and suppose that

$$\sum_{k \in \mathbb{Z}} \left|\hat{f}(k)\right| < +\infty.$$

Then, the partial sums $S_N f$ converge uniformly to a continuous function $S f$ which is equal to f almost everywhere.

Proof. We have

$$\left| \sum_{M \geqslant |k| > N} \hat{f}(k) e^{2i\pi kx} \right| \leqslant \sum_{M \geqslant |k| > N} |\hat{f}(k)|,$$

which proves that $(S_N f)_N$ is a Cauchy sequence in $C^0_\sharp$. It is therefore convergent in $C^0_\sharp$. We denote its limit by Sf, and it remains to show that this limit is identical to f almost everywhere. We show that, in fact, f is in $L^2_\sharp$. Using eqn (7.1.9), we obtain the following inequality:

$$\sum_{|k| \leqslant N} |\hat{f}(k)|^2 \leqslant \sum_{|k| \leqslant N} \|f\|_1 |\hat{f}(k)|.$$

Consequently,

$$\sum_{k \in \mathbb{Z}} |\hat{f}(k)|^2 \leqslant \|f\|_1 \sum_{k \in \mathbb{Z}} |\hat{f}(k)|.$$

It follows, from Theorem 7.1.11, that f is in $L^2_\sharp$ and that

$$\lim_{N \to \infty} \|f - S_N f\|_2 = 0.$$

Moreover,

$$\|S_N f - Sf\|_2 \leqslant \max_{x \in \mathbb{R}} |S_N f(x) - Sf(x)|.$$

Using the triangle inequality and passing to the limit as $N \to \infty$, it can be deduced that $f = Sf$ almost everywhere on $\mathbb{R}$. $\qquad \square$

The following is a useful corollary of this result:

Corollary 7.1.14. Let $f \in L^1_\sharp$. Then f vanishes almost everywhere if and only if all of its Fourier coefficients are zero.

Proof. If f vanishes almost everywhere, it is clear that all its Fourier coefficients are zero. Conversely, if all the Fourier coefficients of f are zero, then we have the case of Lemma 7.1.13: the series of Fourier coefficients of f is absolutely convergent and, therefore, f is the uniform limit of its partial Fourier sums, which are all zero. $\qquad \square$

We present some examples of the application of Lemma 7.1.13. If f has the property

$$(7.1.20) \qquad \sum_{k \in \mathbb{Z}} (1 + k)^2 |\hat{f}(k)|^2 < +\infty,$$

then the Fourier series of f converges uniformly to f. Indeed, in this case the application of the Cauchy–Schwarz inequality gives

$$\sum_{|k|\leqslant N} |\hat{f}(k)| = \sum_{|k|\leqslant N} \frac{\left(1+k^2\right)^{1/2} |\hat{f}(k)|}{\left(1+k^2\right)^{1/2}}$$

$$\leqslant \left(\sum_{|k|\leqslant N} (1+k^2)|\hat{f}(k)|^2\right)^{1/2} \left(\sum_{|k|\leqslant N} \frac{1}{1+k^2}\right)^{1/2},$$

which is bounded independently of N by virtue of the hypotheses on f and the convergence of the series of the general term $(1+k^2)^{-1}$.

We have an estimate of the type given in eqn (7.1.20) if a function f belongs to $L^2_\sharp$ and is the primitive of a function f_1 also belonging to $L^2_\sharp$ in the following sense:

$$f(y) - f(x) = \int_x^y f_1(t)\,\mathrm{d}t, \quad \forall x,\ \forall y \in \mathbb{R} \text{ such that } y > x.$$

The reader may verify that, in this case,

$$\sum_{k\in\mathbb{Z}} (1+4\pi^2 k^2)|\hat{f}(k)|^2 = \|f\|_2^2 + \|f_1\|_2^2.$$

Here is another case of absolute convergence of the series of Fourier coefficients of a function $f \in L^1_\sharp$: suppose that f is the primitive of a function $f_1 \in L^1_\sharp$, which is itself the primitive of a function $f_2 \in L^1_\sharp$ in the following sense:

$$f(y) - f(x) = \int_x^y f_1(t)\,\mathrm{d}t, \quad \forall x,\ \forall y \in \mathbb{R} \text{ such that } y > x$$

and

$$f_1(y) - f_1(x) = \int_x^y f_2(t)\,\mathrm{d}t, \quad \forall x,\ \forall y \in \mathbb{R} \text{ such that } y > x.$$

The reader may verify that in this case, we have the estimate

$$(1+4\pi^2 k^2)|\hat{f}(k)| \leqslant \|f\|_1 + \|f_2\|_1, \quad \forall k \in \mathbb{Z}.$$

7.2. From convolution to pointwise convergence of Fourier series

7.2.1. Convolution

The convolution has already been employed for the proof of Theorem 7.1.5. Generally, if f and g belong to $C^0_\sharp$, the function $y \mapsto f(x-y)g(y)$ is also in $C^0_\sharp$, and we let

$$(f * g)(x) = \int_\sharp f(x-y)\,g(y)\,\mathrm{d}y.$$

The convolution is commutative in $C_\sharp^0$. Indeed,

$$\int_\sharp f(x-y)\,g(y)\,dy = \int_0^1 f(x-y)\,g(y)\,dy,$$

and by the change of variable $t = x - y$, this last expression becomes

$$\int_{x-1}^x f(t)\,g(x-t)\,dt = \int_\sharp f(t)\,g(x-t)\,dt.$$

The convolution is also associative in $C_\sharp^0$, as we see by changing the order of the integration in the following relations:

$$
\begin{aligned}
[(f * g) * h](x) &= \int_\sharp \left(\int_\sharp f(x-y)\,g(y-z)\,dy \right) h(z)\,dz \\
&= \int_0^1 \int_0^1 f(x-y)\,g(y-z)\,h(z)\,dy\,dz \\
&= \int_0^1 \left(\int_0^1 g(y-z)\,h(z)\,dz \right) f(x-y)\,dy \\
&= [f * (g * h)](x).
\end{aligned}
$$

Finally, it is distributive with respect to addition. We verify that the vector space $C_\sharp^0$ is equipped with an algebraic structure by the convolution, and that the norm $L_\sharp^1$ is compatible with the convolution on $C_\sharp^0$, that is,

$$\|f * g\|_1 \leqslant \|f\|_1 \|g\|_1, \quad \forall f, g \in C_\sharp^0.$$

By the density of $C_\sharp^0$, we see that we can extend the convolution to all of $L_\sharp^1$. In fact, the Fubini–Lebesgue theorem, proved in all integration courses, allows us to state a more precise result:

Theorem 7.2.1. For any functions f and g in $\mathcal{L}_\sharp^1$ and almost every x, the function

$$y \mapsto f(x-y)\,g(y)$$

is in $\mathcal{L}_\sharp^1$, and the function $f * g$, which is defined almost everywhere by

$$(7.2.1) \qquad (f * g)(x) = \int_\sharp f(x-y)\,g(y)\,dy$$

is in $\mathcal{L}_\sharp^1$. Furthermore, we have the inequality

$$(7.2.2) \qquad \|f * g\|_1 \leqslant \|f\|_1 \|g\|_1.$$

On $L_\sharp^1$, the convolution is commutative, associative, and distributive with respect to addition. ◇

The proof of this result is found, for example, in the book by P. Malliavin and H. Airault [60, Chapter III] (a high level book) and in the book by J. Dieudonné [24, Chapter XIV] (the book gives results still more general than the preceding one). In fact, all good treatments of integration give the elements needed to prove this theorem, which is an exercise in the application of the Fubini–Lebesgue theorem.

7.2.2. Regularization

Convolution allows us to regularize:

Theorem 7.2.2. Let f and g be in $L_\sharp^1$. If f is C^m, $f * g$ is C^m, and

$$\frac{\mathrm{d}^k}{\mathrm{d}x^k} (f * g) = \frac{\mathrm{d}^k f}{\mathrm{d}x^k} * g, \quad \forall k \leqslant m.$$

Furthermore,

$$(7.2.3) \qquad \max_x \left| \frac{\mathrm{d}^k}{\mathrm{d}x^k} (f * g)(x) \right| \leqslant \max_x \left| \frac{\mathrm{d}^k f}{\mathrm{d}x^k} (x) \right| \|g\|_1, \quad \forall k \leqslant m.$$

If f and g are in $L_\sharp^2$, then $f * g$ is almost everywhere equal to a function belonging to $C_\sharp^0$; we identify $f * g$ to this function and we have the inequality

$$(7.2.4) \qquad \max_x |(f * g)(x)| \leqslant \|f\|_2 \|g\|_2.$$

If f is in $L_\sharp^1$, and g is in $L_\sharp^2$, $f * g$ is in $L_\sharp^2$. Furthermore, we have the inequality

$$(7.2.5) \qquad \|f * g\|_2 \leqslant \|f\|_1 \|g\|_2. \qquad \qquad \diamond$$

Proof. Let x_k be a sequence tending to x, and let f be a C^m periodic function of period 1. It is clear that, as k tends to infinity, the sequence of functions

$$h_k : y \mapsto f(x_k - y)$$

tends in C^m to the function

$$h : y \mapsto f(x - y).$$

It will therefore be a consequence of Lebesgue's theorem, in relation to the continuity and differentiability of integrals dependent on a parameter, that

$$x \mapsto \int_\sharp f(x - y) g(y) \, \mathrm{d}y$$

is C^m and that its derivatives have the given expression. As for the inequality (7.2.3), this immediately follows since, if f is in $C_\sharp^0$,

$$\left| \int_\sharp f(x - y) g(y) \, \mathrm{d}y \right| \leqslant \max_x |f(x)| \|g\|_1.$$

Let f be in $L^2_\sharp$. Then, for every sequence x_k tending to x, the sequence of functions

$$h_k : y \mapsto f(x_k - y)$$

tends in $L^2_\sharp$ to the function

$$h : y \mapsto f(x - y),$$

as k tends to infinity. Indeed, if we let $(A_k f)(y) = f(x_k - y)$, we can apply Theorem 7.1.9, with $E = F = L^2_\sharp$, and $D = C^0_\sharp$. We then see that, if g is in $L^2_\sharp$,

$$(h_k, g)_\sharp \to (h, g)_\sharp,$$

which proves the continuity of $f * g$. Inequality (7.2.4) comes from the Cauchy–Schwartz inequality:

$$|(f * g)(x)| \leqslant \int_\sharp |f(x - y)| \, |g(y)| \, dy$$

$$\leqslant \left(\int_\sharp |f(x - y)|^2 \, dy \right)^{1/2} \left(\int_\sharp |g(y)|^2 \, dy \right)^{1/2} = \|f\|_2 \, \|g\|_2 \, .$$

Let f be in $C^0_\sharp$ and g in $L^2_\sharp$. Let h be some element of $L^2_\sharp$ and let

$$\check{h}(x) = h(-x).$$

The convolution of three factors $f * g * \check{h}$ is well defined. From the first assertion of the theorem, $f * g$ is continuous, and therefore belongs to $L^2_\sharp$, and $f * g * \check{h}$ is also continuous. On the one hand,

$$\left(f * g * \check{h} \right)(0) = \int_\sharp (f * g)(y) \, \check{h}(-y) \, dy = \left(f * g, \bar{h} \right)_\sharp.$$

On the other hand,

$$\left(f * g * \check{h} \right)(0) = \int_\sharp f(y) \, (g * \check{h})(-y) \, dy$$

and from the first two assertions of the theorem

$$\left| [f * g * \check{h}](0) \right| \leqslant \|f\|_1 \max_x |(g * \check{h})(-x)| \leqslant \|f\|_1 \, \|g\|_2 \, \|h\|_2 \, ,$$

since $\|h\|_2 = \|\check{h}\|_2$. We have therefore obtained the following estimate, valid for every h in $L^2_\sharp$:

(7.2.6) $$|(f * g, h)_\sharp| \leqslant \|f\|_1 \, \|g\|_2 \, \|h\|_2 \, .$$

We deduce from inequality (7.2.6), by replacing h by $f * g$, that

$$\|f * g\|_2 \leqslant \|f\|_1 \, \|g\|_2 \, .$$

The linear mapping $f \mapsto f * g$ is a continuous mapping from $C_\#^0$ equipped with the norm $L_\#^1$ to $L_\#^2$. Therefore, there exists a unique analytic continuation of this mapping to all $L_\#^1$. We argue as in the proof of Theorem 7.1.9 to make this continuation, and it satisfies inequality (7.2.5) by continuity. $\square$

7.2.3. Constructive density results

Convolution allows us to prove many constructive density results:

Lemma 7.2.3. Let f_n be a sequence of functions belonging to $L_\#^1$, which have the following properties:

$$\text{(7.2.7)} \qquad \int_\# f_n \, dx = 1, \qquad \forall n \in \mathbb{N};$$

$$\text{(7.2.8)} \qquad \|f_n\|_1 \leqslant K, \qquad \forall n \in \mathbb{N};$$

$$\text{(7.2.9)} \qquad \lim_{n \to \infty} \int_\alpha^{1-\alpha} |f_n(x)| \, dx = 0, \qquad \forall \alpha > 0.$$

Then, for any g in $L_\#^1$, $f_n * g$ tends to g in $L_\#^1$. If g is in $L_\#^2$, $f_n * g$ tends to g in $L_\#^2$. If g is continuous, $f_n * g$ tends to g in $C_\#^0$.

Proof. This proof is obtained by repeated application of Theorem 7.1.9. We begin with the last assertion: we prove essentially the same result as for Theorem 7.1.5, except that this time the f_n are not positive:

$$(f_n * g)(x) - g(x) = \int_\# f_n(y) g(x - y) \, dy - \int_\# f_n(y) g(x) \, dy.$$

Let ω be the modulus of continuity of g. Then,

$$|(f_n * g)(x) - g(x)| \leqslant 2 \left(\int_\alpha^{1-\alpha} |f_n(y)| \, dy \right) \max_x |g(x)|$$
$$+ \left(\int_0^\alpha |f_n(y)| \, dy + \int_{1-\alpha}^1 |f_n(y)| \, dy \right) \omega(\alpha).$$

Given $\epsilon > 0$, if we fix α such that $K\omega(\alpha) \leqslant \epsilon/2$ and take n sufficiently large, so that

$$2 \left(\int_\alpha^{1-\alpha} |f_n|(y) \, dy \right) \max_x |g(x)| \leqslant \frac{\epsilon}{2},$$

we see that $\|f_n * g - g\|_\infty$ tends to 0 and, therefore, $f_n * g$ tends to g.

Now, if g is in $L_\#^1$, we let, with the notation of Theorem 7.1.9, $E = F = L_\#^1$, $A_n g = f_n * g$, and the dense subset D is $C_\#^0$. Theorem 7.2.1 allows us to confirm that

$$\|A_n g\|_1 \leqslant \|f_n\|_1 \|g\|_1 \leqslant K \|g\|_1$$

and we can conclude the required result.

If g is in $L^2_\sharp$, we take $E = F = L^2_\sharp$, A_n and D as above, and Theorem 7.2.2 provides the estimate

$$\|A_n g\|_2 \leqslant \|f_n\|_1 \|g\|_2 \leqslant K \|g\|_2.$$

Then, again, we can apply Theorem 7.1.9. □

Lemma 7.2.3 allows us to see that we can construct approximations to functions which are continuous, integrable, and square-integrable using a sequence of C^∞ functions. The approximation converges, respectively, uniformly, in the mean, and in the quadratic mean.

7.2.4. Convolution and Fourier series

For integrable f and g, we easily calculate the Fourier coefficients of $f * g$:

Lemma 7.2.4. Let f and g be members of $L^1_\sharp$. Then,

$$\widehat{f * g}(k) = \hat{f}(k)\,\hat{g}(k).$$

Proof. Define a function e_k by

$$e_k(x) = \mathrm{e}^{2i\pi kx}.$$

This function is C^∞ and of period 1. Then, if g belongs to $L^1_\sharp$,

$$
\begin{aligned}
(g * e_k)(x) &= \int_\sharp g(y)\,e_k(x - y)\,\mathrm{d}y = \int_\sharp g(y)\,\mathrm{e}^{-2i\pi ky}\,\mathrm{d}y \; \mathrm{e}^{2i\pi kx} \\
&= \hat{g}(k)\,e_k(x).
\end{aligned}
$$

From the associativity of the convolution,

$$
\begin{aligned}
\widehat{f * g}(k) &= (f * g * e_k)(0) = [f * (g * e_k)](0) \\
&= [f * (\hat{g}(k)\,e_k)](0) \\
&= \hat{g}(k)\,(f * e_k)(0) \\
&= \hat{g}(k)\,\hat{f}(k)
\end{aligned}
$$

and we have the result claimed. □

The difficulty of the summation of Fourier series can be understood by introducing a kernel which is defined as follows: If f belongs to $L^1_\sharp$, we have

$$S_N f(x) = \sum_{|k| \leqslant N} \left(\int_\sharp f(y)\,\mathrm{e}^{-2i\pi ky}\,\mathrm{d}y \right) \mathrm{e}^{2i\pi kx} = \int_\sharp f(y) \sum_{|k| \leqslant N} \mathrm{e}^{2i\pi k(x-y)}\,\mathrm{d}y.$$

Let

(7.2.10)
$$D_N(x) = \sum_{|k| \leqslant N} e^{2i\pi k x}.$$

We thus have

$$S_N f = f * D_N.$$

The kernel D_N is the Dirichlet kernel, and it may be expressed as

(7.2.11) $D_N(0) = 2N + 1,$ $D_N(x) = \dfrac{\sin(\pi(2N+1)x)}{\sin(\pi x)},$ $x \notin \mathbb{Z},$

as can be shown by explicit calculation. The Dirichlet kernel has the property (7.2.7) but neither the property (7.2.8) nor the property (7.2.9). This is why the summation of Fourier series is a difficult problem.

7.2.5. Convergence of Fourier series as a local phenomenon

We do, however, have some results on pointwise convergence on the condition of having some precise information. In particular, if the function is piecewise continuously differentiable (in a sense which we will clarify), the partial sums $S_N f(x)$ tend towards the half-sum of the values to the right and left of the function.

First of all, we show that the convergence of a Fourier series is a local phenomenon.

Lemma 7.2.5. Let g belong to $L_\sharp^1$. Suppose that g vanishes almost everywhere in an interval $]a, b[\subset \mathbb{R}$. Then the partial Fourier sums of g uniformly converge to zero on every compact sub-interval of $]a, b[$.

Proof. Without loss of generality, we can suppose that g vanishes almost everywhere on an interval of length strictly less than 1. If not, we can already conclude the required result due to Corollary 7.1.14. We can make a translation to move us to the interval $]-a, a[$, with $|a| < 1/2$. We have

$$S_N g(x) = \int_\sharp g(x - y) D_N(y) \, dy,$$

that is,

$$S_N g(x) = -\frac{i}{2} \int_\sharp g(x - y) \frac{e^{i\pi y}}{\sin \pi y} e^{2i\pi N y} \, dy + \frac{i}{2} \int_\sharp g(x - y) \frac{e^{-i\pi y}}{\sin \pi y} e^{-2i\pi N y} \, dy.$$
(7.2.12)

We can limit ourselves to studying the convergence of the first of the two integrals in the second term of eqn (7.2.12), the other integral being analogous. We note that the function

$$y \mapsto \phi(x, y) = g(x - y) \frac{e^{i\pi y}}{\sin \pi y}$$

is integrable, provided that $\sin \pi y$ is bounded below on the complement of the set where $y \mapsto g(x - y)$ vanishes. If we choose x such that $|x| \leqslant a - \alpha$, then $|\sin \pi y| \geqslant \sin \pi \alpha$ on this set. It therefore follows, from the Riemann–Lebesgue lemma, that $S_N g(x)$ tends to 0 on every compact set included in $]-a, a[$.

We have to show that this convergence is uniform. To do this, we will approximate g by a sequence of C^1 functions, as follows: Let $f \in C_0^1$ be a continuously differentiable function on $\mathbb{R}$, with support included in $[-1, 1]$ and integral 1. For $n \geqslant 2$, we define a function $f_n \in C_\sharp^1$ by its restriction to the interval $[-1/2, 1/2]$ which must be equal to $x \mapsto n f(nx)$. It is clear that the sequence f_n has the properties (7.2.7)–(7.2.9). Lemma 7.2.3 implies that the sequence $g_n = f_n * g$ tends to g in $L_\sharp^1$, and Theorem 7.2.2 says that g_n is C^1. Furthermore, g_n is identically zero on $[-a + \alpha, a - \alpha]$, provided that $n \geqslant 1/\alpha$.

We let

$$\phi_n (x, y) = g_n (x - y) \frac{e^{i\pi y}}{\sin \pi y},$$

and we suppose, from now on, that $\alpha < a/6$ is fixed and that n is greater than $1/\alpha$.

Let $h \in C_\sharp^1$ be a function which coincides with $e^{i\pi y} / \sin \pi y$ if $|y|$ is included between 2α and $1/2$, and which is zero if $|y| \leqslant \alpha$. If $|x| \leqslant a - 3\alpha$ then,

$$\phi_n (x, y) = g_n (x - y) h (y),$$

since if $|x| \leqslant a - 3\alpha$ and $|y| \leqslant 2\alpha$, then $g_n(x - y)$ vanishes.

We can now bound from above the first integral in (7.2.12) for $|x| \leqslant a - 3\alpha$ by decomposing it as

$$\int_\sharp (g - g_n) (x - y) h (y) e^{2i\pi N y} \, dy + \frac{i}{2} \int g_n (x - y) h (y) e^{2i\pi N y} \, dy.$$

The first integral is bounded by $\|g - g_n\|_1 / \sin(2\pi\alpha)$, and we use Lemma 7.1.12 for the second:

$$\left| \int \phi_n (x, y) e^{2i\pi N y} \, dy \right| \leqslant \frac{\|(\partial\phi_n/\partial y) (x, \cdot)\|_1}{2N\pi},$$

which is bounded independently of x by

(7.2.13) $$\left(\frac{\max |f_n| \max |h'| + \max |f_n'| \max |h|}{2\pi N} \right) \|g\|_1 .$$

Fix an $\epsilon > 0$ and choose n such that $\|g - g_n\|_1 / \sin(2\pi\alpha) \leqslant \epsilon/2$. We can then choose N to be large enough so that expression (7.2.13) is less than $\epsilon/2$. □

7.2.6. Pointwise convergence of partial Fourier sums of absolutely continuous functions

We are now going to show that the partial Fourier sums of an integrable function, which is itself the integral of an integrable function, converge pointwise. To this end, we introduce a function Δ_N defined by

$$\Delta_N(x) = \begin{cases} \int_0^x D_N(y)\,dy & \text{if } x \geqslant 0; \\ -\int_x^0 D_N(y)\,dy & \text{if } x < 0. \end{cases}$$

The properties of Δ_N are summarized by the following lemma:

Lemma 7.2.6. The function Δ_N is odd and uniformly bounded on the interval $[-1/2, 1/2]$ independently of N. Furthermore, we have the following relations:

$$(7.2.14) \qquad\qquad \Delta_N\left(\frac{1}{2}\right) = \frac{1}{2},$$

$$(7.2.15) \qquad\qquad \lim_{N\to\infty} \Delta_N(x) = \frac{1}{2}, \quad \forall x > 0,$$

$$(7.2.16) \qquad \lim_{N\to\infty} \Delta_N\left(\frac{1}{2N+1}\right) = \int_0^1 \frac{\sin(\pi y)}{\pi y}\,dy = I \sim 0.589489.$$

Proof. Δ_N is odd, since it is the primitive of the even function D_N which vanishes at 0. If we use the definition of D_N given in eqn (7.2.10), we see that

$$\int_0^{1/2} D_N(y)\,dy = \frac{1}{2} - 2\sum_{k=1}^{N} \left.\frac{\sin(2\pi kx)}{2\pi k}\right|_{x=0}^{x=1/2} = \frac{1}{2},$$

which proves the relation (7.2.14).

Let a and b be two numbers between $1/(2N+1)$ and $1/2$ with $a < b$. We integrate by parts to estimate the following integral:

$$\int_a^b D_N(y)\,dy = -\left.\frac{\cos[(2N+1)\pi y]}{(2N+1)\pi\sin(\pi y)}\right|_a^b - \int_a^b \frac{\cos[(2N+1)\pi y]\cos(\pi y)}{(2N+1)\sin^2(\pi y)}\,dy.$$

Our hypothesis on a and b implies that $(2N+1)\pi\sin(\pi y)$ is bounded below on $[a, b]$ independently of N by a certain κ. Consequently, the integrated term is bounded independently of N:

$$(7.2.17) \qquad \left|-\left.\frac{\cos[(2N+1)\pi y]}{(2N+1)\pi\sin(\pi y)}\right|_a^b\right| \leqslant \frac{2}{\kappa}.$$

Moreover, by the concavity of the sine over $[0, \pi/2]$, there exists a constant γ such that, for every y between 0 and $1/2$,

$$\sin(\pi y) \geqslant \gamma y.$$

We bound the integral from above as follows:

$$(7.2.18) \qquad \left|\int_a^b \frac{\cos[(2N+1)\pi y]\cos(\pi y)}{(2N+1)\sin^2(\pi y)}\,dy\right| \leqslant \frac{1}{\gamma^2(2N+1)}\int_a^b \frac{dy}{y^2}$$

$$\leqslant \frac{1}{\gamma^2(2N+1)}\left(\frac{1}{a} - 2\right).$$

Therefore, there exists a constant C such that, for every N, and for every a, b such that $1/(2N+1) \leqslant a \leqslant b \leqslant 1/2$,

$$(7.2.19) \qquad \left| \int_a^b D_N(y)\, dy \right| \leqslant C.$$

Furthermore, if a is fixed as strictly positive, the upper bounds, given in eqns (7.2.17) and (7.2.18), tend to 0 as N tends to infinity. This proves eqn (7.2.15).

We make the change of variables $t = (2N+1)y$ to estimate the integral appearing in eqn (7.2.16):

$$\int_0^{1/(2N+1)} \frac{\sin\left[(2N+1)\pi y\right]}{\sin(\pi y)}\, dy = \int_0^1 \frac{\sin(\pi t)}{(2N+1)\sin(\pi t/(2N+1))}\, dt.$$

As N tends to infinity, this last integral tends to

$$I = \int_0^1 \frac{\sin(\pi t)}{\pi t}\, dt,$$

by virtue of Lebesgue's theorem, and a numerical calculation gives the value in eqn (7.2.16). To see that Δ_N is bounded on $[-1/2, 1/2]$, it suffices to show that it is bounded on $[0, 1/2]$. If $x \leqslant 1/(2N+1)$, we bound $\Delta_N(x) \geqslant 0$ from above by $\Delta_N(1/(2N+1))$, which is bounded, since Δ_N increases over $[0, 1/(2N+1)]$. If $x \geqslant 1/(2N+1)$, we bound $|\Delta_N(x)|$ from above by $\Delta_N(1/(2N+1)) + |\Delta_N(x) - \Delta_N(1/(2N+1))|$, which is bounded as a result of eqns (7.2.16) and (7.2.19). $\qquad \square$

We can now show the following result:

Lemma 7.2.7. Let $f \in L_\sharp^1$, and suppose that there exists a function $f_1 \in L_\sharp^1$ such that, for every x and y, where $y > x$,

$$f(y) - f(x) = \int_x^y f_1(t)\, dt.$$

In this case, the function f is said to be absolutely continuous. Then, for any x, $S_N f(x)$ tends to $f(x)$.

Proof. A function f which satisfies the conditions of the lemma is necessarily continuous. Since we can translate the variable x, it suffices to show that $S_N f(0)$ tends to $f(0)$ as N tends to infinity. We have

$$S_N f(0) = \int_{-1/2}^{1/2} f(x)\, D_N(-x)\, dx.$$

Taking account of Lemma 7.2.6, we integrate by parts, justifying it by a density argument, and we get

$$S_N f(0) = \frac{f(1/2) + f(-1/2)}{2} - \int_{-1/2}^{1/2} f_1(x)\, \Delta_N(x)\, dx.$$

Lemma 7.2.6 allows us to see that, due to Lebesgue's theorem,

$$\lim_{N\to\infty} \int_{-1/2}^{1/2} f_1(x)\,\Delta_N(x)\,\mathrm{d}x = \frac{f(1/2) - f(0)}{2} - \frac{f(0) - f(-1/2)}{2}.$$

The conclusion of the lemma is then clear.   □

7.2.7. Pointwise convergence of partial Fourier sums of piecewise absolutely continuous functions

To treat the case of functions having a finite number of discontinuities, we introduce a sawtooth function which we define on a period by

$$s(t) = \begin{cases} -x - (1/2) & \text{if } x \in [-1/2, 0]; \\ -x + (1/2) & \text{if } x \in [0, 1/2]. \end{cases}$$

We immediately calculate the Fourier coefficients of s:

$$\hat{s}(k) = \begin{cases} 1/(2\mathrm{i}\pi k) & \text{if } k \neq 0; \\ 0 & \text{if } k = 0. \end{cases}$$

The partial Fourier sums of s are given by

$$S_N s(x) = \sum_{k=1}^{N} \frac{\sin(2\pi k x)}{\pi k}.$$

It is obvious that $S_N s(0)$ tends to 0 as N tends to infinity. We can also note that

(7.2.20) $\Delta_N(x) = S_N s(x) + x.$

Theorem 7.2.8 (Dirichlet). Let g be a function of period 1, which has discontinuities at the points x_j, $1 \leqslant j \leqslant m$ (and at all their translations $x_j + k$, $k \in \mathbb{Z}$). Suppose that there exists a function g_1 such that, for every x and y, $x < y$ for which $]x, y[$ is included in an interval which does not contain a point of discontinuity, we then have

$$g(y) - g(x) = \int_{x}^{y} g_1(t)\,\mathrm{d}t.$$

Then, for any x

$$\lim_{N\to\infty} S_N g(x) = \frac{1}{2}\left[g(x+0) + g(x-0)\right].$$ ◇

Proof. Without loss of generality, we can restrict ourselves to the case of a single discontinuity, due to Lemma 7.2.5. We can also suppose, in return for a translation, that $x_1 = 0$. The function

$$h = g - [g\,(0+0) - g\,(0-0)]\,s$$

is continuous, as we can verify by passing to the limit as x tends to 0. Its value at 0 is

$$h\,(0) = \frac{1}{2}\,[g\,(x+0) + g\,(x-0)]\,.$$

We see that h is a primitive of $h_1(y) = g_1(y) + \big(g(0+0) + g(0-0)\big)y/2$ on $[-1/2, 1/2]$. We can therefore apply Lemma 7.2.7, which implies that $S_N h(0)$ tends to $h(0)$. As $S_N s(0)$ tends to 0, we can conclude the result.   □

7.2.8. Gibbs phenomenon

The convergence of the Fourier series of a discontinuous function, such as in the statement of Theorem 7.2.8, is not uniform, and this is known as the Gibbs

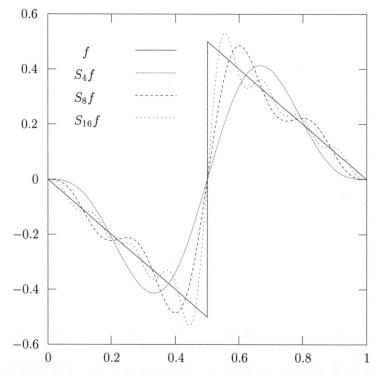

Figure 7.1: Approximation of a function by a partial Fourier sum for $N = 4, 8, 16$.

phenomenon. A graphical illustration of this phenomenon may be found in Figures 7.1 and 7.2, where the partial sums of f are represented for $N = 2^k$, $k = 2, \ldots, 7$.

Theorem 7.2.9. Under the hypotheses of Lemma 7.2.7, as N tends to infinity we have, for every j,

$$S_N g\left(x_j + \frac{1}{2N+1}\right) - g\left(x_j + \frac{1}{2N+1}\right) \to \left(I - \frac{1}{2}\right)\left[g\left(x_j + 0\right) - g\left(x_j - 0\right)\right],$$

$$S_N g\left(x_j - \frac{1}{2N+1}\right) - g\left(x_j - \frac{1}{2N+1}\right) \to -\left(I - \frac{1}{2}\right)\left[g\left(x_j + 0\right) - g\left(x_j - 0\right)\right].$$

Here, I is defined by eqn (7.2.16).   ◇

Proof. Again, using the notation of Theorem 7.2.8, we get back to the case of

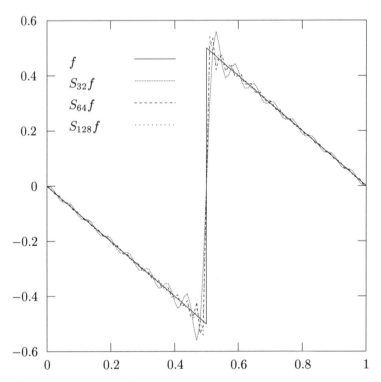

Figure 7.2: Approximation of a function by a partial Fourier sum for $N = 32, 64, 128$.

single discontinuity situated at 0, and we note that

$$S_N h \left(\frac{1}{2N+1} \right) = h \left(\frac{1}{2N+1} - x \right) \Delta_N (x) \Big|_{-1/2}^{1/2}$$

$$- \int_{-1/2}^{1/2} \Delta_N \left(\frac{1}{2N+1} - x \right) h_1 (x) \, dx.$$

Lebesgue's theorem shows us that $S_N h(1/(2N+1))$ tends to $h(0)$ as N tends to infinity. We can easily analyse the behaviour of $S_N s(1/(2N+1))$ as N tends to infinity, due to Lemma 7.2.7 and eqn (7.2.16):

$$\lim_{N \to \infty} S_N s \left(\frac{1}{2N+1} \right) = I.$$

We then have

$$\lim_{N \to \infty} S_N g \left(\frac{1}{2N+1} \right) = \frac{1}{2} \left[g (0+0) + g (0-0) \right] + I \left[g (0+0) - g (0-0) \right].$$

This allows us to conclude the required result. □

7.3. Exercises from Chapter 7

7.3.1. Elementary exercises on Fourier series

We call an expression

$$S \sim \sum_{k \in \mathbb{Z}} a_k e^{2i\pi kx},$$

where the a_k are complex numbers, a formal Fourier series. The word formal signifies that we do not ask any questions about convergence. In particular, two formal series are equal if and only if their coefficients of index k are equal for any k. We equip the vector space of formal Fourier series with its natural vector space structure. The zero element of this space will be the formal series whose coefficients are all zero.

We define the conjugate formal series by

$$\tilde{S} \sim \sum_{k \in \mathbb{Z}} -ia_k \, \text{sgn} (k) \, e^{2i\pi kx},$$

where the function sgn is defined by

$$\text{sgn}(k) = \begin{cases} 1 & \text{if } k > 0; \\ -1 & \text{if } k < 0; \\ 0 & \text{if } k = 0. \end{cases}$$

Exercise 7.3.1. Show that we can write

$$S \sim \frac{A_0}{2} + \sum_{j=1}^{\infty} \left(A_j \cos \left(2\pi j x \right) + B_j \sin \left(2\pi j x \right) \right).$$

Calculate the A_j and the B_j as functions of the a_k.

Exercise 7.3.2. Let f be in $L_\sharp^1$, $a_k = \hat{f}(k)$. Show that if f has real values, A_j and B_j are real for all j.

Exercise 7.3.3. Let f be in $L_\sharp^1$, $a_k = \hat{f}(k)$. Show that if f is even (respectively, odd) then, for every j, B_j (respectively, A_j) is zero.

Exercise 7.3.4. Show that if

$$S \sim \sum_{k=0}^{\infty} A_j \cos \left(2\pi j x \right),$$

then

$$\tilde{S} \sim \sum_{k=0}^{\infty} A_j \sin \left(2\pi j x \right).$$

Exercise 7.3.5. Let f be in $L_\sharp^1$ and let $P \in \mathbb{T}_N$ be the trigonometric polynomial

$$P \left(x \right) = \sum_{|k| \leqslant N} b_k e^{2i\pi k x}.$$

Calculate the Fourier coefficients of the product fP.

Exercise 7.3.6. Let f be in $L_\sharp^1$, and let m be an integer which is strictly greater than 1. Let

$$f_m \left(x \right) = f \left(m x \right).$$

Verify that f_m is in $L_\sharp^1$, and calculate the Fourier coefficients of f_m as functions of the Fourier coefficients of f.

7.3.2. Féjer, La Vallée Poussin, and Poisson kernels

We define a function $K_N(x)$, called a Féjer kernel of order N, by

$$(7.3.1) \qquad K_N \left(x \right) = \sum_{|k| \leqslant N} \left(1 - \frac{|k|}{N+1} \right) e^{2i\pi k x}.$$

Exercise 7.3.7. Show that, if $x \notin \mathbb{Z}$,

$$K_N \left(x \right) = \frac{1}{N+1} \left[\frac{\sin \left[\left(N + 1 \right) \pi x \right]}{\sin \left(\pi x \right)} \right]^2,$$

and that $K_N(0) = N + 1$.

Exercise 7.3.8. Show that K_N has the following two properties:

$$\int_\sharp K_N\,(x)\,\mathrm{d}x = 1,$$

$$\lim_{N\to\infty} \int_\alpha^{1-\alpha} K_N\,(x)\,\mathrm{d}x = 0, \quad \forall \alpha > 0.$$

Exercise 7.3.9. For every f in $L_\sharp^1$, we define the *Féjer sum* of f by

$$(\sigma_N)\,f\,(x) = \sum_{|k|\leqslant N} \left(1 - \frac{|k|}{N+1}\right) \hat{f}\,(k)\,\mathrm{e}^{2\mathrm{i}\pi kx}.$$

Show that σ_N is the arithmetic mean of the partial Fourier sums $S_k f$ for $0 \leqslant k \leqslant N$.

Exercise 7.3.10. Show that, for every f in $L_\sharp^1$, $\sigma_N f$ tends to f in $L_\sharp^1$. Show that if, in addition, f is in $C_\sharp^0$ then $\sigma_N f$ tends to f in $C_\sharp^0$.

Exercise 7.3.11. We define the following kernels:

(i) La Vallée Poussin kernel

$$V_N\,(x) = 2K_{2N+1}\,(x) - K_N\,(x), \quad N \in \mathbb{N};$$

(ii) Poisson kernel

$$P\,(x,r) = 1 + 2\sum_{k=1}^\infty r^k \cos\,(2\pi kx), \quad 0 \leqslant r < 1.$$

Extend Exercise 7.3.10 to the kernel V_N.

Exercise 7.3.12. Show that, for $r < 1$, we have

$$P\,(x,r) = \frac{1-r^2}{1 - 2r\cos\,(2\pi x) + r^2}.$$

Exercise 7.3.13. For every f in $L_\sharp^1$, calculate $f * P\,(\cdot,r)$ as a function of the Fourier coefficients of f. We study the convergence of $f * P\,(\cdot,r)$ to f as r tends to 1 from below, where $f * P\,(\cdot,r)$ is defined by

$$(f * P\,(\cdot,r))\,(x) = \int_\sharp f\,(x-y)\,P\,(y,r)\,\mathrm{d}y.$$

What can we say when f is in $L_\sharp^2$, or in $C_\sharp^0$?

Exercise 7.3.14. Verify Corollary 7.1.14 by convolution: use a sequence of square-integrable functions f_n having the properties (7.2.7) to (7.2.9), and give a proof which is independent of Lemma 7.1.13.

7.3.3. There exists an integrable function whose Fourier coefficients decrease arbitrarily slowly to 0

Let a_k be a sequence of positive or zero numbers such that $a_k = a_{-k}$ tends to 0 as k tends to infinity. Furthermore, suppose that

$$a_{k-1} + a_{k+1} - 2a_k \geqslant 0, \quad \forall k > 0.$$

Exercise 7.3.15. Show that $a_k - a_{k+1}$ decreases for $k > 0$.

Exercise 7.3.16. Show that $(k + 1)a_k - ka_{k-1}$ decreases for $k > 0$. From this, deduce that

$$\lim_{k \to \infty} k\,(a_k - a_{k+1}) = 0,$$

noting that, if this limit is strictly positive, a_k will be bounded below by the sum of a harmonic series. Show that

$$\lim_{N \to \infty} \sum_{k=1}^{N} k\,(a_{k-1} + a_{k+1} - 2a_k) = a_0.$$

Exercise 7.3.17. With K_k as the Féjer kernel of order k (see eqn (7.3.1)), we let

$$f(x) = \sum_{k=1}^{\infty} k\,(a_{k-1} + a_{k+1} - 2a_k)\,K_{k-1}(x).$$

Show that this series converges in $L_\sharp^1$ and that its limit, denoted by f, is positive or zero.

Exercise 7.3.18. Calculate the Fourier coefficients of f.

Exercise 7.3.19. From this, deduce that the Fourier coefficients of an integrable function can tend to 0 arbitrarily slowly.

7.3.4. The existence of sequences of numbers a_k tending to 0 as $|k|$ tends to infinity which are not the Fourier coefficients of any integrable function

Exercise 7.3.20. Let f be in $L_\sharp^1$. Show that, if $\hat{f}(0) = 0$, then

$$F(x) = \int_0^x f(y)\,dy$$

is in $C_\sharp^0$.

Exercise 7.3.21. Furthermore, we suppose that, for every k,

$$\hat{f}(|k|) = -\hat{f}(-|k|) \geqslant 0.$$

Using the convergence of $(K_N * F)(0)$ to $F(0)$ (see Exercise 7.3.10), show that

$$\sum_{k \neq 0} \frac{1}{k} \hat{f}(k) < +\infty.$$

Exercise 7.3.22. Show that there exist sequences a_k tending to 0 such that a_k is not the k-th Fourier coefficient of a function $f \in L^1_\sharp$.

Exercise 7.3.23. Show that we can choose $f \in L^1_\sharp$ in a way that its conjugate Fourier series is not the Fourier series of any integrable function.

7.3.5. Discrete least-squares approximation by trigonometric polynomials

Exercise 7.3.24. Let N be an integer which is greater than or equal to 1, and let $x_\ell = \pi \ell / N$ for ℓ varying from 1 to $2N$. Calculate

$$\sum_{\ell=1}^{2N} e^{ikx_\ell}.$$

Exercise 7.3.25. We define a bilinear form of the space $C([0, 2\pi]) = F$ of continuous real-valued periodic functions of period 2π, by

$$\langle f, g \rangle = \frac{1}{N} \sum_{\ell=1}^{2N} f(x_\ell) g(x_\ell).$$

Show that the functions

$$\frac{1}{\sqrt{2}}, \quad \sin x, \quad \cos x, \quad \dots, \quad \sin(N-1)x, \quad \cos(N-1)x$$

are relatively orthonormal to this bilinear form.

Exercise 7.3.26. We define a semi-norm of F by

$$|f| = \sqrt{\langle f, f \rangle}.$$

Let V_N be the subspace of F generated by $1/\sqrt{2}, \sin x, \cos x, \dots, \sin(N-1)x$, $\cos(N-1)x$. For $f \in F$, we say that the function $\phi \in V_N$ is the discrete least-squares trigonometric polynomial approximation of f if

$$|f - \phi| = \min_{\psi \in V_N} |f - \psi|.$$

Show that, for every f in F, there exists a unique ϕ which is the discrete least-squares trigonometric approximation.

Exercise 7.3.27. Write this ϕ as

$$\phi(x) = \frac{a_0}{2} + \sum_{k=1}^{N-1} (a_k \cos kx + b_k \sin kx).$$

Calculate the a_j and b_j as functions of f.

Exercise 7.3.28. Suppose that f has a uniformly convergent Fourier series:

$$f(x) = \frac{\alpha_0}{2} + \sum_{k=1}^{\infty} (\alpha_k \cos kx + \beta_k \sin kx).$$

Calculate the a_j and b_j as functions of α_j and β_j.

Exercise 7.3.29. We replace the Fourier coefficients α_j and β_j by their approximations using the left rectangle formula 8.1.3 with equidistant points. Show that we can choose the discretization step in such a way that the approximations thus obtained agree with the a_j and b_j.

8

Quadrature

This chapter of the book is dedicated to numerical integration, and it is in preparation for the last chapter of the book, which will be on differential equations. The word quadrature has become celebrated by the problem of the quadrature of a circle. This involves a geometric construction with a ruler and compass to find a square whose area is equal to that of a given circle. That this is impossible is the consequence of two results of the nineteenth century. In 1837, Wantzel showed that the numbers which can be constructed by ruler and compass are algebraic. More precisely, they are obtained by solving a finite sequence of quadratic equations with integer coefficients. In 1882, Lindemann showed that the number π is transcendental, which means that it is not the solution of any polynomial equation with integer coefficients. The reader may wonder: are there many transcendental numbers? It is not difficult to see that the set of algebraic numbers is denumerable: there are as many algebraic numbers as there are rational numbers and integers; therefore, almost all numbers are transcendental. However, it is often extremely difficult to show that a given specific number is transcendental.

This does not prevent anyone from doing quadratures, that is, from calculating areas or integrals. We manipulate integrals, whether we have an explicit expression for them or not, or whether this expression makes use of rational or irrational numbers. However, the effective numerical calculation of integrals, or the numerical approximation of the solutions of differential systems, becomes an interesting problem because most functions do not have a primitive which may be expressed in terms of elementary functions, and most differential systems do not have such solutions.

We therefore seek numerical methods which will allow us to approximate the objects which we cannot, generally, calculate explicitly.

8.1. Numerical integration

Numerically solving a differential equation amounts to finding a numerical approximation of the following problem:

$$(8.1.1) \qquad \frac{du}{dt}(t) = f(t, u(t)), \quad u(t_0) = u_0,$$

where f is a given function from $[t_1, t_2] \times \mathbb{R}^n$ to $\mathbb{R}^n$, t_0 belongs to the interval $[t_1, t_2]$, and u_0 is given in $\mathbb{R}^n$. Before solving this general problem under adequate conditions on f, we consider the simpler differential equation

$$\frac{du}{dt}(t) = f(t), \quad u(t_0) = u_0,$$

where f is an integrable scalar function. The solution is given by

$$(8.1.2) \qquad u(t) = u_0 + \int_{t_0}^{t} f(s)\,\mathrm{d}s.$$

It will be useful to see how to approximate weighted integrals of the form

$$\int_a^b w(x) f(x)\,\mathrm{d}x,$$

where w is an integrable function which is strictly positive almost everywhere, as in the study of the polynomial least-squares approximation (see Chapter 5). The function w is called the weight.

An approximation formula for the integral of a function f on an interval is called a numerical integration formula, or a quadrature formula.

Common sense says that we have little chance of succeeding in numerically approximating eqn (8.1.1) if we do not know how to numerically approximate eqn (8.1.2). Conversely, if we know how to numerically approximate eqn (8.1.2) this will aid us, as we will see later, in constructing schemes to numerically approximate eqn (8.1.1).

One last common sense remark: if f is only integrable in eqn (8.1.2) then the problem is pathological from the numerical point of view as we do not know how to discretize functions which are defined almost everywhere without regularizing them. If we have good reasons for dealing with functions which are only integrable, we are then led to either consider them as linear forms on a space of test functions (as in the measure theory of Radon, or, more generally, the theory of distributions) or to work with their local means. In any case, we are departing from the scope of this book.

8.1.1. Numerical integration for dummies

In all that follows, we assume that the functions which we are integrating numerically are continuous on a compact interval $[a, b]$. Anyone who has studied elementary integration theory (integration of continuous functions is sufficient) has done numerical integration, just as M. Jourdain used to write prose. Indeed, if f is continuous on the compact interval $[a, b]$ and if we have some points

$$a = x_0 < x_1 < x_2 < \cdots < x_{n-1} < x_n = b,$$

the left rectangle formula is written

$$(8.1.3) \qquad I_n^l(f) = \sum_{j=0}^{n-1} f(x_j)(x_{j+1} - x_j).$$

Here we have a Riemann sum of f, and it is shown in every integration course that, when $\max_j(x_{j+1} - x_j)$ tends to 0, the sequence of numbers $I_n^l(f)$ converges to a number which is the integral of f between a and b. Geometrically, we replace the function f by a staircase function having values $f(x_j)$ on the interval $[x_j, x_{j+1}]$, and we trivially integrate the staircase function, see Figure 8.1. The error made is shown by the shaded region.

In the same way, the right rectangle formula is given by

$$(8.1.4) \qquad I_n^r(f) = \sum_{j=0}^{n-1} f(x_{j+1})(x_{j+1} - x_j).$$

This is equivalent to replacing f by a staircase function which has values $f(x_{j+1})$ on the interval $[x_j, x_{j+1}]$. The integration is as easy as the preceding one and the convergence result is the same. The error made is shown by the shaded region in Figure 8.2. Observe that the sign has to be taken into account.

We can also use the midpoint formula:

$$(8.1.5) \qquad I_n^m(f) = \sum_{j=0}^{n-1} f\left(\frac{x_j + x_{j+1}}{2}\right)(x_{j+1} - x_j).$$

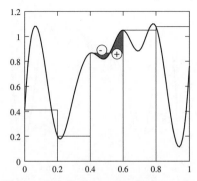

Figure 8.1: Left rectangle formula.

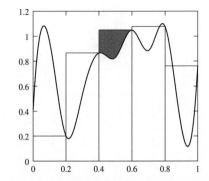

Figure 8.2: Right rectangle formula.

In this case we take the value of the function on $[x_j, x_{j+1}]$ to be the value of f in the middle of the interval. Referring to Figure 8.3 below, we see that if f is sufficiently regular then there is some cancellation of the signs of the error, so the total error made should be less for the midpoint formula than for the left and right rectangle formulae. The convergence result for the Riemann sums can again be applied.

The trapezium formula is given by

$$(8.1.6) \qquad I_{n+1}^t (f) = \sum_{j=0}^{n-1} \frac{f(x_j) + f(x_{j+1})}{2} (x_{j+1} - x_j).$$

This time we have replaced f by a piecewise linear function, which coincides with f at the points x_j, $0 \leqslant j \leqslant n$. As the replacement is more accurate, we hope that the error will be smaller with trapeziums than with rectangles (left or right), although the modification only affects the first and last terms of the quadrature formula. Figure 8.4 shows the greater accuracy, though this higher precision is not true on each interval.

Finally, in a first year course we generally meet Simpson's rule given by

$$(8.1.7) \quad I_{2n+1}^s (f) = \sum_{j=0}^{n-1} \frac{f(x_j) + 4f((x_j + x_{j+1})/2) + f(x_{j+1})}{6} (x_{j+1} - x_j).$$

We see that this is a linear combination of the midpoint formula and the trapezium formula, and we will show that it is more accurate than either of these two formulae. Geometrically, it consists of integrating a function which interpolates f with a second degree polynomial in each interval $[x_j, x_{j+1}]$, with knots at the two end-points and the middle of the interval. Formulae (8.1.3) to (8.1.7) are called composite formulae since they are formed from the juxtaposition on a given interval of formulae on small intervals. These formulae are obtained by a change of variable from a simple formula with weight 1.

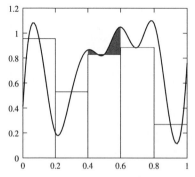

Figure 8.3: Midpoint formula.

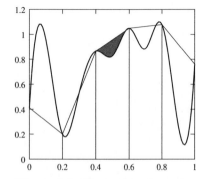

Figure 8.4: Trapezium formula.

8.2. The analysis of quadrature formulae

In numerical integration, we will pose the following questions:

(i) How do we construct quadrature formulae? We have two classes of formulae, namely

simple formulae;

composite formulae.

(ii) What is the order of a formula? How do we estimate the quadrature error in a simple formula?

(iii) How do we estimate the quadrature error in a composite formula?

(iv) Do the geometric symmetries, such as periodicity, provide an advantage with regard to an error compensation process?

(v) Can we find simple formulae of maximal order? More generally, can we tailor integration formulae to given requirements?

In general, a quadrature formula is an expression of the form

(8.2.1)
$$\sum_{j=0}^{n} \lambda_j f(x_j),$$

where the points x_j are $n+1$ pairwise distinct points in the interval $[a, b]$ and the scalars λ_j are chosen in such a way that the quadrature error

$$e_n(f) = \int_a^b f(x) w(x) \, dx - \sum_{j=0}^{n} \lambda_j f(x_j)$$

is not too large in a sense which we will clarify later. We will see later why it is interesting to use a weight w, which we suppose to be integrable and strictly positive almost everywhere on $[a, b]$.

8.2.1. Order of a quadrature formula

To begin with, we look for which classes of functions the rectangle, midpoint, and Simpson formulae are exact. The right and left formulae are exact for constant functions. The midpoint formula is exact for constant functions, but also for linear functions, as shown in Figure 8.5 which demonstrates the error cancellations. The trapezium formula is exact for linear functions, by construction. Finally, the geometric interpretation of Simpson's rule shows that it is exact for polynomials of up to second order. We therefore have the following definition of the order of a formula:

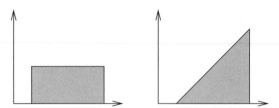

Figure 8.5: The midpoint formula is exact on affine functions: the two shaded areas are equal.

Definition 8.2.1. We say that the quadrature formula (8.2.1) is of order m if m is the largest integer such that the formula is exact on $\mathbb{P}_m$, the vector space of polynomials of degree at most m.

By this definition, the right and left rectangle formulae are of order 0, the trapezium and midpoint formulae are of order 1 (check this) and Simpson's rule is of order at least 2 (and we shall see, in fact, that it is of order 3).

Given the pairwise distinct knots x_j, $0 \leqslant j \leqslant n$, we seek the relations which must be satisfied by the scalars λ_j so that formula (8.2.1) is of order m. For each power x^k, $0 \leqslant k \leqslant m$, we must have the following equalities:

$$(8.2.2) \qquad \sum_{j=0}^{n} \lambda_j x_j^k = \int_a^b x^k w(x)\, dx, \quad \forall k \in \{0, \ldots, m\}.$$

System (8.2.2) is a system of $m+1$ equations for $n+1$ unknowns. To ensure a solution, we need at least as many unknowns as equations. Therefore, we suppose that $m \leqslant n$. We look for its rank: the sub-matrix formed from the first $m+1$ columns is the matrix

$$\begin{pmatrix} 1 & 1 & \cdots & 1 \\ x_0^1 & x_1^1 & \cdots & x_m^1 \\ \vdots & \vdots & & \vdots \\ x_0^m & x_1^m & \cdots & x_m^m \end{pmatrix}.$$

We recognize this to be the matrix of the system (4.1.1) for Lagrange interpolation: it is the Vandermonde matrix, which is invertible. Consequently, system (8.2.2) is of rank $m+1$. In other words, the dimension of the image of the matrix of system (8.2.2) is $m+1$, which is equal to the number of rows of the system or, again, the dimension of the image space. Irrespective of the right-hand side, system (8.2.2) has a solution. In fact, it has an affine space of solutions, of dimension $n - m$. If we fix $n - m$ scalars λ_j, we can find the remaining $m+1$ scalars by solving a system of the type occurring in interpolation. Indeed, if we suppose that the λ_j are given for $j \geqslant m+1$, we use the basis of $\mathbb{P}_m$ formed from the functions ϕ_j, defined in eqn (4.1.1). As the quadrature formula is exact on polynomials of degree at most m, it is, in particular, exact on the ϕ_p:

$$\sum_{j=0}^{n} \lambda_j \phi_p(x_j) = \int_a^b w(x) \phi_p(x) \, dx.$$

Hence, we obtain the value of λ_p for $0 \leqslant p \leqslant m$:

$$\lambda_p = \int_a^b w(x) \phi_p(x) \, dx - \sum_{j=m+1}^{n} \lambda_j \phi_p(x_j).$$

A particularly important case is that in which $m = n$. In this case the system (8.2.2) has a unique solution, which is

(8.2.3)
$$\lambda_j = \int_a^b \phi_j(x) w(x) \, dx.$$

Formula (8.2.1) can then be written as

$$
\begin{aligned}
I_n(f) &= \sum_{j=0}^{n} \left(\int_a^b \phi_j(x) w(x) \, dx \right) f(x_j) \\
&= \int_a^b \sum_{j=0}^{n} f(x_j) \phi_j(x) w(x) \, dx \\
&= \int_a^b P(x) w(x) \, dx,
\end{aligned}
$$

where P is the Lagrange interpolation polynomial of f at the points x_j. In this case the quadrature formula can be interpreted as follows: We interpolate f at the knots $(x_j)_{0 \leqslant j \leqslant n}$ with a polynomial $P \in \mathbb{P}_n$, and we replace the integral of f by the integral of P. Thus, we have obtained a *quadrature formula by interpolation*. Hence, we are assured that, for every choice of $n + 1$ knots, there exists at least one quadrature formula of order n, namely the quadrature by interpolation formula.

8.2.2. On the practical interest of weighted formulae

We consider the weight $1/\sqrt{x}$ on the interval $[0, 1]$ and we consider an interpolation formula with the knots 0 and 1. We have

$$\phi_1(x) = \frac{x - x_1}{x_0 - x_1} = 1 - x \quad \text{and} \quad \phi_2(x) = \frac{x - x_0}{x_1 - x_0} = x.$$

Consequently,

$$\int_0^1 \frac{\phi_1(x)}{\sqrt{x}} \, dx = \int_0^1 \left[\frac{1}{\sqrt{x}} - \sqrt{x} \right] \, dx = \left[2\sqrt{x} - \frac{2}{3} x^{3/2} \right]_0^1 = \frac{4}{3} = \lambda_1.$$

Furthermore,

$$\int_0^1 \frac{\phi_2(x)}{\sqrt{x}}\,dx = \int_0^1 \sqrt{x}\,dx = \left[\frac{2}{3}x^{3/2}\right]_0^1 = \frac{2}{3} = \lambda_2.$$

Therefore, we have obtained the integration formula

$$\int_0^1 \frac{f(x)}{\sqrt{x}}\,dx \simeq \frac{4}{3}f(0) + \frac{2}{3}f(1).$$

We test this formula by taking $f(x) = x^\alpha$ for $\alpha > 1/2$ and compare it with the trapezium formula: the value obtained by the trapezium formula is $1/2$ and the value obtained by our formula is $2/3$. The exact value of the integral is $2/(1+2\alpha)$. We see that the error made by the weighted formula is less in absolute value than the error made by the trapezium formula if

$$\frac{2}{1+2\alpha} \geqslant \frac{1}{2}\left(\frac{1}{2} + \frac{2}{3}\right)$$

or

$$\alpha \leqslant \frac{14}{17}.$$

The error made by the weighted formula is therefore less than the error made by the trapezium formula if $1/2 \leqslant \alpha \leqslant 14/17$. If $-1/2 < \alpha \leqslant 1/2$, then the trapezium formula gives us nothing. Therefore, the weighted formula allows us to integrate singularities better, when they are integrable.

8.2.3. Examples of simple formulae

The simple left and right rectangle formulae on $[0,1]$ are given by

$$I_1^l(f) = f(0) \quad \text{and} \quad I_1^r(f) = f(1),$$

respectively. These are interpolation formulae. The simple midpoint formula is given by

$$I_1^m(f) = f\left(\frac{1}{2}\right).$$

This is also an interpolation formula, as is the simple trapezium formula given by

$$I_2^t(f) = \frac{f(0) + f(1)}{2}.$$

Finally, the simple Simpson's rule is given on $[0,1]$ by

$$I_3^s(f) = \frac{f(0) + 4f(1/2) + f(1)}{6}.$$

We will now calculate its order. It is clear that $I_3^S(1) = 1$. We have

$$I_3^s(x) = \frac{1}{6}\left(4 \times \frac{1}{2} + 1\right) = \frac{1}{2},$$

$$I_3^s(x^2) = \frac{1}{6}\left(4 \times \frac{1}{2^2} + 1\right) = \frac{1}{3},$$

$$I_3^s(x^3) = \frac{1}{6}\left(4 \times \frac{1}{2^3} + 1\right) = \frac{1}{4},$$

and finally,

$$I_3^s(x^4) = \frac{1}{6}\left(4 \times \frac{1}{2^4} + 1\right) = \frac{5}{24}.$$

Consequently, Simpson's rule is of exactly order 3 and, in particular, it is an interpolation formula.

A notable category of quadrature by interpolation formulae is that of the Newton–Cotes formulae. These are the formulae which are obtained with equidistant knots and weight 1. The closed formulae for $n + 1$ points are obtained by including the end-points, and therefore taking as knots

$$x_j = a + \frac{j(b-a)}{n}, \quad 0 \leqslant j \leqslant n.$$

The open formulae for n points are obtained by excluding the end-points, and therefore taking as knots

$$x_j = a + \frac{j(b-a)}{n+1}, \quad 1 \leqslant j \leqslant n.$$

The simple trapezium formula is a closed 2-point Newton–Cotes formula. The simple Simpson's rule is a closed 3-point Newton–Cotes formula. The simple midpoint formula is an open 1-point Newton–Cotes formula. The Newton–Cotes formulae are tabulated in numerous works. We find the most common ones in [19, 51]. Their coefficients are calculated using eqn (8.2.3). We can easily obtain them by using symbolic manipulation software.

8.2.4. Composite formulae

Suppose that we have a quadrature formula with weight 1 on the interval $[0, 1]$:

$$I_n(f) = \sum_{j=0}^{n} \lambda_j f(x_j).$$

By a transformation $x \mapsto a + x(b-a)$ we can deduce from this a quadrature formula on any interval $[a, b]$. Indeed, if g is a continuous function on $[a, b]$, then $x \mapsto g(a + x(b-a))$ is a continuous function on $[0, 1]$, and since

$$\int_a^b g(y)\,\mathrm{d}y = (b-a)\int_0^1 g(a + x(b-a))\,\mathrm{d}x,$$

we have the quadrature formula

$$(8.2.4) \qquad \int_a^b g\left(y\right) \mathrm{d}y \simeq \left(b-a\right) \sum_{j=0}^{n} \lambda_j g\left(a + \left(b-a\right) x_j\right).$$

A composite formula on an interval $[a, b]$ is constructed as follows: We begin with a simple formula I_n on the interval $[0, 1]$. We then subdivide the interval $[a, b]$ by defining a sequence of points

$$a = a_0 < a_1 < \cdots < a_p = b.$$

On each of these intervals $[a_j, a_{j+1}]$ we perform a quadrature of the type given in eqn (8.2.4) to obtain the formula

$$I_{n,p}\left(f\right) = \sum_{i=0}^{p-1} \left(a_{i+1} - a_i\right) \sum_{j=0}^{n} \lambda_j f\left(a_i + x_j \left(a_{i+1} - a_i\right)\right).$$

It goes without saying that the order of the composite formula $I_{n,p}$ is at least equal to the order of the simple formula I_n from which it came.

8.3. The Peano kernel and error estimates

8.3.1. Definition of the Peano kernel

Just as we introduced the Dirichlet kernel in Subsection 7.2.4 to represent the action of the partial sum operator of a Fourier series, here we introduce a kernel, called the Peano kernel, which describes the quadrature error.

Theorem 8.3.1. Let a quadrature formula of order m on $[a, b]$ be denoted by

$$(8.3.1) \qquad I_n\left(f\right) = \sum_{j=0}^{n} \lambda_j f\left(x_j\right).$$

Let $t_+ = \max(t, 0)$ and define $t_+^0 = 1$ if $t > 0$ and $t_+^0 = 0$ if $t \leqslant 0$. Define a function G by

$$G\left(y\right) = \int_a^b \left(x - y\right)_+^m w\left(x\right) \mathrm{d}x - \sum_{j=0}^{n} \lambda_j \left(x_j - y\right)_+^m.$$

Then, for any f in C^{m+1} on $[a, b]$, we have

$$\int_a^b f\left(y\right) w\left(y\right) \mathrm{d}y - I_n\left(f\right) = \frac{1}{m!} \int_a^b f^{(m+1)}\left(y\right) G\left(y\right) \mathrm{d}y.$$

The function G is called the Peano kernel.   ◇

Proof. First of all, we make a preliminary remark: $G(y)$ is the quadrature error made by replacing the integral on $[a, b]$ of $x \mapsto (x - y)_+^m$ by $I_n(x \mapsto (x - y)_+^m)$. We write f with the aid of the Taylor formula with integral remainder:

$$f(x) = P(x) + R(x),$$

where

$$P(x) = \sum_{j=0}^{m} \frac{f^{(j)}(a)}{j!} (x - a)^j$$

and

$$R(x) = \frac{1}{m!} \int_a^x f^{(m+1)}(y)(x - y)^m \, dy$$

is the integral remainder. As we assume the formula to be of order m,

$$\int_a^b P(x) w(x) \, dx = I_n(P),$$

since the formula is exact for polynomials up to order m.

Note that

$$\int_a^x f^{(m+1)}(y)(x - y)^m \, dy = \int_a^b f^{(m+1)}(y)(x - y)_+^m \, dy.$$

We are therefore able to change the order of integration when we integrate R:

$$\int_a^b w(x) R(x) \, dx = \int_a^b \frac{1}{m!} w(x) \int_a^x f^{(m+1)}(y)(x - y)^m \, dy \, dx$$

$$= \int_a^b \frac{1}{m!} w(x) \int_a^b f^{(m+1)}(y)(x - y)_+^m \, dy \, dx$$

$$= \int_a^b \frac{1}{m!} \left(\int_a^b (x - y)_+^m \, w(x) \, dx \right) f(y) \, dy.$$

In the same way, we exchange the summation and integration when we perform the quadrature of R:

$$\sum_{j=0}^{n} \lambda_j R(x_j) = \sum_{j=0}^{n} \lambda_j \frac{1}{m!} \int_a^x f^{(m+1)}(y)(x_j - y)^m \, dx$$

$$= \sum_{j=0}^{n} \lambda_j \frac{1}{m!} \int_a^b f^{(m+1)}(y)(x_j - y)_+^m \, dx$$

$$= \frac{1}{m!} \int_a^b \left(\sum_{j=0}^{n} \lambda_j (x_j - y)_+^m \right) f^{(m+1)}(y) \, dy.$$

Consequently,

$$\int_a^b f(x) \, w(x) \, \mathrm{d}x - I_n(f) = \int_a^b R(x) \, w(x) \, \mathrm{d}x - I_n(R)$$

$$= \int_a^b \frac{1}{m!} f^{(m+1)}(y) \left(\int_a^b w(x)(x-y)_m^+ \, \mathrm{d}x - \sum_{j=0}^n \lambda_j \, (x_j - y)_+^m \right) \mathrm{d}y$$

$$= \frac{1}{m!} \int_a^b f^{(m+1)}(y) \, G(y) \, \mathrm{d}y,$$

which concludes the proof.   □

The Peano kernel allows us to make error estimates. The first of these is given by the following lemma:

Lemma 8.3.2. Let I_n be a quadrature formula which we suppose to be of order m, and let f be a C^{m+1} function on $[a, b]$. Then,

$$(8.3.2) \qquad \left| \int_a^b w(x) f(x) \, \mathrm{d}x - I_n(f) \right| \leqslant \frac{1}{m!} \max_{[a,b]} \left| f^{(m+1)}(x) \right| \int_a^b |G(y)| \, \mathrm{d}y.$$

Proof. The proof is immediate.   □

The estimate that we have just made assumes no sign information on G. If we have some sign information we can do better, indeed, the second mean value theorem states that if f is continuous, and if g is integrable and positive or zero almost everywhere on $[a, b]$, then there exists a real number $\xi \in [a, b]$, such that

$$\int_a^b f(x) \, g(x) \, \mathrm{d}x = f(\xi) \int_a^b g(x) \, \mathrm{d}x.$$

This classic result (but often unknown to degree students) is simply proved as follows. If g is identically zero on $[a, b]$, the result is clear. If not, we have the following inequalities:

$$g(x) \min_{y \in [a,b]} f(y) \leqslant f(x) g(x) \leqslant g(y) \max_{y \in [a,b]} f(y),$$

which we integrate over $[a, b]$, thus obtaining

$$\min_{y \in [a,b]} f(y) \int_a^b g(x) \, \mathrm{d}x \leqslant \int_a^b f(x) g(x) \, \mathrm{d}x \leqslant \int_a^b f(x) g(x) \, \mathrm{d}x.$$

Consequently, the ratio

$$\int_a^b f(x) \, g(x) \, \mathrm{d}x \left/ \int_a^b g(x) \, \mathrm{d}x \right.$$

lies between the minimum and the maximum of f on the interval $[a, b]$. By the continuity of f, there exists a ξ such that $f(\xi)$ is equal to this ratio.

We therefore have an error estimate which is a little less naïve, namely

Lemma 8.3.3. Let I_n be a quadrature formula of order m, and let f be a C^{m+1} function on $[a, b]$. Then, if G, the Peano kernel of I_n, does not change sign on $[a, b]$, there exists a $\xi \in [a, b]$ such that

(8.3.3)
$$\int_a^b w(x) f(x) \, dx - I_n(f)$$
$$= \frac{1}{(m+1)!} f^{(m+1)}(\xi) \left(\int_a^b w(y) y^{m+1} \, dy - I_n(y^{m+1}) \right).$$

Proof. The second mean value formula requires the existence of a ξ such that

(8.3.4)
$$\int_a^b f(x) \, dx - I_n(f) = \frac{1}{m!} f^{(m+1)}(\xi) \int_a^b G(y) \, dy.$$

However, the integral of G over $[a, b]$ can be calculated as follows: the quadrature error on $y \mapsto y^{m+1}$ is given by

$$\int_a^b w(y) y^{m+1} \, dy - I_n(y^{m+1}) = \frac{1}{m!} \int_a^b (m+1)! G(y) \, dy$$
$$= (m+1) \int_a^b G(y) \, dy,$$

which implies

$$\int_a^b G(y) \, dy = \frac{1}{m+1} \left(\int_a^b w(y) y^{m+1} \, dy - I_n(y^{m+1}) \right)$$

and proves the lemma. $\qquad \square$

Examples of Peano kernels

We will calculate some Peano kernels explicitly, beginning with the rectangle formula on $[0, 1]$:
$$I_1(f) = f(c).$$

Suppose that c belongs to $[0, 1]$ and is different to $1/2$, so that the formula is of order 0. We have

$$G_c(y) = \int_0^1 (x - y)_+^0 \, dx - (c - y)_+^0$$
$$= \int_0^1 1_{[y,1]}(x) \, dx - 1_{[0,c]}(y) = 1 - y - 1_{[0,c]}(y).$$

In Figure 8.6 we give the graphical representation of the Peano kernel G_c. We see that it does not change sign on $[0, 1]$ only if $c = 0$ or 1.

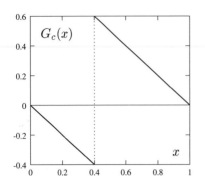
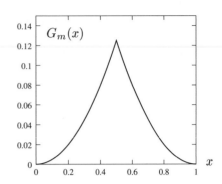

Figure 8.6: The Peano kernel for the rectangle formula (left), with integration knot $c = 0.4$, and for the midpoint formula (right). Note the different scales of the figures.

Now, if $c = 1/2$, the formula is of order 1: the midpoint formula. The calculation of the Peano kernel is therefore a little different:

$$
G_M(y) = \int_a^b (x - y)_+^1 \, dx - \left(\frac{1}{2} - y\right)_+^1
$$

$$
= \left[\frac{x^2}{2} - xy\right]_{x=y}^{x=1} - \left(\frac{1}{2} - y\right)_+
$$

$$
= \frac{1 - y^2}{2} - (1 - y)\, y - \left(\frac{1}{2} - y\right)_+.
$$

Two different cases present themselves. If $y < 1/2$,

$$
G_M(y) = \frac{1 - y^2}{2} - (1 - y)\, y - \frac{1}{2} + y = \frac{y^2}{2}.
$$

If $y \geqslant 1/2$,

$$
G_M(y) = \frac{1 - y^2}{2} - (1 - y)\, y = \frac{(1 - y)^2}{2}.
$$

The Peano kernel of the midpoint formula does not change sign. Consequently, the error incurred by using the midpoint formula is given by eqn (8.3.3):

$$
(8.3.5) \qquad \int_0^1 f(x)\, dx - f\left(\frac{1}{2}\right) = \frac{f''(\xi)}{2!}\left[\int_0^1 y^2\, dy - \frac{1}{4}\right] = \frac{f''(\xi)}{24}.
$$

The graphical representation of this Peano kernel is given in Figure 8.6.

We find the proof of the following results on the order of the Newton–Cotes formulae and their Peano kernels in [51, Chapter 7].

Theorem 8.3.4. A closed Newton–Cotes formula on an even number of intervals (that is, $2p+1$ knots and $2p$ intervals) is of order $2p+1$. ◇

Theorem 8.3.4 leads us to only use Newton–Cotes formulae on an even number of intervals, since we then gain an order with respect to that predicted by the general theory of quadrature by interpolation. We make an exception for the trapezium formula, which is often used. The first closed formula on an even number of intervals is Simpson's rule.

Theorem 8.3.5. An open Newton–Cotes formula on an even number of intervals (that is, $2p-1$ knots and $2p$ intervals) is of order $2p-1$. ◇

Theorem 8.3.5 leads us to only use Newton–Cotes formulae on an even number of intervals. The first open formula on an even number of intervals is the midpoint formula.

Theorem 8.3.6. The Peano kernel for the Newton–Cotes formulae has constant sign. ◇

8.3.2. Quadrature error in composite formulae

Given an interval $[0,1]$ and a simple quadrature formula of order m and weight 1:

$$(8.3.6) \qquad I_n(f) = \sum_{j=0}^{n} \lambda_j f(x_j).$$

We have seen that the composite formula on $[a,b]$ is given by the decomposition of $[a,b]$ into sub-intervals $[a_i, a_{i+1}]$ on which we define new quadrature formulae based on the elementary one. We thus have

$$(8.3.7) \qquad I_{n,p}(f) = \sum_{i=0}^{p-1} (a_{i+1} - a_i) \sum_{j=0}^{n} \lambda_j f(a_i + x_j(a_{i+1} - a_i)).$$

The quadrature error in a composite formula is estimated by means of the following result:

Theorem 8.3.7. Let $I_{n,p}$ be a composite quadrature formula on $[a,b]$ given by eqn (8.3.7) and defined by the elementary formula (8.3.6) of order m on $[0,1]$. Let $h = \max_i(a_{i+1} - a_i)$. Then,

$$(8.3.8) \qquad \left| \int_a^b f(x)\,\mathrm{d}x - I_{n,p}(f) \right| \leqslant \frac{(b-a)h^{m+1}}{m!} \max_{x \in [a,b]} \left| f^{(m+1)}(x) \right| \int_0^1 |G(y)|\,\mathrm{d}y,$$

where G is the kernel of the formula (8.3.6). ◇

Proof. Generally, the formula obtained on an interval $[\alpha, \beta]$, based on the formula on the interval $[0,1]$, is

$$(\beta - \alpha) \sum_{j=0}^{n} \lambda_j f(\alpha + (\beta - \alpha) x_j).$$

Calculating the Peano kernel $\tilde{G}$ of this formula:

$$\tilde{G}\left(y\right) = \int_{\alpha}^{\beta} \left(x - y\right)_{+}^{m} \, \mathrm{d}x - \left(\beta - \alpha\right) \sum_{j=1}^{n} \lambda_{j} \left(\alpha + x_{j} \left(\beta - \alpha\right) - y\right)_{+}^{m}.$$

We make the natural change of variables

$$x = \alpha + \xi \left(\beta - \alpha\right), \quad y = \alpha + \eta \left(\beta - \alpha\right),$$

to obtain

$$\tilde{G}\left(y\right) = \left(\beta - \alpha\right)^{m+1} G\left(\frac{y - \alpha}{\beta - \alpha}\right).$$

Now let $\alpha = a_i$ and $\beta = a_{i+1}$. On each interval $[a_i, a_{i+1}]$, we have the error

$$\frac{1}{m!} \left(a_{i+1} - a_i\right)^{m+1} \int_{a_i}^{a_{i+1}} f^{(m+1)}\left(y\right) G\left(\frac{y - a_i}{a_{i+1} - a_i}\right) \mathrm{d}y.$$

Now

$$\int_{a_i}^{a_{i+1}} \left|G\left(\frac{y - a_i}{a_{i+1} - a_i}\right)\right| \mathrm{d}y = \left(a_{i+1} - a_i\right) \int_{0}^{1} \left|G\left(\eta\right)\right| \mathrm{d}\eta,$$

and therefore, the error on each interval $[a_i, a_{i+1}]$ is bounded above by

$$(8.3.9) \qquad \frac{\left(a_{i+1} - a_i\right)^{m+2}}{m!} \max_{x \in [a,b]} \left|f^{(m+1)}\left(x\right)\right| \left(\int_{0}^{1} \left|G\left(\eta\right)\right| \mathrm{d}\eta\right).$$

Let $h = \max_i (a_{i+1} - a_i)$. We note that

$$(8.3.10) \qquad \sum_{i=0}^{p-1} \left(a_{i+1} - a_i\right)^{m+2} \leqslant h^{m+1} \sum_{i=0}^{p-1} \left(a_{i+1} - a_i\right) = h^{m+1} \left(b - a\right).$$

By adding all the error bounds (8.3.9) and using eqn (8.3.10), we deduce estimate (8.3.8). □

The preceding analysis showed that there is no advantage in doing a numerical integration by a high-order formula on functions which are not very regular. The precision of a formula is strictly limited by the regularity of the function which we are integrating.

8.4. Gaussian quadrature

In this section, we call n the number of knots, because it makes the results easier to remember. Let $[a, b]$ be a compact interval and let w be an integrable weight which is strictly positive almost everywhere on $[a, b]$. Given n knots $x_1, \ldots, x_n$, there exists a quadrature formula of order at least $n - 1$, namely the interpolation formula, and we have seen that it is unique. Observe that if the weight is non-negative almost everywhere, and does not vanish almost everywhere, the uniqueness still holds. This comes from the fact that, since a polynomial is determined by a finite number of coefficients, when the integral of its square multiplied by the weight vanishes, the polynomial itself vanishes, provided that the weight does not vanish almost everywhere. However, such weights are of little practical importance, and we will continue with the simpler hypothesis. We now pose the following problem: how do we determine the knots x_j and the weights λ_j so that the quadrature formula is of the highest possible order, that is

$$\int_a^b f(x) w(x) \, dx - \sum_{j=1}^n \lambda_j f(x_j)$$

vanishes on $\mathbb{P}_k$ for k as large as possible. We already know that $k \geqslant n - 1$, therefore the λ_j are uniquely determined from the x_j by means of the formula

$$(8.4.1) \qquad \lambda_j = \int_a^b w(x) \prod_{\ell : \ell \neq j} \frac{x - x_\ell}{x_j - x_\ell} \, dx.$$

If P belongs to $\mathbb{P}_k$, we must have

$$(8.4.2) \qquad \int_a^b P(x) w(x) \, dx = \sum_{j=1}^n \lambda_j P(x_j).$$

We introduce the polynomial

$$(8.4.3) \qquad p(x) = \prod_{j=1}^n (x - x_j).$$

The Euclidean division of P by p has quotient q and a remainder r, which is a polynomial of degree at most $n - 1$:

$$(8.4.4) \qquad P = pq + r.$$

If $k = n - 1$, the quotient q is zero. Since P is some polynomial in $\mathbb{P}_k$, q is some polynomial in $\mathbb{P}_{k-n}$. We can rewrite relation (8.4.2) as follows:

$$\int_a^b p(x) q(x) w(x) \, dx + \int_a^b r(x) w(x) \, dx = \sum_{j=1}^n \lambda_j p(x_j) q(x_j) + \sum_{j=1}^n \lambda_j r(x_j).$$

Noting that this formula is of order at least $n - 1$ and that $p(x_j)$ is zero for every $j = 1, \ldots, n$, we see that

$$\int_a^b p(x) q(x) w(x) \, dx = 0.$$

In other words, the scalar product of p and q with weight w must be zero:

$$(8.4.5) \qquad (p,q) = 0, \quad \forall q \in \mathbb{P}_{k-n}.$$

The preceding analysis leads us now to present and prove the theorem which describes the maximal-order formulae, known as Gaussian formulae:

Theorem 8.4.1. The unique n-point formula of maximal order is the interpolation formula constructed by taking as knots the zeros of the n-th orthogonal polynomial with respect to the weight w. By convention, the n-th orthogonal polynomial is of degree n, which implies that we begin numbering them from 0. The formula thus determined, is exactly of order $2n - 1$ and is called a Gaussian quadrature formula.   ◇

Proof.   Suppose that P_n is the orthogonal polynomial of degree n with respect to the weight w. We have shown in Theorem 5.5.1 that the zeros of P_n are simple and are all situated in the open interval $]a, b[$. If we make a Euclidean division of P, belonging to $\mathbb{P}_{2n-1}$, by P_n we obtain

$$P = P_n q + r.$$

Let the x_j be the zeros of P_n. By definition of orthogonal polynomials

$$(P_n, q) = 0,$$

that is, eqn (8.4.5). The formula, thus constructed, is certainly of degree at least $2n - 1$. It is not of degree $2n$, indeed

$$\int_a^b P_n(x)^2 \, dx - \sum_{j=1}^n \lambda_j P_n(x_j)^2 = \int_a^b P_n(x)^2 \, dx \neq 0.$$

Conversely, if we have a quadrature formula of order $k \geqslant 2n - 1$,

$$\sum_{j=1}^n \mu_j f(y_j),$$

this formula is clearly an interpolation formula, since $2n - 1 \geqslant n - 1$. We must therefore have

$$(p,q) = 0, \quad \forall q \in \mathbb{P}_{k-n}.$$

Since $k - n \geqslant (2n - 1) - n = n - 1$, we must, in particular, have

$$(p,q) = 0, \quad \forall q \in \mathbb{P}_{n-1}.$$

For the nonzero polynomial $p \in \mathbb{P}_n$ to be orthogonal to $\mathbb{P}_{n-1}$, it is necessary and sufficient that p be a multiple of the n-th orthogonal polynomial with respect to w. The y_j are therefore the x_j and, since the formula is from interpolation, the μ_j are identical to the λ_j.   □

8.5. Numerical integration of periodic functions over a period: Fourier analysis

The error analysis of composite formulae in Theorem 8.3.7 does not take into account possible global compensation from one interval to another. In the case of a periodic function (to fix ideas, of period 1), these global compensations are very interesting, to the point that it is not necessary to employ any other formula than the rectangle formula. Observe that the left and the right rectangle formulae give identical results.

Theorem 8.5.1. Let f be a periodic function of period 1 on $\mathbb{R}$. If f is C^m on $\mathbb{R}$, and if we apply the rectangle formula to it with n points uniformly distributed on the interval $[0, 1]$, we have the following error estimate

$$(8.5.1) \qquad \left| \int_0^1 f(x)\,\mathrm{d}x - \sum_{j=0}^{n-1} \frac{1}{n} f\left(\frac{j}{n}\right) \right| \leqslant \frac{C(m, f)}{n^m}.$$

In other words, the rectangle formula is of infinite order when applied to periodic functions. ◇

Proof. The idea is to see what the rectangle formula gives for trigonometric polynomials. We therefore calculate

$$e_{k,n} = \int_0^1 e^{2i\pi kx}\,\mathrm{d}x - \frac{1}{n} \sum_{j=0}^{n-1} e^{2i\pi kj/n}.$$

Let

$$S = \sum_{j=0}^{n-1} e^{2i\pi kj/n}.$$

This is the sum of a geometric series with ratio $e^{2i\pi k/n}$. If $e^{2i\pi k/n} = 1$, that is, if n is a divisor of k, $S = n$. In the converse case

$$S = \frac{e^{2\pi ik} - 1}{e^{2i\pi k/n} - 1} = 0.$$

We therefore have

$$S = \begin{cases} n & \text{if } n \text{ divides } k; \\ 0 & \text{otherwise.} \end{cases}$$

Moreover,

$$\int_0^1 e^{2i\pi kx}\,\mathrm{d}x = \delta_{0k}.$$

We deduce the value of $e_{k,n}$:

$$(8.5.2) \qquad e_{k,n} = \begin{cases} -1 & \text{if } n \text{ divides } k \text{ and } k \neq 0; \\ 0 & \text{otherwise.} \end{cases}$$

Consequently, if P is a trigonometric polynomial

$$P(x) = \sum_{|k| \leqslant N} a_k e^{2i\pi kx},$$

we have

$$\int_0^1 P(x)\,dx - \frac{1}{n}\sum_{j=0}^{n-1} P\left(\frac{j}{n}\right) = \sum_{|k| \leqslant N} a_k e_{k,n} = -\sum_{l:0<|ln|\leqslant N} a_{ln}.$$

Suppose now that f is C^m and take m to be at least equal to 2, since the case $m = 1$ is a consequence of estimate (8.3.8). We denote by $\hat{f}(k)$ the k-th Fourier coefficient of f. Since f is at least C^2, the Fourier series converges uniformly and by inverting the order of the summation and integration, which is valid by virtue of the uniform convergence, we can write

$$\int_0^1 f(x)\,dx - \sum_{j=0}^{n-1} \frac{1}{n} f\left(\frac{j}{n}\right) = \int_0^1 \sum_{k \in \mathbb{Z}} \hat{f}(k)\, e^{2i\pi kx}\,dx - \sum_{j=0}^{n-1}\frac{1}{n}\sum_{k \in \mathbb{Z}} \hat{f}(k)\, e^{2i\pi kj/n}$$

$$= \sum_{k \in \mathbb{Z}} \hat{f}(k)\, e_{k,n}$$

$$= -\sum_{\ell \in \mathbb{Z}\setminus\{0\}} \hat{f}(\ell n).$$

From the estimate of Lemma 7.1.12

$$\left|\hat{f}(k)\right| \leqslant \frac{C_m}{|k|^m}, \quad \forall k \neq 0,$$

we deduce that

$$\left|\int_0^1 f(x)\,dx - \sum_{j=0}^{n-1}\frac{1}{n} f\left(\frac{j}{n}\right)\right| \leqslant \frac{2C_m}{n^m}\sum_{\ell=1}^{\infty}\frac{1}{\ell^m}.$$

As the series

$$\sum_{\ell=1}^{\infty}\frac{1}{\ell^m}$$

converges for $m \geqslant 2$, we have proved the result claimed. $\square$

8.6. From Bernoulli to Euler and MacLaurin: the delights of integration by parts

We now present another approach to the error analysis for the trapezium formula with uniformly distributed knots. This approach consists of representing the quadrature error in the elementary trapezium formula by means of successive integration by parts, using derivatives of the function being integrated which are of arbitrarily high order. We again find a formula for which the kernel (we are therefore on familiar ground) is a multiple of a Bernoulli polynomial.

The Bernoulli family provided a good dozen scientists in the seventeenth and eighteenth centuries, of whom at least three were very prominent. Amongst these were the mathematicians Jacques (1654–1705) and Jean (1667–1748).

On the subject of the personalities and relations between the brothers Jacques and Jean Bernoulli, consult the book by Stefan Hildebrandt and Anthony Tromba [46, pp. 64–7]. At a time when only a few people in the world understood the infinitesimal calculus of Leibniz, the elder of the brothers had taught mathematical analysis to the younger, who had a temperament which was, to say the least, competitive. Lively scientific conflicts between the two brothers were played out under the watch of contemporary European scholars. It was Leonhard Euler (1707–1783), a student of Jean Bernoulli, who carried the mathematical tradition of the Bernoullis the furthest, adding to it his own originality. Euler was blessed with numerous descendants, and legend says that he did mathematics with his children playing all around him. At the end of his life he lost his sight, which did not prevent him from continuing to work.

I strongly recommend reading the original works of the great authors. Reading the Introduction of Euler's *Introductio in Analysin Infinitorum* is an enthusing experience which is entirely accessible to degree level students. It is good to note that Euler hardly concerned himself with the convergence of the objects on which he worked. However, his supreme intuition traced the paths for contemporary mathematicians, be it as a precursor to non-standard analysis or when he summed divergent series. The phrase 'Euler equation' refers to at least two different objects, one in fluid mechanics and the other in the calculus of variations, both of which are the objects of very active study.

The book by Hildebrandt and Tromba is also very good, and abounds with beautiful pictures and stimulating questions.

The polynomials mentioned here were brought to prominence by Jacques Bernoulli, who introduced and studied them for discrete values of their argument in his work *Ars Conjectandi*, published posthumously in 1713.

8.6.1. Detailed analysis of the trapezium formula

First of all, note that for a periodic function of period 1, the rectangle formula and the trapezium formula are equivalent. We are therefore going to study, in detail, the error for the elementary trapezium formula. Suppose, first of all, that f is C^1. Then,

$$\int_0^1 f(x)\,dx - \frac{1}{2}\left[f(0) + f(1)\right]$$

$$= \frac{1}{2}\int_0^1 (f(x) - f(0))\,dx + \frac{1}{2}\int_0^1 (f(x) - f(1))\,dx$$

$$= \frac{1}{2}\int_0^1 \left(\int_0^x f'(y)\,dy\right)dx - \frac{1}{2}\int_0^1 \left(\int_x^1 f'(y)\,dy\right)dx.$$

Changing the order of the integrations, we see that

$$\int_0^1 \left(\int_0^x f'(y)\,dy\right)dx = \int_0^1 \left(\int_y^1 dx\right)f'(y)\,dy = \int_0^1 (1 - y)\,f'(y)\,dy.$$

In the same way,

$$\int_0^1 \left(\int_x^1 f'(y)\,dy\right)dx = \int_0^1 y f'(y)\,dy.$$

Consequently,

$$\int_0^1 f(x)\,dx - \frac{1}{2}\left[f(0) + f(1)\right] = \int_0^1 \left(\frac{1}{2} - y\right)f'(y)\,dy.$$

We are now going to generalize this result:

Lemma 8.6.1. Let f be a C^{m+1} function on the interval $[0,1]$. Then,

(8.6.1)
$$\int_0^1 f(x)\,dx - \frac{1}{2}\left[f(0) + f(1)\right]$$
$$= \sum_{j=1}^n \alpha_j \int_0^1 f^{(j)}(x)\,dx + \int_0^1 P_{n+1}(x)\,f^{(n+1)}(x)\,dx.$$

The real numbers α_j and the polynomials P_j are defined by

(8.6.2)
$$P_1(x) = \frac{1}{2} - x, \qquad\qquad\qquad \alpha_1 = 0,$$

$$P_{n+1}(x) = \int_0^x (\alpha_n - P_n(y))\,dy, \qquad \alpha_{n+1} = \int_0^1 P_{n+1}(y)\,dy$$

Furthermore, the α_{2j+1} are zero and the P_j satisfy the symmetry relation

$$P_j(x) = (-1)^j P_j(1 - x).$$

Proof. Formula (8.6.1) is true for order 1. Suppose that it is true for order n. We note that

$$P_n = \alpha_n - P_{n+1}',$$

and consequently, if we integrate

$$\int_0^1 P_n(x) f^{(n)}(x) \, dx$$

by parts, we find

$$\int_0^1 (\alpha_n - P'_{n+1}(x)) f^{(n)}(x) \, dx = \alpha_n \int_0^1 f^{(n)}(x) \, dx - \Big[P_{n+1}(x) f^{(n)}(x) \Big]_{x=0}^{x=1}$$
$$+ \int_0^1 f^{(n+1)}(x) P_{n+1}(x) \, dx.$$

By construction, $P_{n+1}(0) = P_{n+1}(1) = 0$, which verifies eqn (8.6.1).

Suppose that P_n satisfies $P_n(1-x) = (-1)^n P_n(x)$. Then, we can write

$$P_{n+1}(1-x) = \int_0^{1-x} (\alpha_n - P_n(y)) \, dy$$
$$= \alpha_n(1-x) + (-1)^{n+1} \int_0^{1-x} P_n(1-y) \, dy$$
$$= \alpha_n(1-x) + (-1)^{n+1} \int_x^1 P_n(y) \, dy$$
$$= \alpha_n(1-x) + (-1)^{n+1} \left(\int_0^1 P_n(y) \, dy - \int_0^x P_n(y) \, dy \right)$$
$$= \alpha_n(1-x) + (-1)^{n+1} \alpha_n + (-1)^{n+1} P_{n+1}(x) + (-1)^n \alpha_n x$$
$$= \alpha_n(1-x) \left(1 + (-1)^{n+1} \right) + (-1)^{n+1} P_{n+1}(x).$$

If $P_{2j+1}(1-x) = -P_{2j+1}(x)$, the integral on $[0, 1/2]$ cancels the integral on $[1/2, 1]$, which implies that $\alpha_{2j+1} = 0$. The above calculation shows us that

$$P_{2j+2}(1-x) = P_{2j+2}(x).$$

Consequently, as $1 + (-1)^{2j+3} = 0$, we see that

$$P_{2j+3}(1-x) = -P_{2j+3}(x).$$

As the symmetry property holds for P_1, the lemma is completely proved. □

8.6.2. The Bernoulli polynomials

The Bernoulli polynomials are defined by means of a generator function:

$$(8.6.3) \qquad t \mapsto \phi(x,t) = \frac{t e^{tx}}{e^t - 1}.$$

This has a power series expansion with respect to t of the form

(8.6.4)
$$\frac{te^{tx}}{e^t - 1} = \sum_{n=0}^{\infty} \frac{B_n(x)}{n!} t^n.$$

The radius of convergence of this expansion for fixed x is equal to 2π, since the numerator and denominator are entire functions of t and the denominator vanishes at $2ik\pi$, $k \in \mathbb{Z}$, but the singularity in the denominator at $t = 0$ is compensated by the singularity in the numerator.

It is clear that ϕ is C^∞ with respect to $t \in \mathbb{R}$ and $x \in \mathbb{R}$, and the radius of convergence of the Taylor series in t is 2π. The partial derivative of ϕ, with respect to x, is

$$\frac{\partial \phi}{\partial x}(x,t) = t\phi(x,t) = \sum_{j=1}^{\infty} \frac{B_{n-1}(x) t^n}{(n-1)!}.$$

We can also differentiate series (8.6.4) term by term (prove it!), which gives us, after equating the terms of equal power in t

(8.6.5)
$$B'_n(x) = nB_{n-1}(x).$$

Since $B_0(x) = 1$, we see that all the B_n are polynomials of exactly degree n. Furthermore,

$$\int_0^1 \phi(x,t)\, dx = \frac{t}{e^t - 1} \int_0^1 e^{tx}\, dx = 1.$$

Consequently, for every $n \geqslant 1$,

$$\int_0^1 B_n(x)\, dx = 0.$$

The polynomials B_n are Bernoulli polynomials.

We are now going to show that, for every $n \geqslant 1$,

(8.6.6)
$$B_n(x) = (-1)^n\, n!\, (\alpha_n - P_n(x)).$$

Explicit calculation gives

$$B_1(x) = x - \frac{1}{2}.$$

Assume that eqn (8.6.6) holds for rank $n - 1$. Then, due to eqn (8.6.5),

$$B'_n(x) = n(-1)^{n-1}(n-1)!\,(\alpha_{n-1} - P_{n-1}(x)),$$

and due to eqn (8.6.2), $B'_n + (-1)^n n! P'_n$ is a constant. By construction,

$$\int_0^1 (\alpha_n - P_n(x))\, dx = 0,$$

and we have seen that B_n has zero integral on $[0,1]$. Consequently, we have verified formula (8.6.6).

8.6.3. The Euler–MacLaurin formula

We are now going to give the Euler–MacLaurin formula which describes the error in the composite trapezium rule. We extend P_n periodically over all of $\mathbb{R}$, and let $\tilde{P}_n$ be the periodic function of period 1, thus obtained. For $n \geqslant 2$, $\tilde{P}_n$ is a continuous function. The recurrence relation (8.6.2) shows that $\tilde{P}_n$ is C^{n-2}. If the function f is defined on the interval $[0, k]$ and is C^{2m+2} on this interval, then

$$
\int_0^k f(x)\,\mathrm{d}x - \frac{1}{2}\sum_{j=0}^{k-1}(f(j) + f(j+1))
$$

$$
= \sum_{j=1}^m \alpha_{2j}\int_0^k f^{(2j)}(x)\,\mathrm{d}x + \int_0^k \tilde{P}_{2m+2}(x)\,f^{(2m+2)}(x)\,\mathrm{d}x.
$$

By a change of variable, we are going to get back to the trapezium formula on $[a, b]$: let $x = a + th$ with $h = (b-a)/k$. We then get

$$
\int_a^b f(x)\,\mathrm{d}x = h\left[\frac{f(a) + f(b)}{2} + \sum_{j=1}^{k-1} f(a+jh)\right] + \sum_{j=1}^m \alpha_{2j}h^{2j}\int_a^b f^{(2j)}(x)\,\mathrm{d}x
$$

(8.6.7)

$$
+ h^{2m+2}\int_a^b \tilde{P}_{2m+2}\left(\frac{x-a}{h}\right) f^{(2m+2)}(x)\,\mathrm{d}x.
$$

Formula (8.6.7) is called the Euler–MacLaurin formula. It allows us to analyse quadrature formulae, as well as to precisely approximate sums of the type

$$
\sum_{j=1}^k f(j),
$$

by comparison with the integral of f from 0 to k and the accurate estimation of the remainder.

We clearly find the results already claimed for the integration of periodic functions, since all the terms $\int_a^b f^{(2j)}(x)\,\mathrm{d}x$ disappear if $b-a$ is a period of f. In particular, the trapezium formula is of exactly infinite order for functions with compact support, integrated over an interval containing their support.

8.7. Discrete Fourier and fast Fourier transforms

We calculate the Fourier coefficients of a periodic function of period 1, by applying the rectangle formula at equidistant points. As the rectangle formula is of infinite order for periodic functions, as we saw in the preceding section, it is pointless to use a more sophisticated formula. In its naïve version, this calculation requires $O(N^2)$ complex multiplications for the calculation of N Fourier coefficients. J. W. Cooley and J. W. Tukey proposed a remarkable algorithm for fast Fourier transforms (FFT) [17], but it seems that it had been discovered previously by Danielson and Lanczos in 1942 and that it was also known to Gauss. To understand the FFT algorithm, it is first of all necessary to note that the rearrangement of the rows or columns highlights self-similarity properties, which are due, on the one hand, to the formulae consisting of many multiplications by 1 or by -1, and moreover, to the (more or less elementary) arithmetic properties of the integer which defines the number of points.

In the simplest case, where N is a power of 2, the FFT algorithm requires $O(N \log_2 N)$ real operations (multiplications or additions) to arrive at the result.

There exist numerous generalizations of FFT, including implementations which work for numbers N which are the products of powers of 2, 3, 5, and 7. The FFT algorithm is a typical example of a recursive numerical algorithm, although this is not necessarily the best way to program it. From its multiscale analysis nature, it is also the ancestor of modern multigrid methods [41] and multiscale wavelet analysis [61].

8.7.1. Discrete Fourier transforms

The Fourier coefficients of a continuous periodic function f, of period 1, are given by

$$\hat{f}(k) = \int_0^1 f(x) \, e^{-2i\pi k x} \, dx.$$

The discretization, by the rectangle formula with equidistant points, is written

$$U_k = \frac{1}{N} \sum_{j=0}^{N-1} f\left(\frac{j}{N}\right) e^{-2i\pi k j/N}.$$

Note that this formula produces at most N distinct complex numbers. Indeed,

$$\begin{aligned}
U_{k+N} &= \frac{1}{N} \sum_{j=0}^{N-1} f\left(\frac{j}{N}\right) e^{-2i\pi(k+N)j/N} \\
&= \frac{1}{N} \sum_{j=0}^{N-1} f\left(\frac{j}{N}\right) e^{-2i\pi k j/N} \\
&= U_k.
\end{aligned}$$

We therefore define the discrete Fourier transform F_N as the linear operator which associates the sequence of U_k, defined by

$$U_k = \sum_{j=0}^{N-1} u_j e^{-2i\pi k j/N},$$

to a finite sequence $(u_j)_{0\leqslant j\leqslant N-1}$ of complex numbers. The matrix of F_N is complex symmetric (warning, it is not Hermitian!). The conjugate linear operator is denoted by $\bar{F}_N$, and we have the following elementary and essential result which implies that, to within a constant factor, F_N is unitary:

Lemma 8.7.1. For all N, we have the following identities:

$$(8.7.1) \qquad\qquad F_N \circ \bar{F}_N = NI = \bar{F}_N \circ F_N.$$

Proof. Let

$$u_j = \sum_{k=0}^{N-1} U_k e^{2i\pi jk/N}.$$

We calculate

$$\sum_{j=0}^{N-1} u_j e^{-2i\pi j\ell/N}.$$

We have

$$\sum_{j=0}^{N-1} u_j e^{-2i\pi j\ell/N} = \sum_{j=0}^{N-1} e^{-2i\pi j\ell/N} \sum_{k=0}^{N-1} U_k e^{2i\pi jk/N}$$

$$= \sum_{k=0}^{N-1} \left(\sum_{j=0}^{N-1} e^{2i\pi j(k-\ell)/N} \right) U_k.$$

Now

$$\sum_{j=0}^{N-1} e^{2i\pi j(k-\ell)/N} = \begin{cases} N & \text{if } N \text{ divides } k - \ell; \\ 0 & \text{otherwise.} \end{cases}$$

As k and ℓ vary from 0 to $N-1$, N can divide $k-\ell$ only if $k = \ell$. Consequently,

$$\sum_{j=0}^{N-1} u_j e^{-2i\pi j\ell/N} = Nu_\ell.$$

The second equality follows immediately from the first by conjugation. $\qquad\square$

Since F_N is a linear operator from $\mathbb{C}^N$ to itself, *a priori* it will need N^2 complex multiplications to calculate the U_k as functions of the u_j. We are going to see that it is nothing of the sort.

8.7.2. Principle of the fast Fourier transform algorithm

We write the matrices of F_2 and F_4 explictly:

$$F_2 = \begin{pmatrix} 1 & 1 \\ 1 & -1 \end{pmatrix} \quad \text{and} \quad F_4 = \begin{pmatrix} 1 & 1 & 1 & 1 \\ 1 & -i & -1 & i \\ 1 & -1 & 1 & -1 \\ 1 & i & -1 & -i \end{pmatrix}.$$

If we calculate $(U_j)_{0 \leqslant j \leqslant 3}$, we can therefore write

(8.7.2)
$$
\begin{aligned}
U_0 &= u_0 + u_1 + u_2 + u_3, \\
U_1 &= u_0 - iu_1 - u_2 + iu_3, \\
U_2 &= u_0 - u_1 + u_2 - u_3, \\
U_3 &= u_0 + iu_1 - u_2 - iu_3.
\end{aligned}
$$

We will rewrite this and change the order of the u_j as follows:

$$
\begin{aligned}
U_0 &= u_0 + u_2 + u_1 + u_3, \\
U_1 &= u_0 - u_2 - iu_1 + iu_3, \\
U_2 &= u_0 + u_2 - u_1 - u_3, \\
U_3 &= u_0 - u_2 + iu_1 - iu_3.
\end{aligned}
$$

We can therefore put this in the form

(8.7.3)
$$
\begin{aligned}
\begin{pmatrix} U_0 \\ U_1 \end{pmatrix} &= F_2 \begin{pmatrix} u_0 \\ u_2 \end{pmatrix} + \begin{pmatrix} 1 & 0 \\ 0 & -i \end{pmatrix} F_2 \begin{pmatrix} u_1 \\ u_3 \end{pmatrix}, \\
\begin{pmatrix} U_2 \\ U_3 \end{pmatrix} &= F_2 \begin{pmatrix} u_0 \\ u_2 \end{pmatrix} - \begin{pmatrix} 1 & 0 \\ 0 & -i \end{pmatrix} F_2 \begin{pmatrix} u_1 \\ u_3 \end{pmatrix}.
\end{aligned}
$$

We see, therefore, that the calculation of F_4 needs the calculation of two transformations F_2 and a multiplication by a 2×2 diagonal matrix.

We can also rewrite eqn (8.7.2), changing the order of the rows, as follows:

$$
\begin{aligned}
U_0 &= u_0 + u_1 + u_2 + u_3, \\
U_2 &= u_0 - u_1 + u_2 - u_3, \\
U_1 &= u_0 - iu_1 - u_2 + iu_3, \\
U_3 &= u_0 + iu_1 - u_2 - iu_3.
\end{aligned}
$$

We then have

(8.7.4)
$$
\begin{aligned}
\begin{pmatrix} U_0 \\ U_2 \end{pmatrix} &= F_2 \left[\begin{pmatrix} u_0 \\ u_1 \end{pmatrix} + \begin{pmatrix} u_2 \\ u_3 \end{pmatrix} \right], \\
\begin{pmatrix} U_1 \\ U_3 \end{pmatrix} &= F_2 \left[\begin{pmatrix} 1 & 0 \\ 0 & i \end{pmatrix} \begin{pmatrix} u_0 - u_2 \\ u_1 - u_3 \end{pmatrix} \right].
\end{aligned}
$$

In this case, we first of all multiply by the 2×2 diagonal matrix, and then make the two transformations F_2.

The two ideas above generalize to the case of $N = 2^n$, which is the most important thing in practice. In the first case, we say that we have an algorithm with decimation-in-time, since the u_j were originally seen as states dependent on discrete time j, and in the second case decimation-in-frequency, as the U_k are modes.

8.7.3. FFT algorithm: decimation-in-frequency

We present the details of the fast Fourier transform in the case of decimation-in-frequency. The decimation-in-time case is completely analogous. We will estimate the number of operations necessary.

We rewrite F_N, firstly grouping all the even modes and then all the odd modes. Letting $M = N/2$, we have:

$$U_{2k} = \sum_{j=0}^{N-1} u_j e^{-4i\pi kj/N} = \sum_{j=0}^{M-1} u_j e^{-2i\pi kj/M} + \sum_{j=M}^{N-1} u_j e^{-2i\pi kj/M}.$$

Note that,

$$(8.7.5) \qquad e^{-2i\pi kj/M} = e^{-2i\pi k(j-M)/M}, \quad \forall j \in \{M, \ldots, N-1\},$$

and consequently,

$$(8.7.6) \qquad \begin{pmatrix} U_0 \\ U_2 \\ \vdots \\ U_{N-2} \end{pmatrix} = F_M \begin{pmatrix} u_0 + u_M \\ u_1 + u_{M+1} \\ \vdots \\ u_{M-1} + u_{N-1} \end{pmatrix}.$$

In the same way, the odd modes are given by

$$U_{2k+1} = \sum_{j=0}^{N-1} u_j e^{-2i\pi(2k+1)j/N}$$

$$= \sum_{j=0}^{M-1} e^{-2i\pi kj/M} e^{-2i\pi j/N} u_j + \sum_{j=M}^{N-1} e^{-2i\pi kj/M} e^{-2i\pi j/N} u_j.$$

Using eqn (8.7.5) and the relation

$$e^{-i\pi j/M} = -e^{-i\pi(j-M)/M},$$

we obtain

$$\begin{pmatrix} U_1 \\ U_3 \\ \vdots \\ U_{N-1} \end{pmatrix} = F_M \begin{pmatrix} u_0 - u_M \\ e^{-i\pi/M}(u_1 - u_{M+1}) \\ \vdots \\ e^{-i\pi(M-1)/M}(u_{M-1} - u_{N-1}) \end{pmatrix}.$$

Let

$$u_{\mathrm{I}} = \begin{pmatrix} u_0 \\ u_1 \\ \vdots \\ u_{M-1} \end{pmatrix}, \quad u_{\mathrm{II}} = \begin{pmatrix} u_M \\ u_{M+1} \\ \vdots \\ u_{N-1} \end{pmatrix}$$

and

$$U_{\text{even}} = \begin{pmatrix} U_0 \\ U_2 \\ \vdots \\ U_{N-2} \end{pmatrix}, \quad U_{\text{odd}} = \begin{pmatrix} U_1 \\ U_3 \\ \vdots \\ U_{N-1} \end{pmatrix}.$$

If $P_M = P_{N/2}$ denotes the diagonal matrix given by

$$(P_M)_{jj} = e^{-i\pi j/M},$$

then we have the block factorization

$$(8.7.7) \qquad \begin{pmatrix} U_{\text{even}} \\ U_{\text{odd}} \end{pmatrix} = \begin{pmatrix} F_{N/2} & 0 \\ 0 & F_{N/2} \end{pmatrix} \begin{pmatrix} I_{N/2} & 0 \\ 0 & P_{N/2} \end{pmatrix} \begin{pmatrix} u_{\text{I}} + u_{\text{II}} \\ u_{\text{I}} - u_{\text{II}} \end{pmatrix}.$$

To recover the components of U in the right order, we now need to make a permutation σ_n of the rows. This is the permutation of 2^n objects, numbered from 0 to $2^n - 1$, for which the matrix $P_{\sigma_n^{-1}}$ (convention of p. 221) is the matrix of the transformation

$$\begin{pmatrix} U_{\text{even}} \\ U_{\text{odd}} \end{pmatrix} \mapsto \begin{pmatrix} U_{\text{I}} \\ U_{\text{II}} \end{pmatrix}.$$

If j varies from 0 to $2^n - 1$, we associate with it its representation in binary

$$j = \sum_{k=0}^{n-1} 2^k d_k = \overline{d_{n-1}d_{n-2}\cdots d_1 d_0}.$$

We then verify that

$$(8.7.8) \qquad \sigma_n\left(\overline{d_{n-1}d_{n-2}\cdots d_0}\right) = \overline{d_{n-2}\cdots d_0 d_{n-1}}.$$

Indeed, we see how we pass from the sequence

$$0, 1, \ldots, 2^n - 1$$

to the sequence

$$0, 2, \ldots, 2p, \ldots, 2^n - 2, 1, 3, \ldots, 2p + 1, \ldots, 2^n - 1.$$

If j is even, we associate $j/2$ to it, that is

$$j = \sum_{k=0}^{n-1} 2^k d_k,$$

with $d_0 = 0$. Therefore,

$$\frac{j}{2} = \sum_{k=1}^{n-1} 2^{k-1} d_k = \sum_{k=0}^{n-2} 2^k d_{k+1} = \overline{d_0 d_{n-1}\cdots d_1}.$$

If j is odd, we associate $2^{n-1} + (j-1)/2$ to it. We therefore have

$$j = \sum_{k=0}^{n-1} 2^k d_k,$$

with $d_0 = 1$. Therefore,

$$\frac{j}{2} - 1 = \sum_{k=0}^{n-2} 2^k d_{k+1},$$

and hence,

$$\frac{j}{2} - 1 + 2^{n-1} = 2^{n-1} + \sum_{k=0}^{n-2} d_{k+1} = \overline{d_0 d_{n-1} \cdots d_1}.$$

In other words,

$$\sigma_n^{-1} \left(\overline{d_{n-1} \cdots d_1 d_0} \right) = \overline{d_0 d_{n-1} \cdots d_1},$$

which implies eqn (8.7.8). To take account of the structure of the discrete Fourier transform, we define the permutation matrices $P_{\sigma_p,k}$ of 2^k objects by

$$P_{\sigma_p,p} = P_{\sigma_p},$$

for $p \geqslant 1$, and

$$P_{\sigma_p,k+1} = \begin{pmatrix} P_{\sigma_p,k} & 0 \\ 0 & P_{\sigma_p,k} \end{pmatrix},$$

for $k \geqslant p \geqslant 1$. At the binary representation level, it is clear that

$$(8.7.9) \qquad \sigma_{p,k} \left(\overline{d_{k-1} \cdots d_p d_{p-1} d_{p-2} \cdots d_0} \right) = \overline{d_{k-1} \cdots d_p d_{p-2} \cdots d_0 d_{p-1}}.$$

From eqn (8.7.7) we deduce that

$$F_N u = P_{\sigma_n^{-1}} \begin{pmatrix} F_{N/2} & 0 \\ 0 & F_{N/2} \end{pmatrix} \begin{pmatrix} I_{N/2} & 0 \\ 0 & P_{N/2} \end{pmatrix} \begin{pmatrix} u_{\mathrm{I}} + u_{\mathrm{II}} \\ u_{\mathrm{I}} - u_{\mathrm{II}} \end{pmatrix}.$$

Obviously, $F_{N/2}$ has the same structure.

Thus, we can graphically represent the fast Fourier transform by a scheme consisting of lattices known as butterflies and horizontal arrows. The butterfly represents the mapping

$$\begin{pmatrix} u \\ v \end{pmatrix} \mapsto \begin{pmatrix} u + v \\ u - v \end{pmatrix}$$

from $\mathbb{C}^2$ to itself. The horizontal arrow simply symbolizes a transfer of data, and the arrow labelled with a complex number symbolizes multiplication by this complex number.

We refer to Figure 8.7 for the case $N = 8$, $n = 3$ with $w = \exp(-i\pi/2)$. At the end, it is necessary to make the matrix permutation

$$P_{\sigma_{2,3}^{-1}} P_{\sigma_3^{-1}} = P_{\sigma_{2,3}^{-1} \circ \sigma_3^{-1}},$$

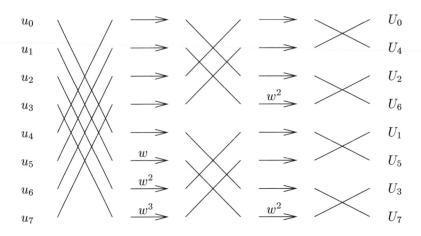

Figure 8.7: The FFT butterflies.

to recover the components of U in the natural order. It results from eqns (8.7.8) and (8.7.9) that

$$\sigma_3 \circ \sigma_{2,3} \left(\overline{d_2 d_1 d_0}\right) = \sigma_3 \left(\overline{d_2 d_0 d_1}\right) = \overline{d_0 d_1 d_2}.$$

Generally, we can check that

$$\sigma_n \circ \sigma_{n-1,n} \circ \cdots \circ \sigma_{2,n} \left(\overline{d_{n-1} d_{n-2} \cdots d_1 d_0}\right) = \overline{d_0 d_1 \cdots d_{n-2} d_{n-1}}$$

is the permutation τ_n, which swaps element p with element $\tau_n(p)$, whose binary expansion is obtained by reversing the order of the digits of the binary expansion of p. This explains the order of the elements in the last column of Figure 8.7 and shows the effects of the structure of decimation-in-frequency: after the algebraic part of the FFT, represented by the operations appearing in eqn (8.7.7), it is advisable to de-interlace the frequencies by permuting them by τ_n.

Operation Count 8.7.2. Let $N = 2^n$. An FFT on N complex numbers demands at most $5N \log_2 N$ real operations (additions or multiplications).

Proof. We agree to count real multiplications and additions. The addition of two complex numbers requires two real additions, and the multiplication of two complex numbers requires two real additions and four real multiplications. Let a_n be the number of real operations necessary for an FFT algorithm on $N = 2^n$ modes. We let $M = N/2$. From eqn (8.7.7), to obtain $u_{\mathrm{I}} + u_{\mathrm{II}}$ and $u_{\mathrm{I}} - u_{\mathrm{II}}$, we need $2M$ complex additions or $2N$ real operations.

The matrix multiplication of $P_M(u_{\mathrm{I}} - u_{\mathrm{II}})$ by the diagonal matrix P_M requires M complex multiplications or $6M = 3N$ real operations. To make the two fast Fourier transforms in dimension M, we need $2a_{n-1}$ real operations. We therefore have the inequality

$$a_n \leqslant 2a_{n-1} + 5 \times 2^n.$$

By induction, and observing that F_1 requires no operations, we have

$$a_n \leqslant 5n2^n.$$

We have obtained the estimate

$$a_n \leqslant 5N \log_2 N.$$

Note that the classic count, which only counts complex multiplications, gives $N2^{-1} \log_2 N$ operations, as can be checked by the reader. ☐

8.8. Exercises from Chapter 8

8.8.1. Summation of series with Bernoulli numbers and polynomials

This subsection allows us to study the error made in a numerical integration formula with equidistant knots. All the necessary calculations can be made by hand.

Bernoulli numbers

Recall the definition (8.6.4) of Bernoulli polynomials.

Exercise 8.8.1. Show that, for all $j \geqslant 1$, $B_{2j+1}(0) = 0$.

Exercise 8.8.2. The Bernoulli numbers are defined by the formula

$$b_{2j} = B_{2j}(0).$$

Calculate b_2 and b_4.

Exercise 8.8.3. Show that, for all $j \geqslant 1$,

$$b_{2j} = -\int_0^1 B_{2j}(x)\,\mathrm{d}x.$$

Exercise 8.8.4. Calculate B'_j in terms of B_{j-1} and the Bernoulli numbers.

Error estimates and formal computations

In this subsection, we will prove an error formula relating to the trapezium rule. This formula will be studied for polynomials, by operator calculus.

Only the last two exercises depend on the first section.

We let $\mathbb{P}$ be the space of polynomials of a real variable, and $\mathbb{P}_n$ be the space of polynomials of degree at most n. h is a fixed strictly positive real number.

We define the following linear operators on $\mathbb{P}$:

$$(If)(x) = f(x),$$
$$(Ef)(x) = f(x+h),$$
$$(\Delta f)(x) = f(x+h) - f(x),$$
$$(Df)(x) = \frac{df}{dx}.$$

We say that two operators A and B are equal if, for any f in $\mathbb{P}$, $Af = Bf$.

Exercise 8.8.5. Show that, for every f in $\mathbb{P}$, there exists m such that $D^m f = \Delta^m f = 0$.

Exercise 8.8.6. We let

$$(8.8.1) \qquad\qquad e^{hD} = \sum_{n=1}^{\infty} \frac{h^n D^n}{n!}.$$

Show that the sum defining $e^{hD} f$ is finite for all f in $\mathbb{P}$ and that

$$(8.8.2) \qquad\qquad e^{hD} = E.$$

Exercise 8.8.7. Show that, for every $y \in \mathbb{R}$, we can define linear operators Γ_y and C_y on $\mathbb{P}$ by

$$\Delta\Gamma_y f = f, \qquad (\Gamma_y f)(y) = 0,$$

and

$$hDC_y f = f, \qquad (C_y f)(y) = 0.$$

Use the functions

$$g_k(x) = \frac{(x-y)(x-y-h)\cdots(x-y-(k-1)h)}{k!}$$

and apply Δ to them to solve the first part of the question.

Show that $\Gamma_y \Delta f - f$ and $C_y D f - f$ are of degree zero. Calculate $\Delta C_y h D \Gamma_y$.

Exercise 8.8.8. Calculate $\Delta C_y f$ as a function of J, defined by

$$(Jf)(x) = \int_x^{x+h} f(t)\,dt.$$

Exercise 8.8.9. With ϕ being the function given by eqn (8.6.3) and ψ defined by

$$\psi(t) = \phi(0, t),$$

we let

$$(8.8.3) \qquad\qquad A = hD\Gamma_y - \sum_{j=0}^{\infty} \psi^{(j)}(0) \frac{(hD)^j}{j!}.$$

Show, by using Exercise 8.8.3, that A does not depend on y. Calculate $A\Delta$ by using eqn (8.8.2). What is the value of A?

Exercise 8.8.10. Show that

$$Jf(x) - h\frac{f(x) + f(x+h)}{2} = -\sum_{j=1}^{\infty} \frac{b_{2j}h^{2j}}{(2j)!}\left[f^{(2j-1)}(x+h) - f^{(2j-1)}(x)\right],$$

(8.8.4)

by applying ΔC_y to A.

Application to the summation of series

In this section, we evaluate the error term in eqn (8.8.3), if we truncate the infinite sum at order m.

Exercise 8.8.11. Let f be a $(2n+2)$ times differentiable function on $[0,1]$. Show that the Euler–MacLaurin formula can be rewritten as

$$\int_0^1 f(x)\,\mathrm{d}x - \frac{1}{2}(f(0) + f(1)) = -\sum_{j=1}^{m} \frac{b_{2j}}{(2j)!}\left[f^{(2j-1)}(1) - f^{(2j-1)}(0)\right]$$

(8.8.5)

$$+ \frac{1}{(2m+2)!}\int_0^1 B_{2m+2}(x)\,f^{(2m+2)}(x)\,\mathrm{d}x.$$

We extend B_j into a periodic function of period 1. Show that, if f is $(2m+2)$ times differentiable on $[a,b]$ and $h = (b-a)/n$, we have

$$\int_a^b f(x)\,\mathrm{d}x - h\Big[\frac{1}{2}f(a) + \sum_{j=1}^{n-1} f(a+jh) + \frac{1}{2}f(b)\Big]$$

(8.8.6)

$$= -\sum_{j=1}^{m} \frac{B_j}{(2j)!}h^{2j}\left[f^{(2j-1)}(b) - f^{(2j-1)}(a)\right]$$

$$+ \frac{h^{2m+2}}{(2m+2)!}\int_a^b \bar{B}_{2m+2}\left(\frac{x-a}{h}\right)f^{(2m+2)}(x)\,\mathrm{d}x.$$

Exercise 8.8.12. Application of eqn (8.8.6). Let

$$f(x) = \frac{1}{(x+10)^{3/2}}.$$

By letting $m = 1$ and with the aid of formula (8.8.6), evaluate the error incurred by replacing

$$\sum_{n=10}^{\infty} \frac{1}{n^{3/2}}$$

by an integral expression, plus the first error term. We give

$$B_4(x) = x^2(1-x)^2 + b_4.$$

Exercise 8.8.13. How many terms of the series must be added to obtain an accuracy which is at least as good?

8.8.2. The Fredholm integral equation of the first kind

Let K be a continuous mapping from $[0, 1] \times [0, 1]$ to $\mathbb{R}$. We define an operator $\mathcal{K}$ on $E = C^0([0, 1])$ by

$$(8.8.7) \qquad (\mathcal{K}u)(x) = \int_0^1 K(x, y) u(y) \, dy.$$

Recall the definition of a modulus of continuity: if a function f is uniformly continuous from the convex part of a vector space metric F, provided with a distance d_F, to values in a metric space G, provided with a distance d_G, there exists an increasing continuous function ω_f from $\mathbb{R}^+$ to itself, vanishing at zero, such that, for every x and x' in F, we have

$$(8.8.8) \qquad d_G(f(x), f(x')) \leqslant \omega_f(d_F(x, x')).$$

Exercise 8.8.14. Show that the image of E by $\mathcal{K}$ is included in E. Given a distance on $[0, 1] \times [0, 1]$ use the modulus of continuity ω_K of K.

Exercise 8.8.15. We provide E with the maximum norm, denoted $\| \cdot \|$. Show that, for every u in E,

$$(8.8.9) \qquad \|\mathcal{K}u\| \leqslant \|u\| \max_{x \in [0,1]} \int_0^1 |K(x, y)| \, dy.$$

Exercise 8.8.16. Show that, if K is at least m times continuously differentiable with respect to its first variable, $\mathcal{K}u$ is C^m.

Exercise 8.8.17. Let

$$\Lambda = \left(\max_{x \in [0,1]} \int_0^1 |K(x, y)| \, dy \right)^{-1}.$$

Show that, for every f in E and for every λ in $]-\Lambda, \Lambda[$, there exists a unique u such that

$$(8.8.10) \qquad u - \lambda \mathcal{K}u = f.$$

Exercise 8.8.18. Let $v \in E$. Given $h = 1/n$, we approximate the integral of v on $[0, 1]$ by the trapezium formula with equidistant knots $y_k = kh$, $0 \leqslant k \leqslant n$. Write down this formula, and estimate the difference between the integral of v and its numerical approximation using the modulus of continuity of v. Improve this estimate by supposing that v is C^2.

Exercise 8.8.19. We want to numerically approximate the solution of eqn (8.8.10). To do this, we replace the integral with respect to y by a quadrature, by means of the trapezium method, with the same knots as in the preceding question. We denote this expression by

$$h \sum_{j=0}^{n} \gamma_j(x) u(jh).$$

Evaluate, for every u in E,

$$\left| (Ku)(x) - h \sum_j \gamma_j(x) u(jh) \right|,$$

supposing, first of all, that K is continuous, and then that K is C^2 with respect to x.

Exercise 8.8.20. Show that, if K is continuous,

$$\lim_{h \to 0} \max_{0 \leqslant k \leqslant n} \left[h \sum_{j=0}^{n} |\gamma_j(kh)| \right] = \Lambda^{-1}.$$

Exercise 8.8.21. We define an operator $\mathcal{K}^h$ from $\mathbb{R}^{n+1}$ to itself by

$$\left(\mathcal{K}^h U \right)_k = h \sum_{j=0}^{n} \gamma_j(kh) U_j, \quad 0 \leqslant k \leqslant n.$$

Providing $\mathbb{R}^{n+1}$ with the norm $\|U\| = \max_j |U_j|$, evaluate the norm of the operator of $\mathcal{K}^h$.

Exercise 8.8.22. Show that, for every λ in $]-\Lambda, \Lambda[$ and for every h less than a certain $h_0(|\lambda|)$ which should be specified, the problem

$$U - \lambda \mathcal{K}^h U = F$$

possesses a solution, for any F in $\mathbb{R}^{n+1}$.

Exercise 8.8.23. Let $F^h = \left(f(kh) \right)_{0 \leqslant k \leqslant n}$. Let U^h be the solution of

$$U^h - \lambda \mathcal{K}^h U^h = F^h.$$

In order to evaluate the error committed by the process of numerical approximation we denote by V^h the vector defined by

$$\left(V^h \right)_k = u(kh),$$

where u is the solution of eqn (8.8.10). Show that

$$\left[U^h - V^h - \lambda \mathcal{K}^h U^h + \lambda \mathcal{K}^h V^h \right]_k = \lambda \left(\mathcal{K}^h V^h \right)_k - \lambda (Ku)(kh).$$

Evaluate the second term of the above expression using Exercise 8.8.17. Show that $(I - \lambda \mathcal{K}^h)^{-1}$ exists and is uniformly bounded as h tends to 0. Deduce from this that

$$\lim_{h \to 0} \max_{1 \leqslant k \leqslant n} \left| \left(U^h \right)_k - u(kh) \right| = 0.$$

Make this estimate precise if we assume that K is C^2 with respect to x and y.

8.8.3. Towards Franklin's periodic wavelets

Diagonalization of cyclic matrices

We say that a square matrix A is cyclic if it is of the form

$$A = \begin{pmatrix} a_0 & a_1 & a_2 & \cdots & a_{N-2} & a_{N-1} \\ a_{N-1} & a_0 & a_1 & \cdots & a_{N-3} & a_{N-2} \\ a_{N-2} & a_{N-1} & a_0 & \cdots & a_{N-4} & a_{N-3} \\ & & & \ddots & & \\ a_2 & a_3 & a_4 & \cdots & a_0 & a_1 \\ a_1 & a_2 & a_3 & \cdots & a_{N-1} & a_0 \end{pmatrix}.$$

If necessary, we use the notation

$$A = \text{circu}\,(a_0, a_1, \ldots, a_{N-1}),$$

and denote by Γ_N the set of $N \times N$ cyclic matrices. If necessary, we identify the finite sequence $(a_j)_{0 \leqslant j \leqslant N-1}$ and the infinite periodic sequence defined by

$$a_{j+kN} = a_j, \quad \forall k \in \mathbb{Z},\ \forall j \in \{0, \ldots, N-1\}.$$

The vector space of periodic sequences of period N on $\mathbb{Z}$ is denoted by $\ell_{\sharp,N}$.

Then, the Hermitian scalar product of two sequences $a, b \in \ell_{\sharp,N}$ is given by

$$(a, b)_{\sharp,N} = \sum_{j=0}^{N-1} a_j \bar{b}_j.$$

Exercise 8.8.24. Show that Γ_N is closed under addition, the multiplication by a scalar, and under matrix multiplication. Use the convolution of the elements $\ell_{\sharp,N}$ defined by

$$(a * b)_j = \sum_{k=0}^{N-1} a_k b_{j-k}.$$

Exercise 8.8.25. Let ω be an N-th root of unity and let x be the vector

$$x = \begin{pmatrix} 1 \\ \omega \\ \vdots \\ \omega^{N-1} \end{pmatrix}.$$

Calculate Ax for A in Γ_N.

Exercise 8.8.26. Calculate all the eigenvectors and eigenvalues of the cyclic matrices

$$\text{circu}\,(a_0, a_1, \ldots, a_{N-1})$$

as functions of the coefficients a_j, $0 \leqslant j \leqslant N-1$.

Exercise 8.8.27. Show that every cyclic matrix can be diagonalized on an orthogonal basis. Give a necessary and sufficient condition so that a cyclic matrix is Hermitian.

Properties of certain function spaces

We denote by $L^1_\sharp$ the space of functions from $\mathbb{R}$ to itself which are periodic of period 1 and whose restriction to every compact subset of $\mathbb{R}$ is integrable. The integral of the functions $f \in L^1_\sharp$ is defined by

$$\int_\sharp f(x)\,\mathrm{d}x = \int_a^{a+1} f(x)\,\mathrm{d}x,$$

where a is any real number.

We denote by $L^2_\sharp$ the space of functions from $\mathbb{R}$ to itself which are periodic of period 1, measurable, and whose restriction to every subspace of $\mathbb{R}$ is square-integrable. The scalar product between functions of $L^2_\sharp$ is defined by

$$(f,g)_\sharp = \int_\sharp f(x)\,g(x)\,\mathrm{d}x.$$

The integer N is fixed and greater than or equal to 1, and we let

$$h = \frac{1}{N}.$$

The space V_N is defined as the subspace of $L^1_\sharp$ formed from all the continuous functions whose restriction to each interval $[kh, (k+1)h]$ is a polynomial of at most degree 1.

Exercise 8.8.28. Show that the mapping, which to $f \in V_N$ associates the collection of its values at the points jh, $j \in \mathbb{Z}$, is an isomorphism Φ_N from V_N on $\ell_{\sharp,N}$.

Exercise 8.8.29. Let e_0 be a periodic sequence of period N defined by

$$(e_0)_j = \delta_{0j}, \quad 0 \leqslant j \leqslant N.$$

Give the function $\phi = \Phi_N^{-1}(e_0)$ and sketch its behaviour, where $N = 8$.

Exercise 8.8.30. Let τ be the translation defined on V_N by

$$(\tau f)(x) = f(x - h).$$

With an abuse of the notation, we will also write for the sequences belonging to $\ell_{\sharp,N}$

$$(\tau a)_j = a_{j-1}.$$

Show that the family of $(\tau^m \phi)_{0 \leqslant m \leqslant N-1}$ form a basis of V_N.

Exercise 8.8.31. For every integer $m \in \mathbb{Z}$, calculate

$$(\phi, \tau^m \phi)_\sharp .$$

For every j and m, deduce from this the value of

$$\left(\tau^j \phi, \tau^m \phi\right)_\sharp .$$

Exercise 8.8.32. For f and g in V_N, let

$$\Phi_N (f) = (a_j)_{j \in \mathbb{Z}} \quad \text{and} \quad \Phi_N (g) = (b_j)_{j \in \mathbb{Z}} .$$

Write $(f, g)_\sharp$ as a function of the sequences a_j and b_j. Show that the matrix A, a bilinear form on $\ell_{\sharp, N}$, defined by

$$(a, b) \mapsto \left(\Phi_N^{-1} (a), \Phi_N^{-1} (b)\right)_\sharp$$

is cyclic, real symmetric, and positive definite.

Exercise 8.8.33. We intend to show that there exists an orthogonal basis of V_N for the scalar product $(\cdot, \cdot)_\sharp$ formed from functions $(\tau^m \psi)_{0 \leqslant m \leqslant N-1}$, where we want to determine the real even function ψ. Let

$$\Phi_N (\psi) = c = (c_j)_{j \in \mathbb{Z}} .$$

Show that the block column matrix

$$C^* = \begin{pmatrix} c & \tau c & \cdots & \tau^{N-1} c \end{pmatrix}$$

is identical to

$$\mathrm{circu} \, (c_0, c_1, \ldots, c_{N-1}) .$$

Exercise 8.8.34. Show that ψ answers the question posed if and only if

$$C^* A C = I.$$

Using the first part of the problem, give all the functions ψ which satisfy the question.

Exercise 8.8.35. Suppose that N is even, and let $N = 2M$. In this question we use the notation ϕ_M and ψ_M in place of ϕ and ψ, respectively, and we introduce the functions ϕ_M and ψ_M similarly in V_M. We denote by W_M the orthogonal complement of V_M in V_N. Construct the function χ_M belonging to W_M such that the $\tau^{2m} \chi_M$ form a basis of W_M. We can reduce this to considering functions χ_M with support in $[-2h, 2h]$.

Exercise 8.8.36. Reasoning as in Exercise 8.8.34, show that there exists an orthonormal basis of W_M formed from functions $(\tau^{2m} \sigma_M)_{0 \leqslant m \leqslant M-1}$ and determine σ_M.

Remark 8.8.37. If $N = 2^n$, the successive functions σ_M, with $M = 2^j$ form periodic Franklin wavelets.

Part III

Numerical linear algebra

I used to think, when I was a student, that linear algebra is this boring subject where you prove only fairly obvious things, and then you have these gigantic and stupid calculations. The non-constructive approach to linear algebra has little interest, because it does not tackle the most important question, which is the effective construction of the objects whose existence we prove. But the constructive approach is absolutely fascinating. Take an easy example: suppose that A is a square positive definite matrix, and that we want to solve the linear system

$$Ax = b.$$

We know that A can be diagonalized in an orthonormal basis, and in that basis, the above equation reduces to a diagonal system, i.e. a completely trivial question. Fine, but how do we get, practically, the orthonormal basis? It turns out that getting the orthonormal basis is much more difficult than solving the original equation. This problem is treated, at least partially in Chapter 13, and it is a highly nonlinear problem, as we shall see. Therefore, one must find efficient practical methods for solving linear systems which do not rely on determining a good basis by unspecified means.

In Chapter 9, we treat the so-called direct methods of resolution of linear systems; in Chapter 10, we define a number of analytical tools in order to be able to take up iterative methods in Chapter 11. There, we find that iterative methods are very efficient, even for solving problems which admit a solution in finite terms.

In Chapter 12, we use orthogonality-related constructions to devise other methods for solving linear systems. These methods also come as a preparation for finding the eigenvalues and eigenvectors of a matrix, and they have a strong Lie group-theoretical flavour, but we are not supposed to say so, lest some pure mathematician might have heard us and be offended that creepy applied mathematicians might walk on their turf, and lest some applied mathematician might hear it and scream that highbrow pure mathematics is irrelevant in the realm of numerical analysis and should be shunned as useless theoretical gobbledygook. But of course, here we are only with friends, and if we do not use big words and show that the methods are efficient, who cares what they are called?

9

Gauss's world

We consider the system of n equations and n unknowns of the form

$$(9.0.1) \qquad Ax = b,$$

where A is a square matrix and x, b belong to $\mathbb{K}^n$. More explicitly, the system may be written as

$$(9.0.2) \qquad \begin{aligned} a_{11}x_1 + a_{12}x_2 + \ldots + a_{1n}x_n &= b_1, \\ a_{21}x_1 + a_{22}x_2 + \ldots + a_{2n}x_n &= b_2, \\ &\vdots \\ a_{n1}x_1 + a_{n2}x_2 + \ldots + a_{nn}x_n &= b_n. \end{aligned}$$

The objective is to find an equivalent system to eqn (9.0.1), that is, one which has the same set of solutions as (9.0.1), but is triangular. As we saw in Section 3.1, such systems with triangular matrices are very easy to solve.

9.1. The Gaussian elimination algorithm without pivoting

The title of this section will become clear in the lines which follow. We are concerned with an algorithm for which success is not guaranteed. Since we discover the criteria for success as we go along, the only thing to do is to try it. It is, therefore, an algorithm for the lucky.

9.1.1. Just elimination

Let us suppose that the coefficient $a_{11} = \pi_1$ in the system (9.0.2) is not zero—this is the first pivot. We are going to fill the whole of the first column under a_{11} with zeros, by making linear combinations of row i, $i = 2, \ldots, n$ with row

1. We must, therefore, subtract a_{i1}/π_1 times the first row from the i-th row to achieve our objective. The i-th row then becomes

$$\left(a_{i1} - \frac{a_{i1}a_{11}}{\pi_1}\right)x_1 + \ldots + \left(a_{in} - \frac{a_{i1}a_{1n}}{\pi_1}\right)x_n = b_i - \frac{a_{i1}b_1}{\pi_1}.$$

It is clear that the coefficient of x_1 in this row vanishes. We keep intact the first row of the system (9.0.2). We now have a new system which is written as

$$(9.1.1) \quad \begin{pmatrix} \pi_1 & a_{12} & \cdots & a_{1n} \\ 0 & a_{22} - \dfrac{a_{21}a_{12}}{\pi_1} & \cdots & a_{2n} - \dfrac{a_{21}a_{1n}}{\pi_1} \\ \vdots & \vdots & & \vdots \\ 0 & a_{n2} - \dfrac{a_{n1}a_{12}}{\pi_1} & \cdots & a_{nn} - \dfrac{a_{n1}a_{1n}}{\pi_1} \end{pmatrix} x = \begin{pmatrix} b_1 \\ b_2 - \dfrac{a_{21}b_1}{\pi_1} \\ \vdots \\ b_n - \dfrac{a_{n1}b_1}{\pi_1} \end{pmatrix}.$$

The transformation from the system (9.0.2) to the system (9.1.1) can be denoted in terms of matrices, since we are merely making linear combinations of the rows. Let us introduce some new notation:

$$(9.1.2) \qquad a' = \begin{pmatrix} a_{21} \\ \vdots \\ a_{n1} \end{pmatrix}, \quad p' = \frac{a'}{\pi_1}, \quad A = \begin{pmatrix} \pi_1 & \ell' \\ a' & A'_1 \end{pmatrix}, \quad b = \begin{pmatrix} b_1 \\ b' \end{pmatrix}.$$

Let $\widehat{M}$ and $\widehat{L}_1$ be the matrices

$$(9.1.3) \qquad \widehat{M} = \begin{pmatrix} 1 & 0 \\ -p' & I_{n-1} \end{pmatrix} \quad \text{and} \quad \widehat{L}_1 = \begin{pmatrix} 1 & 0 \\ p' & I_{n-1} \end{pmatrix}.$$

The inverse of $\widehat{M}$ is $\widehat{L}_1$. Indeed, by block multiplication:

$$\begin{aligned} \widehat{L}_1\widehat{M} &= \begin{pmatrix} 1 & 0 \\ p' & I_{n-1} \end{pmatrix}\begin{pmatrix} 1 & 0 \\ -p' & I_{n-1} \end{pmatrix} \\ &= \begin{pmatrix} 1 \times 1 + 0 \times (-p') & 1 \times 0 + 0 \times I_{n-1} \\ p' \times 1 + I_{n-1} \times (-p') & p' \times 0 + I_{n-1} \times I_{n-1} \end{pmatrix} \\ &= \begin{pmatrix} 1 & 0 \\ 0 & I_{n-1} \end{pmatrix} = I_n. \end{aligned}$$

An elementary calculation shows that the system (9.1.1) is equivalent to

$$(9.1.4) \qquad\qquad \widehat{M}Ax = \widehat{M}b.$$

Another block calculation gives

$$\widehat{M}A = \begin{pmatrix} 1 & 0 \\ -p' & I_{n-1} \end{pmatrix}\begin{pmatrix} \pi_1 & \ell' \\ a' & A'_1 \end{pmatrix} = \begin{pmatrix} \pi_1 & \ell' \\ 0 & -p'\ell' + A'_1 \end{pmatrix}$$

and

$$\widehat{M}b = \begin{pmatrix} b_1 \\ b' - p'b_1 \end{pmatrix}.$$

If we let

(9.1.5) $$\widehat{A}_1 = \widehat{M}A = \begin{pmatrix} \pi_1 & \ell' \\ 0 & A_1 \end{pmatrix},$$

we see that, if we know how to solve the system comprising the last $n - 1$ rows, the first row gives x_1 very simply. Moreover, the set of solutions of the system (9.1.4) is identical to the set of solutions of the system (9.0.1), since we pass from one to the other by a premultiplication with a regular matrix.

The system (9.1.4) is, therefore, equivalent to the system (9.0.1), since we have not eliminated haphazardly, but have used a strategy which does not destroy information.

The new matrix $\widehat{M}A$ has the following form (block notation):

$$\widehat{M}A = \begin{pmatrix} \pi_1 & * \\ 0 & A' \end{pmatrix},$$

where the asterisk denotes coefficients whose precise expressions do not interest us for the moment. The matrix A' has $n - 1$ rows and $n - 1$ columns. If the element in the first column and the first row of A' is not zero (this is the second pivot), then we can apply the same algorithm to A' that we applied to A and fill the second column of $\widehat{M}A$ with zeros from the third row, whilst retaining an equivalent system. This procedure does not change the first or second row of the matrix $\widehat{M}A$, since we add a multiple of the second row to the j-th row of the matrix for $j \geqslant 3$.

By induction, provided that we do not meet a zero pivot, we obtain an equivalent system for which the matrix is given by

$$\begin{pmatrix} \text{pivot} & * & \cdots & * \\ 0 & \text{pivot} & \cdots & * \\ \vdots & & & \vdots \\ 0 & \cdots & 0 & \text{pivot} \end{pmatrix}.$$

We have therefore described Gaussian elimination. It remains to interpret it precisely in terms of matrices. This is the object of the next section.

9.1.2. Matrix interpretation of Gaussian elimination

In this section, we describe precisely the matrices which are involved in the elimination algorithm.

Theorem 9.1.1. Let A be an $n \times n$ matrix. Suppose that, in the course of the process of Gaussian elimination, no pivot is zero. Then, there exists a lower

triangular matrix L, with ones on the diagonal, and an invertible upper triangular
matrix U such that matrix A may be decomposed as

$$A = LU.$$

Furthermore, such a decomposition of A is unique. It is called the LU decomposition of A.   ◇

   It is important to be familiar with the following proofs. They give subtle
insights into matrices and provide excellent practice at block multiplication.

Proof of Theorem 9.1.1. We prove the result by induction on the spatial
dimension. If $n = 1$, it suffices to take $L = (1)$ and $U = A$. Assume the
statement to be true in dimension $n - 1$, i.e., for every matrix A of size $n - 1$
for which the elimination does not produce vanishing pivots, a decomposition
$A = LU$ of the stated type exists.

   The analysis of Subsection 9.1.1 shows that a matrix of size n satisfying the
conditions of the statement can be written as

$$(9.1.6) \qquad\qquad A = \widehat{L}_1 \widehat{A}_1,$$

with $\widehat{L}_1$ given by eqn (9.1.3) and $\widehat{A}_1$ given by eqn (9.1.5). The induction hypothesis then implies that

$$A_1 = L_1 U_1.$$

However,

$$\begin{pmatrix} \pi_1 & \ell' \\ 0 & L_1 U_1 \end{pmatrix} = \begin{pmatrix} 1 & 0 \\ 0 & L_1 \end{pmatrix} \begin{pmatrix} \pi_1 & \ell' \\ 0 & U_1 \end{pmatrix}$$

and therefore

$$A = \begin{pmatrix} 1 & 0 \\ p' & I_{n-1} \end{pmatrix} \begin{pmatrix} 1 & 0 \\ 0 & L_1 \end{pmatrix} \begin{pmatrix} \pi_1 & \ell' \\ 0 & U_1 \end{pmatrix}.$$

Thus, the theorem is proved with

$$L = \begin{pmatrix} 1 & 0 \\ p' & L_1 \end{pmatrix} \quad \text{and} \quad U = \begin{pmatrix} \pi_1 & \ell' \\ 0 & U_1 \end{pmatrix},$$

which gives the existence. Assume now that there exist the two decompositions

$$A = L_1 U_1 = L_2 U_2$$

satisfying the stated conditions. Then, it is possible to write

$$L_2^{-1} L_1 = U_2 U_1^{-1}.$$

In consequence, the upper triangular matrix $U_2 U_1^{-1}$ would be equal to the lower
triangular matrix $L_2^{-1} L_1$ having only ones on the diagonal. Therefore, they are
both equal to the identity, which proves uniqueness.   □

The matrix L contains a great deal of information that we are now going to examine.

Lemma 9.1.2. The element L_{ij} of L, for $1 \leqslant j < i \leqslant n$, is the value by which the i-th row must be multiplied at the j-th stage of the Gaussian elimination.

Proof. If the dimension n is equal to 1, nothing needs to be proved. Let us show first that, for $n > 1$, the $n \times n$ matrix A can be written in the form

$$(9.1.7) \qquad A = \widehat{L}_1 \cdots \widehat{L}_{n-1} U,$$

with an upper triangular matrix U and lower triangular matrices $\widehat{L}_j$, having ones on their diagonal and of the form

$$\widehat{L}_j = \begin{pmatrix} I_{j-1} & 0 \\ 0 & L'_j \end{pmatrix},$$

with L'_j being a square matrix of size $(n+1-j)$. Assume that this decomposition holds for the matrices of size at most $n-1$, and let A be a matrix of size n satisfying the conditions of the lemma. Due to the induction assumption, the matrix A_1 of eqn (9.1.5) can be written in the form

$$(9.1.8) \qquad A_1 = \widetilde{L}_2 \cdots \widetilde{L}_{n-1} \widetilde{U},$$

with an upper triangular matrix U and lower triangular matrices $\widetilde{L}_j$, having ones on their diagonal and of the form

$$\widetilde{L}_j = \begin{pmatrix} I_{j-2} & 0 \\ 0 & L'_j \end{pmatrix}.$$

We observe that

$$\begin{pmatrix} \pi_1 & \ell' \\ 0 & A_1 \end{pmatrix} = \begin{pmatrix} 1 & 0 \\ 0 & \widetilde{L}_1 \cdots \widetilde{L}_{n-2} \end{pmatrix} \begin{pmatrix} \pi_1 & \ell' \\ 0 & \widetilde{U} \end{pmatrix}$$

and that

$$\begin{pmatrix} 1 & 0 \\ 0 & \widetilde{L}_2 \cdots \widetilde{L}_{n-1} \end{pmatrix} = \begin{pmatrix} 1 & 0 \\ 0 & \widetilde{L}_2 \end{pmatrix} \cdots \begin{pmatrix} 1 & 0 \\ 0 & \widetilde{L}_{n-1} \end{pmatrix}.$$

Consequently, by letting

$$\widehat{L}_j = \begin{pmatrix} 1 & 0 \\ 0 & \widetilde{L}_j \end{pmatrix}$$

and using the relation (9.1.6), expression (9.1.7) is proved. Observe that

$$\widehat{L}_{k+1} \cdots \widehat{L}_{n-1} = \begin{pmatrix} I_k & 0 \\ 0 & L_k \end{pmatrix},$$

with a lower triangular matrix $\bar{L}_k$ having ones on its diagonal. Therefore, if we let

$$U = \begin{pmatrix} U_k & V_k \\ 0 & W_k \end{pmatrix},$$

the upper left block being of size $k \times k$, we will have

$$\widehat{L}_{k+1} \cdots \widehat{L}_{n-1} U = \begin{pmatrix} U_k & V_k \\ 0 & \bar{L}_k W_k \end{pmatrix}.$$

Thus, letting $A_k = \bar{L}_k W_k$, we may rewrite eqn (9.1.7) as follows:

$$A = \widehat{L}_1 \cdots \widehat{L}_k \widehat{A}_k, \quad \widehat{A}_k = \begin{pmatrix} U_k & V_k \\ 0 & A_k \end{pmatrix}.$$

The analysis of Subsection 9.1.1 implies that

(9.1.9) $$A_k = \begin{pmatrix} 1 & 0 \\ p'_{k+1} & I_{n-k-1} \end{pmatrix} \begin{pmatrix} \pi_{k+1} & \ell'_{k+1} \\ 0 & A'_{k+1} - p'_{k+1}\ell'_{k+1} \end{pmatrix}$$

and, by uniqueness, we see that

(9.1.10) $$L'_{k+1} = \begin{pmatrix} 1 & 0 \\ p_{k+1} & I_{n-k-1} \end{pmatrix}, \quad A_{k+1} = A'_{k+1} - p'_{k+1}\ell'_{k+1}.$$

On the other hand, if $\tilde{p}'_1$ denotes the first element of p'_1 and $\tilde{p}'_1$ is the column vector made out of its remaining $n-2$ elements, it is possible to write

$$\widehat{L}_1 \widehat{L}_2 = \begin{pmatrix} 1 & 0 & 0 \\ \tilde{p}'_1 & 1 & 0 \\ \tilde{p}'_1 & 0 & I_{n-2} \end{pmatrix} \begin{pmatrix} 1 & 0 & 0 \\ 0 & 1 & 0 \\ 0 & p'_2 & I_{n-2} \end{pmatrix} = \begin{pmatrix} 1 & 0 & 0 \\ \tilde{p}'_1 & 1 & 0 \\ \tilde{p}'_1 & p'_2 & I_{n-2} \end{pmatrix}.$$

A simple induction now shows that L, which is the product of the matrices $\widehat{L}_j$, is a matrix having ones on its diagonal, zeros above and below, and the p'_j aligned in order of increasing j.   □

Remark 9.1.3. Multiplication between $\widehat{L}_1$ and $\widehat{L}_2$ is not commutative (exercise). Neither is that between $\widehat{M}_1$ and $\widehat{M}_2$. Consequently, we cannot read the coefficients of the linear combinations which are used in the elimination from L^{-1}.

9.2. Putting it into practice: operation counts

9.2.1. The madness of Cramer's rule

By way of a comparison, we begin by evaluating the number of operations necessary to solve a system by Cramer's rule. From any first year maths course, recall that

$$x_j = \frac{D_j}{D_0}, \quad j = 1, \ldots, n,$$

where each of the D_j, $1 \leqslant j \leqslant n$, is the determinant of a matrix of n rows and n columns. More precisely, D_0 is the determinant of the matrix A, and D_j is the determinant of the matrix obtained from A by replacing the j-th column of A by b. The cost of the calculation is therefore the same for all the determinants. The formula for the expansion of a determinant gives

$$D_0 = \det A = \sum_{\sigma} \epsilon(\sigma) \prod_{i=1}^{n} a_{i,\sigma(i)}.$$

Here, σ runs through the set of permutations of n objects and $\epsilon(\sigma)$ is the signature of the permutation σ which has the value ± 1. It is known that there are exactly $n!$ distinct permutations of n objects. Each product of n factors requires $n-1$ operations. Therefore, we must make $n!(n-1)$ multiplications, and add the $n!$ products obtained, giving in total

$$n!(n-1) + n! - 1 = n(n!) - 1 \sim n(n!) \text{ floating-point operations.}$$

As we have $n+1$ determinants to calculate, the number of operations necessary to solve a system by Cramer's rule is of order

$$n((n+1)!)$$

as n tends to infinity.

Let us evaluate this quantity when $n = 100$. We can evaluate the factorial by using Stirling's formula

$$n! \sim n^{n+(1/2)} e^{-n} \sqrt{2\pi},$$

which is very accurate. We calculate the second term in this expression as $e^{-100} \simeq 10^{-43.43}$, since $\log_{10} e \simeq 0.4343$, and therefore

$$
\begin{aligned}
100 \times 101! &= 101 \times 100 \times 100! \\
&\simeq 100 \times 101 \times 100^{100.5} \times 10^{-43.43} \times \sqrt{2\pi} \\
&\simeq 10^{205-44} 10^{0.57} \sqrt{2\pi} \\
&\simeq 9.4 \times 10^{161}.
\end{aligned}
$$

With a computer processing at 100 megaflops (10^8 floating-point operations per second), we can do

$$10^8 \times 365 \times 86\,400 \text{ operations per year,}$$

which we round to 3×10^{15} operations. Therefore, about 3×10^{146} years are required to solve our system. Taking into account that the universe is 15 billion years old, we would need, therefore, at least 10^{135} times the age of the universe to solve our system.

9.2.2. Putting elimination into practice

In order to solve a linear system in practice, we calculate the LU decomposition of the matrix of the system, supposing that it is possible, and then solve the following two systems:

$$(9.2.1) \qquad\qquad Ly = b \quad \text{and} \quad Ux = y.$$

Since the two matrices L and U are invertible, the systems (9.2.1) are equivalent to the system (9.0.1). Each of the two systems (9.2.1) are very easy to solve by successive substitution. We will see later that triangular systems require few operations. We point out that the construction of the matrices L and U is an intermediate result in the process of elimination. Consequently, it is necessary to store the LU decomposition when we have to calculate several systems with matrix A, so as to avoid pointless recalculation.

9.2.3. Operation counts for elimination

For the moment, we count the operations for the LU decomposition. The cost of constructing $\widehat{L}_j$ is the same as the cost of constructing L'_j. This cost is precisely the cost of calculating p'_j, a column vector of $n - j$ rows. We therefore have to make $n - j$ divisions, since the elements of p'_j are the $(A_{j-1})_{k1} / \pi_j$, for k from $j + 1$ to n. Consequently, the total cost of constructing L is

$$(9.2.2) \qquad\qquad \sum_{j=1}^{n-1}(n-j) = \sum_{k=1}^{n-1} k = \frac{n(n-1)}{2} \sim \frac{n^2}{2}.$$

As the relations (9.1.9) and (9.1.10) show, the transformation from A_j to A_{j+1} is made by modifying all the elements of A_j except the first row, which does not change, and the first column, set to zero except for its first element. From eqn (9.1.10), each element of A_{j+1} requires a multiplication and a subtraction. We have, therefore, $2(n - j - 1)^2$ operations to perform to construct $\widehat{A}_{j+1}$ from $\widehat{A}_j$.

In total, constructing $\widehat{A}_{n-1}$ requires

$$\sum_{j=0}^{n-1} 2(n-j-1)^2 = 2\sum_{k=1}^{n-1} k^2 = 2\frac{(n-1)n(2n-1)}{6} \sim \frac{2n^3}{3}.$$

If we compare this last estimate with the estimate (9.2.2), we obtain the following result:

Operation Count 9.2.1. The number of operations necessary to decompose an $n \times n$ matrix into the form $A = LU$ is of order $2n^3/3$, for large n.

Let us move on to the operation count necessary to solve the systems (9.2.1). The system $Ly = b$ may be written as

$$
\begin{aligned}
y_1 && &= b_1, \\
L_{21}y_1 + \quad y_2 && &= b_2, \\
&& &\vdots \\
L_{n1}y_1 + L_{n2}y_2 + \ldots + y_n &= b_n.
\end{aligned}
$$

To solve the first equation requires 0 operations, for the second, two operations, and for the j-th it is necessary to perform $j - 1$ products of 2 factors, $j - 2$ additions, and a subtraction, giving $2(j - 1)$ floating-point operations. In total, the solution of $Ly = b$ costs

$$
2\sum_{j=1}^{n-1} j = 2\frac{(n-1)\,n}{2} = n\,(n-1) \text{ operations.}
$$

We solve the system $Ux = y$ in the same way, rewriting it in the form

$$
\begin{aligned}
U_{nn}x_n &= y_n, \\
U_{n-1,n-1}x_{n-1} + U_{n-1,n}x_n &= y_{n-1}, \\
&\vdots \\
U_{11}x_1 + \ldots + U_{1,n-1}x_{n-1} + U_{1n}x_n &= y_1.
\end{aligned}
$$

We see that the solution of row n costs one operation, row $n - 1$ costs three, and row $n - j$ needs $2j - 1$. Consequently, the cost of solving this triangular system is equal to the cost of the preceding one plus one operation per row, giving

$$
(n - 1)\,n + n = n^2.
$$

In total, we obtain the following result:

Operation Count 9.2.2. The total cost of solving the two triangular systems (9.2.1) of n rows and n columns is of order $2n^2$, for large n.

If we want to solve a 100×100 system by Gaussian elimination it would cost us $2 \times 100^3/3 + 2 \times 100^2 \sim 6.6 \times 10^5$ floating-point operations. On the same computer as before, that would take $6.6 \times 10^5 \times 10^{-8}$ seconds, or less than 7 thousandths of a second. It could not be done by hand, but it is certainly completely within reach, even on a PC, which would take a few tens of seconds.

9.2.4. Inverting a matrix: putting it into practice and the operation count

We now ask another practical question: how do we calculate the inverse of a matrix A? From Chapter 3, the columns v^j of A^{-1} are the images by A^{-1} of the canonical basis vectors e^j and so

$$
Av^j = e^j.
$$

We are going to exploit the LU decomposition of A to calculate the v^j. This is done in the following two stages:

$$Lw^j = e^j \quad \text{and} \quad Uv^j = w^j.$$

From the preceding results, we would therefore need $n \times 2n^2$ operations, or $2n^3$ operations, to solve the two systems. But this count is too large. Indeed, the triangular system $Lw^j = e^j$ restricted to its first $j-1$ rows and columns has the zero solution. Therefore, we just have to solve

$$w_j^j = 1,$$
$$L_{j+1,j}w_j^j + w_{j+1}^j = 0,$$
$$\vdots$$
$$L_{nj}w_j^j + L_{n,j+1}w_{j+1}^j + \ldots + w_n^j = 0.$$

The cost of the solution of this triangular system of $n - j + 1$ rows and $n - j + 1$ columns is of order $(n-j+1)^2$ operations. Consequently, the cost of constructing w^j is equivalent to

$$\sum_{j=1}^{n} (n - j + 1)^2 = \sum_{k=1}^{n} k^2 \sim \frac{n^3}{3}.$$

On the other hand, we cannot hope for the same type of economy in the construction of the v^j, since there is no reason why the last components of the w^j should be zero. We summarize this result as follows:

Operation Count 9.2.3. Let A be an $n \times n$ matrix having an LU decomposition. Then, $4n^3/3$ floating-point operations are required to construct A^{-1} once the LU decomposition is known, giving in total $2n^3/3 + 4n^3/3 = 2n^3$ operations.

9.2.5. Do we need to invert matrices?

We have just shown that it costs about 3 times as many operations to invert a matrix than to solve a linear system. However, we may imagine that this loss is compensated by an economy of scale if we need to solve many linear systems with the same matrix A. Let us see if this is right.

Suppose that we must solve

$$Ax^k = b^k,$$

where k goes from 1 to K and K is large compared with 1. If we store the LU decomposition then we would have to perform

$$\frac{2n^3}{3} + 2Kn^2 \text{ operations}$$

to solve all of these systems.

Now let us consider the other hypothesis. The construction of A^{-1} has cost us $2n^3$ operations, including the LU decomposition. The calculation of each of the n elements of $A^{-1}b^k$ requires n multiplications and $n-1$ subtractions, giving $n(2n-1) \sim 2n^2$ floating-point operations for the product of a matrix with a vector. In total, this approach would require

$$2n^3 + 2Kn^2 \text{ operations,}$$

which exceeds the result found previously by $4n^3/3$.

Consequently, we do not need to calculate the inverse of a matrix to solve linear systems. It is only when we need the inverse of a matrix explicitly that we should calculate it, and we use the LU decomposition for this.

For reference, let us calculate the cost of the product of two $n \times n$ matrices. This is the same as doing n matrix–vector multiplications, giving $2n^3$ operations. This result is remarkable and serves almost as a conclusion to Section 9.2:

Operation Count 9.2.4. The number of operations necessary to multiply together two $n \times n$ matrices is of order $2n^3$. It is equal to the number of operations required to invert an $n \times n$ matrix.

This would be the conclusion if we limited ourselves to the case of sequential machines. In the case of parallel machines, it could be more advantageous to calculate the inverse of a matrix once and for all when faced with a repetitive calculation. It is very easy indeed to parallelize a matrix multiplication algorithm. On the other hand, the algorithm for solving a triangular system by elimination is not parallelizable, since, at each step, it makes use of the result of the previous calculation.

The choice of an efficient algorithm is dependent upon the computer technology available. It is therefore susceptible to evolution. Furthermore, there is not, generally, a simple choice of algorithm, since each problem suggests different choices. It is necessary, therefore, to try to understand the principles of the algorithms, with the aim of taking as sensible a decision as possible. It is not, however, necessary to program the algorithms which we could need in practice, since there exist many program libraries, which are available on all sizes of machines. These are often free, although sometimes, particularly for large machines, they cost money.

9.3. Elimination with pivoting

9.3.1. The effect of a small pivot

If, in the course of the elimination without pivoting algorithm, we find a zero pivot, the process stops. If a pivot is very small, without being zero, the process does not stop, but the result can be tainted by considerable errors, as we are going to see in the example below.

Let ϵ be a small real number and consider the following system of two equations and two unknowns:

(9.3.1)
$$\epsilon x + y = 1,$$
$$x + y = 2.$$

Using elimination without pivoting, we obtain the following equivalent system:

$$\epsilon x + y = 1,$$
$$\left(1 - \frac{1}{\epsilon}\right) y = 2 - \frac{1}{\epsilon}.$$

Therefore, we immediately have the following value of y:

(9.3.2)
$$y = \frac{1 - 2\epsilon}{1 - \epsilon} \simeq 1.$$

By substitution, we obtain

(9.3.3)
$$x = \frac{1 - y}{\epsilon} = \frac{1}{1 - \epsilon} \simeq 1.$$

Suppose that the number ϵ is small enough for information to be lost in the following floating-point operations:

$$1 \ominus \epsilon = 1, \quad 1 \ominus (2 \otimes \epsilon) = 1.$$

This will be the case if the mantissa of the floating-point numbers has k significant figures (in internal representation) and if $\epsilon < \beta^{-k-1}$, where β denotes the base of the representation of the numbers in the machine being considered. We will suppose also that $1/\epsilon$ does not exceed the capacity of the machine.

In this case,

$$1 \ominus (1 \oslash \epsilon) = -(1 \oslash \epsilon)$$

and

$$2 \ominus (1 \oslash \epsilon) = -(1 \oslash \epsilon).$$

Consequently, the calculation gives

$$y = 1.$$

The substitution gives

$$x = (1 \ominus y) \oslash \epsilon = 0.$$

The error committed in y was acceptable, since it was carried in the last decimal of this number. The error in x is obviously unacceptable.

Let us try the elimination on the following system, obtained after exchanging the rows of the system (9.3.1):

$$x + y = 2,$$
$$\epsilon x + y = 1.$$

The same process of elimination as before leads to the following equivalent system:

$$x + y = 2,$$
$$(1 - \epsilon)\, y = 1 - 2\epsilon,$$

which becomes, in floating-point arithmetic,

$$x + y = 2,$$
$$y = 1.$$

The solution, which is completely acceptable this time, is

$$x = 1 \quad \text{and} \quad y = 1.$$

The error that we made with the first elimination came from dividing a number by a small pivot ϵ (see eqn (9.3.3)), which considerably amplifies the errors.

In the same way, we could have swapped the order of the variables in eqns (9.3.1) to obtain the system

$$y + \epsilon x = 1,$$
$$y + x = 2.$$

We improve the results by this process.

9.3.2. Partial pivoting and total pivoting: general description and cost

We now describe the pivoting algorithms precisely.

In the partial pivoting by column algorithm, one stage of elimination consists of selecting from the first column the element a_{i1} of maximum absolute value, then exchanging row i with row 1, if $i \neq 1$, and, finally, filling the first column (except the first element) with zeros, using the process of elimination described in Section 9.1.

In the partial pivoting by row algorithm, we select the element a_{1j} of maximum absolute value from the first row, we exchange the first column with column j, if $j \neq 1$, and we do an elimination step. To exchange column 1 with column j comes down to making a change of variables

$$y_j = x_1, \quad y_1 = x_j, \quad y_k = x_k \quad \text{for} \quad k \neq 1, j.$$

Finally, in the total pivoting algorithm we combine the two preceding strategies by searching in the whole matrix for the element a_{ij} of maximum absolute value. After exchanging the first row with the i-th row as well as the first and j-th columns, we do an elimination step.

From the point of view of operation counts (and therefore the complexity of the process), a pivot search by column or by row requires a search on n elements and a total pivot search requires a search on n^2 elements. A recursive search algorithm on $N = 2^p$ elements costs $p2^{p-1}$ tests and permutations. The total cost of partial pivoting in comparisons and exchanges is

$$\sum_{j=2}^{n} \frac{1}{2} j \log_2 j.$$

We find the order of this quantity by writing

$$\sum_{j=2}^{n} \frac{1}{2} j \log_2 j = \sum_{j=2}^{n} \frac{1}{2\ln 2} j \ln j$$

$$\sim \frac{1}{2\ln 2} \int_{2}^{n} x \ln x \, dx$$

$$= \frac{1}{2\ln 2} \left[\frac{x^2}{2} \ln x \Big|_{x=2}^{x=n} - \int_{2}^{n} \frac{x}{2} \, dx \right]$$

$$\sim \frac{n^2 \log_2 n}{4}.$$

An analogous calculation in the case of total pivoting gives

$$\sum_{j=2}^{n} \frac{1}{2} j^2 \log_2 j^2 = \sum_{j=2}^{n} j^2 \log_2 j$$

$$= \sum_{j=2}^{n} \frac{1}{\ln 2} j^2 \ln j$$

$$\sim \frac{1}{\ln 2} \int_{2}^{n} x^2 \ln x \, dx$$

$$= \frac{1}{\ln 2} \left[\frac{x^3}{3} \ln x \Big|_{x=2}^{x=n} - \int_{2}^{n} \frac{x^2}{3} \, dx \right]$$

$$\sim \frac{n^3 \log_2 n}{3}.$$

The cost of total pivoting is an order of magnitude larger than that of partial pivoting. According to the implementation, comparisons and exchanges are more or less rapid relative to floating-point operations, but it should be remembered that total pivoting is much slower than partial pivoting. In fact, partial pivoting is used more often than total pivoting.

9.3.3. Aside: permutation matrices

To interpret partial pivoting and total pivoting from a matrix perspective, we need to express row and column exchanges in terms of matrix multiplications. To this end, we define and study permutation matrices.

Let σ be a permutation of n objects. The matrix P_σ associated with the permutation σ is the $n \times n$ matrix defined by

$$(P_\sigma)_{ij} = \delta_{i\sigma(j)},$$

where δ_{pq} denotes the Kronecker delta.

How do we compose permutation matrices? Let σ and σ' be two permuation matrices. Then

$$(P_{\sigma'} P_\sigma)_{ik} = \sum_{j=1}^{n} \delta_{i\sigma'(j)} \delta_{j\sigma(k)}.$$

The nonzero terms in this sum must satisfy $i = \sigma'(j)$ and $j = \sigma(k)$. Consequently, this sum has the value 1 if $i = \sigma' \circ \sigma(k)$ and zero otherwise. We see then that

(9.3.4) $$P_{\sigma'} P_\sigma = P_{\sigma' \circ \sigma}.$$

Note that

$$(P_{\sigma^{-1}})_{ij} = \delta_{i\sigma^{-1}(j)} = \delta_{\sigma(i)j}.$$

Consequently,

(9.3.5) $$P_{\sigma^{-1}} = (P_\sigma)^* = (P_\sigma)^{-1}.$$

Regrouping (9.3.4) and (9.3.5), we see that the matrices P_σ are orthogonal, which is obvious *a priori* from geometric considerations: changing the order of the components of a vector in $\mathbb{R}^n$, or in $\mathbb{C}^n$, does not change its Euclidean length.

In other words, the mapping $\sigma \mapsto P_\sigma$ defines a mapping from the permutation group on n objects into the group of orthogonal matrices O_n. This mapping respects the structure of the group and, therefore, it is called a group homomorphism.

What is the effect of a left or right multiplication of a permutation matrix on some $n \times n$ matrix A? We have

$$(P_\sigma A)_{ik} = \sum_{j=1}^{n} \delta_{i\sigma(j)} A_{jk} = A_{\sigma^{-1}(i)k}.$$

Consequently, the multiplication $P_\sigma A$ is equivalent to operating the permutation σ^{-1} on the rows of A.

In the same way,

$$(AP_\sigma)_{ik} = \sum_{j=1}^{n} A_{ij}\delta_{j\sigma(k)} = A_{i\sigma(k)}.$$

The multiplication AP_σ is equivalent to applying the permutation σ on the columns of A.

9.3.4. Matrix interpretation of partial and total pivoting

Elimination with partial column pivoting is interpreted as a matrix decomposition by the following theorem:

Theorem 9.3.1. Let A be an $n \times n$ matrix. Then, it is invertible if and only if we can find a permutation σ, a lower triangular matrix L with 1 on the diagonal, and an invertible upper triangular matrix U such that

(9.3.6) $P_\sigma A = LU.$

Decomposition (9.3.6) is not generally unique.   ◇

Proof. It is obvious that, if P_σ, L, and U are as given in the theorem, $A = P_\sigma^* LU$ is invertible.

Conversely, we argue by induction on the spatial dimension. If $n = 1$, nothing needs to be proved. Assume that the statement is true for every square matrix of dimension at most $n - 1$, and let A be a square matrix of dimension n. Since A is invertible, its first column does not vanish identically. Let $\pi_1 = a_{i1}$ be an element of maximal absolute value in the first column of A, and let τ be the transposition which exchanges 1 and i. As the square of a transposition is the identity, we perform an elimination on $P_\tau A$. As in eqn (9.1.6), we let

$$P_\tau A = \widehat{L}_1 \widehat{A}_1,$$

with $\widehat{A}_1$ given by eqn (9.1.5). We infer from the induction assumption that there exists a permutation matrix P_{σ_1}, an upper triangular matrix U_1, and a lower triangular matrix L_1 having ones on the diagonal, such that

$$P_{\sigma_1} A_1 = L_1 U_1.$$

We may now write

$$P_\tau A = \widehat{L}_1 \begin{pmatrix} \pi_1 & \ell' \\ 0 & P_{\sigma_1}^{-1} L_1 U_1 \end{pmatrix}.$$

However, we have the identity

(9.3.7) $P_\tau A = \begin{pmatrix} 1 & 0 \\ p' & I_{n-1} \end{pmatrix} \begin{pmatrix} 1 & 0 \\ 0 & P_{\sigma_1}^{-1} \end{pmatrix} \begin{pmatrix} 1 & 0 \\ 0 & L_1 \end{pmatrix} \begin{pmatrix} \pi_1 & \ell' \\ 0 & U_1 \end{pmatrix}.$

If σ_1 is a permutation on $n - 1$ objects, the permutation ρ on n objects which leaves the first one invariant and acts as σ_1 on the $n - 1$ following objects has the matrix

$$\begin{pmatrix} 1 & 0 \\ 0 & P_{\sigma_1} \end{pmatrix}.$$

It is also the permutation matrix of P_ρ. We premultiply the identity (9.3.7) by P_ρ and we then obtain

$$P_\rho P_\tau A = LU,$$

with

$$L = \begin{pmatrix} 1 & 0 \\ P_{\sigma_1} p' & L_1 \end{pmatrix}, \quad U = \begin{pmatrix} \pi_1 & \ell' \\ 0 & U_1 \end{pmatrix},$$

which proves the theorem.

The decomposition (9.3.6) is not unique, since the choice of transposition τ_j is not generally unique at each step of the elimination. $\qquad\square$

Elimination with partial row pivoting is interpreted by the following analogous theorem:

Theorem 9.3.2. Let A be an $n \times n$ matrix. Then, it is invertible if and only if we can find a permutation σ, a lower triangular matrix L with ones on the diagonal, and an invertible upper triangular matrix U such that

$$(9.3.8) \qquad\qquad\qquad AP_\sigma = LU.$$

The decomposition (9.3.8) is not generally unique. $\qquad\diamond$

Proof. Let $B = A^*$. From Theorem 9.3.1, B is invertible if and only if it has the decomposition $P_\sigma B = LU$. Consequently, A is invertible if and only if it has the following decomposition, obtained on passing to the adjoint:

$$A (P_\sigma)^* = U^* L^*.$$

We have seen that $(P_\sigma)^* = P_{\sigma^{-1}}$. Moreover, let D be the diagonal of U^* and let

$$L' = U^* D^{-1}, \quad U' = DL^*, \quad \sigma' = \sigma^{-1}.$$

Then, A is invertible if and only if it has the decomposition

$$AP_{\sigma'} = L'U'.$$

This is exactly what we wanted to prove. $\qquad\square$

Finally, the total pivoting strategy is interpreted by a decomposition

$$P_\sigma AP_{\sigma'} = LU,$$

the proof of which is left to the reader.

9.3.5. The return of the determinant

The determinant is not used to calculate the solutions of linear systems; it is linear systems which are used to calculate the determinant! More precisely, we obtain the value of the determinant of an invertible square matrix A from the decomposition

$$P_\sigma A = LU,$$

which always exists. By calculating the determinant of each term of this identity, we obtain

$$\det (P_\sigma) \det (A) = \det (L) \det (U) .$$

The determinant of P_σ is equal to the signature of the permutation σ. We determine an integer $m(\sigma)$ as follows: initially, m equals 0, and each time that we actually make a row permutation, in the search for the maximal pivot by column, we add 1 to m. The value of m at the end of the process is $m(\sigma)$. We then have

$$\det (P_\sigma) = (-1)^{m(\sigma)} .$$

The calculation of

$$\det (U) = \prod_{j=1}^{n} U_{jj}$$

is completely trivial and explicit, and $\det(L)$ equals 1. Therefore, we see that

(9.3.9) $$\det (A) = (-1)^{m(\sigma)} \prod_{j=1}^{n} U_{jj}.$$

9.3.6. Banded matrices

A matrix is said to be full if it has 'few' zero coefficients. A matrix which is not full is said to be sparse. The most simple example of a sparse matrix is a banded matrix. This is a matrix which only has nonzero elements on a certain number of diagonals centred around the principal diagonal. We will see in this section that the solution of a linear system for which the matrix is banded is often easier than the solution of a system for which the matrix is full.

We begin with the following definition:

Definition 9.3.3. We say that the matrix A of n rows and n columns is banded if there exists an integer $q \geqslant 1$ such that

$$|i - j| \geqslant q \quad \Longrightarrow \quad A_{ij} = 0.$$

In this case, if we want to specify the integer q, we say that A is a band-q matrix.

If $q = 1$, A is a diagonal matrix. If $q = 2$, A is a tridiagonal matrix of the form

$$A = \begin{pmatrix} A_1 & C_1 & 0 & 0 & \cdots & 0 \\ B_2 & A_2 & C_2 & 0 & & 0 \\ 0 & B_3 & A_3 & C_3 & & 0 \\ \vdots & & \ddots & \ddots & \ddots & \vdots \\ 0 & & & B_{n-1} & A_{n-1} & C_{n-1} \\ 0 & & \cdots & & B_n & A_n \end{pmatrix}.$$

In general, a band-q matrix has at most $2q - 1$ nonzero diagonals.

Such matrices can arise when we discretize partial or ordinary differential equations. We consider an example in which we want to solve

$$-u''(x) = f(x), \quad x \in]0, 1[, \quad u(0) = u(1) = 0.$$

To discretize this problem, we look for an approximation to u at the points jh, with $h = 1/n$ and j going from 1 to $n - 1$. We will denote this approximation to $u(jh)$ by U_j. We replace the second derivative $u''(jh)$ by the finite difference

$$\frac{U_{j+1} - 2U_j + U_{j-1}}{h^2},$$

which is justified by the truncated expansion

$$u(x + h) - 2u(x) + u(x - h) = h^2 u''(x) + O(h^4).$$

We are therefore going to solve the linear problem

$$-\frac{U_{j+1} - 2U_j + U_{j-1}}{h^2} = f(jh), \quad j = 1, \ldots, n - 1.$$

The matrix of this problem is

$$\begin{pmatrix} 2 & -1 & 0 & 0 & \cdots & 0 \\ -1 & 2 & -1 & 0 & & 0 \\ 0 & -1 & 2 & -1 & & 0 \\ \vdots & & \ddots & \ddots & \ddots & \vdots \\ 0 & & & -1 & 2 & -1 \\ 0 & & \cdots & 0 & -1 & 2 \end{pmatrix},$$

which is tridiagonal and symmetric. It has been shown in Subsection 4.6.2 that this matrix is positive definite. Another proof of this fact is given in Chapter 11.

The essential property of banded matrices, from the point of view of LU decomposition, is stated in the following theorem:

Theorem 9.3.4. Let A be a band-q matrix. If A has the decomposition $A = LU$ then L and U are band-q. Due to their particular structure, L and U each have at most q nonzero diagonals. ◇

Proof. It suffices to reread the proof of Theorem 9.1.1, counting the nonzero elements. The matrix $\widehat{L}_1$ is given by

$$\widehat{L}_1 = \begin{pmatrix} 1 & 0 \\ p' & I_{n-1} \end{pmatrix},$$

where p' is defined by

$$p' = \frac{1}{\pi_1} \begin{pmatrix} A_{21} \\ \vdots \\ A_{q1} \\ 0 \\ \vdots \\ 0 \end{pmatrix},$$

and the matrix $\widehat{L}_1$ is therefore banded. In the same way,

$$\ell' = \begin{pmatrix} A_{12} & \cdots & A_{1q} & 0 & \cdots & 0 \end{pmatrix}.$$

As A_1' is band-q and $p'\ell'$ has nonzero elements only for the indices $i, j \leqslant q - 1$, we see that

$$A_1 = A_1' - p'\ell'$$

is also band-q. The new matrix $\widehat{A}_1$ is band-q. By induction, the result is clear. □

Let us look at the advantage, in terms of operation counts, of banded matrices.

Operation Count 9.3.5. Let A be a band-q matrix admitting an LU decomposition. Suppose that $n \gg q$ (which implies, in particular, that $n \gg 1$). The number of operations necessary to construct the LU decomposition is of order $n(2(q-1)^2 + q - 1)$, and the number of operations necessary to solve the two systems

$$Ly = b \quad \text{and} \quad Ux = y$$

is of order $n(4q - 3)$.

Proof. The generic elimination step requires the construction of p^j, which requires $q - 1$ divisions, and the construction of $p^j \ell^j$, which requires $(q - 1)^2$ multiplications, the other elements of this matrix being zero. It is then necessary to subtract $p^j \ell^j$ from $\widehat{A}^{j-1}$, which requires $(q - 1)^2$ subtractions. The construction of LU requires, therefore, at most $(q - 1) + 2(q - 1)^2$ floating-point operations per row. There are at most q non-generic rows, and these require the least number of operations. Since $n \gg q$, the number of operations necessary to construct LU is of order $n((q - 1) + 2(q - 1)^2)$.

The generic row of the system $Ly = b$ has q nonzero coefficients, one of which is equal to 1. Redoing Operation Count 9.2.3, the solution of this system requires

$2q - 2$ operations. Similarly, the generic row of the system $Ux = y$ involves q nonzero coefficients and therefore requires $2q - 1$ floating-point operations for its solution. We thus obtain the count stated.   □

Remark 9.3.6. The preceding operation count shows that about $8n$ operations are required to solve a tridiagonal system A. We can compare this estimate to the number of operations needed to multiply a tridiagonal matrix by a vector x. The j-th component of the result is equal to $B_j x_{j-1} + A_j x_j + C_j x_{j+1}$, which requires 5 floating-point operations. It needs around $5n$ operations to calculate Ax, which is only a little less than to calculate $A^{-1}x$.

Remark 9.3.7. The inverse of a banded matrix A is generally not a banded matrix (exercise). In the case where a band-q matrix admits LU decomposition, it is enormously more advantageous to solve the systems $Ax^k = b^k$, exploiting the LU decomposition, than to calculate A^{-1}. Each calculation of $A^{-1}b^k$ requires $O(2n^2)$ operations, while each solution of $Ly^k = b^k$ and $Ux^k = y^k$ requires $O(n(4q - 3))$ operations. The difference is an order of magnitude.

9.4. Other decompositions: *LDU* and Cholesky

9.4.1. The *LDU* decomposition

We can give a more symmetric character to the LU decomposition:

Definition 9.4.1. A matrix A of n rows and n columns admits an LDU decomposition if there exists a lower triangular matrix L with ones on the diagonal, an invertible diagonal matrix D, and an upper triangular matrix U with ones on the diagonal such that $A = LDU$.

We then have the following easy result:

Lemma 9.4.2. A matrix A admits the LDU decomposition if and only if it admits the LU decomposition. The LDU decomposition is unique.

Proof. The proof of this result is completely parallel to that of Theorem 9.1.1, the only difference being that at each step the diagonal of U is normalized. Details are left to the reader.   □

From the algorithmic point of view, the construction of the LDU decomposition is quite analogous to the construction of the LU decomposition, and the reader can easily modify the constructions of Section 9.1. With the notation (9.1.2) and (9.1.3), we see that the first elimination step consists of writing

$$A = \widehat{L}_1 \begin{pmatrix} \pi_1 & 0 \\ 0 & I_{n-1} \end{pmatrix} \begin{pmatrix} 1 & \pi_1^{-1}\ell' \\ 0 & -p'\ell' + A_1' \end{pmatrix}.$$

With respect to the LU decomposition, the extra cost is the computation of

$\pi_1^{-1}\ell'$, that is $n - 1$ operations. Therefore, the total overhead is

$$\sum_{j=1}^{n-1} j \sim \frac{n^2}{2} \text{ floating-point operations,}$$

which is negligible relative to $2n^3/3$.

When the matrix A is Hermitian and admits an LDU decomposition, we can write

$$LDU = A = A^* = U^*D^*L^*,$$

and deduce from the uniqueness of this decomposition that $L = U^*$ and D is real.

In this case, we calculate about half as many coefficients, and the cost of this decomposition is of order $n^3/3$, for large n. Nevertheless, we have no guarantee that decomposition without pivoting will succeed. An example of a symmetric matrix not possessing an LU decomposition is given by

$$A = \begin{pmatrix} 0 & 1 \\ 1 & 0 \end{pmatrix}.$$

If we use partial pivoting we destroy the symmetric structure of the matrix. If we use total pivoting this will cost a lot more operations. In the section which follows, we consider a case in which we show that it is useless to proceed by pivoting.

9.4.2. The Cholesky method

Recall that a Hermitian matrix A is said to be positive definite if

$$x^*Ax \geqslant 0, \quad \forall x \qquad \text{and} \qquad x^*Ax = 0 \quad \Longrightarrow \quad x = 0.$$

If A is a positive definite matrix, it is invertible: if $x \in \ker A$ then $x^*Ax = 0$ and therefore $x = 0$. The classic decomposition of a symmetric positive definite matrix is described in the following theorem:

Theorem 9.4.3. Let A be a positive definite Hermitian matrix. Then, there exists a unique upper triangular matrix C with positive diagonal such that

$$A = C^*C. \qquad\qquad \diamond$$

Remark 9.4.4. This result shows us that A possesses an LDL^* decomposition. Indeed, let Δ be the diagonal matrix whose diagonal is equal to that of C, and let $L = C^*\Delta^{-1}$. As C has a positive diagonal, Δ is Hermitian. Let $D = \Delta^2$. Then

$$LDL^* = C^*\Delta^{-1}\Delta^2\Delta^{-1}C = C^*C = A.$$

Proof of Theorem 9.4.3. The proof is by induction on the spatial dimension n. For $n = 1$, the matrix A is a single element a, and the vector x a single component ξ. Consequently, on identifying $(\xi)^* = (\bar{\xi})$,

$$x^* A x = (\xi)^* (a) \xi = a |\xi|^2.$$

If A is positive definite, then $a > 0$ and we have a Cholesky decomposition with $C = (\sqrt{a})$.

Suppose that the result is true up to dimension $n - 1$. Let A be an $n \times n$ positive definite Hermitian matrix. Then, it is of the form

$$A = \begin{pmatrix} \alpha & \ell \\ \ell^* & \widehat{A} \end{pmatrix}.$$

It is clear that $\widehat{A}$ is also positive, definite, and Hermitian. We look for an upper triangular matrix C with positive diagonal of the form

$$C = \begin{pmatrix} \beta & m \\ 0 & B \end{pmatrix},$$

where, of course, B will be upper triangular with a positive diagonal. We must then have the identity

$$\begin{pmatrix} \alpha & \ell \\ \ell^* & \widehat{A} \end{pmatrix} = A = C^* C = \begin{pmatrix} \beta^2 & \beta m \\ \beta m^* & B^* B + m^* m \end{pmatrix}.$$

Consequently, we must solve the following nonlinear system in β, m, and B:

(9.4.1) $$\beta^2 = \alpha,$$
(9.4.2) $$\beta m = \ell,$$
(9.4.3) $$m^* m + B^* B = \widehat{A}.$$

We first verify that $\alpha > 0$. Let e_1 be the first vector of the canonical basis. Then, with the block decomposition

$$e_1 = \begin{pmatrix} 1 \\ 0 \end{pmatrix},$$

we have

$$e_1^* A e_1 = \begin{pmatrix} 1 & 0 \end{pmatrix} \begin{pmatrix} \alpha & \ell \\ \ell^* & \widehat{A} \end{pmatrix} \begin{pmatrix} 1 \\ 0 \end{pmatrix} = \begin{pmatrix} \alpha & \ell \end{pmatrix} \begin{pmatrix} 1 \\ 0 \end{pmatrix} = \alpha.$$

We can therefore solve eqn (9.4.1) by choosing

(9.4.4) $$\beta = \sqrt{\alpha}.$$

This relation implies that

(9.4.5) $$m = \frac{\ell}{\sqrt{\alpha}},$$

from eqn (9.4.2). To solve eqn (9.4.3), it suffices to show that $\widehat{A} - m^*m$ is a positive definite Hermitian matrix, and then to use the induction hypothesis. It is clear that we have a Hermitian matrix. To show that it is positive definite, we will show that, for all $\widehat{x} \in \mathbb{C}^{n-1}$ (respectively, $\mathbb{R}^{n-1}$), there exists a $\xi \in \mathbb{C}$ (respectively, $\mathbb{R}$) such that

$$(9.4.6) \qquad \widehat{x}^* \left(\widehat{A} - m^*m \right) \widehat{x} = \begin{pmatrix} \xi^* & \widehat{x}^* \end{pmatrix} A \begin{pmatrix} \xi \\ \widehat{x} \end{pmatrix}.$$

Here, we have identified ξ^* with the complex conjugate of ξ. The left-hand side of eqn (9.4.6) is equal to

$$\widehat{x}^* \widehat{A} \widehat{x} - \widehat{x}^* m^* m \widehat{x}$$

and the right-hand side of eqn (9.4.6) is

$$\alpha \xi^* \xi + \xi^* \ell \widehat{x} + \widehat{x}^* \ell^* \xi + \widehat{x}^* \widehat{A} \widehat{x}.$$

Equating these two expressions, we obtain

$$(9.4.7) \qquad \alpha \left| \xi \right|^2 + \xi^* \ell \widehat{x} + \xi \widehat{x}^* \ell^* + \widehat{x}^* m^* m \widehat{x} = 0.$$

Noting that $(\ell \widehat{x})^* = \widehat{x}^* \ell^* = \overline{\ell \widehat{x}}$, we use the relations (9.4.4) and (9.4.5) to see that eqn (9.4.7) may be written as

$$\left| \alpha \xi + \ell \widehat{x} \right|^2 = 0.$$

Consequently, the choice of ξ is given by

$$\xi = -\frac{\ell \widehat{x}}{\alpha}.$$

It is now clear that eqn (9.4.6) holds for such a ξ, and we deduce immediately that $\widehat{A} - m^*m$ is a positive definite Hermitian matrix. Due to the induction hypothesis, we see that we can find an upper triangular Hermitian matrix B with positive diagonal that satisfies eqn (9.4.3).

We now verify uniqueness. If $A = C_1^* C_1 = C_2^* C_2$, it follows from Remark 9.4.4 that A admits the following two decompositions:

$$A = L_1 D_1 L_1^* = L_2 D_2 L_2^*.$$

Matrix D_1 is the square of the diagonal Δ_1 of C_1 and matrix D_2 is the square of the diagonal Δ_2 of C_2. Furthermore,

$$L_1 = C_1^* \Delta_1^{-1} \quad \text{and} \quad L_2 = C_2^* \Delta_2^{-1}.$$

The uniqueness of the LDL^* decomposition implies that $\Delta_2^2 = \Delta_1^2$ and, as Δ_1 and Δ_2 are diagonal with positive coefficients, they are equal. As $L_1 = L_2$, we see immediately that $C_1 = C_2$. We have therefore shown uniqueness. □

9.4.3. Putting the Cholesky method into practice and operation counts

One step of the Cholesky decomposition consists of calculating

$$\beta = \sqrt{\alpha},$$
$$m = \frac{1}{\beta} \begin{pmatrix} A_{12} & \cdots & A_{1n} \end{pmatrix},$$
$$B^*B = \widehat{A} - m^*m.$$

The calculation of the first row requires the taking of a square root, the calculation of the second requires $n - 1$ divisions, and the calculation of the third requires $(n - 1)^2$ multiplications and $(n - 1)^2$ subtractions. However, there are some redundancies, since $\widehat{A}$ and m^*m are Hermitian. Therefore, we only need to calculate the $n(n - 1)/2$ lower triangular terms. Consequently, a Cholesky step requires the taking of a square root and $(n - 1) + n(n - 1)$ arithmetic operations, giving n^2 arithmetic operations. The final count is described as follows, after summation over all the steps:

Operation Count 9.4.5. The Cholesky decomposition for a matrix A of n rows and n columns requires n square roots and a number of arithmetic operations of order $n^3/3$.

The Cholesky method for solving a linear system with positive definite Hermitian matrix consists of determining the Cholesky decomposition of the matrix and then solving the two triangular systems, which has negligible cost compared with the decomposition, at least if the matrix is not a narrow-banded matrix. In the case of a banded matrix, we can combine the advantages of a Cholesky decomposition with the particular properties of banded matrices.

Remark 9.4.6. The Cholesky decomposition presents two advantages. We are assured that a Hermitian matrix admits a decomposition as a product of two triangular matrices without pivoting. Furthermore, the method requires almost two times fewer arithmetic operations than the non-Hermitian case. The 'almost' corresponds to the cost of the n square roots, which take more or less time according to the implementation.

9.5. Exercises from Chapter 9

9.5.1. Exercises on the rank of systems of vectors

Exercise 9.5.1. What is the rank of the system of vectors:

$$\begin{pmatrix} 1 \\ 2 \\ -1 \end{pmatrix}, \quad \begin{pmatrix} 3 \\ 6 \\ -3 \end{pmatrix}, \quad \begin{pmatrix} 3 \\ 9 \\ 3 \end{pmatrix}, \quad \begin{pmatrix} 2 \\ 5 \\ 0 \end{pmatrix}.$$

Exercise 9.5.2. Give the kernel and the image of the matrix A which has the preceding vectors as columns.

Hint: the aim of the exercise is to find a method which does not use determinants. It is particularly recommended to reason geometrically.

Exercise 9.5.3. Let U be the subspace of $\mathbb{R}^5$ generated by the vectors

$$\begin{pmatrix} 1 \\ 3 \\ -2 \\ 2 \\ 3 \end{pmatrix}, \quad \begin{pmatrix} 1 \\ 4 \\ -3 \\ 4 \\ 2 \end{pmatrix}, \quad \begin{pmatrix} 2 \\ 3 \\ -1 \\ -2 \\ 9 \end{pmatrix},$$

and V the subspace of $\mathbb{R}^5$ generated by the vectors

$$\begin{pmatrix} 1 \\ 3 \\ 0 \\ 2 \\ 1 \end{pmatrix}, \quad \begin{pmatrix} 1 \\ 5 \\ -6 \\ 6 \\ 3 \end{pmatrix}, \quad \begin{pmatrix} 2 \\ 5 \\ 3 \\ 2 \\ 1 \end{pmatrix}.$$

Find a basis of U, V, $U + V$, and $U \cap V$.

Hint: the beginning of this exercise is identical to the preceding one. To find a basis of $U + V$ and a basis of $U \cap V$, geometric reasoning is required.

9.5.2. Echelon matrices and least-squares

In this section in three parts, we generalize the *LDU* decomposition to the case of rectangular matrices and we use this generalization to solve general linear systems by the least-squares method.

Let $V = \mathbb{C}^n$, $W = \mathbb{C}^m$, and A be a matrix of m rows and n columns. We denote by A^* the adjoint of A defined by

$$(A^*)_{ij} = \bar{A}_{ji}.$$

We define the following vector subspaces:

(9.5.1)
$$\begin{aligned} V_1 &= \ker A &&= \{x \in V : Ax = 0\}, \\ V_2 &= \operatorname{Im} A^* &&= A^*W, \\ W_1 &= \operatorname{Im} A &&= AV, \\ W_2 &= \ker A^* &&= \{x \in W : A^*x = 0\}. \end{aligned}$$

We equip V and W with their respective canonical scalar products, that is,

$$(x, x') = \sum_{i=1}^{n} x_i \bar{x}'_i \quad \text{if } x, x' \in V,$$

$$(x, x') = \sum_{i=1}^{m} x_i \bar{x}'_i \quad \text{if } x, x' \in W.$$

These scalar products are denoted identically, but are distinct.

First part: elementary linear algebra

Exercise 9.5.4. Show that V is the direct orthogonal sum of V_1 and V_2, and that W is the direct orthogonal sum of W_1 and W_2.

Exercise 9.5.5. Give a necessary and sufficient condition on the spaces (9.5.1) so that the system of linear equations

$$(9.5.2) \qquad\qquad Ax = b$$

possesses at least one solution.

Exercise 9.5.6. Give a necessary and sufficient condition on the spaces (9.5.1) so that system (9.5.2) has at most one solution.

Exercise 9.5.7. We define a linear mapping A_0 from V_2 to W_1 by

$$(9.5.3) \qquad\qquad A_0 x = Ax, \quad \forall x \in V_2.$$

Find the kernel and the image of A_0.

Exercise 9.5.8. Compare the kernel of A and the kernel of A^*A, and deduce that

$$\dim \operatorname{Im} A = \dim \operatorname{Im} A^*A.$$

The common value of these two dimensions is the rank of the matrix A and it will be denoted by r. Warning! Matrix A is generally not square.

Second part: echelon matrices and the generalization of Gaussian elimination

We call a matrix A echelon if it has the following properties:

$$A_{ij} = 0 \quad \text{if } j < f(i), \qquad A_{i,f(i)} \neq 0 \quad \text{for all } i \text{ such that } f(i) \leqslant n.$$

Here, f is a function from $\mathbb{N}$ into $\mathbb{N}$ which satisfies

$$f(i+1) \geqslant f(i) + 1,$$

and $f(1)$ is arbitrary. We say that an element $A_{i,f(i)}$ is a pivot (nonzero by definition) of A.

Exercise 9.5.9. Give an example of a nonzero rectangular ($m \neq n$) echelon matrix.

Exercise 9.5.10. Count the maximum number of nonzero coefficients that row i could have. Show that, if A has k nonzero rows, these k nonzero rows are the first k rows of the matrix and they are independent.

Exercise 9.5.11. How many independent columns does A possess? (Use Exercise 9.5.8). Determine a system of independent columns of maximum dimension. You can make use of the pivots.

Exercise 9.5.12. Let A be some matrix. Show that there exists a permutation σ_1 on $\{1, \ldots, m\}$ such that

$$P_{\sigma_1} A = G_1 D_1,$$

where the matrix P_{σ_1} of the permutation is defined by

$$(P_{\sigma_1})_{ij} = \delta_{i, \sigma_1(j)}$$

and the matrices G_1 and D_1 are $m \times m$ and $m \times n$ matrices, respectively, of the form

$$
G = \begin{pmatrix}
1 & 0 & 0 & 0 & \cdots & 0 \\
* & 1 & 0 & 0 & & 0 \\
* & 0 & 1 & 0 & & 0 \\
* & 0 & 0 & 1 & & 0 \\
\vdots & \vdots & & & & \vdots \\
* & 0 & & & \cdots & 1
\end{pmatrix}, \quad
D = \begin{pmatrix}
d_{11} & * & \cdots & * \\
0 & * & \cdots & * \\
\vdots & \vdots & \cdots & \vdots \\
0 & * & \cdots & *
\end{pmatrix}.
$$

Here, d_{11} is zero or nonzero, and the asterisks represent arbitrary numbers.

Exercise 9.5.13. Using Exercise 9.5.11 as the first stage of an induction, show that there exists a permutation τ of $\{1, \ldots, m\}$ such that

$$(9.5.4) \qquad\qquad\qquad P_\tau A = LU,$$

where L is a square $m \times m$ matrix with ones on the diagonal and U is an $m \times n$ echelon matrix. Show how this decomposition permits the solution of the system (9.5.2) when the condition of Exercise 9.5.5 is satisfied.

Exercise 9.5.14. Show that there exist matrices $\hat{L}$ and $\hat{U}$ of dimension $m \times r$ and $r \times n$, respectively, such that

$$(9.5.5) \qquad\qquad\qquad P_\tau A = \hat{L}\hat{U}.$$

Here, r is the rank of A.

Exercise 9.5.15. Show that $\hat{U}\hat{U}^*$ and $\hat{L}^*\hat{L}$ are invertible.

Exercise 9.5.16. We define $B = \hat{L}(\hat{L}^*\hat{L})^{-1}\hat{L}^*$. Show that

$$B^2 = B, \qquad ((I - B)x, Bx) = 0, \quad \forall x,$$

and conclude that B is the orthogonal projection on Im $\hat{L}$.

Exercise 9.5.17. Show that Im $\hat{L} = $ Im $P_\tau A$.

Third part: solution of a linear system in the least-squares sense

Exercise 9.5.18. Suppose that the sufficient condition for existence in Exercise 9.5.5 is not satisfied. We then seek to minimize

$$\phi(x) = \|Ax - b\|^2.$$

Show that, if x minimizes ϕ on V, then

(9.5.6) $\Re(Ax, Ay) - \Re(b, Ay) = 0, \quad \forall y \in V.$

Derive a linear system satisfied by x. You can replace y by iy in eqn (9.5.6).

Exercise 9.5.19. Give a necessary and sufficient condition such that the preceding linear system has a solution.

Exercise 9.5.20. Show that, even if the system (9.5.6) possesses more than one solution, Ax is unique.

Exercise 9.5.21. If the system does not satisfy the uniqueness condition of Exercise 9.5.6, let $\hat{x}$ be such that

$$\|\hat{x}\| = \min\{\|x\| : \ x \text{ minimizes } \phi \text{ on } V\}.$$

Show that this $\hat{x}$ satisfies

$$A\hat{x} \in \operatorname{Im} A \quad \text{and} \quad \hat{x} \in \operatorname{Im} A^*.$$

Express $\hat{x}$ as a function of B, the projection of $\operatorname{Im} A$ to W, and A_0.

Exercise 9.5.22. Suppose that $P_r = I$ in the decomposition (9.5.5). Show that

$$\hat{x} = \hat{U}^* \left(\hat{U}\hat{U}^*\right)^{-1} \left(\hat{L}^*\hat{L}\right)^{-1} \hat{L}^* b$$

satisfies the conditions (9.5.6).

Exercise 9.5.23. Let

$$A^+ = \hat{U}^* \left(\hat{U}\hat{U}^*\right)^{-1} \left(\hat{L}^*\hat{L}\right)^{-1} \hat{L}^*.$$

What are the values of AA^+ and A^+A?

9.5.3. The conditioning of a linear system

We consider the linear system

$$Ax = b,$$

with A an invertible square matrix. We perturb the data A and b by δA and δb, respectively, which results in x being perturbed by δx. We then have

$$(A + \delta A)(x + \delta x) = b + \delta b.$$

In the following we denote a vector norm by $|\cdot|$ and the subordinate operator norm of this vector norm by $\|\cdot\|$; see Section 10.3.4 for a definition.

We define the condition number $\kappa(A)$ by writing

$$\kappa(A) = \|A\|\,\|A^{-1}\|.$$

Exercise 9.5.24. Show that, if $\|\delta A\| \leqslant 1/\|A^{-1}\|$, then

$$\frac{|\delta x|}{|x|} \leqslant \frac{\kappa(A)}{1 - \kappa(A)\,\|\delta A\|\,/\,\|A\|}\left(\frac{|\delta b|}{|b|} + \frac{\|\delta A\|}{\|A\|}\right).$$

Exercise 9.5.25. Show that the condition number of A is bounded below by the ratio $|\lambda_n/\lambda_1|$, where λ_n is the largest eigenvalue of A in modulus and λ_1 is the smallest eigenvalue in modulus.

Exercise 9.5.26. Show that, if $|\cdot|$ is the Euclidean norm (or Hermitian norm, in the complex case) and the matrix A is normal, then the condition number of A for this norm is precisely the ratio between the absolute maximum and minimum eigenvalues.

Exercise 9.5.27. Give examples of matrices of condition number 1.

Exercise 9.5.28. Give an example which proves that the condition number depends on the chosen vector norm. Compare the condition number of a rotation matrix in $\mathbb{R}^3$ for the norms 1, 2, and ∞. These norms are defined in eqns (10.2.3), (10.2.4), and (10.2.5), respectively.

Exercise 9.5.29. Let

$$A = \begin{pmatrix} 1 & 1 \\ 1 & 1.0001 \end{pmatrix}.$$

Calculate the eigenvalues of A and give $\kappa(A)$ for a norm of your choice, which you should define carefully. Solve the two systems

$$Ax = \begin{pmatrix} 2 \\ 2 \end{pmatrix}, \quad A(x + \delta x) = \begin{pmatrix} 2 \\ 2.0001 \end{pmatrix}.$$

Compare the relative variation of the solution $|\delta x|/|x|$ with that of the right-hand sides of the systems. What do you conclude?

Exercise 9.5.30. Let

$$A = \begin{pmatrix} 1 & 100 \\ 0 & 1 \end{pmatrix}.$$

Compare the solutions of the systems

$$Ax = \begin{pmatrix} 100 \\ 1 \end{pmatrix} \quad \text{and} \quad A(x + \delta x) = \begin{pmatrix} 100 \\ 0 \end{pmatrix}.$$

What is the amplication factor of the error?

9.5.4. Inverting persymmetric matrices

We denote the transpose of some matrix B by $B^\top$.

In the following, $\mathcal{M}_n$ denotes the space of real square matrices of n rows and n columns. The identity matrix of $\mathcal{M}_n$ will be written I_n.

Exercise 9.5.31. Let E_n be the element of $\mathcal{M}_n$ defined by

$$(E_n)_{ij} = \delta_{n+1-i,j}$$

or

$$E_n = \begin{pmatrix} 0 & 0 & 0 & 0 & \cdots & 0 & 1 \\ 0 & 0 & 0 & 0 & \cdots & 1 & 0 \\ \vdots & & & & \reflectbox{$\ddots$} & & \vdots \\ 0 & 0 & 0 & 1 & \cdots & 0 & 0 \\ 0 & 0 & 1 & 0 & \cdots & 0 & 0 \\ 0 & 1 & 0 & 0 & \cdots & 0 & 0 \\ 1 & 0 & 0 & 0 & \cdots & 0 & 0 \end{pmatrix}.$$

We say that an element B of $\mathcal{M}_n$ is persymmetric if it satisfies

$$B = E_n B^\top E_n.$$

Show that persymmetric matrices form a vector subspace of $\mathcal{M}_n$, denoted $\mathcal{P}_n$. What is the dimension of this subspace of $\mathcal{M}_n$? Show that the inverse of a regular persymmetric matrix is also persymmetric. Does the subspace $\mathcal{P}_n$ form an algebra for the multiplication of matrices?

Exercise 9.5.32. We say that a matrix is Toeplitz if it is of the form

$$A = \begin{pmatrix} r_0 & r_1 & r_2 & \cdots & r_{n-1} \\ r_{-1} & r_0 & r_1 & \cdots & r_{n-2} \\ r_{-2} & r_{-1} & r_0 & \cdots & r_{n-3} \\ \vdots & \vdots & \vdots & \ddots & \vdots \\ r_{-n+1} & r_{-n+2} & r_{-n+3} & \cdots & r_0 \end{pmatrix}.$$

Show that the inverse of a Toeplitz matrix is generally not a Toeplitz matrix. To do this, consider the matrix of finite differences given by

$$A_n = \begin{pmatrix} 2 & -1 & 0 & \cdots & 0 & 0 \\ -1 & 2 & -1 & \cdots & 0 & 0 \\ 0 & -1 & 2 & \cdots & 0 & 0 \\ \vdots & & & \ddots & & \vdots \\ 0 & 0 & 0 & \cdots & 2 & -1 \\ 0 & 0 & 0 & \cdots & -1 & 2 \end{pmatrix}.$$

Show that there does not exist a Toeplitz matrix B such that

$$B A_n = I_n.$$

We denote by $\mathcal{T}_n$ the set of positive definite symmetric Toeplitz matrices and we only consider elements of $\mathcal{T}_n$ in the following. We propose to solve a system whose matrix is an element of $\mathcal{T}_n$. Without loss of generality, we can suppose that $r_0 = 1$. We obtain a series $(1, r_1, \ldots, r_{n-1})$ of real numbers such that the corresponding symmetric Toeplitz matrix T_n is positive definite. We denote by T_k the element of $\mathcal{T}_k$ obtained from T_n by taking the first k columns and the first k rows. This is, therefore, the symmetric Toeplitz matrix corresponding to the series $(1, r_1, \ldots, r_{k-1})$.

Exercise 9.5.33. Let
$$R_k = (r_1, r_2, \ldots, r_k)^\top .$$

Suppose that we know how to solve the system
$$T_k y_k = -R_k.$$

Calculate the block product
$$\begin{pmatrix} I_k & E_k y_k \\ 0 & 1 \end{pmatrix}^\top \begin{pmatrix} T_k & E_k R_k \\ R_k^\top E_k & 1 \end{pmatrix} \begin{pmatrix} I_k & E_k y_k \\ 0 & 1 \end{pmatrix}$$

and deduce that $1 + R_k^\top y_k$ is strictly positive, by using the fact that T_{k+1} is positive definite.

Exercise 9.5.34. Calculate the solution $(z^\top \ \alpha_k)^\top$ of the system
$$\begin{pmatrix} T_k & E_k R_k \\ R_k^\top E_k & 1 \end{pmatrix} \begin{pmatrix} z \\ \alpha_k \end{pmatrix} = \begin{pmatrix} -R_k \\ -r_{k+1} \end{pmatrix}.$$

We can express z as a function of y_k and α_k, and then substitute into the equation in α_k.

Exercise 9.5.35. Show that the calculations of the preceding question need $O(k)$ floating-point operations.

Exercise 9.5.36. Let
$$\beta_k = 1 + R_k^\top y_k.$$

Show that
$$\beta_k = \left(1 - \alpha_{k-1}^2\right)\beta_{k-1}.$$

Exercise 9.5.37. Give an algorithm, using the β_k, to calculate the solution of the problem
$$T_n y_n = -R_n$$

in $O(n^2)$ multiplications or divisions.

Exercise 9.5.38. We propose now to solve the system having the matrix T_n and an arbitrary right-hand side b. Suppose that we possess the solutions of the systems
$$T_k x_k = B_k \quad \text{and} \quad T_k y_k = -R_k,$$

where B_k denotes a column vector formed from the first k rows of the column vector b. Calculate the solution of

$$\begin{pmatrix} T_k & E_k R_k \\ R_k^\top E_k & 1 \end{pmatrix} \begin{pmatrix} \nu_k \\ \mu_k \end{pmatrix} = \begin{pmatrix} B_k \\ b_{k+1} \end{pmatrix},$$

whose right-hand side is the vector B_{k+1}.

Exercise 9.5.39. Give an algorithm based on the solution 'in parallel' of the two systems

$$T_k x_k = B_k = (b_1, \ldots, b_k)^\top \quad \text{and} \quad T_k y_k = -R_k = -(r_1, \ldots, r_k)^\top.$$

Exercise 9.5.40. Show that this algorithm leads to a solution in $2n^2$ multiplications or divisions, plus lower-order terms.

10

Theoretical interlude

The rest of the book requires some supplementary knowledge of matrix analysis. This chapter contains various information which is of a more analytical than algebraic nature: properties of eigenvalues, matrix norms, spectral radius etc.

Just as the do-it-yourself expert finds himself with dozens of tools spilled over the ground to do even the simplest repair, the mathematician needs an entire workshop to understand matrices.

We are therefore on holiday (provisionally) from numerical analysis and, one by one, we will construct the tools which will allow us to return to it.

10.1. The Rayleigh quotient

We begin with a definition:

Definition 10.1.1. Let A be a Hermitian matrix in a space of dimension n. For $x \neq 0$, let

$$(10.1.1) \qquad r_A(x) = \frac{x^* A x}{x^* x}.$$

The function r_A is called the Rayleigh quotient associated with A.

The Rayleigh quotient of A is linked to the spectrum of A. This relation is expressed by the two following theorems:

Theorem 10.1.2. Suppose that the eigenvalues of the Hermitian matrix A are arranged in increasing order:

$$\lambda_1 \leqslant \cdots \leqslant \lambda_n.$$

Then,

$$\max_{x \neq 0} r_A(x) = \lambda_n \quad \text{and} \quad \min_{x \neq 0} r_A(x) = \lambda_1. \qquad \diamond$$

240

Proof. Let $(e_j)_{1 \leqslant j \leqslant n}$ be an orthonormal basis of eigenvectors of A corresponding to the eigenvalues λ_j. In this basis, x has the decomposition

$$x = \sum_{j=1}^{n} x_j e_j$$

and we can therefore write

$$r_A(x) = \frac{\sum_{j=1}^{n} \lambda_j x_j^2}{\sum_{j=1}^{n} x_j^2}.$$

It is evident that we have

$$\lambda_1 \sum_{j=1}^{n} x_j^2 \leqslant \sum_{j=1}^{n} \lambda_j x_j^2 \leqslant \lambda_n \sum_{j=1}^{n} x_j^2$$

and, therefore, we have the inequalities

$$\lambda_1 \leqslant r_A(x) \leqslant \lambda_n, \quad \forall x \neq 0.$$

If we take $x = e_1$, then $x^* A x = \lambda_1$, and if we take $x = e_n$, then $x^* A x = \lambda_n$. $\square$

The second theorem allows us to get all the eigenvalues of A from the minimax properties of the Rayleigh quotient.

Theorem 10.1.3. Under the conditions of Theorem 10.1.2 we have

(10.1.2) $$\min_{\dim W = k} \max_{x \in W \setminus \{0\}} r_A(x) = \lambda_k,$$

(10.1.3) $$\max_{\dim W = k} \min_{x \in W \setminus \{0\}} r_A(x) = \lambda_{n-k+1}.$$ ◇

Proof. The second formula comes from the first on passing from A to $-A$, which swaps the maximum and minimum and requires the renumbering of the eigenvalues. We will therefore content ourselves with proving the first formula. First of all, let $\bar{W}$ be the vector space generated by $e_1, \ldots, e_k$. It is of dimension k and, by application of Theorem 10.1.2, we see that

$$\max_{x \in \bar{W} \setminus \{0\}} r_A(x) = \lambda_k.$$

This shows us that

$$\inf_{\dim W = k} \max_{x \in W \setminus \{0\}} r_A(x) \leqslant \lambda_k.$$

Conversely, let W be some subspace of dimension k. If Z is the space of dimension $n - k + 1$ which is generated by $e_k, \ldots, e_n$, the intersection of W and Z cannot be reduced to zero. If it was reduced to zero then W and Z would be in direct sum, and the sum of their dimensions is greater than the dimension of the space, which is absurd. Therefore, let $z \in W \cap Z$, $z \neq 0$. It is clear that

$$r_A(z) \geqslant \min_{z \in Z \setminus \{0\}} r_A(z) = \lambda_k.$$

Consequently,

$$\max_{x \in W \setminus \{0\}} r_A(x) \geqslant r_A(y) \geqslant \lambda_k.$$

Since this property holds for every W, we see that

$$\inf_{\dim W = k} \max_{x \in W \setminus \{0\}} r_A(x) = \lambda_k.$$

It follows, from the beginning of the proof, that this lower bound is attained. $\square$

10.2. Spectral radius and norms

10.2.1. Spectral radius

Definition 10.2.1. The spectral radius $\rho(A)$ of a matrix A is the maximum modulus of its eigenvalues.

The following result is a consequence of this definition and of the theorems relating to the Rayleigh quotient:

Lemma 10.2.2. Let A be a Hermitian matrix. Then,

$$(10.2.1) \qquad\qquad \rho(A) = \max_{x \neq 0} |r_A(x)|.$$

The proof is left to the reader.

A classic and subtle exercise consists of showing that, for every square $n \times n$ matrices A and B,

$$(10.2.2) \qquad\qquad \rho(AB) = \rho(BA).$$

It goes without saying, that we make no commutativity hypothesis on A and B. We are going to show a stronger result in fact: the spectrum of AB is identical to the spectrum of BA. Indeed, let (λ, x) be a pair of eigenvalue and eigenvector of AB. We therefore have $ABx = \lambda x$, or, on pre-multiplying this relation with B, $BABx = \lambda Bx$. Suppose, first of all, that $Bx = y \neq 0$. Then y is an eigenvector of BA for the eigenvalue λ. If $Bx = 0$ then from the relation $ABx = \lambda x$ we deduce that $\lambda x = 0$, and therefore, B is not invertible. This implies that BA is not invertible: the rank of BA is $\dim \operatorname{Im} BA \leqslant \dim \operatorname{Im} B < N$. Therefore 0 is in the spectrum of BA. We have therefore shown that the spectrum of BA contains the spectrum of AB. The converse is obvious.

10.2.2. Norms of vectors, operators, and matrices

Let V be a vector space of dimension n over $\mathbb{K}$.

Definition 10.2.3. The norm $|x|$ of a vector x is a mapping from V to $\mathbb{R}^+$ which satisfies the following properties:

  (i) If $|x| = 0$, then $x = 0$;

 (ii) $|\lambda x| = |\lambda|\,|x|$, $\forall x \in V$, $\forall \lambda \in K$ (homogeneity);

(iii) $|x + y| \leqslant |x| + |y|$, $\forall x, y \in V$ (triangle inequality).

We give some examples of norms: if x is a vector with components $(x_j)_{1 \leqslant j \leqslant n}$ we define

$$(10.2.3) \qquad |x|_1 = \sum_{j=1}^{n} |x_j|,$$

which has already been defined in Exercise 2.6.2,

$$(10.2.4) \qquad |x|_2 = \left[\sum_{j=1}^{n} |x_j|^2 \right]^{1/2},$$

which is the Euclidean norm, and

$$(10.2.5) \qquad |x|_\infty = \max_{1 \leqslant j \leqslant n} |x_j|,$$

which is the maximum norm, or ℓ^∞ norm. It is simple to show that these expressions all define norms.

Let $p \in [1, \infty[$. The expression

$$(10.2.6) \qquad |x|_p = \left[\sum_{j=1}^{n} |x_j|^p \right]^{1/p}$$

is a norm. See the exercises for a proof.

Another example is constructed from a Hermitian positive definite matrix A. Let

$$(10.2.7) \qquad |x|_A = (x^* A x)^{1/2}.$$

This norm is deduced from a scalar product; it can be written as the sum of the square of the coordinates, with respect to a basis which is orthogonal in the canonical basis but usually not orthonormal. If A coincides with the identity matrix, then we fall back on the usual Euclidean norm. We now move on to the definition of a matrix norm.

Definition 10.2.4. A matrix norm is a norm $\| \cdot \|$ on a vector space of square matrices, which satisfies the following algebra property:

$$(10.2.8) \qquad \|AB\| \leqslant \|A\|\,\|B\|.$$

10.3. Topology and norms

10.3.1. Topology refresher

We assume familiarity with the definition of a metric space, and the essential topological properties of a metric space: open and closed sets, neighbourhoods, continuity, compactness, Cauchy sequences, and complete metric spaces.

Let E_j be spaces equipped with a distance d_j ($1 \leqslant j \leqslant n$). The topology on the product space $E = \prod_{j=1}^{n} E_j$ is defined by the product distance below

$$d\left(x,y\right) = \max_{1 \leqslant j \leqslant n} d_j\left(x_j, y_j\right).$$

A product of complete metric spaces is complete (for the product distance). A product of compact sets is compact.

It is obvious that a norm $\|\cdot\|$ on a vector space defines a distance by

$$d\left(x,y\right) = \|x - y\|.$$

On the field $\mathbb{K} = \mathbb{R}$ (respectively, $\mathbb{K} = \mathbb{C}$) the distance between x and y is the absolute value (respectively, the modulus) of $x - y$. The product topology on $V = \mathbb{K}^n$ is given by

$$d\left(x,y\right) = \max_{1 \leqslant j \leqslant n} |x_j - y_j|.$$

Therefore, the product topology on V is defined by the norm $|x|_\infty$, and this is the only topology which we will consider from now on.

Let $\|\cdot\|$ be some norm on V. We have the following inequalities, where $(e_j)_{1 \leqslant j \leqslant n}$ is the canonical basis of V:

$$(10.3.1) \qquad \|x\| = \left\| \sum_{j=1}^{n} x_j e_j \right\| \leqslant \sum_{j=1}^{n} |x_j| \, \|e_j\| \leqslant |x|_\infty \sum_{j=1}^{n} \|e_j\|.$$

We immediately deduce from this that all norms are continuous.

10.3.2. Equivalence of norms

Recall that a subset C of V is compact if and only if it is closed and bounded. This remark allows us to deduce the following essential result:

Lemma 10.3.1. All norms on $V = \mathbb{K}^n$ are equivalent. In other words, if N_1 and N_2 are two norms on $V = \mathbb{K}^n$, there exist constants $\gamma > 0$ and $\Gamma \geqslant \gamma$ such that

$$(10.3.2) \qquad \gamma N_1\left(x\right) \leqslant N_2\left(x\right) \leqslant \Gamma N_1\left(x\right), \quad \forall x \in \mathbb{K}^n.$$

Proof. Let S be the unit sphere for the norm $|\cdot|_\infty$; it is closed and bounded for the topology of V, and hence compact. Relation (10.3.1) shows that N_1 is continuous over V. Therefore, it attains its minimum and its maximum over S:

$$\gamma_1 \leqslant N_1(x) \leqslant \Gamma_1, \quad \forall x \in S.$$

Furthermore, γ_1 cannot be zero, since there exists $y \in S$ such that $\gamma_1 = N_1(y)$. If γ_1 were zero, then the definition of the norm would imply that y would be zero, and we could not have $|y|_\infty = 1$. If $x \neq 0$, we see that

$$\gamma_1 \leqslant N_1\left(\frac{x}{|x|_\infty}\right) \leqslant \Gamma_1.$$

Consequently,

$$\gamma_1 |x|_\infty \leqslant N_1(x) \leqslant \Gamma_1 |x|_\infty.$$

Similarly, we have the following inequality for N_2:

$$\gamma_2 |x|_\infty \leqslant N_2(x) \leqslant \Gamma_2 |x|_\infty.$$

If we let $\gamma = \gamma_2/\Gamma_1$ and $\Gamma = \Gamma_2/\gamma_1$, we obtain the conclusion of the lemma. $\square$

We give a geometric interpretation of this result. Note that the unit ball for a norm N on $\mathbb{K}^n$ is a convex closed bounded subset of $\mathbb{K}^n$ which is invariant under the transformations $x \mapsto \lambda x$ if $|\lambda| = 1$. In Figure 10.1 we represent the unit ball for some norm in $\mathbb{R}^2$.

It is possible to prove that, for every convex closed bounded subspace C of non-empty interior, which is invariant under the transformation $x \mapsto \lambda x$ for every $|\lambda| = 1$, there exists a norm N such that

$$C = \{x \in \mathbb{K}^n : N(x) \leqslant 1\}.$$

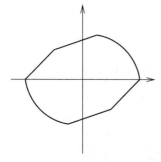

Figure 10.1: An arbitrary unit ball.

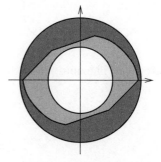

Figure 10.2: The previous ball is included in a round ball and contains a round ball.

This norm is given by the relation

$$N\left(x\right) = \inf\left\{r \in \mathbb{R}^+ \,:\, x \in rC\right\}.$$

The equivalence of all norms means that we can insert the unit ball of a norm N between scalings of the unit ball of another norm, as is indicated in Figure 10.2. We also give some representations in $\mathbb{R}^2$ of unit balls for the classic norms. See Figures 10.3 to 10.6.

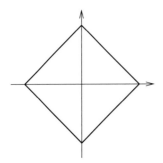

Figure 10.3: The unit ball of ℓ^1.

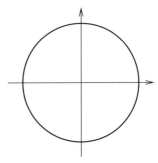

Figure 10.4: The unit ball of ℓ^2.

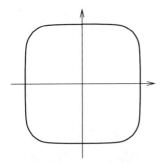

Figure 10.5: The unit ball of ℓ^p, $p = 5$.

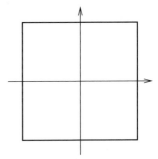

Figure 10.6: The unit ball of ℓ^∞.

Figure 10.7: The unit ball for the norm $|\cdot|_A$.

Finally, to represent the norm $|\cdot|_A$ we let A be the diagonal matrix

$$A = \begin{pmatrix} \lambda_1 & 0 \\ 0 & \lambda_2 \end{pmatrix}.$$

The perimeter of the unit ball for this norm is an ellipse of semi-axes $1/\lambda_1$ and $1/\lambda_2$. Figure 10.7 is representative of the case where $\lambda_2 \gg \lambda_1$.

10.3.3. Linear mappings: continuity, norm

Let V be a vector space of dimension n over $\mathbb{K}$. V is isomorphic to $\mathbb{K}^n$: if we choose a basis e_j in V, the isomorphism is a mapping ϕ which associates a vector x to the n-tuple of its coordinates in the basis e_j. We transport the topology of $\mathbb{K}^n$ to V by this isomorphism, by defining a norm N on V:

$$N(x) = |\phi(x)|_\infty.$$

It is immediately clear that N is a norm. With the basis $(f_k)_{1 \leqslant k \leqslant n}$ in V, we define a different isomorphism ψ, which will correspond to a different norm,

$$M(x) = |\psi(x)|_\infty.$$

Since all norms are equivalent on $\mathbb{K}^n$, these two norms M and N are equivalent, and the topology of V does not depend on the choice of basis which is used to make the isomorphism.

We give a concrete example of this phenomenon: let V be the vector space of real polynomials of degree at most n. We describe two different coordinate systems. The first consists of taking for a basis the monomials $1, X^1, X^2, \ldots, X^n$. We obtain the coordinates of a polynomial P in this basis by the following formula:

$$x_j = \frac{P^{(j)}(0)}{j!}, \quad j = 0, 1, \ldots, n,$$

which is simply Taylor's formula. To make the second coordinate system, we fix n pairwise distinct real points, $\xi_0, \xi_1, \ldots, \xi_n$. In the second coordinate system,

$$y_j = P(\xi_j), \quad j = 0, 1, \ldots, n.$$

The corresponding basis of V is given by

$$\phi_k(X) = \prod_{j \neq k} \frac{X - \xi_j}{\xi_k - \xi_j},$$

which are the classical Lagrange interpolation polynomials. We easily verify that

$$\phi_k(\xi_j) = \delta_{jk}.$$

If we equip $\mathbb{R}^n$ with the maximum norm, we thus construct two equivalent but different norms. The first is

$$\max_{0\leqslant j\leqslant n}\left|\frac{P^{(j)}(0)}{j!}\right|$$

and the second is

$$\max_{0\leqslant j\leqslant n}\left|P(\xi_j)\right|.$$

Lemma 10.3.2. All linear mappings from a finite-dimensional space over $\mathbb{K}$ to a finite-dimensional space over $\mathbb{K}$ are continuous.

Proof. From the preceding discussion, the study of the continuity of linear mappings in finite dimensions is equivalent to the study of the continuity of $x \mapsto Ax$ from $\mathbb{K}^m$ to $\mathbb{K}^n$, where A is an $m \times n$ matrix. We have the following obvious inequality:

$$|Ax|_\infty = \max_{1\leqslant i\leqslant m}\left|\sum_{j=1}^{n}A_{ij}x_j\right|$$

$$\leqslant \max_{1\leqslant i\leqslant m}\sum_{j=1}^{n}|A_{ij}|\,|x_j| \leqslant |x|_\infty \max_{1\leqslant i\leqslant m}\sum_{j=1}^{n}|A_{ij}|.$$

Therefore, there exists a constant C such that

$$|Ax|_\infty \leqslant C\,|x|_\infty\,, \quad \forall x \in \mathbb{K}^m.$$

We immediately deduce continuity.   □

10.3.4. Subordinate norms

In the preceding proof, the choice of norm $|\cdot|_\infty$ is a question of convenience. It is clear that, if M and N are norms given on the finite-dimensional spaces V and W, respectively, then, for every linear mapping f from V to W, there exists a constant C such that

$$N\left(f\left(x\right)\right) \leqslant CM\left(x\right).$$

Theorem 10.3.3. Let V and W be vector spaces of finite dimension over $\mathbb{K}$, and let M and N be the norms on V and W, respectively. Then, the lower bound $\|f\|_{M,N}$ of numbers C such that

$$N\left(f\left(x\right)\right) \leqslant CM\left(x\right)$$

is a norm on the vector space of linear mappings from V to W and, furthermore, it satisfies

$$(10.3.3) \quad \|f\|_{M,N} = \max_{x\neq 0}\frac{N\left(f\left(x\right)\right)}{M\left(x\right)} = \max_{M(x)\leqslant 1} N\left(f\left(x\right)\right) = \max_{M(x)=1} N\left(f\left(x\right)\right).$$

We say that the norm $\|f\|_{M,N}$ is the operator norm of f, defined from the norms M and N.   ◇

Proof. We begin with the last expression: let $S(V)$ be the unit sphere in V for the norm M:

$$S(V) = \{x \in V : M(x) = 1\}.$$

As we saw previously, $S(V)$ is a compact subset of V. Therefore, the continuous function $x \mapsto N(f(x))$ attains its maximum on $S(V)$. Let Γ be this maximum. It follows that

$$\max_{M(x)=r} N(f(x)) = \Gamma r, \quad \forall r \in \mathbb{R}^+,$$

which is a consequence of the homogeneity of the norm. From this we deduce that

$$\sup_{M(x)\leqslant 1} N(f(x)) = \sup_{r\in[0,1]} \sup_{M(x)=r} N(f(x)) = \Gamma.$$

It is clear that this upper bound is attained on $S(V)$. Then, if $x \neq 0$, $N(f(x))/M(x) = N\big(f(x/M(x))\big)$. The element $x/M(x)$ is in $S(V)$ and we again have

$$\Gamma = \max_{x\neq 0} \frac{N(f(x))}{M(x)}.$$

It is obvious that $N(f(x)) \leqslant \Gamma M(x)$ for every x in V and therefore, $\|f\|_{M,N} \leqslant \Gamma$. Conversely, since the upper bounds are attained, there exists a $y \neq 0$ such that $N(f(y)) = \Gamma M(y)$ and therefore, $\|f\|_{M,N} \geqslant \Gamma$.

It remains to verify that $\|f\|_{M,N}$ is a norm. Positivity and homogeneity are obvious. The triangle inequality is true as

$$\max_{M(x)=1} N(f(x) + g(x)) \leqslant \max_{M(x)=1} N(f(x)) + \max_{M(x)=1} N(g(x)).$$

Finally, if $\|f\|_{M,N} = 0$, $N(f(x)) = 0$ for every x, and therefore, $f = 0$. □

In the particular case of an endomorphism f of V (that is, a linear mapping from V into itself), we say that the norm of $\|f\|_{M,M}$ defined from a vector norm M is subordinate to the norm M. For simplicity, we will denote this operator norm $\|f\|_M = \|f\|_{M,M}$.

Lemma 10.3.4. Let $\mathcal{M}_n$ be the n^2-dimensional vector space of $n \times n$ matrices, and let N be some vector norm on $\mathbb{K}^n$. Then the operator norm $\|\cdot\|_N$, subordinate to the vector norm N, is a matrix norm.

Proof. It suffices to verify that if $A, B \in \mathcal{M}_n$, then

$$(10.3.4) \qquad \|AB\|_N \leqslant \|A\|_N \|B\|_N.$$

Now, we have

$$N(ABx) \leqslant \|A\|_N N(Bx) \leqslant \|A\|_N \|B\|_N N(x).$$

The result is therefore clear. □

10.3.5. Examples of subordinate norms

We will calculate the corresponding subordinate norms for some of the vector norms defined previously:

Lemma 10.3.5. Let A be an $n \times n$ real or complex matrix. The subordinate operator norms $\|A\|_p$ to the norm ℓ^p, for $p = 1$, 2, and ∞, are given by

$$(10.3.5) \qquad \|A\|_1 = \max_{1 \leqslant j \leqslant n} \sum_{i=1}^{n} |A_{ij}|,$$

$$(10.3.6) \qquad \|A\|_\infty = \max_{1 \leqslant i \leqslant n} \sum_{j=1}^{n} |A_{ij}|,$$

and

$$(10.3.7) \qquad \|A\|_2 = \sqrt{\rho\,(A^*A)} = \sqrt{\rho\,(AA^*)}.$$

Proof. By definition

$$\|A\|_1 = \max_{|x|_1 = 1} |Ax|_1 \,.$$

Now,

$$\begin{aligned}
|Ax|_1 &= \sum_{i=1}^{n} \left| \sum_{j=1}^{n} A_{ij} x_j \right| \\
&\leqslant \sum_{i=1}^{n} \sum_{j=1}^{n} |A_{ij}|\,|x_j| \\
&\leqslant \sum_{j=1}^{n} |x_j| \max_k \sum_{i=1}^{n} |A_{ik}| \,.
\end{aligned}$$

Therefore, we see that

$$\|A\|_1 \leqslant \max_j \sum_{i=1}^{n} |A_{ij}| \,.$$

In fact, we have equality here: choose k such that

$$\sum_{i=1}^{n} |A_{ik}| = \max_j \sum_{i=1}^{n} |A_{ij}|$$

and $x_j = \delta_{jk}$. Then,

$$\|Ax\|_1 = \sum_{i=1}^{n} \left| \sum_{j=1}^{n} A_{ij} \delta_{jk} \right| = \sum_{i=1}^{n} |A_{ik}| \,.$$

We have therefore shown that

$$\|A\|_1 = \max_{1 \leqslant j \leqslant n} \sum_{i=1}^{n} |A_{ij}|.$$

Similarly,

$$\|A\|_\infty = \max_{|x|_\infty = 1} |Ax|_\infty.$$

We easily verify that

$$\|A\|_\infty \leqslant \max_{1 \leqslant i \leqslant n} \sum_{j=1}^{n} |A_{ij}|.$$

Conversely, let k be such that

$$(10.3.8) \qquad \sum_{j=1}^{n} |A_{kj}| = \max_{1 \leqslant i \leqslant n} \sum_{j=1}^{n} |A_{ij}|.$$

Suppose that the quantity (10.3.8) is not zero. We choose x such that

$$x_j = \begin{cases} |A_{kj}|/A_{kj} & \text{if } A_{kj} \neq 0; \\ 0 & \text{otherwise.} \end{cases}$$

Then, the reader may verify that, for such an x, $|x|_\infty = 1$ and

$$|Ax|_\infty = \sum_{j=1}^{n} |A_{kj}|.$$

This shows us that

$$\|A\|_\infty = \max_{1 \leqslant i \leqslant n} \sum_{j=1}^{n} |A_{ij}|.$$

The case of the Euclidean norm is a little different, since we do not generally have an explicit expression for $\|A\|_2$. Indeed,

$$\|A\|_2^2 = \max_{x \neq 0} \frac{|Ax|_2^2}{|x|_2^2}.$$

But $|Ax|_2^2 = (Ax)^* Ax = x^* A^* Ax$ and, furthermore, $|x|_2^2 = x^* x$. We recognize a Rayleigh quotient. To get its value it would be necessary to calculate the largest eigenvalue of $A^* A$. It is immediate that $A^* A$ is Hermitian, positive, and semi-definite and, therefore,

$$\|A\|_2^2 = \rho(A^* A),$$

from Lemma 10.2.2. Using the equality $\rho(A^* A) = \rho(AA^*)$, we see that

$$\|A\|_2 = \sqrt{\rho(A^* A)} = \sqrt{\rho(AA^*)},$$

which proves the lemma. □

10.3.6. The Frobenius norm is not subordinate

We have seen that all operator norms are matrix norms (which have the algebra property). There are matrix norms which are not subordinate to any vector norm. An example is the Frobenius norm of a matrix, defined by

$$F(A) = \sqrt{\operatorname{trace}(A^*A)}.$$

Recall that the trace of a square matrix B is the sum of the diagonal elements. We have

$$(A^*A)_{ii} = \sum_{j=1}^{n} (A^*)_{ij} A_{ji} = \sum_{j=1}^{n} \bar{A}_{ji} A_{ji} = \sum_{j=1}^{n} |A_{ji}|^2.$$

Consequently,

$$F(A)^2 = \sum_{i,j=1}^{n} |A_{ji}|^2.$$

The Frobenius norm is, therefore, nothing other than the Euclidean norm of A seen as a vector of $\mathbb{K}^{n^2}$. It is therefore clear that this is a norm. We verify the algebra property:

$$F(AB)^2 = \sum_{i,k} |(AB)_{ik}|^2 = \sum_{i,k} \left| \sum_j A_{ij} B_{jk} \right|^2 = \sum_{i,k} \sum_{j,\ell} A_{ij} B_{jk} \overline{A_{i\ell}}\, \overline{B_{\ell k}}.$$

Let

$$u_{j\ell} = \sum_i A_{ij} \bar{A}_{i\ell} \quad \text{and} \quad v_{j\ell} = \sum_k B_{jk} \overline{B_{\ell k}}.$$

Then, by virtue of the Cauchy–Schwartz inequality,

$$F(AB)^2 = \sum_{j,\ell} u_{j\ell} v_{j\ell} \leqslant \left(\sum_{j,\ell} |u_{j\ell}|^2 \right)^{1/2} \left(\sum_{j,\ell} |v_{j\ell}|^2 \right)^{1/2}.$$

But

$$|u_{j\ell}|^2 = \left| \sum_i A_{ij} \overline{A_{i\ell}} \right|^2 \leqslant \sum_i |A_{ij}|^2 \sum_p |A_{p\ell}|^2.$$

Consequently,

$$\sum_{j,\ell} |u_{j\ell}|^2 \leqslant \sum_{i,j,\ell,p} |A_{ij}|^2 \sum_p |A_{p\ell}|^2 = F(A)^4.$$

We can derive an analogous formula for B, and we see that

(10.3.9) $$F(AB)^2 \leqslant F(A)^2 F(B)^2.$$

We will show that the Frobenius norm is not an operator norm (subordinate to a vector norm), indeed, the Frobenius norm of the identity has value $\sqrt{n}$. On the other hand, for any vector norm N, $\|I\|_N = 1$. Therefore, if $n \geqslant 2$, the Frobenius norm is not an operator norm. However, the Frobenius norm gives a useful estimate of $\|A\|_2$, and it is convenient because it is explicit. Indeed, we know that if A^*A has eigenvalues μ_j which are all positive or zero and arranged in increasing order

$$\mu_1 \leqslant \cdots \leqslant \mu_n,$$

then

$$\operatorname{trace}(A^*A) = \sum_{j=1}^{n} \mu_j$$

and

$$\rho(A^*A) = \mu_n.$$

We therefore have the estimate

$$\rho(A^*A) \leqslant \operatorname{trace}(A^*A) \leqslant n\rho(A^*A),$$

which implies that

(10.3.10) $$\|A\|_2 \leqslant F(A) \leqslant \sqrt{n}\,\|A\|_2.$$

It remains to see a simple and useful application of the equivalence of norms to the convergence of sequences of matrices.

Lemma 10.3.6. The following assertions are equivalent for a sequence of matrices B_k belonging to $\mathcal{M}_{m,n}$:

(i) $(B_k)_{k \in \mathbb{N}}$ converges in $\mathcal{M}_{m,n}$ equipped with some norm N;

(ii) Each of the sequences $(B_k)_{ij}$ converges in $\mathbb{K}$;

(iii) For every $x \in \mathbb{K}^n$, $B_k x$ converges in $\mathbb{K}^m$.

Proof. All vector norms are equivalent on a finite-dimensional vector space. The space $\mathcal{M}_{m,n}$ is equipped with a norm N and we let

$$\|B\|_{\ell^\infty} = \max_{\substack{1 \leqslant i \leqslant m \\ 1 \leqslant j \leqslant n}} |B_{ij}|,$$

which is also a norm on $\mathcal{M}_{m,n}$. Therefore, there exists a constant $\gamma > 1$ such that

$$\gamma^{-1} N(B) \leqslant \|B\|_{\ell^\infty} \leqslant \gamma N(B).$$

It is therefore clear that (i) is equivalent to (ii). To show that (ii) implies (iii) we note that, if B_k tends to a certain limit B, there is convergence for the norm $\|\cdot\|_\infty$, in particular. Then,

$$\left|B_k x - Bx\right|_\infty \leqslant \|B_k - B\|_\infty \left|x\right|_\infty$$

and, therefore, $B_k x$ tends towards Bx.

Conversely, from (iii) we deduce that $B_k e_j$ tends to a certain limit f_j, for every element e_j of the canonical basis of $\mathbb{K}^n$. As $B_k e_j$ is the j-th column of B_k, we see that $(B_k)_{ij}$ converges. We can therefore apply (ii), which is equivalent to (i). $\qquad\square$

10.4. Exercises from Chapter 10

10.4.1. Continuity of the eigenvalues of a matrix with respect to itself

We say that a norm $\| \cdot \|$ on $\mathbb{K}^n$ ($\mathbb{K} = \mathbb{R}$ or $\mathbb{C}$) satisfies the Δ property if the operator norm that is subordinate to it satisfies the following condition: for every diagonal matrix of the form

$$
D = \operatorname{diag}(\lambda_1, \lambda_2, \ldots, \lambda_n) = \begin{pmatrix} \lambda_1 & 0 & & \cdots & & 0 \\ 0 & \lambda_2 & 0 & & & \\ \vdots & & \ddots & \ddots & \ddots & \vdots \\ & & & 0 & \lambda_{n-1} & 0 \\ 0 & & \cdots & & 0 & \lambda_n \end{pmatrix},
$$

the norm of D is given by

$$
\|D\| = \max_{1 \leqslant i \leqslant n} |\lambda_i| .
$$

All of the matrices considered in this section are square and of order n.

Exercise 10.4.1. We denote by x_i the i-th component of the vector $x \in \mathbb{K}^n$. Show that the following norms satisfy the Δ condition:

$$
\|x\|_p = \left(\sum_{i=1}^{p} |x_i|^p \right)^{1/p} , \quad 1 \leqslant p < \infty,
$$

$$
\|x\|_\infty = \max_{1 \leqslant i \leqslant n} |x_i| .
$$

In all that follows, the norm $\| \cdot \|$ satisfies the Δ property.

Exercise 10.4.2. Calculate the operator norm of

$$
(\mu I_n - \operatorname{diag}(\lambda_1, \lambda_2, \ldots, \lambda_n))^{-1}
$$

when μ is not equal to any of the λ_i.

Exercise 10.4.3. Let A be a diagonalizable matrix, whose eigenvalues are denoted by λ_i, with i going from 1 to n, and let P be the corresponding transformation matrix

$$
P^{-1} A P = \operatorname{diag}(\lambda_1, \lambda_2, \ldots, \lambda_n) = D.
$$

Let E be some matrix. Show that, for every eigenvalue μ of $A + E$, we have the inequality

$$\min_{1 \leqslant i \leqslant n} |\mu - \lambda_i| \leqslant \|P\| \, \|P^{-1}\| \, \|E\|.$$

To do this, denote an eigenvector, associated to the eigenvalue λ of $A + E$, by x and bound from below the norm of $(\mu - A)^{-1} E$, by studying the action of this operator on x.

Exercise 10.4.4. Deduce from the preceding question that the eigenvalues of a matrix are continuous with respect to the matrix. Give as precise a formulation as possible of this assertion.

Exercise 10.4.5. Suppose that the matrix A is normal. Show that we can choose a norm satisfying the Δ property such that, for every eigenvalue μ of $A + E$,

$$\min_{1 \leqslant i \leqslant n} |\mu - \lambda_i| \leqslant \|E\|.$$

10.4.2. Various questions on norms

Exercise 10.4.6. Let $p \in \,]1, \infty[$, and let q be defined by

$$\frac{1}{p} + \frac{1}{q} = 1.$$

Show that, for every α and β in $\mathbb{R}^+$,

$$\alpha\beta \leqslant \frac{\alpha^p}{p} + \frac{\beta^q}{q}.$$

Exercise 10.4.7. For x in $\mathbb{K}^n$, let

(10.4.1) $$|x|_p = \left(\sum_{j=1}^n |x_j|^p \right)^{1/p}.$$

Deduce, from the preceding equation, the Hölder inequality

$$\sum_{j=1}^n |x_j y_j| \leqslant |x|_p \, |y|_q.$$

Exercise 10.4.8. Show that we have the Minkowski inequality

$$|x + y|_p \leqslant |x|_p + |y|_p.$$

Verify that eqn (10.4.1) defines a norm on $\mathbb{K}^n$.

Exercise 10.4.9. Let $\mathbb{P}_n$ be the vector space of polynomials of degree at most n with coefficients in $\mathbb{K}$. Given $m \geqslant n + 1$ distinct points $(x_j)_{1 \leqslant j \leqslant m}$ of $\mathbb{R}$. Show that

$$|P|_{\text{inter}} = \max_{1 \leqslant j \leqslant m} |P(x_j)|$$

defines a norm on $\mathbb{P}_n$.

Exercise 10.4.10. Let $\mathbb{T}_N$ be the space of complex trigonometric polynomials of degree at most N and period 1. This is formed of functions from $\mathbb{R}$ to $\mathbb{C}$ of the form

$$f(x) = \sum_{|k| \leqslant N} a_j e^{2i\pi jx}.$$

We equip $\mathbb{T}_N$ with the norms

(10.4.2) $$|f|_k = \left(\sum_{|j| \leqslant N} |a_j|^2 \sum_{p=0}^{k} (2\pi j)^{2p} \right)^{1/2},$$

for $k \geqslant 0$. Quickly check that the expressions (10.4.2) do define norms. We denote by D the differentiation operator from $\mathbb{T}_N$ to itself. Calculate the norms $\|D\|_{k+1,k}$ and $\|D\|_{k,k}$, for every positive or zero k.

11

Iterations and recurrence

We know, from Chapter 9, that elimination can be used to solve linear systems in a finite number of machine operations. However, there are at least two non-elementary difficulties. The first one is that the LU decomposition is approximately as bulky as the original matrix, and the Cholesky is about half as bulky as the original decomposition. However, when discretizing partial differential equations, we do not even create the matrix as an object in the computer memory. We are content to describe it by its action on vectors, and for that we do not need the whole matrix in memory, we just need an algorithm.

The second reason is that, even when it is convenient to create the matrix on the computer, either as an array of numbers or as a sparse matrix, i.e., by giving only the indices and values of its nonzero coefficients, we have to deal with another fact: solving a triangular system is not very efficient on parallel machines, since we need all the previous results at any given step before we go on to the next step. But the essential factor which slows down parallel computations is the communication time between processors. Solving a triangular system is an essentially sequential task and, therefore, it creates a bottle-neck on a parallel machine. Hence, it is important to have alternative ways of solving linear systems.

The alternative is to devise an iterative solution, i.e., to replace a process in finite terms by a process in infinite terms. This looks like a terribly awkward thing to do. We lose the safety of algebra to go into the realm of (approximative) analysis. In fact, this process is applied mainly to very large matrices, which come usually from an ordinary or partial differential system: we know that we have committed an error when discretizing our problem; therefore, if we do not solve our problem exactly, but within an acceptable error, we can safely assume that this is enough for all of the purposes which we have in mind.

It remains to see how to construct such methods.

The following methods which are presented in this chapter: Jacobi, Gauss–Seidel, and over-relaxation are somewhat *passé*, since, nowadays, the favoured iterative method for solving a linear system is the conjugate gradient in the

Hermitian case, and its generalizations in other cases. And, to be completely true, the conjugate gradient method is a method in finite terms, which is used as an iterative method: if n denotes the spatial dimension, it can be proved that the conjugate gradient method stops after n iterations; however, in practice, only $m \ll n$ iterations are used.

However, there are two important ideas which are—still—widely used in practice. Firstly, multigrid approximations are based on refined versions of Jacobi's method. The second is that, with the use of a pre-conditioner, i.e., an approximate inverse, iterative methods can be made extremely efficient.

Therefore, some of the more refined and modern iterative methods are treated in the problems section at the end of the chapter: Richardson's pre-conditioned method, gradient and conjugate gradient methods, and an initiation to multi-grids. A much more extensive treatment is given in Canuto *et al.* [14].

Another reason for studying iterative methods is that they are discrete models of differential equations, and they display, indeed, many of the phenomena found in differential equations, including exponential growth or decay. Then, they are termed linear recurrences, and they are, in fact, the key to the understanding of numerical schemes for ordinary differential equations and, in particular, of multistep methods. Finally, they give us a few explicit, or almost explicit, solutions of linear difference equations, which are extremely useful to understand what is going on in the discretization of linear partial differential equations. Therefore, they are extremely important building blocks.

11.1. Iterative solution of linear systems

Let A be an $n \times n$ matrix. Suppose that

$$(11.1.1) \qquad\qquad A = M - N,$$

where M is an invertible matrix. In practice, we assume that the system with matrix M is easy to solve, for example, if M is diagonal, tridiagonal or triangular. We define a sequence of vectors x^k by a given initial vector x^0 and a recurrence relation:

$$(11.1.2) \qquad\qquad M x^{k+1} = N x^k + b.$$

Suppose that the sequence x^k is convergent. Then, if $x^\infty = \lim_{k\to\infty} x^k$, we have the relation

$$M x^\infty = N x^\infty + b \quad \Longleftrightarrow \quad A x^\infty = b.$$

In other words, if the sequence of x^k converges, then its limit is the solution of the linear system $Ax = b$.

We will ask ourselves several questions about this sequence. Is it convergent? At what speed does it converge? What choices can we make when decomposing A in the form of eqn (11.1.1)?

Two examples of iterative methods

For the following two examples we suppose that A is an invertible matrix which has no zero diagonal element.

The i-th row of the Jacobi method is written

$$(11.1.3) \quad a_{i1}x_1^k + \ldots + a_{i,i-1}x_{i-1}^k + a_{ii}x_i^{k+1} + a_{i,i+1}x_{i+1}^k + \ldots + a_{in}x_n^k = b_i.$$

Knowing x^k, we find the value of x^{k+1} by solving each row. We can even solve these n equations in parallel.

The Gauss–Seidel method is a modification of the Jacobi method which consists of using the values $x_1^{k+1}, \ldots, x_{i-1}^{k+1}$ calculated previously in the i-th equation. Consequently, this i-th row is written

$$(11.1.4) \quad a_{i1}x_1^{k+1} + \ldots + a_{i,i-1}x_{i-1}^{k+1} + a_{ii}x_i^{k+1} + a_{i,i+1}x_{i+1}^k + \ldots + a_{in}x_n^k = b_i.$$

We will determine the matrices M and N in each of the cases examined. For this we define the following decomposition of A:

$$(11.1.5) \qquad\qquad A = D - E - F,$$

where D, E, and F are given by

$$(11.1.6) \qquad\qquad D_{ij} = \delta_{ij}A_{jj},$$

$$(11.1.7) \qquad\qquad E_{ij} = \begin{cases} -A_{ij} & \text{if } i > j; \\ 0 & \text{otherwise,} \end{cases}$$

and

$$(11.1.8) \qquad\qquad F_{ij} = \begin{cases} -A_{ij} & \text{if } i < j; \\ 0 & \text{otherwise.} \end{cases}$$

Thus D is a diagonal matrix, whose diagonal elements are those of A, $-E$ is lower triangular with 0 on the diagonal (in a way, this is the lower triangular part of A), and $-F$ is upper triangular with 0 on the diagonal (the upper triangular part of A). With this notation the Jacobi method is written as

$$Dx^{k+1} - (E + F)x^k = b,$$

which corresponds to the choice $M = D$ and $N = E + F$. The Gauss–Seidel method is written as

$$(D - E)x^{k+1} = Fx^k + b,$$

which corresponds to the choice $M = D - E$ and $N = F$.

Elementary theory of the convergence of iterative methods

We begin with the following elementary result:

Theorem 11.1.1. Let an iterative method be defined by eqns (11.1.1) and (11.1.2). Suppose that there exists a vector norm $|\cdot|$ such that, for the corresponding subordinate norm denoted $\|\cdot\|$, we have

(11.1.9) $$\|M^{-1}N\| < 1.$$

Then, for every initial x^0, sequence (11.1.2) converges.   ◇

Proof. Let $B = M^{-1}N$ and $c = M^{-1}b$, and let $k \geqslant 0$. Then it is equivalent to write eqn (11.1.2) as

(11.1.10) $$x^{k+1} = Bx^k + c.$$

Consequently, we have

$$x^{k+1} - x^k = Bx^k + c - Bx^{k-1} - c = \cdots = B^r \left(x^{k+1-r} - x^{k-r} \right)$$

for any $r \leqslant k$. It follows that

$$x^{k+1} - x^k = B^k \left(x^1 - x^0 \right).$$

Consequently, for $k > \ell$,

$$x^k - x^\ell = \sum_{j=\ell}^{k-1} x^{j+1} - x^j = \sum_{j=\ell}^{k-1} B^j \left(x^1 - x^0 \right),$$

where by convention, $B^0 = I$. The triangle inequality allows us to write

$$\left| x^k - x^\ell \right| \leqslant \sum_{j=\ell}^{k-1} \|B\|^j \left| x^1 - x^0 \right| \leqslant \frac{\|B\|^\ell}{1 - \|B\|} \left| x^1 - x^0 \right|.$$

This proves that $(x^k)_{k \geqslant 0}$ is a Cauchy sequence, and therefore that it converges.
      We can also deduce the convergence result from the fixed point theorem. The mapping $x \mapsto Bx + c$ from $\mathbb{K}^n$ to itself, equipped with the norm $|\cdot|$ is a strict contraction. Therefore, it has a unique fixed point which is obtained as the limit of the sequence of iterations (11.1.2).   □

Note that we have obtained a convergence result, which is a result of a topological nature, from a hypothesis on the norm of the matrix A, which is a hypothesis of a metrical nature. This situation is not, in itself, scandalous, but we are going to show in what follows that we can link topological information to information which does not depend on the norm.

Detailed theory of the convergence of iterative methods

The answer to the question posed at the start of this section is summarized in the following result:

Theorem 11.1.2. The iterative method (11.1.2) converges for every initial x^0 and for every vector b if and only if $\rho(M^{-1}N) < 1$. ◇

This result is often not very applicable since the norms appearing in Theorem 11.1.1 are a lot easier to calculate than the spectral radius. Nevertheless, Theorem 11.1.2 is a notable theorem. It is a corollary to the theorem given below:

Theorem 11.1.3. Let A be an $n \times n$ matrix with coefficients in $\mathbb{C}^n$. The following two assertions are equivalent:

(i) For any x, $A^k x$ tends to 0 as k tends to infinity;

(ii) The spectral radius $\rho(A)$ is strictly less than 1. ◇

Proof. We show first that (i) implies (ii). Let λ be an eigenvalue of A and let x be a corresponding eigenvector (of course, x is not a null vector). We have

$$A^k x = \lambda^k x.$$

It is clear that $|\lambda| < 1$.

Conversely, recall the following first year result: for every square matrix A there exists an invertible matrix P, a diagonal matrix D, and an upper triangular matrix N with zero diagonal and commuting with D, such that

$$A = P^{-1}(D + N)P.$$

It is clear that the diagonal of the matrix D consists only of the eigenvalues of A. Calculating A^k:

$$A^k = \underbrace{\left(P^{-1}(D+N)P\right)\left(P^{-1}(D+N)P\right)\cdots\left(P^{-1}(D+N)P\right)}_{k \text{ identical factors}}$$

$$= P^{-1}(D+N)^k P.$$

Therefore, it suffices to calculate $(D+N)^k$, which the binomial formula enables us to do as follows, due to the commutation hypothesis:

(11.1.11)
$$(D + N)^k = \sum_{j=0}^{k} C_k^j\, D^{k-j} N^j.$$

The reader may verify that N is nilpotent: $N^n = 0$, since at each multiplication the number of zero diagonals increases by one. Consequently, the sum in eqn (11.1.11) is a sum of at most n terms for $k \geqslant n$:

$$\sum_{j=0}^{n-1} C_k^j\, D^{k-j} N^j.$$

Let $\|\cdot\|$ be a matrix norm such that $\|D\| = \rho(D) = \rho(A)$. This is the case for the norms subordinate to $|\cdot|_p$ for $1 \leqslant p \leqslant \infty$. We therefore have the upper bound

$$(11.1.12) \qquad \left\|(D+N)^k\right\| \leqslant \sum_{j=0}^{n-1} C_k^j \, \rho\,(D)^{k-j} \, \|N\|^j \,.$$

It is then clear that, if $\rho(D) < 1$, $(D+N)^k$ tends to zero as k tends to infinity. In particular, examination of the inequality (11.1.12) shows that there exists a constant $C < 1$ and a constant D, such that

$$(11.1.13) \qquad \left\|A^k\right\| \leqslant C^k D. \qquad\qquad \square$$

From this result, we immediately deduce Theorem 11.1.2:

Proof of Theorem 11.1.2. Suppose that $\rho(M^{-1}N) < 1$. Then, with the notation of Theorem 11.1.1, $B = M^{-1}N$ and $c = M^{-1}b$. We have

$$x^k = B^k x^0 + \sum_{j=0}^{k-1} B^j c.$$

Relation (11.1.13) shows that $B^k x^0$ tends to 0 geometrically and that the sequence of partial sums $\sum_{j=0}^{n-1} B^j c$ tends to the sum of the convergent series $\sum_{j=0}^{\infty} B^j c$. To show the converse, we first of all consider the complex case. If $\rho(A) \geqslant 1$, there exists an $x \neq 0$ such that $Ax = \lambda x$ with $|\lambda| = \rho(A)$. If we begin with $x^0 = 0$ and $c = x$, the sequence of x_k is given by

$$x_k = \left(\sum_{j=0}^{k-1} \lambda^k\right) x,$$

which diverges.

If we consider the real case, two possibilities present themselves: either A possesses a real eigenvalue λ, whose absolute value is equal to $\rho(A)$, in which case we return to the preceding situation, or every eigenvalue of A of modulus $\rho(A)$ is complex. In the latter case, they appear as complex conjugate pairs. We define an operator $\tilde{A}$ on $\mathbb{C}^n$ by the following process of complexification[1]: if $x \in \mathbb{R}^n$ and $y \in \mathbb{R}^n$ we let $z = x + iy$ which we identify with the vector $(x_j + iy_j)_{1 \leqslant j \leqslant n}$ from $\mathbb{C}^n$. The operator $\tilde{A}$ operates on $\mathbb{C}^n$ by

$$\tilde{A}\,(x + iy) = Ax + iAy.$$

The reader may verify, as an exercise, that $\tilde{A}$ is $\mathbb{C}$-linear. The characteristic polynomial of A is the same as that of $\tilde{A}$, and therefore, they have the same

---

[1] For the reader familiar with the notion of complexifying a real vector space, we return to the preceding case by placing ourselves in the complexified space, see [59].

eigenvalues. Let $\lambda = \mu + i\nu = re^{i\theta}$, $\theta \neq k\pi$ be an eigenvalue of A such that $\rho(A) = |\lambda|$. Therefore, there exists a vector $z = x + iy \neq 0$, such that

$$\tilde{A}(x + iy) = (\mu + i\nu)(x + iy).$$

We have

$$\tilde{A}^k(x + iy) = \lambda^k z.$$

We deduce, from this, that

$$A^k x = \Re\big(\tilde{A}^k(x + iy)\big) = \frac{1}{2}\big(\lambda^k z + \bar{\lambda}^k \bar{z}\big).$$

Similarly,

$$A^k y = \Im\big(\tilde{A}^k(x + iy)\big) = \frac{1}{2i}\big(\lambda^k z - \bar{\lambda}^k \bar{z}\big).$$

Since $\Im\lambda \neq 0$, we verify that $y \neq 0$. If we let $c = y$ and $x^0 = 0$, we see that

$$x^k = \sum_{j=0}^{k-1} A^j y = \frac{1}{2i}\sum_{j=0}^{k-1}\lambda^j z - \bar{\lambda}^j \bar{z},$$

which diverges as n tends to infinity. $\qquad\square$

Comparison of the norm of a matrix and its spectral radius

We begin by bounding below the norms of matrices by means of the spectral radius:

Lemma 11.1.4. Let A be an $n \times n$ matrix with complex coefficients and let $\|\cdot\|$ be some matrix norm on $\mathcal{M}_n(\mathbb{C})$. Then $\rho(A) \leqslant \|A\|$.

Proof. Since we are considering the complex case, we know that A possesses an eigenvector for every eigenvalue λ whose modulus is equal to $\rho(A)$. Let x be such an eigenvector. If the norm that we are considering is a norm which is subordinate to N, it is sufficient to write

$$\rho(A)N(x) = N(Ax) \leqslant \|A\|_N N(x),$$

and therefore, $\|A\|_N \geqslant \rho(A)$. We are going to prove this result for a matrix norm in the following way: the product xx^* is a nonzero $n \times n$ matrix. We have

$$Ax = \lambda x,$$

and therefore, on right multiplying by x^*,

$$Axx^* = \lambda xx^*.$$

Hence, on using the algebra property,

$$|\lambda|\,\|xx^*\| \leqslant \|A\|\,\|xx^*\|.$$

This gives us the desired result. $\qquad\square$

To bound matrix norms from above we restrict ourselves to subordinate norms and we begin with a result pertaining to matrices with a spectral radius less than 1:

Lemma 11.1.5. Let A be an $n \times n$ matrix with coefficients in $\mathbb{K}$ such that $\rho(A) < 1$. For all vector norms M, it is possible to construct a vector norm N, dependent on M and A, such that

$$\|A\|_N < 1.$$

Proof. Let

$$(11.1.14) \qquad\qquad N(x) = \sum_{j=0}^{\infty} M\left(A^j x\right).$$

This expression is well defined for all x by virtue of relation (11.1.13), which assures the geometric convergence of the series which defines N. We verify that we have really defined a norm: it is immediate that $N(x) \geqslant 0$ for every x. Let λ be a scalar, then

$$N(\lambda x) = \sum_{j=0}^{\infty} M\left(A^j \lambda x\right) \sum_{j=0}^{\infty} |\lambda| \, M\left(A^j x\right) = |\lambda| \, N(x).$$

Finally, if x and y are vectors,

$$N(x+y) = \sum_{j=0}^{\infty} M\left(A^j (x+y)\right) \leqslant \sum_{j=0}^{\infty} M\left(A^j x\right) + M\left(A^j y\right) = N(x) + N(y).$$

If $N(x)$ vanishes, then the first term in the series defining N vanishes, and therefore, $M(x) = 0$, which implies that x vanishes. We therefore see that N is a norm on $\mathbb{K}^n$. For $N(x) = 1$, we calculate $N(Ax)$:

$$(11.1.15) \qquad N(Ax) = \sum_{j=0}^{\infty} M\left(A^{j+1} x\right) = \sum_{j=1}^{\infty} M\left(A^j x\right) = 1 - M(x).$$

As a result

$$(11.1.16) \qquad\qquad \|A\|_N = \max_{N(x)=1} N(Ax) = \max_{N(x)=1} (1 - M(x)).$$

The function M attains its minimum on the compact set $\{x : \ N(x) = 1\}$ and this minimum is not zero. Consequently, we have proved that $\|A\|_N < 1$. $\qquad\square$

From the preceding result, we are going to obtain a precise lower bound on the spectral radius in terms of a well-chosen subordinate norm:

Theorem 11.1.6. For any matrix $A \in \mathcal{M}_n(\mathbb{K})$ and any $\epsilon > 0$, there exists a vector norm N dependent on A and on ϵ, such that

$$(11.1.17) \qquad\qquad \|A\|_N \leqslant \rho(A) + \epsilon. \qquad\qquad\qquad \diamond$$

Proof. Let $B = A(\rho(A) + \epsilon)^{-1}$. It is clear that $\rho(B) < 1$. We can therefore apply the preceding theorem to B and there exists an N, such that

$$\|B\|_N < 1.$$

Consequently,

$$\|A\|_N = \|B\|_N \left(\rho(A) + \epsilon\right) < \rho(A) + \epsilon.$$

This gives us the result. $\qquad\qquad\qquad\qquad\qquad\qquad\qquad\qquad\qquad\square$

We will now construct an example of a vector norm N for which $\|A\|_N < 1$ when

$$A = \begin{pmatrix} 1/2 & 10^{25} \\ 0 & 1/2 \end{pmatrix}.$$

We choose

$$N(x) = a\,|x_1| + |x_2|.$$

Let $(1/2) + \epsilon = \mu \in \,]1/2, 1[$ and seek an a, such that

$$a\left|\frac{x_1}{2} + 10^{25}x_2\right| + \left|\frac{x_2}{2}\right| \leqslant \mu\left(a\,|x_1| + |x_2|\right).$$

It suffices to verify that

$$\frac{a}{2}\,|x_1| + 10^{25}a\,|x_2| + \frac{|x_2|}{2} \leqslant \left(\frac{a}{2} + \epsilon a\right)|x_1| + \left(\frac{1}{2} + \epsilon\right)|x_2|.$$

We see that it is sufficient that

$$10^{25}a \leqslant \epsilon.$$

If we take $\epsilon = 10^{-1}$, for instance, which is not very small, we are led to choose $a = 10^{-26}$. The unit ball for this norm N is, therefore, the set

$$\left\{x \in \mathbb{R}^2 : \ 10^{-26}\,|x_1| + |x_2| \leqslant 1\right\}.$$

Geometrically, this unit ball is a lozenge whose diagonals are the axes. The diagonal along x_1 has 10^{26} times the length of the diagonal along x_2. It is quite difficult to draw since 10^{26} is a very large number: recall that Avogadro's number is about 6×10^{23} and this is the number of molecules contained in about 22 litres of the air which we breathe. Suppose that the small diagonal of our lozenge measures 1 micron. Then the large diagonal has a length of about ten light years. We would need the aid of the drawing instruments of that celebrated cinema hero ET to be able to draw this type of figure.

Finally, we prove a limit theorem which allows us to recover the spectral radius of a matrix from any norm:

Theorem 11.1.7. Let A be an $n \times n$ matrix with complex coefficients. Then, for every norm $\| \cdot \|$ on $\mathcal{M}_n(\mathbb{C})$,

$$(11.1.18) \qquad \lim_{k \to \infty} \left\| A^k \right\|^{1/k} = \rho(A). \qquad \diamond$$

Proof. For every vector norm M, $\|A\|_M \geqslant \rho(A)$, and for every $\epsilon > 0$ there exists N_ϵ such that $\|A\|_{N_\epsilon} \leqslant \rho(A) + \epsilon$. We easily verify that $\rho(A^k) = \rho(A)^k$ by referring to the Jordan form of A. Consequently, we have the inequalities

$$(11.1.19) \qquad \rho(A)^k \leqslant \left\| A^k \right\|_{N_\epsilon} \leqslant (\rho(A) + \epsilon)^k.$$

For the first inequality we have used Lemma 11.1.4 applied to A^k and for the second the algebra property of subordinate norms has been employed.

If $\| \cdot \|$ designates some norm (not necessarily a matrix norm) on $\mathcal{M}_n(\mathbb{C})$, there exists a constant C_ϵ for every $\epsilon > 0$, such that, for every matrix B

$$C_\epsilon^{-1} \|B\|_{N_\epsilon} \leqslant \|B\| \leqslant C_\epsilon \|B\|_{N_\epsilon}.$$

Using the inequalities (11.1.19), we obtain

$$C_\epsilon^{-1} \rho(A)^k \leqslant C_\epsilon^{-1} \left\| A^k \right\|_{N_\epsilon} \leqslant \left\| A^k \right\| \leqslant C_\epsilon \left\| A^k \right\|_{N_\epsilon} \leqslant C_\epsilon (\rho(A) + \epsilon)^k.$$

We raise these relations to the power $1/k$ and we get

$$(11.1.20) \qquad C_\epsilon^{-1/k} \rho(A) \leqslant \left\| A^k \right\| \leqslant C_\epsilon^{1/k} (\rho(A) + \epsilon).$$

Given $\alpha > 0$ we choose $\epsilon = \alpha/2$. There exists a C_ϵ for which eqn (11.1.20) holds for every $k \geqslant 1$. Since $C_\epsilon^{1/k}$ tends to 1 as k tends to infinity, we can find an $\ell(\alpha)$ such that

$$C_\epsilon^{1/k} (\rho(A) + \epsilon) \leqslant \rho(A) + \alpha, \quad \forall k \geqslant \ell(\alpha) \quad \text{and} \quad C_\epsilon^{-1/k} \rho(A) \geqslant \rho(A) - \alpha.$$

This proves the desired estimate. $\qquad \qquad \square$

Remark 11.1.8. It is not necessary to work in $\mathcal{M}_n(\mathbb{C})$. The result is still true in $\mathcal{M}_n(\mathbb{R})$ but needs a more delicate argument using norms on the complexified space.

Some sufficient conditions on the convergence or divergence of iterative methods

We begin with a theorem on iterative methods for Hermitian matrices:

Theorem 11.1.9. Let A be a Hermitian positive definite matrix, with the decomposition $A = M - N$, where M is invertible. If $M + N^*$ (which is still Hermitian) is positive definite, then the iterative method (11.1.2) converges. $\qquad \diamond$

Proof. First of all, we verify that $M + N^*$ is still Hermitian:

$$M + N^* = M + (M - A)^* = M + M^* - A,$$

which is Hermitian. We equip $\mathbb{C}^n$ with the norm $|x|_A = (x^*Ax)^{1/2}$ and this (astute) choice allows us to prove the theorem. Note that $M^{-1}N = I - M^{-1}A$ and therefore,

$$\left\| M^{-1}N \right\|_A = \max\left\{ |x - M^{-1}Ax|_A / |x|_A = 1 \right\}.$$

But, if we let $M^{-1}Ax = y$ which is equivalent to $Ax = My$ and $x^*A = y^*M^*$, then

$$
\begin{aligned}
(x^* - y^*)\, A\, (x - y) &= x^*Ax - x^*Ay - y^*Ax + y^*Ay \\
&= 1 - y^*M^*y - y^*My + y^*Ay \\
&= 1 - y^* \left(M + M^* - A \right) y.
\end{aligned}
$$

The hypothesis that $M^* + N$ is positive definite implies that

$$\min_{\substack{|x|_A = 1 \\ y = M^{-1}Ax}} y^* \left(M + M^* - A \right) y > 0.$$

This proves the theorem. $\qquad\square$

We are going to apply this convergence criterion to the Gauss–Seidel method and to a more general method, called the relaxation method. Suppose that $A = D - E - F$. The Gauss–Seidel method involves writing the iterations in the form

$$(D - E)\, x^{k+1} = Fx^k + b.$$

We see that all of the matrix D acts on the vector x^{k+1}. We introduce a parameter α in a way so that a part of D acts on the vector x^{k+1} and the rest on x^k. We thus write

$$(\alpha D - E)\, x^{k+1} + ((1 - \alpha)\, D - F)\, x^k = b.$$

Classically, we denote $\alpha = 1/\omega$ and the relaxation method is written as

$$\left(\frac{D}{\omega} - E \right) x^{k+1} - \left(\frac{1 - \omega}{\omega} D + F \right) x^k = b.$$

The matrix M is equal to $(D/\omega) - E$ and the matrix N is equal to $((1-\omega)D/\omega) + F$. The matrix of the relaxation method is

$$\mathcal{L}_\omega = \left(\frac{D}{\omega} - E \right)^{-1} \left(\frac{1 - \omega}{\omega} D + F \right).$$

We note that for $\omega = 1$ the relaxation method is equivalent to the Gauss–Seidel method.

We have, up till now, supposed D to be the diagonal of matrix A, but we can decompose A by blocks of size $n_i \times n_j$, with

$$\sum_{j=1}^{N} n_j = n.$$

In this case, matrix D is block diagonal and has as diagonal blocks the diagonal blocks of A. If $A = A^*$ then $D = D^*$ because the same is true for each of the blocks A_{ii}. If A is also positive definite, the same is true of the blocks A_{ii}. Indeed, let

$$x = \begin{pmatrix} x_1 \\ \vdots \\ x_N \end{pmatrix}$$

be the decomposition of vector x by blocks of n_i rows. If all the blocks of x are zero except the i-th then

$$x^* A x = x_i^* A_{ii} x_i,$$

which shows that A_{ii} is positive definite and so is D.

It goes without saying that, in this case, $-E$ is formed from blocks situated under the diagonal of A, and that $-F$ is the matrix formed from blocks situated above the diagonal of A.

With a decomposition such as $A = D - E - F$ from the block form of A, we define the block Jacobi method, the block Gauss–Seidel method, and the block relaxation method in a completely analogous way to the element-wise methods of the same name.

We can now state a result on the convergence of the relaxation method, by elements or by blocks, when A has the right properties:

Theorem 11.1.10. Let A be a positive definite Hermitian matrix. If $\omega \in\ ]0,2[$, the element-wise or block relaxation method converges.   ◇

Proof. We have

$$M + N^* = \frac{D}{\omega} - E + \frac{(1-\omega)\,D^*}{\omega} + F^*.$$

By construction, $F^* = E$ and $D^* = D$. Consequently,

$$M + N^* = \frac{(2-\omega)\,D}{\omega}.$$

For $M + N^*$ to be positive definite, it is necessary and sufficient that $((2-\omega)/\omega)x^* D x$ is strictly positive if and only if x is nonzero. Under the stated hypotheses, D is positive definite and therefore, $M + N^*$ is positive definite if and only if $(2-\omega)/\omega$ is strictly positive, that is, if $\omega \in\ ]0,2[$. We therefore conclude the required result with the aid of Theorem 11.1.9.   □

It is generally more difficult to apply Theorem 11.1.9 to Jacobi's method. The proof of convergence often uses the particular properties of A.

We now give a lower bound for the spectral radius of the matrix $\mathcal{L}_\omega$, which leads to a sufficient condition for divergence.

Theorem 11.1.11. For any $\omega \neq 0$, we have

$$\rho\left(\mathcal{L}_\omega\right) \geqslant |\omega - 1| . \qquad \diamond$$

Proof. The determinant of a block triangular matrix B of size $n_i \times n_j$, is equal to the product of the determinants of its diagonal blocks. This result is shown in Exercise 3.3.4.

We apply this result to the calculation of $\det \mathcal{L}_\omega$:

$$\det \mathcal{L}_\omega = \frac{\det\left(\dfrac{(1-\omega)D}{\omega} + F\right)}{\det\left(\dfrac{D}{\omega} - E\right)} = \frac{\left(\dfrac{1-\omega}{\omega}\right)^n \det D}{\left(\dfrac{1}{\omega}\right)^n \det D} = (1-\omega)^n .$$

Moreover, since the determinant is the product of the eigenvalues of a matrix, we see that

$$\rho\left(\mathcal{L}_\omega\right)^n \geqslant |\det\left(\mathcal{L}_\omega\right)| .$$

We conclude that $\rho(\mathcal{L}_\omega)^n \geqslant |1-\omega|^n$ and we obtain the conclusion of the theorem. $\qquad \square$

We now cite two results which we will not prove. For their proofs we refer to the book of P. G. Ciarlet [16, pp. 105–9]:

Theorem 11.1.12. Let A be a block tridiagonal matrix. Then the spectral radii of block Jacobi matrices and the corresponding Gauss–Seidel matrices are linked by the relation

$$\rho\left(\mathcal{L}_1\right) = \rho\left(J\right)^2 ,$$

from which we see that the two methods converge or diverge simultaneously. When convergent, the Gauss–Seidel method converges more rapidly than the Jacobi method. $\qquad \diamond$

We note that there exist matrices A for which the Jacobi method converges and the Gauss–Seidel method does not converge.

We can also compare the Jacobi method and the relaxation method. Again, from [16] we have

Theorem 11.1.13. Let A be a block tridiagonal matrix such that every eigenvalue of the corresponding block Jacobi matrix is real. Then the block Jacobi method and the block relaxation method diverge or converge simultaneously for $0 < \omega < 2$. When they converge, the function $\omega \in \,]0, 2[\, \mapsto \rho(\mathcal{L}_\omega)$ has the shape given by Figure 11.1, where the optimal parameter of relaxation ω_0 is given by

$$(11.1.21) \qquad \omega_0 = \frac{2}{1 + \sqrt{1 - \rho\left(J\right)^2}} . \qquad \diamond$$

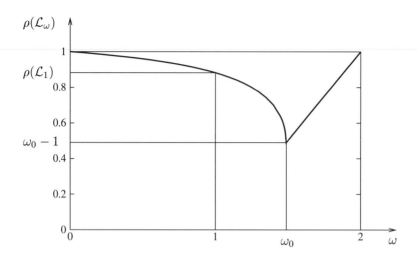

Figure 11.1: The spectral radius of the relaxation matrix $\mathcal{L}_\omega$, as a function of ω.

Figure 11.1 allows us to see that the optimal parameter ω_0 is strictly greater than 1. It is because of this that the relaxation method is often called the *over-relaxation method*. The shape of this figure shows that it is better to overestimate the relaxation parameter than to underestimate it.

The combination of Theorems 11.1.10 and 11.1.13 confirm that, for a positive definite block tridiagonal Hermitian matrix A, the methods of Jacobi, Gauss–Seidel, and relaxation for $0 < \omega < 2$, converge. Furthermore, the optimal relaxation parameter is given by formula (11.1.21). If $\rho(J) > 0$, then $\rho(\mathcal{L}_{\omega_0}) = \omega_0 - 1 < \rho(\mathcal{L}_1) < \rho(J)$.

We will show later the advantage of the over-relaxation method compared to the Jacobi method and the Gauss–Seidel method.

11.2. Linear recurrence and powers of matrices

In the vector space $\mathbb{C}^n$ we consider the linear recurrence

$$(11.2.1) \qquad \sum_{j=0}^{p} A_j x^{j+p} = 0,$$

where the matrices A_j are given in $\mathcal{M}_n(\mathbb{C})$ and the linear recurrence is initialized by giving p vectors x^j for $j = 0, \ldots, p-1$. We suppose, furthermore, that the matrix A_p is invertible. If we let $B_j = -A_p^{-1} A_j$, and if we define a vector y^j in $C\mathrm{e}^{pn}$ by

$$y^j = \begin{pmatrix} x^j \\ x^{j+1} \\ \vdots \\ x^{j+p-1} \end{pmatrix},$$

we see that recurrence (11.2.1) can be put in the form

$$y^{j+1} = Cy^j,$$

where the matrix C is given by the following block decomposition:

$$(11.2.2) \qquad C = \begin{pmatrix} 0 & I & 0 & \cdots & & 0 \\ 0 & 0 & I & & & \\ \vdots & & & \ddots & \ddots & \\ & & & & 0 & I \\ B_0 & B_1 & \cdots & & & B_{p-1} \end{pmatrix}.$$

Matrix C is called the companion matrix of the recurrence (11.2.1). The solution of relation (11.2.1) is given by

$$y^k = C^k y^0.$$

The choice of $\mathbb{C}$ as the number field is justified by the use of a Jordan decomposition; however, all the results are also true for real matrices. Using the Jordan form of C we have, at least theoretically, all the solutions of the recurrence. In particular, seeking the eigenvectors of C is equivalent to seeking vectors $x^0, \ldots, x^{p-1}$, such that

$$x^1 = \lambda x^0,$$
$$x^2 = \lambda x^1,$$

$$\vdots$$

$$x^{p-1} = \lambda x^{p-2},$$

$$\sum_{j=0}^{p-1} B_j x^j = \lambda x^{p-1}.$$

It is equivalent to find a vector x_0, such that

$$\left(\sum_{j=0}^{p} \lambda^j A_j \right) x_0 = 0.$$

The equation

$$(11.2.3) \qquad \det \left(\sum_{j=0}^{p} \lambda^j A_j \right) = 0$$

is the characteristic equation of the recurrence (11.2.1). Therefore, it is this which gives us the eigenvalues of C. The reader can verify that

$$\det\left(\sum_{j=0}^{p} \lambda^j A_j\right) = \gamma \det\left(\lambda I - C\right),$$

where γ is a constant. If all the eigenvalues of C are pairwise distinct and if $n = 1$, then the solution of recurrence (11.2.1) is a linear combination of the sequences $\left(\lambda^k\right)_{k\geqslant 0}$, where λ runs through the eigenvalues of C.

If $n = 1$ we consider the decomposition of C in Jordan blocks. We recall that there exists a transformation matrix P such that $C = P^{-1}JP$, where J is a block diagonal matrix, each of the diagonal blocks being of the form

$$J\left(\lambda, m\right) = \begin{pmatrix} \lambda & 1 & 0 & \cdots & 0 \\ 0 & \lambda & 1 & & 0 \\ \vdots & \ddots & \ddots & \ddots & \vdots \\ 0 & & 0 & \lambda & 1 \\ 0 & \cdots & & 0 & \lambda \end{pmatrix}.$$

If C has Jordan blocks of size strictly greater than 1, then our procedure for constructing particular solutions does not provide us with enough of them which are independent. It is simple to calculate the powers of $J(\lambda, m)$:

$$J\left(\lambda, m\right) = \lambda I_m + J\left(0, m\right).$$

We therefore have

(11.2.4) $$J\left(\lambda, m\right)^k = \sum_{\ell=0}^{\min(m-1,k)} C_k^\ell \lambda^{k-\ell} J\left(0, m\right)^\ell.$$

Each of these powers $J(0, m)$ is given by

(11.2.5) $$\left(J\left(0, m\right)^\ell\right)_{ij} = \delta_{j-i,\ell}.$$

In this case, the examination of formula (11.2.5) shows that it is necessary to add the particular solutions of the form $\left(k^j \lambda^k\right)_{k\geqslant 0}$, where j is an integer between 1 and $m - 1$, and m is the multiplicity of a Jordan block associated with λ.

Thus, we can answer some questions concerning the recurrence (11.2.1). If we demand that the solution x^k tends to 0 for all initial data as k tends to infinity, the study of Section 11.1 shows that it is necessary and sufficient that $\rho(C) < 1$. If we demand only that the solution x^k to the recurrence (11.2.1) remains bounded for all initial data as k tends to infinity, we can state the necessary and sufficient condition, which will be proved in Section 17.3, Lemma 17.3.1 $\rho(C) \leqslant 1$ and, for every eigenvalue λ of modulus 1, the corresponding Jordan blocks must be of dimension 1. This same result is also proved in Subsection 3.3.6.

11.2.1. The spectrum of a finite difference matrix

We have already met, in Subsection 4.6.2, the following matrix which arises from a finite difference problem:

$$
A = \begin{pmatrix}
2 & -1 & 0 & 0 & \cdots & 0 \\
-1 & 2 & -1 & 0 & & 0 \\
0 & -1 & 2 & -1 & & 0 \\
\vdots & & \ddots & \ddots & \ddots & \vdots \\
0 & & & -1 & 2 & -1 \\
0 & & \cdots & 0 & -1 & 2
\end{pmatrix}.
$$

Seeking an eigenvalue of this matrix corresponds to writing, for a scalar λ,

$$
\begin{aligned}
2x_1 - x_2 &= \lambda x_1, \\
-x_1 + 2x_2 - x_3 &= \lambda x_2, \\
&\vdots \\
-x_{j-1} + 2x_j - x_{j+1} &= \lambda x_j, \\
&\vdots \\
-x_{n-1} + 2x_n &= \lambda x_n.
\end{aligned}
$$

(11.2.6)

These relations fall into the class of linear recurrences, with the condition that $x_0 = 0$ and there existing a λ such that $x_{n+1} = 0$. Matrix A is real and symmetric, and, therefore, all of its eigenvalues are real. Hence, we will only concern ourselves with real λ. The recurrence (11.2.6) can be rewritten as

$$
x_{j+1} + (\lambda - 2) x_j + x_{j-1} = 0
$$

and its characteristic equation is

(11.2.7) $$\rho^2 + (\lambda - 2) \rho + 1 = 0.$$

The discriminant of this second-order equation is

$$
\Delta = (\lambda - 2)^2 - 4.
$$

If $\lambda \notin \,]0,4[$, Δ is positive and eqn (11.2.7) has two distinct real roots whose product is equal to 1. We denote the root with the largest absolute value by ρ_+ and the other by ρ_-. We have

$$
x_j = a\rho_+^j + b\rho_-^j.
$$

The initial condition x_0 forces $a = -b$. The condition $x_{n+1} = 0$ is satisfied if $a(\rho_+^{n+1} - \rho_-^{n+1}) = 0$. This is only possible if $a = 0$, which would imply that $x_j = 0$ for every j. This choice of λ is impossible since a zero vector is not an eigenvector. If $\lambda = 0$ or $\lambda = 4$, the solution of the recurrence is of the form

$$
x_j = a\rho^j + bj\rho^j,
$$

where ρ is the double root of eqn (11.2.7). The initial condition forces $a = 0$ and the final condition forces $b = 0$. We again have a contradiction, and we must choose $\lambda \in \,]0, 4[$. In this case, the roots of eqn (11.2.7) are complex conjugates of modulus 1, which we denote by $\rho = e^{i\theta}$ and $\bar{\rho}$. These solutions have the form

$$x_j = a\rho^j + b\bar{\rho}^j.$$

The initial condition forces

$$a = -b,$$

which implies that

$$x_j = 2ai \sin(j\theta).$$

The final condition forces

$$2ai \sin((n + 1)\theta) = 0,$$

which is possible if we have

$$\theta = \frac{m\pi}{n + 1}.$$

From relation (11.2.7) we get the corresponding value of λ:

$$\lambda = 2 - \frac{\rho^2 + 1}{\rho} = 2 - e^{i\theta} - e^{-i\theta} = 2 - 2\cos\theta$$

$$= 4\sin^2\left(\frac{\theta}{2}\right) = 4\sin^2\left(\frac{m\pi}{2(n + 1)}\right).$$

The sine function is positive and strictly increasing on the interval $[0, \pi/2]$ and we have thus obtained n distinct eigenvalues. So, we know that we have found all the eigenvalues of A.

We will apply this result to the methods of Jacobi, Gauss–Seidel, and relaxation for A. The Jacobi method is given by the decomposition

(11.2.8) $M = D = 2I$ and $N = E + F = 2I - A.$

Consequently, the eigenvalues of the Jacobi matrix $J = M^{-1}N$ are

$$1 - \frac{\lambda_m}{2} = 1 - 2\sin^2\left(\frac{m\pi}{2(n + 1)}\right) = \cos\left(\frac{m\pi}{n + 1}\right).$$

From Definition 10.2.1, the spectral radius of the Jacobi matrix is, therefore,

$$\rho(J) = \max_{1 \leqslant m \leqslant n} \left|\cos\left(\frac{m\pi}{n + 1}\right)\right| = \cos\left(\frac{\pi}{n + 1}\right).$$

Therefore, for large n, we have the truncated expansion

$$\rho(J) = 1 - \frac{\pi^2}{2(n + 1)^2} + O\left(n^{-4}\right).$$

From Theorem 11.1.12, we have

$$\rho\left(\mathcal{L}_1\right) = \rho\left(J\right)^2 = 1 - \frac{\pi^2}{\left(n+1\right)^2} + O\left(n^{-4}\right).$$

The optimal relaxation parameter ω_0, determined in Theorem 11.1.13, is given by

$$\omega_0 = \frac{2}{1 + \sqrt{1 - \rho\left(J\right)^2}} = 2\left[1 + \frac{\pi}{n+1} + O\left(n^{-2}\right)\right]^{-1}$$

$$= 2 - \frac{2\pi}{n+1} + O\left(n^{-2}\right).$$

Consequently, the optimal spectral radius $\rho\left(\mathcal{L}_{\omega_0}\right) = |\omega_0 - 1|$ is

$$\rho\left(\mathcal{L}_{\omega_0}\right) = 1 - \frac{2\pi}{n+1} + O\left(n^{-2}\right).$$

To see the numerical consequences of these different estimates, we estimate the number of iteration steps that are needed to halve the error using any of these methods. With the notation of the proof of Theorem 11.1.2,

$$x^k = B^k x^0 + \sum_{j=0}^{k-1} B^j c$$

and

$$x^\infty = \lim_{k \to \infty} x^k = \sum_{j=0}^{\infty} B^j c.$$

Therefore, the error is given by

$$x^\infty - x^k = \sum_{j=k}^{\infty} B^j c - B^k x^0.$$

For every $\epsilon > 0$, there exists a vector norm N such that $\|B\|_N \leqslant \rho + \epsilon$. We therefore have the estimate

$$N\left(x^\infty - x^k\right) \leqslant \left(\rho\left(B\right) + \epsilon\right)^k N\left(x^0\right) + \sum_{j=k}^{\infty} \left(\rho\left(B\right) + \epsilon\right)^j N\left(c\right)$$

$$= \left(\rho\left(B\right) + \epsilon\right)^k \left[\frac{N\left(c\right)}{1 - \rho\left(B\right) - \epsilon} + N\left(x^0\right)\right].$$

We therefore see that, to divide the error by 2, we must have

$$k \ln \left(\rho\left(B\right) + \epsilon\right) \leqslant -\ln 2$$

and therefore, it is necessary that

$$k \geqslant -\frac{\ln 2}{\ln \rho(B)}.$$

If there exists a norm N such that $\|B\|_N = \rho(B)$ (which is the case when B is Hermitian and the norm N is the Hermitian norm), then this estimate is optimal. We will be content with this estimate in what follows.

We estimate $-\ln \rho(B)$ when $B = J$, $B = \mathcal{L}_1$, or $B = \mathcal{L}_{\omega_0}$:

$$-\ln \rho(J) \sim -\ln\left(1 - \frac{\pi^2}{2(n+1)^2}\right) \sim \frac{\pi^2}{2(n+1)^2},$$

$$-\ln \rho(\mathcal{L}_1) \sim -\ln\left(1 - \frac{\pi^2}{(n+1)^2}\right) \sim \frac{\pi^2}{(n+1)^2},$$

$$-\ln \rho(\mathcal{L}_{\omega_0}) \sim -\ln\left(1 - \frac{2\pi}{n+1}\right) \sim \frac{2\pi}{n+1}.$$

We see that it needs about $2(n+1)^2 \ln 2/\pi^2$ steps of the Jacobi method to divide the error by 2, $(n+1)^2 \ln 2/\pi^2$ steps of the Gauss–Seidel method, and $2(n+1) \ln 2/\pi$ steps in the case of relaxation with optimal parameter. Going from Jacobi to Gauss–Seidel allows us to divide the number of steps needed by a constant factor, whilst going from Gauss–Seidel or Jacobi to relaxation with optimal parameter gains an order of magnitude.

It is important to note that exploiting the structure of the matrix and the astute introduction of parameters in the methods leads to considerable gains in numerical efficiency. The method of (over-)relaxation is only one of the methods which allow such gains. Some other methods are studied in the forthcoming exercises.

11.3. Exercises from Chapter 11

11.3.1. Finite difference matrix of the Laplacian in a rectangle

Consider the system of linear equations

(11.3.1) $-4u_{i,j} + u_{i+1,j} + u_{i,j+1} + u_{i-1,j} + u_{i,j-1} = h^2 f_{ij},$

where i varies between 1 and m and j varies between 1 and n. We let

$$u_{0,j} = u_{m+1,j} = u_{i,0} = u_{i,n+1} = 0,$$

so that the system (11.3.1) is well defined for $i = 1$ or m and for $j = 1$ or n.

Exercise 11.3.1. Explicitly write down the relations satisfied by $u_{i,j}$ when $i = 1$ or m and when $j = 1$ or n.

Exercise 11.3.2. How many unknowns does the system (11.3.1) have, and how many equations?

Exercise 11.3.3. Give a numbering

$$(11.3.2) \qquad k = K(i, j)$$

such that the matrix of the system (11.3.1) is in block tridiagonal form. Determine the blocks explicitly.

Exercise 11.3.4. Show that, for every $(u_{i,j})_{\substack{1 \leqslant i \leqslant m \\ 1 \leqslant j \leqslant n}} \in \mathbb{R}^{mn}$, we have

$$\sum_{\substack{1 \leqslant i \leqslant m \\ 1 \leqslant j \leqslant n}} u_{i,j} \left(-4u_{i,j} + u_{i+1,j} + u_{i,j+1} + u_{i-1,j} + u_{i,j-1} \right) \geqslant 0.$$

For which u does this expression vanish?

Exercise 11.3.5. Choose a direct method to solve this system. Justify your answer by an evaluation of the operation count necessary for its solution.

Exercise 11.3.6. Show that every vector

$$(11.3.3) \qquad u_{i,j} = \sin \alpha i h \sin \beta j h,$$

where h is a real positive number, satisfies

$$(11.3.4) \qquad -4u_{i,j} + u_{i+1,j} + u_{i,j+1} + u_{i-1,j} + u_{i,j-1} = \lambda u_{i,j},$$

for $2 \leqslant i \leqslant m - 1$ and for $2 \leqslant j \leqslant n - 1$. What is the value of λ?

Exercise 11.3.7. Show that, for certain choices of α and β, the vectors $u_{i,j}$, given by eqn (11.3.3), also satisfy eqn (11.3.4) for $i = 1$, $i = m$, $j = 1$, $j = n$.
Hint: to do this, use the expression for λ found previously and substitute into eqn (11.3.4) for the particular values of i and j considered here.

Exercise 11.3.8. Give all the eigenvalues and all the eigenvectors of the matrix A of the system (11.3.1).

Exercise 11.3.9. Calculate the eigenvalues of the Jacobi matrix of the system (11.3.1) from those of A. From this, deduce the spectral radius of J for large m and n.

Exercise 11.3.10. Evaluate the condition number

$$\mathrm{cond}_2(A) = \|A\|_2 \|A^{-1}\|_2.$$

11.3.2. Richardson's and pre-conditioned Richardson's methods

Definition of Richardson's method

Exercise 11.3.11. Let $A = M - N$ be the decomposition given in eqn (11.1.1) and define the residual

$$r^k = b - Ax^k.$$

Show that relation (11.1.2) is equivalent to

$$M\left(x^{k+1} - x^k\right) = r^k.$$

Exercise 11.3.12. We wish to accelerate this iterative method. Let α^k be a real parameter; we shall consider the iterative method

(11.3.5) $$M\left(x^{k+1} - x^k\right) = \alpha^k r^k.$$

This iterative method is called the pre-conditioned Richardson's method—stationary if α^k does not depend on k, dynamical otherwise. If M is the identity, the method is called Richardson's method. The matrix M is called the pre-conditioner. Show that if α^k does not depend on k, then eqn (11.3.5) can be written as

(11.3.6)
$$\begin{aligned}
Mz^k &= r^k, \\
x^{k+1} &= x^k + \alpha z^k, \\
r^{k+1} &= r^k - \alpha A z^k,
\end{aligned}$$

and that the matrix of this iterative method is $I - \alpha M^{-1}A$. Show that the methods of Jacobi, Gauss–Seidel, and the relaxation method are Richardson's methods.

Analysis of the stationary pre-conditioned Richardson's method

Exercise 11.3.13. Define the Cayley transform on matrices L such that $I - L$ is not singular by
$$\Gamma\left(L\right) = \left(I - L\right)^{-1}\left(I + L\right).$$

What is the inverse of $\Gamma(L)$? On which set of matrices is it defined? Show that the spectrum of a matrix L is contained in the open unit disk $\{z :\ |z| < 1\}$ if and only if the spectrum of $\Gamma(L)$ is contained in the open right-hand side plane $\{z :\ \Re z > 0\}$.

Exercise 11.3.14. Find the Cayley transform of $I - \alpha M^{-1}A$.

Exercise 11.3.15. Infer from Exercise 11.3.14 that a necessary and sufficient condition of convergence of the method (11.3.6) is that the eigenvalues λ_j of the matrix $M^{-1}A$ satisfy the following condition:

$$\alpha\left|\lambda_j\right|^2 < 2\Re\lambda_j.$$

Exercise 11.3.16. Show that it is possible to find α such that the method (11.3.6) is a convergent method if and only if the spectrum of $M^{-1}A$ is included in $\{z :\ \Re z > 0\}$.

Exercise 11.3.17. Assume that the spectrum of $M^{-1}A$ is included in $]0, \infty[$. Let its lower bound be $\lambda_{\min}$ and its upper bound be $\lambda_{\max}$. We wish to choose a parameter α for which the convergence is fastest. Show that the best α is given by

$$\alpha^* = \frac{2}{\lambda_{\min} + \lambda_{\max}}.$$

Exercise 11.3.18. Calculate the spectral radius of $I - \alpha^* M^{-1}A$ in terms of $\lambda_{\min}$ and $\lambda_{\max}$.

Exercise 11.3.19. Assume that $M^{-1}A$ is symmetric and positive definite; denote by $\| \cdot \|$ the matrix norm subordinated to the Euclidean norm. Show that

(11.3.7)
$$\|I - \alpha^* M^{-1}A\| = \frac{\lambda_{\max} - \lambda_{\min}}{\lambda_{\max} + \lambda_{\min}}.$$

The right-hand side of eqn (11.3.7) will be called ρ^*.

Remark 11.3.20. It is therefore important that $\lambda_{\max}/\lambda_{\min}$ be as close as possible to 1. This means that if M^{-1} is a good approximation of A^{-1}, we stand to gain a lot by applying pre-conditioned iterative methods.

Exercise 11.3.21. Assume that M and A are symmetric positive definite. Define a scalar product on $\mathbb{R}^d$ by

$$(x, y)_M = x^\top My,$$

and denote by $| \cdot |_M$ and $\| \cdot \|_M$ the corresponding vector and operator norms. Show that $M^{-1}A$ is self-adjoint with respect to this scalar product. Show that

$$\|I - \alpha^* M^{-1}A\|_M = \frac{\lambda_{\max} - \lambda_{\min}}{\lambda_{\max} + \lambda_{\min}}.$$

Analysis of a dynamical Richardson's method: the pre-conditioned steepest gradient method

Exercise 11.3.22. Consider z^{k+1} given by the method (11.3.6) as a function of α; assume M to be symmetric positive definite and A to be regular. Show that if z^k does not vanish, there exists a number α^k at which $\alpha \mapsto |z^{k+1}|_M^2$ reaches its minimum, and give the expression of α^k. Give also the value of z^{k+1} and show that

(11.3.8)
$$\frac{|z^k|_M^2 - |z^{k+1}|_M^2}{|z^k|_M^2} = \frac{\left((z^k)^\top A z^k \right)^2}{\left((z^k)^\top A^\top M^{-1} A z^k \right) |z^k|_M^2}.$$

Conclude that the sequence $|z^k|_M$ converges.

Exercise 11.3.23. Assume that $A + A^\top$ is positive definite, and let $\beta > 0$ be such that

$$z^\top A z \geqslant \beta |z|_M^2.$$

Show that

$$\frac{\left(\left(z^k\right)^\top A z^k\right)^2}{\left(z^k\right)^\top A^\top M^{-1} A z^k} \geqslant \frac{\beta^2 \left|z^k\right|_M^4}{\|A^\top M^{-1} A\| \left|z^k\right|^2} \geqslant c \left|z^k\right|_M^2,$$

and conclude that z^k converges to 0 and hence, that x^k converges to $x = A^{-1}b$.

Exercise 11.3.24. Assume that A is symmetric positive definite, and let B be the M-symmetric positive square root of $M^{-1}A$ (see Subsection 3.3.5), i.e.,

$$B^2 = M^{-1}A \quad \text{and} \quad MB = B^\top M.$$

Show that the right-hand side of eqn (11.3.8) can be rewritten as

$$\frac{\left|Bz^k\right|_M^4}{(M^{-1}ABz^k, Bz^k)_M \, (A^{-1}MBz^k, Bz^k)_M}$$

and infer from the Kantorovich inequality (3.3.10) that

$$\left|z^{k+1}\right|_M \leqslant \rho^* \left|z^k\right|_M.$$

Exercise 11.3.25. Define the Richardson steepest descent method as the following algorithm:

$$r^k = b - Ax^k,$$

$$\alpha^k = \frac{\left(r^k\right)^\top r^k}{\left(r^k\right)^\top A r^k},$$

$$x^{k+1} = x^k + \alpha^k r^k.$$

Show that, if A is symmetric positive definite, then the error $e^k = x^k - x$ satisfies the estimate

$$\left|e^{k+1}\right|_A \leqslant \rho^* \left|e^k\right|_A.$$

Remark 11.3.26. The method of Exercise 11.3.25 is simpler than that of Exercise 11.3.24 with $M = I$ because we have been able to exploit the symmetry of A.

11.3.3. Convergence rate of the gradient method

Exercise 11.3.27. Let A be a symmetric positive definite real $d \times d$ matrix. Show that it is equivalent to solve

(11.3.9) $Ax = b$

and to find a minimizer of the function

$$y \mapsto \frac{1}{2}y^\top Ay - y^\top b = f(y)$$

over $\mathbb{R}^d$.

Exercise 11.3.28. Calculate the gradient $v = \nabla f(y)$ and find the value α which minimizes

$$\alpha \mapsto f(y - \alpha v).$$

Exercise 11.3.29. Consider the algorithm

(11.3.10)
$$r^0 = b - Ax^0,$$
$$\alpha^k = \frac{\left| r^k \right|^2}{\left(r^k \right)^{\mathsf{T}} A r^k},$$
$$x^{k+1} = x^k + \alpha^k r^k,$$
$$r^{k+1} = b - Ax^{k+1}$$

Let x be the solution of eqn (11.3.9). Show that the error $e^k = x^k - x$ satisfies the relation

$$e^{k+1} = \left(I - \alpha^k A \right) e^k.$$

Let $\lambda = \lambda_{\min}$ and $\mu = \lambda_{\max}$ be the smallest and the largest eigenvalues of A, respectively. We will show that the rate of convergence of the algorithm (11.3.10) is no better than

$$\rho^* = \frac{\mu - \lambda}{\mu + \lambda}.$$

Exercise 11.3.30. Let u be an eigenvector of A relative to the eigenvalue λ and let v be an eigenvector of A relative to the eigenvalue μ. We will assume that u and v are of Euclidean norm 1. Show that, if e^0 belongs to the space spanned by u and v, then e^k also belongs to that space, for all integers k.

Exercise 11.3.31. Write

$$e^k = x^k u + y^k v.$$

Assuming that x^0 and y^0 do not vanish, show that

$$\frac{y^{k+1}}{x^{k+1}} = -\frac{\lambda^2}{\mu^2} \frac{x^k}{y^k}.$$

Exercise 11.3.32. Writing, from now on,

$$\frac{y^{2k}}{x^{2k}} = p, \quad \frac{y^{2k+1}}{x^{2k+1}} = q,$$

calculate y^{2k+2}/y^{2k} in terms of λ, μ, and p.

Exercise 11.3.33. Show that this ratio is maximal for the choice $p = \lambda/\mu$ and that it is equal to the square of ρ^*.

Exercise 11.3.34. Show that the method (11.3.10) is a Richardson's steepest descent method with $M = I$, and hence, that the convergence rate is, at most, equal to ρ^*, according to Exercise 11.3.19.

11.3.4. The conjugate gradient

Let A be a positive symmetric definite $d \times d$ real matrix. Let

$$f(y) = \frac{1}{2} y^\top A y - b^\top y.$$

We denote by x the solution of $Ax = b$, where x is the minimizer of f over $\mathbb{R}^d$. The Euclidean scalar product is denoted by $(\cdot, \cdot)$, with corresponding vector norm $|\cdot|$ and operator norm $\|\cdot\|$. The scalar product $(\cdot, \cdot)_A$ is defined by

$$(x, y)_A = x^\top A y,$$

and the corresponding vector and operator norms are denoted by $|\cdot|_A$ and $\|\cdot\|_A$.

Exercise 11.3.35. Given p^0 and x^0 in $\mathbb{R}^d$, we seek a sequence of vectors $p^1, \ldots, p^k$ and a sequence of reals $\alpha^0, \ldots, \alpha^k$ such that, for all $j = 1, \ldots, k$, the minimum of f over the affine space

$$V_j = x^0 + \bigoplus_{i=0}^{j} \mathbb{R} p^j$$

is attained at $x^0 + \alpha^0 p^0 + \ldots + \alpha^j p^j$. Give the value of α^0 in terms of p^0 and $r^0 = b - Ax^0$. Show that the sequence $(p^j)_j$ must satisfy the relations

$$\left(p^j, p^i\right)_A = 0, \quad \forall i, \ \forall j \neq i,$$

and the sequence α^j is given by

$$r^j = b - Ax^j, \quad \alpha^j = \frac{\left(r^j\right)^\top p^j}{|p^j|_A^2}.$$

In order to define an iterative method, we must give an initialization and a way of generating p^{k+1} in terms of p^k. Therefore, we define the conjugate gradient method by the following algorithm:

(11.3.11) $p^0 = r^0 = b - Ax^0,$

and while p^k does not vanish:

(11.3.12) $\alpha^k = \frac{\left(r^k, p^k\right)}{|p^k|_A^2},$

(11.3.13) $x^{k+1} = x^k + \alpha^k p^k,$

(11.3.14) $r^{k+1} = r^k - \alpha^k A p^k,$

(11.3.15) $\beta^{k+1} = -\frac{\left(r^{k+1}, A p^k\right)}{|p^k|^2},$

(11.3.16) $p^{k+1} = r^{k+1} + \beta^{k+1} p^k.$

Conjugate gradient as a direct method

Exercise 11.3.36. Show that eqns (11.3.15) and (11.3.16) imply that

$$\left(p^k, p^{k+1}\right)_A = 0.$$

Show that eqns (11.3.12) and (11.3.13) imply that

$$\left(p^k, r^{k+1}\right) = 0.$$

Show that, when the index k is replaced by $k-1$ in eqn (11.3.16), it implies that

$$\left(p^k, r^k\right)_A = \left|p^k\right|^2_A.$$

Show that eqn (11.3.12) implies that

$$\left(r^{k+1}, r^k\right) = 0.$$

Conclude that we have the following alternative expressions for α^k and β^k:

$$\alpha^k = \frac{\left|r^k\right|^2}{\left|p^k\right|^2_A} \quad \text{and} \quad \beta^k = \frac{\left|r^{k+1}\right|^2}{\left|r^k\right|^2}.$$

Exercise 11.3.37. Assume that p^k vanishes. Then, show that $r^{k-1} = -\beta^k p^{k-1}$ and $|r^k|^2 = -\beta^k(r^{k-1}, r^k)$, so that r^k vanishes.

Exercise 11.3.38. Show, by induction on k, that the following relations hold:

$$\left(r^i, r^j\right) = 0, \quad \left(p^i, p^j\right)_A = 0, \quad \forall i = 0, \ldots, k-1, \quad \forall j = i+1, \ldots, k.$$

Exercise 11.3.39. Show that the conjugate gradient method converges in a finite number of steps. Give this number in terms of the dimension d of the space.

Conjugate gradient as an iterative method

When the matrix A is very large, it does not make much sense to perform the number of steps which are required by the conjugate gradient method to guarantee us an exact solution—up to round-off error. In modern codes, the conjugate gradient is viewed as an iterative algorithm, and its rate of convergence can be analysed with precision.

Exercise 11.3.40. Show that $x^{k+1} - x^0$ is the orthogonal projection of $x - x^0$ onto the space spanned by $p^0, \ldots, p^k$, where the orthogonality is taken with respect to the scalar product $(\cdot, \cdot)_A$.

Exercise 11.3.41. Let $K_{k+1}(r^0)$ be the space spanned by $r^0, Ar^0, \ldots, A^k r^0$. This space is called a Krylov space. Show that the span of $p^0, \ldots, p^k$ is equal to $K_{k+1}(r^0)$.

Exercise 11.3.42. Show that any $y \in K_{k+1}(r^0)$ can be written as

$$y = \sum_{j=1}^{k+1} \gamma_j A^j \left(x - x^0 \right).$$

Exercise 11.3.43. Let $\mathbb{P}^*_{k+1}$ be the subset of $\mathbb{P}_{k+1}$ formed from the polynomials P of degree at most $k+1$ such that $P(0) = 1$. Show that

$$\left| x - x^{k+1} \right|_A = \min \left\{ \left| P(A)(x - x^0) \right|_A :\ P \in \mathbb{P}^*_{k+1} \right\}.$$

Denoting the eigenvalues of A by $\lambda_1, \ldots, \lambda_d$, show that

$$(11.3.17) \qquad \left| p(A)(x - x^0) \right|_A \leqslant \max_{1 \leqslant j \leqslant d} p(\lambda_j) \left| x - x^0 \right|_A, \quad \forall p \in \mathbb{P}^*_{k+1}.$$

Exercise 11.3.44. Let $\lambda_{\min}$ and $\lambda_{\max}$ be the smallest and the largest eigenvalues of A, respectively. Let Q_{k+1} be the Chebyshev polynomial of degree $k+1$ defined in Theorem 5.2.4. Define a polynomial P by

$$P(t) = Q_{k+1} \left(\frac{\lambda_{\max} + \lambda_{\min} - 2t}{\lambda_{\max} - \lambda_{\min}} \right) \Big/ Q_{k+1} \left(\frac{\lambda_{\max} + \lambda_{\min}}{\lambda_{\max} - \lambda_{\min}} \right).$$

Show, with the help of the recurrence relation (5.2.4), that

$$(11.3.18)$$
$$Q_{k+1} \left(\frac{\lambda_{\max} + \lambda_{\min}}{\lambda_{\max} - \lambda_{\min}} \right)$$
$$= 2 \left(1 + \left(\frac{\sqrt{\lambda_{\max}} - \sqrt{\lambda_{\min}}}{\sqrt{\lambda_{\max}} + \sqrt{\lambda_{\min}}} \right)^{2k+2} \right) \left(\frac{\sqrt{\lambda_{\max}} + \sqrt{\lambda_{\min}}}{\sqrt{\lambda_{\max}} - \sqrt{\lambda_{\min}}} \right)^{k+1}.$$

Exercise 11.3.45. Infer, from eqns (11.3.17) and (11.3.18), that the conjugate gradient method converges at a rate which is, at most, equal to

$$\frac{\sqrt{\lambda_{\max}} - \sqrt{\lambda_{\min}}}{\sqrt{\lambda_{\max}} + \sqrt{\lambda_{\min}}}.$$

Pre-conditioned conjugate gradient

An improvement over the conjugate gradient method is the pre-conditioned conjugate gradient: assume that M is a symmetric positive definite matrix such that linear systems with matrix M are easy to solve. Also assume that $M^{-1}A$ is reasonably small. The pre-conditioned conjugate gradient algorithm is defined by

$$(11.3.19) \qquad\qquad r^0 = b - Ax^0, \quad p^0 = z^0 = M^{-1}r^0,$$

and while p^k does not vanish,

$$\alpha^k = \frac{\left(z^k, r^k\right)}{|p^k|^2},$$

$$x^{k+1} = x^k + \alpha^k p^k,$$

$$r^{k+1} = r^k - \alpha^k A p^k,$$

$$M z^{k+1} = r^{k+1},$$

$$\beta^{k+1} = -\frac{\left(z^{k+1}, r^{k+1}\right)}{(z^k, r^k)},$$

$$p^{k+1} = z^{k+1} + \beta^{k+1} p^k.$$

Exercise 11.3.46. Show that the analysis of Exercises 11.3.36 to 11.3.45 can be entirely reproduced in the pre-conditioned case, up to the following changes: the Euclidean scalar product has to be replaced by the scalar product $x, y \mapsto x^T M y = (x, y)_M$ in appropriate places, the matrix A must be replaced by $M^{-1} A$ and the residual r^k by $z^k = M^{-1} r^k$. Show that the convergence rate is then

$$\frac{\sqrt{\lambda_{\max}\left(M^{-1} A\right)} - \sqrt{\lambda_{\min}\left(M^{-1} A\right)}}{\sqrt{\lambda_{\max}\left(M^{-1} A\right)} + \sqrt{\lambda_{\min}\left(M^{-1} A\right)}}.$$

11.3.5. Introduction to multigrid methods

The multigrid method proposes to correct the bad features of classical iterative methods for systems coming from the discretization of partial differential equations. It is a recursive iterative method, the recursion being performed on the scale of the spatial discretization.

It applies to discretizations of elliptic partial differential equations, or problems that can be reduced to them, and more generally to network problems, structural problems and many more. But we shall keep to the simplest case: we just want to find a fast iterative method for solving

$$A_N x = b,$$

with A_N being the $(N - 1) \times (N - 1)$ finite difference matrix given by

$$(11.3.20) \qquad A_N = N^2 \begin{pmatrix} 2 & -1 & & & \\ -1 & 2 & -1 & & \\ & \ddots & \ddots & \ddots & \\ & & -1 & 2 & -1 \\ & & & -1 & 2 \end{pmatrix}.$$

The main idea of the multigrid method is that some classical iterations, such as damped Jacobi iterations (11.3.21), act as low-pass filters: they significantly damp the high frequencies, whilst the low frequencies are not altered much by one iteration step; therefore, if we want to efficiently reduce the residual, after it has been, more or less, cleaned from its high frequency components, we will make a correction on the low frequency modes, using the coarse grid, i.e., a grid having a space step which is twice as large.

In the bigrid method, we suppose that N is even: we do a damped Jacobi iteration with an appropriately chosen parameter on the fine grid ($h = 1/N$); then we lift the residual to the coarse grid ($h = 2/N$); we solve an equation on the coarse grid, and we interpolate the solution on the fine grid to obtain a correction on it.

The beauty of the bigrid method is that the spectral radius of the matrix of the iterations does not depend on the space step. However, we could do more than one sweep of damped Jacobi iterations on the fine grid; we could also replace the coarse grid resolution by some other algorithm, for instance, some iterative method steps.

If we assume that N is a power of 2, the idea of the multigrid method is extremely simple: instead of solving on the coarse grid, we do one, or several, damped Jacobi sweeps, and we correct using an even coarser grid, on which we do a sweep, and so on, until we reach a very simple grid which could have only one point, and on which the resolution is trivial; then, we successively interpolate the corrections on all the finer grids, possibly doing more sweeps at each pass.

Many combinations are possible, and multigrid methods are still an active research subject. They are fascinating objects, and they are very close to the ideas used in a Fast Fourier Transform, and also in wavelet algorithms.

We will only treat the description of the bigrid method and the reason for its remarkable convergence. The multigrid method is even more beautiful: we just give a few indications at the end of the problem, since it is technically heavy to prove anything about multigrids.

Damped Jacobi methods

Let A be an arbitrary $d \times d$ symmetric positive definite matrix. Write $A = D - E - F$, with the notation of eqn (11.1.5). A damped Jacobi iteration is a generalization of the Jacobi iteration, which is written as

$$(11.3.21) \qquad Dx^{k+1} = (1 - \omega) Dx^k + \omega (E + F) x^k + \omega b.$$

Exercise 11.3.47. Give the eigenvalues $\mu_j(\omega)$ of the matrix $J(\omega)$ of a damped Jacobi method, as a function of ω and of the eigenvalues of $J(1) = D^{-1}(E + F)$.

Exercise 11.3.48. When A is equal to A_N given by eqn (11.3.20), calculate

$$\max \{|\mu_j(\omega)| : 1 \leqslant j \leqslant N - 1\},$$

and

$$(11.3.22) \qquad \max \{|\mu_j(\omega)| : N/2 \leqslant j \leqslant N - 1\}.$$

Exercise 11.3.49. Show that the expression (11.3.22) is minimal for $\omega = 2/3$ and give its value.

Exercise 11.3.50. Justify the statement: the damped Jacobi method with $\omega = 2/3$ is a low-pass filter.

Bigrid method

We assume, henceforth, that N is even, and we let $n = N/2$. The subspace V_k, for $1 \leqslant k \leqslant n - 1$, is the subspace of $\mathbb{R}^{N-1}$ spanned by the vectors $u_k = (\sin(jk\pi/N))_{1 \leqslant j \leqslant N-1}$ and $v_k = (\sin(j(N-k)\pi/N))_{1 \leqslant j \leqslant N-1}$; V_n is spanned by $u_n = (\sin(j\pi/2))_{1 \leqslant j \leqslant N-1}$. The space W_k is the subspace of $\mathbb{R}^{n-1}$ spanned by $w_k = (\sin(jk\pi/n))_{1 \leqslant j \leqslant n-1}$. We also define a restriction operator R from $\mathbb{R}^{N-1}$ to $\mathbb{R}^{n-1}$ by

$$(11.3.23) \qquad (Rx)_j = \frac{x_{2j-1} + 2x_{2j} + x_{2j+1}}{4}, \quad 1 \leqslant j \leqslant n - 1,$$

and an interpolation operator S from $\mathbb{R}^{n-1}$ to $\mathbb{R}^{N-1}$ by

$$(11.3.24) \qquad (Sx)_j = \begin{cases} x_{j/2} & \text{if } j \text{ is even;} \\ \left(x_{(j-1)/2} + x_{(j+1)/2}\right)/2 & \text{if } j \text{ is odd.} \end{cases}$$

In this last definition, it is assumed that x_0 and x_n are set equal to 0.

Exercise 11.3.51. Show that R maps V_k to W_k and V_n to 0. Give, in terms of $\kappa = \cos^2(k\pi/2N)$, the matrix of the restriction of R to V_k, equipped with the basis $\{u_k, v_k\}$, and W_k, equipped with the basis w_k.

Exercise 11.3.52. Show that S maps W_k to V_k and give the matrix of the restriction of S to W_k, equipped with the basis w_k, and V_k, equipped with the basis $\{u_k, v_k\}$.

Exercise 11.3.53. We define the bigrid algorithm as follows: $u^0 \in \mathbb{R}^{N-1}$ is the initial guess. We perform a Jacobi sweep:

$$(11.3.25) \qquad u^1 = J\left(\frac{2}{3}\right) u^0 + \frac{2}{3} D^{-1} b;$$

we calculate the residual:

$$(11.3.26) \qquad u^2 = b - A_N u^1;$$

we restrict the residual to the coarse grid:

$$(11.3.27) \qquad u^3 = R u^2;$$

we solve on the coarse grid the following problem:

$$(11.3.28) \qquad A_n u^4 = u^3;$$

we interpolate the result on the fine grid:

$$(11.3.29) \qquad\qquad u^5 = S u^4$$

and we use the result as a correction to u^1:

$$(11.3.30) \qquad\qquad u^6 = u^1 + u^5.$$

Give the matrix of the iteration described by the algorithm (11.3.25) to (11.3.30), in terms of R, S, A_N, A_n, and $J(2/3)$.

Exercise 11.3.54. For $k = 1, \ldots, n$, show that the restriction of the matrix of the iteration to the space W_k, equipped with the basis $\{u_k, v_k\}$, is given by

$$\begin{pmatrix} 1 \\ 1 \end{pmatrix} \begin{pmatrix} 1 - \kappa & \kappa \end{pmatrix} \begin{pmatrix} (4\kappa - 1)/3 & 0 \\ 0 & (3 - 4\kappa)/3 \end{pmatrix}.$$

Calculate the spectral radius of this matrix. Show that this spectral radius is bounded by $1/3$ for all N and k.

Exercise 11.3.55. We modify the algorithm (11.3.25) to (11.3.30) by performing ν Jacobi sweeps instead of 1:

$$u^{0,0} = u^0,$$

$$u^{0,r} = J\left(\frac{2}{3}\right) u^{0,r-1} + \frac{2}{3} D^{-1} b, \quad \forall r = 1, \ldots, \nu,$$

$$u^1 = u^{0,\nu}.$$

What is now the spectral radius ρ_ν of the matrix of the iteration? Show that it satisfies an estimate of the form

$$\rho_\nu \leqslant \frac{C}{\nu}.$$

Informal description of the multigrid method

As we understand what the bigrid method does, it suffices now to sketch what the multigrid method does. Assume that N is equal to 2^q; we replace the step with resolution $A_{N/2}$ by an iterative method. We perform one or several Jacobi sweeps in dimension $(N/2) - 1$, calculate the residual, and then lift it to a grid with $(N/4) - 1$ points. If we had a trigrid method, we would apply a resolution in dimension $(N/4) - 1$, interpolate the residual, so as to generate a correction on the grid of dimension $(N/2) - 1$, and finish as in the bigrid method.

There is clearly a recursive definition of the p-grid method. For $p = 2$, we get an iterative method depending on N and $N/2$. This method can be written as

$$(11.3.31) \qquad x^{k+1} = M\left(q, q - 1, \nu_q\right) x^k + \left(I_{2^q - 1} - M\left(q, q - 1, \nu_q\right)\right) b,$$

where ν_q is the number of Jacobi sweeps on the grid with q points. The p-grid method is defined from the $(p-1)$-grid method as follows: suppose that we have defined $M(q-1, q-2, \ldots, 1, \nu_{q-1}, \ldots, \nu_1, \mu_{q-2}, \ldots, \mu_2)$, where the ν_i describe the number of Jacobi sweeps on each grid and the μ_i describe the number of iterations used to approximate the inverse of A_{2^i}.

We perform ν_q sweeps of the damped Jacobi iteration with initial guess y and data b. We obtain a vector z, we calculate the residual $b - A_{2^q} z$, we restrict it to the grid with $2^{q-1} - 1$ points, and we apply μ_{q-1} iterations of the $(p-1)$-grid method to the restriction, with 0 as an initial guess. Then, we interpolate the result and we add the corresponding correction to z.

Exercise 11.3.56. Give a matrix description of the multigrid algorithm with 3 grids.

Exercise 11.3.57. Describe a bigrid algorithm for the case of finite differences on a rectangle (see Subsection 11.3.1) and calculate the spectral radius of the corresponding iteration matrix.

Exercise 11.3.58. Describe a multigrid algorithm in the case of finite differences on a rectangle.

12

Pythagoras' world

In this chapter, we discuss good old right angles and all the numerical and mathematical marvels they give rise to, as much for the solution of systems of equations as for the various interesting qualitative properties of Hermitian matrices. The reader will, perhaps, have noticed that this subject has already been touched upon in Section 3.1, Lemma 3.1.9, and in Section 10.1 in the study of the Rayleigh quotient.

We have already worked with orthogonality in Chapters 5, 7, and 8 and we will return to it in Chapter 13.

12.1. About orthogonalization

An essential property of unitary matrices is that they have a norm of 1 for the operator norm subordinate to the Hermitian norm. As a result of this, there is no (numeric) difficulty with multiplying by such a matrix, since it does not increase the relative error. We begin by trying to construct an orthonormal basis from some other basis. This is known as the Gram–Schmidt orthonormalization.

12.1.1. The Gram–Schmidt orthonormalization revisited

We are going to show that the Gram–Schmidt orthonormalization is equivalent to a matrix decomposition called QR. More precisely, we are going to recover the Gram–Schmidt orthonormalization process from the Cholesky decomposition:

Theorem 12.1.1. Let A be a matrix belonging to $\mathcal{M}_n(\mathbb{K})$. If A is invertible, there exists a unitary matrix Q and an upper triangular matrix R, which has a positive diagonal, such that

$$(12.1.1) \qquad\qquad A = QR.$$

Furthermore, this decomposition is unique and equivalent to the Gram–Schmidt orthonormalization of the basis formed from the column vectors of A. ◇

Proof.   The matrix A^*A is Hermitian and positive definite, since $x^*A^*Ax = |Ax|_2^2$ is strictly positive and is nonzero if and only if $x \neq 0$. We then know that there exists an upper triangular matrix R, with a positive diagonal, such that A^*A admits the Cholesky decomposition

$$A^*A = R^*R.$$

We let $Q = AR^{-1}$ and calculate Q^*Q:

$$Q^*Q = (R^*)^{-1} A^* AR^{-1} = (R^*)^{-1} R^*RR^{-1} = I_n.$$

Consequently, Q is unitary and we have shown the existence of such a decomposition. Uniqueness is shown as follows: suppose that $A = Q_1 R_1 = Q_2 R_2$, then,

$$A^*A = R_1^*Q_1^*Q_1 R_1 = R_1^*R_1$$

and

$$A^*A = R_2^*Q_2^*Q_2 R_2 = R_2^*R_2.$$

By virtue of the uniqueness of the Cholesky decomposition (Theorem 9.4.3), we see that $R_1 = R_2$, and we immediately deduce that $Q_1 = Q_2$.

Relation (12.1.1) can be written as

$$(12.1.2) \qquad\qquad A_{ik} = \sum_{j=1}^{n} Q_{ij} R_{jk}.$$

Noting that $R_{jk} = 0$ if $j > k$, we can rewrite eqn (12.1.2) as

$$(12.1.3) \qquad\qquad A_{ik} = \sum_{j=1}^{k} Q_{ij} R_{jk}.$$

We denote the column vectors of A by $(f_j)_{1 \leqslant j \leqslant n}$ and the column vectors of Q by $(q_j)_{1 \leqslant j \leqslant n}$. Then, relation (12.1.3) is written vectorially as

$$(12.1.4) \qquad\qquad f_k = \sum_{j=1}^{k} q_j R_{jk}.$$

For $k = 1$, we thus have

$$(12.1.5) \qquad\qquad f_1 = q_1 R_{11},$$

with $R_{11} > 0$. The vector q_1 is of norm 1, since Q is unitary. Therefore, we must have

$$R_{11} = |f_1|,$$

and relation (12.1.5) is the first step of the Gram–Schmidt orthonormalization. For $k > 1$, relation (12.1.4) is interpreted as follows: we choose the coefficients R_{jk}, $j < k - 1$ such that the vector $f_k - \sum_{j=1}^{k-1} q_j R_{jk}$ is orthogonal to q_j, for all $j < k-1$. We then write $R_{kk}q_k = f_k - \sum_{j=1}^{k-1} q_j R_{jk}$, and we choose $R_{kk} > 0$ such that $|q_k| = 1$. This is the k-th step of the Gram–Schmidt orthonormalization. $\square$

A consequence of this result is that we can always complete an independent system of k orthonormal vectors $(q_j)_{1 \leqslant j \leqslant k}$ of $\mathbb{K}^n$ to produce an orthonormal basis. Indeed, there always exists a choice of $(n-k)$ vectors from the canonical basis e_i, denoted $(e_{i_r})_{1 \leqslant r \leqslant n-k}$, such that the family of vectors formed from $(q_j)_{1 \leqslant j \leqslant k}$ and from $(e_{i_r})_{1 \leqslant r \leqslant n-k}$ is a basis of $\mathbb{K}^n$. It suffices then, to orthonormalize the basis thus constructed. The term Gram–Schmidt orthogonalization is also used when an orthogonal basis is constructed by induction from an arbitrary basis, without performing the normalization step.

A very important lemma, which is usually proved in the first year of a degree, is the Schur lemma. We reprove it here using block notation:

Theorem 12.1.2 (Schur's lemma). Let A be a complex $n \times n$ matrix. Then A can be made triangular in an orthonormal basis.   ◇

Proof.   Making A triangular in an orthonormal basis is equivalent to finding a unitary matrix Q such that QAQ^{-1} is triangular. We reason by induction on n, the dimension of the space. If $n = 1$ the result is trivial. Suppose that it is true for n, and that A is an $(n+1) \times (n+1)$ matrix. We know that A possesses at least one eigenvector, which we can assume to have norm 1. We denote this eigenvector by f_1 and let λ be the corresponding eigenvalue. We complete the family consisting of only f_1 by the vectors $f_2, \ldots, f_{n+1}$ in the manner of constructing an orthonormal basis. We therefore have a unitary transformation matrix P such that

$$PAP^{-1} = \begin{pmatrix} \lambda & \ell \\ 0 & B \end{pmatrix},$$

where B is an $n \times n$ matrix. The induction hypothesis tells us that there exists a unitary matrix U and a triangular matrix T such that

$$UBU^{-1} = T.$$

We let

$$V = \begin{pmatrix} 1 & 0 \\ 0 & U \end{pmatrix}.$$

It is immediate that V is unitary and, furthermore,

$$\begin{aligned} VPAP^{-1}V^{-1} &= \begin{pmatrix} 1 & 0 \\ 0 & U \end{pmatrix} \begin{pmatrix} \lambda & \ell \\ 0 & B \end{pmatrix} \begin{pmatrix} 1 & 0 \\ 0 & U^{-1} \end{pmatrix} \\ &= \begin{pmatrix} \lambda & \ell \\ 0 & UB \end{pmatrix} \begin{pmatrix} 1 & 0 \\ 0 & U^{-1} \end{pmatrix} \\ &= \begin{pmatrix} \lambda & \ell U^{-1} \\ 0 & UBU^{-1} \end{pmatrix} \\ &= \begin{pmatrix} \lambda & \ell U^{-1} \\ 0 & T \end{pmatrix}. \end{aligned}$$

We have therefore found a matrix $Q = VP$ such that QAQ^{-1} is triangular.   □

12.1.2. Paths of inertia

From the QR decomposition, we are going to give a topological proof, due to G. Strang [74], of Sylvester's inertia theorem, which states the following:

Theorem 12.1.3. Let A be a Hermitian matrix and C be some invertible matrix. Then, the number of strictly positive (respectively, zero, strictly negative) eigenvalues of C^*AC is equal to the number of strictly positive (respectively, zero, strictly negative) eigenvalues of A. ◇

The proof of this theorem depends on a continuity result about the eigenvalues of Hermitian operators which we will state and prove below; it depends on the following minimax characterization of the eigenvalues of a Hermitian matrix.

Theorem 12.1.4. Let A be a Hermitian matrix, r_A be the Rayleigh quotient associated with A, and $\lambda_p(A)$ be the p-th eigenvalue defined by

$$\lambda_p(A) = \min_{\dim W = p} \max_{x \in W \setminus \{0\}} r_A(x).$$

Then $\lambda_p(A)$ is a continuous function of A. More precisely, for every p and all Hermitian matrices A and B, we have the following inequality:

$$|\lambda_p(A) - \lambda_p(B)| \leqslant \|A - B\|_2. \qquad ◇$$

Proof. For all Hermitian matrices A and B, we have

$$r_A(x) = r_B(x) + r_{A-B}(x).$$

We deduce, from Lemma 10.3.5 and Definition 10.2.1, that

$$\|A - B\|_2 = \max_{x \neq 0} |r_{A-B}(x)|,$$

and consequently,

$$|r_A(x) - r_B(x)| \leqslant \|A - B\|_2, \quad \forall x \in \mathbb{K}^n \setminus \{0\}.$$

The conclusion of the theorem is then immediate. □

We can order Hermitian matrices by deciding that $A \geqslant B$ if and only if $A - B$ is positive or zero (in the sense of sesquilinear forms). We then have the following result which links the order of the matrices with the order of their eigenvalues:

Lemma 12.1.5. Let A and B be two Hermitian matrices such that $A \geqslant B$. Then, for all $p = 1, \ldots, n$, $\lambda_p(A) \geqslant \lambda_p(B)$.

Proof. The hypotheses of the theorem imply that, for all $x \neq 0$,

$$r_A(x) \geqslant r_B(x).$$

The conclusion of the theorem follows immediately from Theorem 10.1.3. □

Proof of Theorem 12.1.3. This is a topological proof, unlike the usual proof which is algebraic.

To begin with, suppose we consider the case of invertible matrices A. Such a matrix has exactly p strictly negative eigenvalues and exactly $n - p$ strictly positive eigenvalues. Let $t \mapsto C(t)$ be a continuous mapping from $[0,1]$ to $GL_n(\mathbb{K})$, the group of invertible matrices with coefficients in $\mathbb{K}$. If we can construct a continuous path of invertible matrices, starting at $C(0) = C$ and ending at a unitary matrix $C(1)$, i.e., a matrix satisfying $C(1)^* = C(1)^{-1}$, we will have succeeded in proving the result.

Indeed, we remark, first of all, that $B(1) = C(1)^* A C(1)$ is similar to A and therefore, has the same eigenvalues as A. In particular, $B(1)$ has exactly p strictly negative eigenvalues and $n - p$ strictly positive eigenvalues. Suppose that the number of strictly negative eigenvalues of the Hermitian matrix $B(t) = C(t)^* A C(t)$ is not constant with respect to t. We denote by $p(t)$ the number of strictly negative eigenvalues of $B(t)$. If $p(t_0) \neq p$, for a certain value $t_0 \in [0,1[$, two symmetric cases are then possible: $p(t_0) > p$ and $p(t_0) < p$.

Consider, for example, the first case. Consequently, $\lambda_{p(t_0)}(B(t)) < 0$ and $\lambda_{p(t_0)}(B(1)) > 0$. From Theorem 12.1.4, we have that $\lambda_{p(t_0)}(B(t_0))$ is a continuous function of $t \in [0,1]$, and therefore, its image is connected. In particular, there will exist a value t_1 of t for which $\lambda_{p(t_0)}(B(t_1))$ vanishes, which is impossible since $B(t)$ is invertible for all t.

Similarly, if we have the case where $p(t_0) < p$, then we consider the continuous function $t \mapsto \lambda_p(B(t))$, which takes a negative value at $t = 1$ and a positive value at $t = t_0$.

We now show that we can construct a path having the required properties. Let $C = QR$ be the decomposition studied in Theorem 12.1.1. Let

$$C(t) = Q\big((1 - t)R + tI\big).$$

It is clear that $C(t)$ is continuous with respect to t and that $C(0) = C$ and $C(1) = Q$, which is unitary. Matrix $C(t)$ is invertible since $(1 - t)R + tI$ is an upper triangular matrix with a strictly positive diagonal for all $t \in [0,1]$. We have therefore proved the theorem in the case where A is invertible.

We pass now to the general case. The eigenvalues of A are

$$\lambda_1 \leqslant \cdots \leqslant \lambda_p < 0 = \lambda_{p+1} = \cdots = \lambda_q < \lambda_{q+1} \leqslant \cdots \leqslant \lambda_n.$$

Let $B(\epsilon) = C^*(A + \epsilon I)C$, and let $\mu_k(\epsilon)$ be the k-th eigenvalue of $B(\epsilon)$. The eigenvalues of $A + \epsilon I$ are

$$\lambda_1 + \epsilon \leqslant \cdots \leqslant \lambda_p + \epsilon < \epsilon = \lambda_{p+1} = \cdots = \lambda_q < \lambda_{q+1} + \epsilon \leqslant \cdots \leqslant \lambda_n + \epsilon.$$

The first part of the proof shows us that, if $\epsilon \in \,]0, -\lambda_p[$, $A + \epsilon I$, and therefore $B(\epsilon)$, have exactly p strictly negative eigenvalues and $n - p$ strictly positive eigenvalues. Furthermore, if $\epsilon \in \,]-\lambda_{q+1}, 0[$, $A + \epsilon I$ and $B(\epsilon)$ have exactly q

strictly negative eigenvalues and $n - q$ strictly positive eigenvalues. In particular, if ϵ is a small positive number, $\mu_q(\epsilon)$ and $\mu_{p+1}(\epsilon)$ are strictly positive, and if ϵ is a small negative number, $\mu_q(\epsilon)$ and $\mu_{p+1}(\epsilon)$ are strictly negative. By continuity, we see that $\mu_q(0) = \mu_{p+1}(0) = 0$. This shows that $B(0)$ has at least as many zero eigenvalues as A. We now have to bound below the number of strictly negative eigenvalues and strictly positive eigenvalues of $B(0)$. For small positive ϵ, $B(\epsilon)$ has exactly p strictly negative eigenvalues. Consequently, from Lemma 12.1.5, $B(\epsilon) - \epsilon C^* C = B(0)$ has at least p strictly negative eigenvalues. We argue in the same manner for the positive eigenvalues. □

12.1.3. Topological properties of the Cholesky and QR decompositions

We will return to the QR decomposition, whose continuity properties we will examine, after having looked at those of the Cholesky decomposition.

Lemma 12.1.6. The Cholesky decomposition defines a continuous mapping from the set of positive definite Hermitian matrices to the set of upper triangular matrices with a strictly positive diagonal.

Proof. We will use the notation of Theorem 9.4.3. Let A be a positive definite Hermitian matrix. If n is equal to 1, $A = (\alpha)$, and as the mapping $\alpha \mapsto \sqrt{\alpha}$ is continuous on $\mathbb{R}_*^+$, we see that the mapping $A \mapsto C = (\sqrt{\alpha})$ is continuous.

Suppose that we have continuity in $n - 1$ dimensions. We write

$$A = \begin{pmatrix} \alpha & \ell \\ \ell^* & \hat{A} \end{pmatrix},$$

and from the proof of Theorem 9.4.3 we have $A = C^* C$ with

$$C = \begin{pmatrix} \beta & m \\ 0 & B \end{pmatrix}$$

and

$$\beta = \sqrt{\alpha}, \quad m = \frac{\ell}{\sqrt{\alpha}}, \quad B^* B = \hat{A} - \frac{\ell^* \ell}{\alpha}.$$

It is clear that β and m are continuous functions of A, mapping to values in $\mathbb{R}^+$ and $(n-1)$-dimensional linear forms, respectively, provided that α is strictly positive. The mapping $A \mapsto \hat{A} - \ell^* \ell / \alpha$ is a continuous mapping to $(n-1) \times (n-1)$ matrices, and we saw in the proof of Theorem 9.4.3 that it maps to positive definite Hermitian matrices. By induction, we know that the mapping $\hat{A} - \ell^* \ell / \alpha \mapsto B$ is continuous and maps to $(n-1) \times (n-1)$ upper triangular matrices with strictly positive diagonals. Consequently, the mapping $A \mapsto C$ is a continuous mapping to values in $n \times n$ upper triangular matrices with strictly positive diagonals. □

Lemma 12.1.7. The QR decomposition defines a continuous mapping from $GL_n(\mathbb{C})$, the group of invertible complex $n \times n$ matrices to the product of $U_n(\mathbb{C})$, the group of unitary matrices, and $T_n^+(\mathbb{C})$, the group of complex upper triangular matrices with strictly positive diagonal.

Proof. We will use the notation of Theorem 12.1.1. Let $A \in GL_n(\mathbb{C})$. The mapping $A \mapsto A^*A$ is clearly continuous, and we have seen that it maps to positive definite Hermitian matrices. It follows, from Lemma 12.1.6, that the mapping $A \mapsto R$, where R is the member of $T_n^+(\mathbb{C})$ defined by

$$A^*A = R^*R,$$

is continuous. Furthermore, $A \mapsto Q = AR^{-1}$ is continuous, and from the proof of Theorem 12.1.1, it maps to the elements in the set of unitary matrices.   □

12.1.4. Operation counts and numeric strategies

The QR decomposition is extremely useful practically for the solution of linear systems and, as we will see later, for the search for the eigenvalues and eigenvectors of a matrix. Unfortunately, it is slow and pretty unstable in its naïve form:

Operation Count 12.1.8. The QR decomposition of an invertible matrix, viewed as a Gram–Schmidt orthogonalization, requires n square roots and of order $2n^3$ arithmetic operations.

Proof. As in Theorem 12.1.1, we denote the column vectors of A by $(f_j)_{1 \leqslant j \leqslant n}$ and the column vectors of Q by $(q_j)_{1 \leqslant j \leqslant n}$. For $k = 1$, we have

$$f_1 = q_1 R_{11}.$$

We therefore need to calculate $R_{11} = \sqrt{|f_1|^2}$, which requires n multiplications, $n - 1$ additions and the taking of a square root. Furthermore, the calculation of q_1 requires n divisions: it is necessary to divide each of the components of f_1 by R_{11}. To calculate q_k, knowing the q_ℓ for $\ell \leqslant k - 1$, we write that

$$f_k = \sum_{j=1}^{k} q_j R_{jk}.$$

Consequently,

$$R_{\ell k} = (f_k, q_\ell)$$

and the calculation of $R_{\ell k}$ requires n multiplications and $n - 1$ additions. In all, it needs $(k - 1)(2n - 1)$ operations to calculate the $R_{\ell k}$ for $\ell \leqslant k - 1$. We have

$$R_{kk} = \left| f_k - \sum_{j=1}^{k-1} q_j R_{jk} \right|.$$

Each of the components of a vector for which we calculate the norm is calculated by means of $k-1$ multiplications and $k-1$ additions or subtractions. It therefore needs $n(2k-2)$ operations to construct this vector. The calculation of the norm of this vector needs n multiplications, $n-1$ additions and a square root. Then, the calculation of q_k needs n divisions. Finally, the calculation of Q_k and $R_{\ell k}$ for $\ell \leqslant k$ requires

$$(k-1)(2n-1) + (2k-2)n + 2n - 1 + n = n(4k-1) - k$$

arithmetic operations and the taking of a square root. By summation we find the number of arithmetic operations is of order $2n^3$ plus n square roots. $\square$

Furthermore, the Gram–Schmidt process is not very stable numerically, as is shown in [69].

12.1.5. Hessenberg form

The most practical method of calculating QR decompositions is to apply a numerical strategy which is dependent on the particular properties of so-called Hessenberg matrices. A square matrix is in upper Hessenberg form if it has the following form:

$$\begin{pmatrix} * & * & \cdots & \cdots & \cdots & * \\ * & * & \cdots & \cdots & \cdots & * \\ 0 & \ddots & \ddots & & & \vdots \\ \vdots & \ddots & \ddots & \ddots & & \vdots \\ \vdots & & \ddots & \ddots & \ddots & \vdots \\ 0 & \cdots & \cdots & 0 & * & * \end{pmatrix}.$$

In other words, all the coefficients A_{ij} of this matrix are zero if $i > j+1$. Obviously, a matrix A will be lower Hessenberg if A^* is upper Hessenberg.

There are two interesting properties of matrices in Hessenberg form. First of all, we can put a matrix into Hessenberg form in $O(n^3)$ operations (which is not much better than the preceding operation), by a procedure which is stable. In addition, the QR decomposition of an upper Hessenberg matrix A produces matrices Q and R by a stable procedure such that Q and RQ are upper Hessenberg.

Let us verify that Q and RQ are, indeed, upper Hessenberg, if A is upper Hessenberg and invertible. As in Theorem 12.1.1, we denote the column vectors of A by $(f_j)_{1 \leqslant j \leqslant n}$ and the column vectors of Q by $(q_j)_{1 \leqslant j \leqslant n}$. We have

$$f_1 = q_1 R_{11}.$$

It is obvious that only the first two components of q_1 can be nonzero. Suppose that up to row $k-1$, only the first $j+1$ components of q_j are nonzero. Then, from

$$f_k = \sum_{j=1}^{k} q_j R_{jk},$$

we deduce that only the first $k+1$ components of q_k can be nonzero. This little induction therefore shows us that Q is upper Hessenberg. As for the product RQ, which we will need later, it is also in upper Hessenberg form. We calculate its coefficients $(RQ)_{ij}$, for $i > j + 1$, as follows:

$$(RQ)_{ij} = \sum_{k} R_{ik} Q_{kj} = \sum_{i \leqslant k \leqslant j+1} R_{ik} Q_{kj} = 0.$$

12.1.6. Householder transformations

To put a matrix in Hessenberg form, we make use of the Householder transformations, see [47], which we are now going to introduce.

We describe these Householder transformations: for every $v \in \mathbb{C}^n \setminus \{0\}$, we write

$$S_n(v)\, x = x - 2\frac{vv^* x}{v^* v}.$$

Geometrically, $S_n(v)$ is the orthogonal reflection with respect to the hyperplane which is orthogonal to v. Indeed, if $x = \lambda v$,

$$S_n(v)\, x = \lambda v - 2\lambda\frac{vv^* v}{v^* v} = -\lambda v = -x.$$

Moreover, if x is orthogonal to v, that is to say $v^* x = 0$, it is clear that

$$S_n(v)\, x = x.$$

The transformation $S_n(v)$ is, at the same time, Hermitian and unitary:

$$S_n(v) = I - 2\frac{vv^*}{v^* v}$$

is clearly Hermitian. Furthermore,

$$S_n(v)^2 = I - 4\frac{v^* v}{v^* v} + 4\frac{vv^* vv^*}{(v^* v)^2} = I - 4\frac{v^* v}{v^* v} + 4\frac{v^* v\, vv^*}{v^* v\, v^* v} = I.$$

Abusing the language a little, we will consider the identity as a Householder transformation. The Householder transformations will be useful by virtue of the following lemma:

Lemma 12.1.9. For all vectors x and y with the same Euclidean norm, there exists a vector v, a Householder transformation $S_n(v)$, and a complex number ω of modulus 1 such that

$$\omega y = S_n(v)\, x.$$

Proof. Suppose, first of all, that $x = y = 0$. The case is trivial, and any v and ω are suitable. Now, if x and y both have nonzero norm but are linearly dependent, we will take the identity as the Householder transformation, and ω such that $x - \omega y = 0$. Such an ω exists since x and y are linearly dependent. We suppose now that x and y are linearly independent. Then we have

$$x - 2\frac{vv^*x}{v^*v} = \omega y,$$

therefore,

$$2vv^*x = (v^*v)(x - \omega y).$$

Consequently, v and $x - \omega y$ are linearly dependent. Since $S_n(v)$ depends only on the direction of v, we may take

$$v = x - \omega y.$$

To determine ω we examine $\omega y - S_n(v)x$:

$$\omega y - S_n(v)x = \omega y - x + 2\frac{vv^*x}{v^*v}$$

$$= v\left(-1 + 2\frac{v^*x}{v^*v}\right)$$

$$= \left[\frac{v}{v^*v}\right]v^*(2x - v)$$

$$= \left[\frac{v}{v^*v}\right](x^* - \bar{\omega}y^*)(x + \omega y).$$

We therefore have, taking account of the equal norms of x and y,

$$\bar{\omega}y^*x = \omega x^*y.$$

It is therefore necessary to choose, $\omega = \pm e^{i\theta}$, if $y^*x \neq 0$, where θ is the argument of the complex number y^*x. If $y^*x = 0$, the choice of θ is immaterial, and we can take $\omega = \pm 1$. $\qquad \square$

In practice, the choice of sign in front of $e^{i\theta}$ is governed by conditioning considerations. We will choose, in preference, the sign which leads to an $x - \omega y$ of largest norm.

12.1.7. QR decomposition by Householder transformations

With a finite series of Householder transformations, we can find the QR decomposition of a matrix:

Theorem 12.1.10. Let A be an invertible matrix. Then there exist $n - 1$ Householder transformations $S_n(v_1), S_n(v_2), \ldots, S_n(v_{n-1})$ such that

$$S_n(v_{n-1}) \cdots S_n(v_1) A = T$$

is upper triangular. In particular, the QR decomposition of A is given by

$$R = D^{-1}T, \quad Q = S_n\left(v_{n-1}\right)\cdots S_n\left(v_1\right)D,$$

where D is a diagonal matrix whose coefficients are all of modulus 1.   ◇

Proof. We are going to argue by induction on the dimension of the space. In one dimension, the result is trivial. Suppose it to be true in dimension $n-1$. Let A be an $n\times n$ invertible matrix. It can be put in the form

$$A = \begin{pmatrix} f_1 & f_2 & \cdots & f_n \end{pmatrix},$$

where the f_j are the column vectors of A. From Lemma 12.1.9, there exists a vector v_1 in $\mathbb{C}^n$ such that

$$S_n\left(v_1\right)f_1 = |f_1|\omega e_1,$$

where e_1 is the first vector in the canonical basis of $\mathbb{C}^n$. We then have

$$S_n\left(v_1\right)A = \begin{pmatrix} |f_1|\omega e_1 & S_n\left(v_1\right)f_2 & \cdots & S_n\left(v_1\right)f_n \end{pmatrix} = \begin{pmatrix} |f_1|\omega & \ell \\ 0 & \hat{A} \end{pmatrix}.$$

From the induction hypothesis, there exist vectors $\hat{v}_2,\ldots,\hat{v}_{n-1}$ in $\mathbb{C}^{n-1}$ such that

$$S_{n-1}\left(\hat{v}_{n-1}\right)\cdots S_{n-1}\left(\hat{v}_2\right)\hat{A} = \hat{T}.$$

Letting

$$v_j = \begin{pmatrix} 0 \\ \hat{v}_j \end{pmatrix}$$

for $2\leqslant j\leqslant n-1$, we see that

$$S_n\left(v_j\right) = \begin{pmatrix} 1 & 0 \\ 0 & S_{n-1}\left(\hat{v}_j\right) \end{pmatrix}$$

and that

$$S_n\left(v_{n-1}\right)\cdots S_n\left(v_1\right)A = \begin{pmatrix} |f_1|\omega & \ell \\ 0 & \hat{T} \end{pmatrix},$$

which proves the possibility of a triangulation by Householder transformations.

The matrix T thus obtained is upper triangular, but its diagonal terms are not necessarily strictly positive, although they are nonzero. If we write

$$D = \text{diag}\left(T_{ii}/|T_{ii}|\right),$$

it is clear that $D^{-1}T$ is upper triangular with a strictly positive diagonal. Moreover, D is unitary, therefore QD is unitary, and we have obtained the QR decomposition of A.   □

It is shown how to put this algorithm into practice in [12, 34].

The advantage of this method of triangulation is that the transformations $S_n(v_j)$ are all unitary and, therefore, they do not change the conditioning of the matrix. This decomposition can be used to solve a linear system. However, the transformation to a triangular system by Householder transformations is more costly than LU Gaussian decomposition. We can show that it requires order $2n^3$ arithmetic operations, and $2n$ square roots. The numerical stability, see [58], can balance the greater cost of the calculations.

On the other hand, if A is an upper Hessenberg matrix, the matrices appearing in the algorithm of Theorem 12.1.10 are all upper Hessenberg, and the operation count is a lot more favourable:

Theorem 12.1.11. For an upper Hessenberg matrix A, the matrices in the expression

$$S_n (v_j) \cdots S_n (v_1) A,$$

described in Theorem 12.1.10, are all upper Hessenberg. ◇

Operation Count 12.1.12. The QR decomposition of an upper Hessenberg matrix requires order $n^3/3$ arithmetic operations, and order n square roots.

Proof. Let $C(n)$ be the number of operations necessary to go from A to the matrix $S_n(v_1)A$. We write $x = f_1$, $v = \hat{v}_1$, and $y = |f_1|e_1$. As f_1 has at most two nonzero components, the calculation of $|f_1|$ demands 2 multiplications, an addition and the taking of a square root. To get ω in the complex case, we must calculate y^*x, which demands a multiplication and the calculation of the modulus of this complex number. We obtain ω by division, which involves two complex operations and the taking of another square root. The calculation of $v = x - \omega y$ demands only a multiplication and a subtraction since y has only one nonzero component. The calculation of v^*v demands 3 complex operations, and the calculation of v^*f_k demands, for its part, $k + 1$ multiplications and k additions, as f_k has at most $k+1$ nonzero components. It needs 2 further arithmetic operations to get $2v^*f_k/v^*v$. Finally, the calculation of $f_k - (2v^*f_k/v^*v)v$ requires 2 subtractions and 2 multiplications since v has only two nonzero components. We also note that $S_n(v)f_k$ has at most its first $k + 1$ components nonzero, which proves that $\hat{A}$ is Hessenberg. In total, we have to make

$$2 + 1 + 2 + 2 + 3 + \sum_{k=2}^{n-1} (2k + 1 + 4) \sim n^2 \text{ arithmetic operations}$$

and 2 square roots. We therefore have

$$C(n) \sim C(n-1) + n^2.$$

In total, summing with respect to n, we find the result claimed. □

12.1.8. Hessenberg form by Householder transformations

Putting a matrix into Hessenberg form is also done by Householder transformations:

Lemma 12.1.13. For every invertible matrix A of n rows and n columns, there exists a family of Householder transformations

$$(S_n(v_j))_{1 \leqslant j \leqslant n-2}$$

such that

$$S_n(v_{n-2}) \cdots S_n(v_1) A S_n(v_1) \cdots S_n(v_{n-2})$$

is upper Hessenberg.

Proof. We will prove by induction on the dimension. In two dimensions the result is obvious, since every 2×2 matrix is upper Hessenberg. Suppose that in $n-1$ dimensions we can find $n-3$ vectors in $\mathbb{C}^{n-1}$, with zero first component, which allow us to put an invertible matrix into Hessenberg form by Householder transformations. Let A be an $n \times n$ invertible matrix. We can put it into the form

$$A = \begin{pmatrix} a_{11} & \ell \\ p & \hat{A}^1 \end{pmatrix}.$$

There exists a vector $\hat{v}_1$ in $\mathbb{C}^{n-1}$ such that

$$S_{n-1}(\hat{v}_1) p = \omega |p| \hat{e}_1,$$

where $\hat{e}_1$ is the first vector in the canonical basis of $\mathbb{C}^{n-1}$. Then, let

$$v_1 = \begin{pmatrix} 0 \\ \hat{v}_1 \end{pmatrix},$$

which implies that

$$S_n(v_1) = \begin{pmatrix} 1 & 0 \\ 0 & S_{n-1}(\hat{v}_1) \end{pmatrix}.$$

Consequently,

$$\begin{aligned}
S_n(v_1) A S_n(v_1) &= \begin{pmatrix} 1 & 0 \\ 0 & S_{n-1}(\hat{v}_1) \end{pmatrix} \begin{pmatrix} a_{11} & \ell \\ p & \hat{A}^1 \end{pmatrix} \begin{pmatrix} 1 & 0 \\ 0 & S_{n-1}(\hat{v}_1) \end{pmatrix} \\
&= \begin{pmatrix} a_{11} & \ell S_{n-1}(\hat{v}_1) \\ S_{n-1}(\hat{v}_1) p & S_{n-1}(\hat{v}_1) \hat{A}^1 S_{n-1}(\hat{v}_1) \end{pmatrix}.
\end{aligned}$$

From the induction hypothesis, there exist $n-3$ vectors $\hat{v}_2, \ldots, \hat{v}_{n-2}$ in $\mathbb{C}^{n-1}$, with zero first component such that

$$\hat{B} = S_{n-1}(\hat{v}_{n-2}) \cdots S_{n-1}(\hat{v}_2) \left[S_{n-1}(\hat{v}_1) \hat{A}^1 S_{n-1}(\hat{v}_1) \right] S_{n-1}(\hat{v}_2) \cdots S_{n-1}(\hat{v}_{n-2})$$

is upper Hessenberg. We then have, on letting

$$v_j = \begin{pmatrix} 0 \\ \hat{v}_j \end{pmatrix}, \quad 2 \leqslant j \leqslant n-2, \qquad m = \ell S_{n-1}(\hat{v}_1) \cdots S_{n-1}(\hat{v}_{n-2}),$$

the relation

$$S_n(v_{n-2}) \cdots S_n(v_2)[S_n(v_1) AS_n(v_1)] S_n(v_2) S_n(v_{n-2}) = \begin{pmatrix} a_{11} & m \\ S_{n-1}(\hat{v}_1)p & \widehat{B} \end{pmatrix},$$

noting that, for $j \geqslant 2$,

$$S_{n-1}(\hat{v}_j) S_{n-1}(\hat{v}_1)p = S_{n-1}(\hat{v}_1)p,$$

since the first component of $\hat{v}_j$ is zero. It is clear that the matrix

$$\begin{pmatrix} a_{11} & m \\ S_{n-1}(\hat{v}_1)p & \widehat{B} \end{pmatrix}$$

is upper Hessenberg $\qquad\qquad\qquad\qquad\qquad\qquad\qquad\qquad\qquad\qquad\square$

The number of operations necessary to put a matrix into upper Hessenberg form is $O(n^3)$, as the reader can calculate. It is sufficient to do this once initially.

12.2. Exercises from Chapter 12

12.2.1. The square root of a Hermitian positive definite matrix

In all of this problem the basis field is $\mathbb{C}$.

We denote the set of $n \times n$ matrices with complex coefficients by $\mathcal{M}$. Let $\mathcal{H}$ be the subset of $\mathcal{M}$ of Hermitian matrices and $\mathcal{P}$ be the subset of $\mathcal{H}$ formed from matrices which are positive definite.

We denote the Euclidean norm on $\mathbb{C}^n$ by $|\cdot|$ and the subordinate operator norm by $\|\cdot\|$.

We define a mapping from $\mathcal{P}$ to $\mathcal{M}$ as follows: if $B \in \mathcal{P}$, the Cholesky decomposition is denoted

(12.2.1) $$B = C^*C$$

and we let

(12.2.2) $$F(B) = C, \quad G(B) = CC^* = F(B) F(B)^*.$$

Exercise 12.2.1. Show that G maps to elements in $\mathcal{P}$.

Exercise 12.2.2. Show that $\|G(B)\| = \|B\| = \|F(B)\|^2$.

Exercise 12.2.3. Let Λ be a diagonal matrix belonging to $\mathcal{P}$. Show that there exists a matrix H belonging to $\mathcal{P}$ such that $H^2 = \Lambda$, and give H explicitly.

Exercise 12.2.4. Let Λ be defined as in the preceding question and let K be a matrix belonging to $\mathcal{P}$ such that $K^2 = \Lambda$. We denote by $\sqrt{\Lambda}$ the matrix $F(\Lambda)$, which is clearly diagonal and in $\mathcal{P}$. Let $K = QR$, with Q unitary and R upper triangular with a strictly positive diagonal.

(i) Show by calculating K^*K that $R = \sqrt{\Lambda}$;

(ii) Show that K can only be positive definite if Q has 1 as its only eigenvalue. Suppose that ω is an eigenvalue of Q distinct from 1, associated to an eigenvector x and obtain a contradiction;

(iii) Conclude that there exists a unique matrix $H \in \mathcal{P}$ such that $H^2 = \Lambda$.

Exercise 12.2.5. Let B be in $\mathcal{P}$. Deduce from the preceding question that there exists a unique matrix H in $\mathcal{P}$ such that $H^2 = B$. This matrix will be denoted $\sqrt{B}$.

Exercise 12.2.6. Show that if B is symmetric positive definite and real, then $\sqrt{B}$ is also symmetric positive definite and real.

Part IV

Nonlinear problems

In this part we treat three different kinds of nonlinear problems: the calculation of eigenvalues and eigenvectors of a matrix, the resolution of nonlinear equations and systems, and the numerical integration of ordinary differential equations.

These three problems are deeply related, and I shall write only a few words about them.

First, for practical applications, it is often necessary to find the modes of a vibrating structure, for instance, earthquake certification of high buildings requires the computation of their eigenmodes.

But, of course, eigenmodes are interesting because they appear as special solutions of the differential equation which governs the motion of a large structure. Currently, nonlinear effects are not taken into account by earthquake certification, but if interaction between the nonlinear ground and the structure is considered, one might have to develop a nonlinear analysis for earthquake certification of large structures.

When integrating nonlinear differential systems, one has often to solve a nonlinear system of equations at each step. Being able to solve these nonlinear systems is then completely crucial in order to obtain reasonable computing times and reasonable accuracy.

13

Spectra

The theoretical problem of finding eigenvectors and eigenvalues of a square matrix A could be considered as being totally solved after a first year course. But the calculation of these famous eigenelements or spectra is another problem entirely!

The numerical computation of eigenvalues of linear operators is a subject which has a bad name, for bad and for good reasons. The readers will decide for themselves after reading this chapter whether the reputation is deserved. But, in order to make the problems more palatable, let me emphasize some essential ideas, which might otherwise be lost in the technicalities.

We learn in any first year course that the eigenvalues of a matrix are the roots of its characteristic polynomial. Assume that we have a good method for computing the characteristic polynomial. If we normalize the leading coefficient X^n to be equal to 1, the information given by the characteristic polynomial is totally contained in the sequence of n real numbers. On the other hand, the matrix we started from contains n^2 numbers.

In some cases, there is much more information in the matrix than in its characteristic polynomial. In particular, if the matrix is self-adjoint, or skew-adjoint, or unitary, which can be tested with a few operations, or is a consequence of the nature of the problem, then it is quite clear that we know much about the eigenvalues and eigenvectors and that this information is not visible in the characteristic polynomial. On the other hand, if A is a companion matrix of the form

$$A = \begin{pmatrix} 0 & 1 & 0 & 0 & \cdots & 0 \\ 0 & 0 & 1 & 0 & \cdots & 0 \\ 0 & 0 & 0 & 1 & \cdots & 0 \\ \vdots & & & & \ddots & \vdots \\ -a_0 & -a_1 & -a_2 & -a_3 & \cdots & -a_{n-1} \end{pmatrix},$$

then the characteristic polynomial of A is

$$\det(XI - A) = X^n + a_{n-1}X^{n-1} + \ldots + a_1 X + a_0.$$

The amount of information contained in the matrix is exactly the same as the amount of information contained in the characteristic polynomial.

The computation of eigenvalues and eigenvectors is a part of linear numerical algebra; but it is *not* a linear problem: the eigenvalues of $A + B$ are not the sum of the eigenvalues of A and the eigenvalues of B. Let us give an elementary combinatorial argument. If A and B have n distinct eigenvalues, then we might pair the sums of eigenvalues in $n!$ different ways so as to obtain $n!$ n-tuples; this is far too many possibilities. The first reason why the computation of eigenelements can be really difficult is the nonlinearity of the problem.

If an eigenvalue of a linear operator is not simple, the problem is usually ill-conditioned. This means that the variation of the eigenelements is very large with respect to perturbations of the elements of the matrix. Let J be an $n \times n$ Jordan block, with zeros on the diagonal, and let $A(\varepsilon)$ be the perturbed matrix

$$(13.0.1) \qquad A(\varepsilon) = \begin{pmatrix} 0 & 1 & 0 & 0 & \cdots & 0 \\ 0 & 0 & 1 & 0 & \cdots & 0 \\ 0 & 0 & 0 & 1 & \cdots & 0 \\ \vdots & & & & \ddots & \vdots \\ 0 & 0 & 0 & 0 & \cdots & 1 \\ \varepsilon & 0 & 0 & 0 & \cdots & 0 \end{pmatrix}.$$

This matrix is a companion matrix, hence its characteristic polynomial is

$$P(X; \varepsilon) = X^n - \varepsilon,$$

and we know the eigenvalues explicitly: they are equal to $\varepsilon^{1/n} \exp(2i\pi k/n)$, with $k = 0, \ldots, n - 1$. This means that a variation of $O(\varepsilon)$ in the elements of the matrix can correspond to a variation of $O(\varepsilon^{1/n})$ in the eigenvalues. Moreover, in general, we will not know that there is a Jordan block of high multiplicity in the matrix: somehow, the program should be written with a warning to alert us that something unusual is going on. But this is difficult to implement.

One could say: alright, the matrix (13.0.1) has been made up to make life really unpleasant; what about a nice 2×2 matrix, with nicely distinct eigenvalues. Surely, nothing bad can happen in such a simple case? Thus, take the matrices

$$(13.0.2) \qquad A = \begin{pmatrix} 1 & 10^4 \\ 0 & 11 \end{pmatrix} \quad \text{and} \quad A' = \begin{pmatrix} 1 & 10^4 \\ -10^{-5} & 11 \end{pmatrix}.$$

The eigenvalues of A are

$$\lambda_1 = 1, \quad \lambda_2 = 11.$$

An explicit calculation gives the eigenvalues of A':

$$\lambda_1' = 6 - \sqrt{24.9} \sim 1.01001, \quad \lambda_2' = 6 + \sqrt{24.9} \sim 10.98999.$$

This means that a change of 10^{-5} in the elements of the matrix, which is also at most 10^{-5} relative to the eigenvalues, brings a perturbation of order $1/100$ in one of the eigenvalues.

It is not sufficient to observe that the eigenvalues can be very sensitive to the perturbations of the elements in the matrix: the dependence is much more dramatic for the eigenvectors. Going back to the matrix (13.0.1): for $\varepsilon > 0$, there are n distinct eigenvalues and, hence, n distinct eigenvectors; for $n = 0$, these n distinct eigenvectors collapse to a single eigenvector corresponding to the eigenvalue 0. It is interesting to get a more precise picture: letting ω be an n-th root of unity, the normalized eigenvectors are all of the form

$$(13.0.3) \qquad \frac{1}{\mu(\varepsilon)} \begin{pmatrix} 1 \\ \varepsilon^{1/n}\omega \\ \vdots \\ \varepsilon^{(n-1)/n}\omega^{n-1} \end{pmatrix}.$$

Here the positive quantity $\mu(\varepsilon)$ is given by

$$\mu(\varepsilon)^2 = 1 + \varepsilon^{2/n} + \ldots + \varepsilon^{2(n-1)/n}.$$

Therefore, the word 'collapsing' describes the phenomenon very well: the basis of eigenvectors, given by expression (13.0.3), for ω running through all the n-th roots of unity, becomes more and more singular as ε tends to 0. Such a situation must be considered as systematic: the continuity of eigenvectors with respect to the elements of the matrix cannot be assumed; we can only expect the continuity of generalized eigenspaces relative to a cluster of eigenvalues converging to a given multiple eigenvalue.

The definition of conditioning is not the same for the resolution of a linear system as for the computation of the eigenelements of a matrix. We have seen in Chapter 5 that the Hilbert matrix is very well conditioned for the second situation and very ill-conditioned for the first one. Conversely, the matrix $I + A(\varepsilon)$, with $A(\varepsilon)$ defined by eqn (13.0.1), is very well conditioned for the second situation and very ill-conditioned for the first one.

In this chapter we will mainly consider two methods: one is the power method and simple modifications to it; the other one is the QR method. In fact, they are basically the same, but more on that later.

The idea of the power method is utterly naïve: if A has an eigenvalue λ whose modulus is strictly larger than the moduli of every other eigenvalue, and if it happens also to be simple, we take any non-vanishing vector x and we repeatedly apply A to x. The components along the eigenvector corresponding to λ will increase fast, relative to the other components. In the limit, the relative importance of the other components tends to 0. This is too naïve to work: if $|\lambda|$ is strictly larger than 1, we expect an overflow; if it is strictly smaller than 1, we expect an underflow. Thus, we must normalize at each step.

Of course, we do not know whether an arbitrary matrix has an eigenvalue possessing the property which we have hypothesized. This is not a problem: numerical analysts are bold souls and they *try* numerical methods, in the hope of detecting from the output whether the result looks correct.

Maybe, we are not interested in the eigenvalue which has the largest modulus, but in the eigenvalue which is smallest in modulus. Then, instead of repeatedly applying A, we repeatedly apply A^{-1} to x; and, since we know better than to invert matrices, we repeatedly solve a linear system whose matrix is A, and we normalize the result. The analysis is the same for this inverse power method as for the direct power method. Geometrically, we have performed an inversion and a conjugation in the complex plane.

Now, we can modify the inverse power method and get much more from it. Intuitively the inverse power method converges the best when the ratio of the smallest eigenvalue to any other eigenvalue is the smallest. If we translate A by μI, choosing μ to be close to the eigenvalue we want to compute, we get a much better convergence for the inverse power method, and the method will give the eigenvalue closest to μ. So, now we can be pedantic and observe that we have performed a homographic transformation in the complex plane completed by ∞, i.e., the Riemann sphere: much ado about nothing.

The QR method can now be explained by waving our hands: we apply the power method simultaneously to all of the basis vectors and we have to renormalize at each step to get something meaningful. The reader is referred to Section 13.4 for the definition of the QR method, and to the sequence of exercises in Subsection 13.5.2. Moreover, there are still many open questions in the mathematical analysis of the QR method.

Time-dependent problems can be solved theoretically by means of decompositions on a basis of eigenvectors, however, this method is rarely used in practice. An important case is the calculation of the motion of a linear vibrating structure. The idea is basically the same as for the argument against inverting matrices in order to solve a linear system: if there is a method which works as well and is faster, we choose the faster method. So, though we are brave numerical analysts, we are not foolhardy: knowing that there are many pitfalls in the computation of eigenelements, we calculate them only if it cannot be avoided.

We need eigenvalues and eigenvectors if we are interested in the vibration modes of structures, acoustic or electromagnetic fields, and so on. And, we should not be surprised to learn that what causes mathematical difficulties can also cause physical difficulties: the higher the degree of degeneracy of an eigenvalue, the larger the instability at resonance. So, if we care for stability, we try to draw eigenvalues apart at the time of design. Conversely, we may be interested in resonance, as in tuning; for example, tuning a bell or any musical instrument, or tuning for receiving electromagnetic waves. In this case, also, we are interested in eigenvalues and eigenvectors.

It would be nice to have an idea of the size of the set of ill-conditioned matrices relative to the computation of their eigenelements. This question has a strong algebraic and geometric flavour: the set of matrices must be stratified into sub-varieties on which the algebraic and geometric multiplicity of the eigenvalues are fixed. Now, how large is the set of matrices which are 'close', in a sense to be defined, to such sub-varieties? The question looks wide open.

Let us conclude: it is necessary to calculate eigenvalues and eigenfunctions of linear operators; in practice, we do use packages, but it is particularly important to understand the underlying mathematical considerations, in order to find out whether the results given by the software have any value. There are all sorts of motivations, coming from fluid and solid mechanics, from physics, from statistics, from economics etc. The matrices are large, the problems are not well conditioned, and all the art of the numerical analyst is required to find a solution.

13.1. Eigenvalues: the naïve approach

13.1.1. Seeking eigenvalues and polynomial equations

There exist algorithms to calculate the characteristic polynomial, but they are little used because the calculation of the roots of a polynomial in the complex plane is often a badly conditioned problem, a problem which is not limited to the case of multiple roots. J. H. Wilkinson [79] has proposed the following example: let

$$P(X) = (X + 1) \cdots (X + 20)$$

and let the perturbation be

$$Q(X) = 2^{-23} X^{19}.$$

The calculation gives the following roots for the perturbed polynomial $P + Q$:

$$-1.000\,000\,000,$$
$$-2.000\,000\,000,$$
$$-3.000\,000\,000,$$
$$-4.000\,000\,000,$$
$$-4.999\,999\,928,$$
$$-6.000\,006\,944,$$
$$-6.999\,697\,234,$$
$$-8.007\,267\,603,$$
$$-8.917\,250\,249,$$
$$-10.095\,266\,145 \pm 0.643\,500\,904\,i,$$
$$-11.793\,633\,881 \pm 1.652\,329\,728\,i,$$
$$-13.992\,358\,137 \pm 2.518\,830\,070\,i,$$
$$-16.730\,737\,466 \pm 2.812\,624\,894\,i,$$
$$-19.502\,439\,400 \pm 1.940\,330\,347\,i,$$
$$-20.846\,908\,101.$$

Note that the perturbation of the zeros is very important. Indeed, this polynomial is badly conditioned for calculating its zeros. This phenomenon is difficult to predict. There exist algorithms to calculate the characteristic polynomial (without Cramer's formula!), but they are not used to find the eigenvalues and eigenvectors of a numeric matrix, except in two dimensions.

13.1.2. The bisection method

We begin with a very simple method for finding the eigenvalues of Hermitian matrices.

Recall the principle of the bisection method for finding the roots of a nonlinear equation $f(x) = 0$. We suppose that f is given on an interval $[a, b]$, is continuous, and satisfies

$$f(a) f(b) < 0.$$

Let $c_1 = (a + b)/2$. Either $f(c_1) = 0$, and in this case we have found a zero and the algorithm stops, or $f(c_1) \neq 0$, and in this case one of the two products $f(a)f(c_1)$ or $f(c_1)f(b)$ is strictly negative. In this case, we denote by $[a_1, b_1]$ the interval $[a, c_1]$ or $[c_1, b]$ such that the product of the function values at the end-points is strictly negative. Having obtained an interval $[a_k, b_k]$ such that $f(a_k)f(b_k) < 0$, we write $c_{k+1} = (a_k + b_k)/2$. If $f(c_{k+1}) = 0$, the algorithm stops. If not, we denote by $[a_{k+1}, b_{k+1}]$ whichever of the intervals $[a_k, c_{k+1}]$ or $[c_{k+1}, b_k]$ is such that the product of the function values of the end-points is strictly negative. Thus, in N steps of bisection, we localize at least one zero of f with a precision of $2^{-N}(b - a)$.

We are going to couple this bisection method (which has a very general use) with Sylvester's inertia theorem. Let A be a Hermitian matrix. Suppose that $A - \mu I$ admits an $LD_\mu L^*$ decomposition, and denote by $p(\mu)$ the number of eigenvalues of A which are strictly greater than μ. It is clear that $p(\mu)$ is the number of strictly positive eigenvalues of D_μ.

For a value $\mu' < \mu$, $A - \mu' I \geqslant A - \mu I$, and consequently, by virtue of Lemma 12.1.5, $\lambda_{p(\mu)}(A - \mu I) \leqslant \lambda_{p(\mu)}(A - \mu' I)$, which shows us that $p(\mu') \geqslant p(\mu)$. Hence, in the interval $]\mu', \mu]$, A has exactly $p(\mu') - p(\mu)$ eigenvalues.

Suppose that we have already determined that A possesses exactly one eigenvalue in the interval $]\mu', \mu]$. Then, by application of the bisection method, we can determine this eigenvalue with a precision of $2^{-N}(\mu - \mu')$ in N calculations of the decomposition LDL^*. We will see later that the inverse power method can be applied to calculate an eigenvalue precisely if we already know a good approximation.

13.2. Resonance and vibration

13.2.1. Galloping Gertie

In 1831, near Manchester, a bridge collapsed whilst being crossed by a military detachment marching in step. Since this time, the military regulations of every country order the infantry to stop marching in step whilst crossing a bridge.

It was common, at the beginning of aviation, for an aeroplane to crash following uncontrollable oscillations of its wings, as a result of the loss of control of the elevators.

The earthquake-proof construction of tall buildings is regulated in several towns in California, by ensuring that the natural frequencies of the buildings are far from the characteristic frequencies of earthquakes. Furthermore, the foundations include vibration-damping elements.

The Ariane rocket is designed in such a way that the vibrations created by the rocket motors do not resonate with the structure (which is a very thin hull filled with liquid).

Here is another story about the destructive effects of resonance, which I have adapted from the autobiography of Theodore von Kármán, [76, pp. 211–15].

The collapse of the bridge over the straits of Tacoma on the 7th of November 1940 in Washington State, USA was due to a subtle resonance created by the interaction between the bridge and the turbulent movements of the air. This 1.6 km long suspension bridge was, at the time, the third longest in the world and was considered at its inauguration to be at the pinnacle of civil engineering. From the begining, the behaviour of the bridge was bizarre. In winds of 7 or 8 km/h, it oscillated with a maximum amplitude surpassing one metre. The movement of this fine steel ribbon was so spectacular that visitors came from far away to cross it and it became nicknamed 'Galloping Gertie'.

The engineers tried in vain to stabilize the bridge by anchoring it with thick cables attached to blocks of concrete. Other procedures were also tried, but nothing worked: Galloping Gertie swayed and for four months they watched its behaviour. Since this did not change with the passage of time, the Washington state authorities began to say that the bridge was safe.

The morning of the collapse, nothing foretold what was going to happen. In spite of a storm during the night, the bridge continued to sway as usual. At ten o'clock in the morning the wind blew at 67 km/h, the strongest that the bridge had ever been subjected to. All of a sudden, a few minutes after ten o'clock in the morning, the movement changed character: the rhythm of the displacement from low to high took place in violent, torsional movements, and as an observer said 'it seemed that the bridge was going to turn over'. The authorities prohibited traffic from the bridge.

In the following minutes, the torsional movement continued more and more violently. At one instant, one end of the road appeared to an observer to be 8.5 m higher than the other. At the following instant, 8.5 m lower. The cables of the main span, instead of ascending and descending together as in their usual spring movement, pulled and twisted in opposite directions, inclining the roadway from one side to the other at 45 degrees. The street lights on the bridge were nearly horizontal. For half an hour, the steel girders, the suspension cables and the concrete road were subjected to these terrible stresses. Finally, at eleven o'clock in the morning, the structure could resist no longer. The street lights began to collapse. The central span exploded and a two hundred metre section detached itself and collapsed into the bottom of the straits with a deafening noise.

Theodore von Kármán analysed the causes of the destruction of the bridge. Working principally on problems of aerodynamics applied to aircraft, he had highlighted, in 1911, the formation of vortices in the wake of an obstacle. These vortices are created alternately from one side of the obstacle and then the other. The vortices are arranged in staggered rows, like the street lamps from one side of a street to the other, and from this we get their name: von Kármán vortex streets. This discovery won him international recognition in aeronautical circles.

The Tacoma bridge had a roadway covered in metal, and the plates forming the wall were pushed by the wind until this formed vortex sheets at their side, which caused the oscillations and the collapse of the bridge.

The Tacoma bridge was reconstructed with openings in the roadway and the lateral walls, and it holds till this day. The other great American bridges were checked and found safe.

13.2.2. Small vibrations

Consider a mechanical system, for which position is described by a point $q \in \mathbb{R}^n$. We will assume that the masses are included in the chosen coordinates. If the point q moves under the action of a force $-\nabla U$, with U a potential depending only on q, the fundamental principle of dynamics implies that

$$(13.2.1) \qquad \ddot{q} = -\nabla U(q).$$

We have used here the notation of mechanics and physics: $\dot{u} = du/dt$ and $\ddot{u} = d^2u/dt^2$ are the derivatives with respect to time and ∇ is the gradient operator. We easily verify that the total energy, which is the sum of the potential energy and kinetic energy

$$(13.2.2) \qquad E(q,p) = U(q) + \frac{|\dot{p}|^2}{2},$$

remains constant when q is a solution to the system of differential eqns (13.2.1). The system has a stable equilibrium q_0 if U has a local minimum at q_0. If U (which we assume to be sufficiently differentiable) has a second derivative $A = D^2U(q_0)$ (the Hessian matrix, or the Hessian) which is positive definite, we know that U has a local minimum at q_0. The small vibration approximation consists of replacing U in the neighbourhood of q_0 by $(q - q_0)^* A(q - q_0)/2$ and thus linearizing the system. Without loss of generality, we can suppose that $q_0 = 0$, and the system becomes

$$(13.2.3) \qquad \ddot{q} = -Aq.$$

We seek solutions of eqn (13.2.3) which are of the form $v \cos(\omega t + \phi)$, where v is a vector in $\mathbb{R}^n$:

$$-\omega^2 v \cos(\omega t + \phi) = -Av \cos(\omega t + \phi).$$

Therefore, we must have

$$Av = \omega^2 v,$$

that is to say, v is an eigenvector of A associated with the positive eigenvalue ω^2, since A is positive definite. The number ω is the angular frequency of the vibration, and the frequency of the vibration is $\omega/2\pi$.

Every solution of eqn (13.2.3) is of the form

$$\sum_{k=1}^{n} v_k \cos(\omega_k t + \phi_k),$$

since the orthogonal projection of system (13.2.3) onto the eigenvector v_k associated with the eigenvalue ω_k^2 leads to the equation

$$(13.2.4) \qquad \ddot{x}_k + \omega_k^2 x_k = 0,$$

where x_k is the component of x on v_k.

Consider a linear equation of angular frequency ω, to which we add a small dissipation coefficient $\epsilon > 0$, and calculate the stationary response of this system to an excitation $e^{i\alpha t}$. We look for a solution of the form $k e^{i\alpha t}$ to the differential equation

$$\ddot{x} + \epsilon \dot{x} + \omega^2 x = e^{i\alpha t}.$$

A simple calculation shows that

$$k = \frac{1}{\omega^2 - \alpha^2 + i\epsilon\alpha}$$

and, therefore, the modulus of k is given by

$$|k| = \left(\left(\omega^2 - \alpha^2 \right)^2 + \epsilon^2 \alpha^2 \right)^{1/2},$$

whose representative curve has the shape indicated in Figure 13.1.

Therefore, we see that an excitation of unit amplitude and of angular frequency α gives a solution whose amplitude is inversely proportional to the coefficient of dissipation ϵ.

This abundantly shows the danger of resonance. When the vibrations of a system are of large amplitude, on the one hand, the linearized system (13.2.3) is no longer a good approximation to the nonlinear system (13.2.4) and, on the other hand, we leave the domain of validity of the physical model.

The search for resonance is one case where we are not much interested in the eigenvectors.

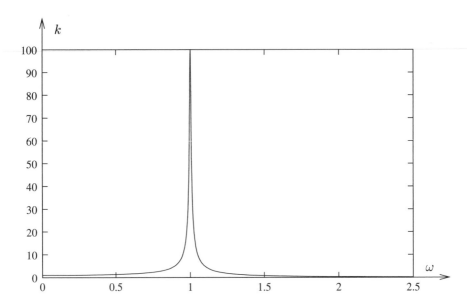

Figure 13.1: Amplitude of the response of an oscillator to a harmonic excitation of pulsation α; $\epsilon = 0.01$, $\omega = 1$.

13.3. Power method

13.3.1. The straightforward case

The power algorithm is described as follows: A is a matrix in $\mathbb{K}^n$. Given a linear form y^* and an initial vector x^0, the vector x^{k+1} is defined from x^k by the following relations:

$$(13.3.1) \qquad x^{k+1/2} = Ax^k, \quad x^{k+1} = \frac{x^{k+1/2}}{y^* x^{k+1/2}}.$$

The convergence properties of the above algorithm are set out as follows:

Theorem 13.3.1. Let A be a matrix in $\mathbb{K}^n$ which possesses a simple eigenvalue λ whose modulus is strictly greater than the moduli of all the other eigenvalues. Then there exists an open set U of $\mathbb{K}^n \times \mathbb{K}^n \approx \mathbb{K}^{2n}$ whose complement has Lebesgue measure zero in $\mathbb{K}^{2n}$ such that, for every pair (y, x^0) in U, the sequence $\left(x^k\right)_{k \in \mathbb{N}}$ has a limit. The limit of the x^k is an eigenvector of A associated with the eigenvalue λ. $\diamond$

Remark 13.3.2. Observe that in the real case, if A has a simple eigenvalue λ whose modulus is larger than the modulus of any other eigenvalue, then λ must be real.

Proof. We note, first of all, that

$$x^k = \frac{A^k x^0}{y^* A^k x^0}.$$

Indeed, this relation is true for $k = 1$. If it holds for k we have

$$x^{k+1/2} = \frac{A^{k+1} x^0}{y^* A^k x^0},$$

consequently,

$$y^* x^{k+1/2} = \frac{y^* A^{k+1} x^0}{y^* A^k x^0},$$

and we see that

$$x^{k+1} = \frac{A^{k+1} x^0}{y^* A^k x^0} \frac{y^* A^k x^0}{y^* A^{k+1} x^0} = \frac{A^{k+1} x^0}{y^* A^{k+1} x^0}.$$

We present the algorithm in the form (13.3.1) with the aim of avoiding the overflows and underflows which would spoil the precision of the calculations. These overflows and underflows can occur if the spectral radius of A is different from 1, as the study in Section 11.1 shows that A^k grows as k tends to infinity like $\rho(A)^k$.

Let v be an eigenvector associated with the eigenvalue λ. There exists a subspace W of $\mathbb{K}^n$ which is invariant to A, that is $AW \subset W$, and such that

$$\mathbb{K}^n = \mathbb{K}v \oplus W.$$

We denote by P the projection in W parallel to $\mathbb{K}v$, and z the vector of $\mathbb{K}^n$ such that

$$x = (z^* x) v + P x, \quad \forall x \in \mathbb{K}^n.$$

The operator B from W to W, defined by

$$Bx = \lambda^{-1} A x, \quad \forall x \in W,$$

has spectral radius less than 1 from the hypotheses of the theorem.

We have

$$A^k x^0 = \lambda^k \left(z^* x^0 \right) v + A^k P x^0 = \lambda^k \left[\left(z^* x^0 \right) v + B^k P x^0 \right].$$

Consequently, if $y^* A^k x^0$ does not vanish for any value of k,

$$(13.3.2) \qquad x^k = \frac{A^k x^0}{y^* A^k x^0} = \frac{\left(z^* x^0 \right) v + B^k P x^0}{y^* \left[\left(z^* x^0 \right) v + B^k P x^0 \right]}.$$

Suppose that $y^* v$ and $x^* x^0$ do not vanish. Then it follows from the study of Section 11.1 that the numerator of eqn (13.3.2) tends towards $(z^* x^0)v$ and that the denominator of eqn (13.3.2) tends towards $(z^* x^0)(y^* v)$. Consequently,

$$x^k \to \frac{v}{y^* v}.$$

We therefore have convergence for all data (y, x^0) not belonging to the set F defined by

$$F_k = \left\{ (y, x) \in \mathbb{K}^{2n} : y^* A^k x = 0 \right\},$$
$$F_\infty = \left(\{y \in \mathbb{K}^n : y^* v = 0\} \times \mathbb{K}^n \right) \cup \left(\mathbb{K}^n \times \{x \in \mathbb{K}^n : z^* x = 0\} \right),$$
$$F = F_1 \cup F_2 \cup \cdots \cup F_\infty.$$

We are going to show now that F is a closed set of measure zero. Note that $\rho(A) = |\lambda| > 0$. Consequently, none of the powers of A is zero. The set F_k is closed, since it is a level set of the continuous function $(y, x) \mapsto y^* A^k x$. It is a set of measure zero. Indeed, let $N_1 = \{x \in \mathbb{K}^n : A^k x = 0\}$. N_1 is the kernel of A^k, which being of co-dimension at least 1, has Lebesgue measure zero in $\mathbb{K}^n$. For $x \notin N_1$,

$$N_x = \left\{ y \in \mathbb{K}^n : y^* A^k x = 0 \right\}$$

is a hyperplane of $\mathbb{K}^n$ and, therefore, the measure of N_x vanishes. From the Fubini–Lebesgue theorem, the set F_k is therefore negligible, since almost all its intersections $F_k \cap (\mathbb{K}^n \times \{x\})$ are negligible relative to the Lebesgue measure in $\mathbb{K}^n$.

The set F_∞ is a union of two products having the same structure: each is a product of $\mathbb{K}^n$ with a hyperplane of $\mathbb{K}^n$. It is therefore negligible in $\mathbb{K}^{2n}$. The set F is a countable union of negligible sets and is also negligible.

It remains to show that F is closed. Let (y^ℓ, x^ℓ) be a convergent sequence of elements of F, which have limit (y, x). We therefore have for each ℓ an index $k(\ell)$ such that $(y^\ell, x^\ell) \in F_{k(\ell)}$. If the number of distinct indices $k(\ell)$ which appear in this sequence is finite, it is clear that the limit of (y^ℓ, x^ℓ) is in F. Suppose for the moment that there appear an infinite number of indices in this sequence. By extracting a subsequence, we can suppose that $k(p) = k(\ell_p)$ tends towards infinity. In this case to simplify the notation we let $\xi_p = x^{\ell_p}$ and $\eta_p = y^{\ell_p}$ and we have

$$\eta_p^* A^{k(p)} \xi_p = 0.$$

Reasoning as in the first part of the proof, we see that

$$\eta_p^* \left[(z^* \xi_p) v + B^{k(p)} P \xi_p \right] = 0.$$

If $z^* x = 0$, the pair (y, x) is in F. If $z^* x \neq 0$, passing to the limit in the above inequality gives

$$(y^* v)(z^* x) = 0$$

and, therefore,

$$y^* v = 0,$$

which implies again that (y, x) is in F. We have therefore shown that F is closed.
□

13.3.2. Modification of the power method

In this subsection, we work under the hypotheses of Theorem 13.3.1: A is an $n \times n$ matrix which has a simple eigenvalue λ whose modulus is strictly greater than all the others. We are going to examine a modification to the power method. First of all, instead of the normalization obtained by dividing $x^{n+1/2}$ by $y^* x^{n+1/2}$, we are going to divide $x^{n+1/2}$ by its norm

$$(13.3.3) \qquad x^{n+1/2} = A x^n, \quad x^{n+1} = \frac{x^{n+1/2}}{|x^{n+1/2}|}.$$

We see then that

$$x^k = \frac{A^k x^0}{|A^k x^0|},$$

and the analysis made in Theorem 13.3.1 shows that

$$A^k x^0 = \lambda^k \left(z^* x^0 \right) v + A^k P x^0.$$

Consequently, if $z^* x^0 \neq 0$,

$$x^k = \frac{A^k x^0}{|A^k x^0|} = \left[\frac{\lambda}{|\lambda|} \right]^k \frac{\left(z^* x^0 \right) v + B^k P x^0}{|\left(z^* x^0 \right) v + B^k P x^0|}.$$

We see that the vector $A^k x^0 / |A^k x^0|$ does not converge if λ is not a strictly positive real number. If $\lambda = r e^{i\theta}$,

$$x^k = e^{ik\theta} \frac{\left(z^* x^0 \right) v + B^k P x^0}{|\left(z^* x^0 \right) v + B^k P x^0|},$$

and, therefore,

$$e^{-ik\theta} x^k \to \frac{\left(z^* x^0 \right) v}{|\left(z^* x^0 \right) v|}.$$

In other words, x^k tends towards the eigensubspace associated with λ, which does not necessarily imply the convergence of x^k. On the other hand, the ratio $(A x^k)_j / x_j^k$ of the j-th component of $A x^k$ and the j-th component of x^k tends towards λ, provided that x_j^k is not too small. Therefore, we can have two phenomena in the real case: x^k converges or $(-1)^k x^k$ converges.

To ensure convergence it is also required that

$$z^* x^0 \neq 0 \quad \text{and} \quad x^0 \notin \bigcup_{k=1}^{\infty} \ker A^k.$$

The union of the kernels of the powers of A is an increasing sequence of vector subspaces which is stationary for sufficiently large k. Jordan's theory of the decomposition of matrices shows that, since A has a nonzero eigenvalue λ, this space cannot be equal to $\mathbb{K}^n$. Consequently, the set of initial conditions for which we do not have convergence is closed and of measure zero, as in Theorem 13.3.1.

Note that the rate of convergence of x^k towards $\ker(A - \lambda I)$ is geometric in $\rho(B)^k$, where $\rho(B) = |\lambda|^{-1} \max_{\lambda_j \neq \lambda} |\lambda_j|$.

The modified power method is particularly interesting when A is Hermitian, $|\cdot|$ being the Euclidean norm. The remarkable point is that the sequence defined by (13.3.3)

$$\mu^k = \left(x^k\right)^* A x^k$$

converges twice as fast towards λ as the ratio $(Ax^k)_j / x^k_j$. Indeed,

$$x^k = \frac{A^k x^0}{|A^k x^0|},$$

which gives

$$\left(x^k\right)^* A x^k = \frac{\left(x^0\right)^* A^{2k+1} x^0}{\left(x^0\right)^* A^{2k} x^0}.$$

Since A is Hermitian, it possesses an eigenvector v of Euclidean norm 1, relative to the eigenvalue λ, and we can write

$$Ax = \lambda \left(v^* x\right) v + APx,$$

where P is the orthogonal projection on the orthogonal supplement of $\mathbb{K}v$. This orthogonal projection is necessarily Hermitian and idempotent ($P^2 = P$). Hence,

$$
\begin{aligned}
\left(x^0\right)^* A^{2k+1} x^0 &= \left[\left(v^* x^0\right) v + Px^0\right]^* A^{2k+1} \left[\left(v^* x^0\right) v + Px^0\right] \\
&= \left[\left(v^* x^0\right) v + Px^0\right]^* \left[\lambda^{2k+1} \left(v^* x^0\right) v + A^{2k+1} Px^0\right] \\
&= \lambda^{2k+1} \left|v^* x^0\right|^2 + \left(x^0\right)^* P A^{2k+1} Px^0.
\end{aligned}
$$

Similarly,

$$\left(x^0\right)^* A^{2k} x^0 = \lambda^{2k} \left|v^* x^0\right|^2 + \left(x^0\right)^* P A^{2k} Px^0.$$

We therefore obtain

$$\mu^k = \lambda \frac{\left|v^* x^0\right|^2 + \lambda^{-2k-1} \left(x^0\right)^* P A^{2k+1} Px^0}{\left|v^* x^0\right|^2 + \lambda^{-2k} \left(x^0\right)^* P A^{2k} Px^0}.$$

The convergence of μ^k to λ is geometric if $v^* x^0 \neq 0$. The rate of convergence is obtained from $\rho(PAP/\lambda)^2 = \rho(B)^2$: we have a convergence which is twice as rapid.

Since A is Hermitian, the hypothesis $v^* x^0 \neq 0$ suffices to ensure that $A^k x^0$ does not vanish for any k.

13.3.3. Inverse power method

We have seen that the power method gives only the eigenvalue–eigenvector pair corresponding to the eigenvalue with the largest modulus. Let $\sigma \in \mathbb{C}$ be such that

$$0 < |\sigma - \lambda_j| < |\sigma - \lambda_k|, \quad \forall k \neq j,$$

which implies that λ_j is simple. Then the largest eigenvalue (in modulus) of $(A - \sigma I)^{-1}$ is $(\lambda_j - \sigma)^{-1}$. Therefore, we have convergence of the sequence x^k when it is defined by either the relations (13.3.1) applied to $(A - \sigma I)^{-1}$, that is,

$$(13.3.4) \qquad (A - \sigma I)\, x^{k+1/2} = x^k \quad \text{and} \quad x^{k+1} = \frac{x^{k+1/2}}{y^* x^{k+1/2}},$$

or when it is defined by the relations (13.3.3) applied to $(A - \sigma I)^{-1}$, that is,

$$(13.3.5) \qquad (A - \sigma I)\, x^{k+1/2} = x^k \quad \text{and} \quad x^{k+1} = \frac{x^{k+1/2}}{\left| x^{k+1/2} \right|}.$$

The convergence of the sequence (13.3.4) holds for (y, x^0) in an open dense set of $\mathbb{K}^n \times \mathbb{K}^n$, whose complement is of measure zero. The convergence of the sequence (13.3.5) holds for x^0 in an open dense set of $\mathbb{K}^n$, whose complement is of measure zero.

Practically, it is not necessary to calculate the inverse of $A - \sigma I$. We store a decomposition $A = LU$ or $P_\tau A = LU$, and each solution of the system requires $O(n^2)$ operations, once the initial investment of $O(n^3)$ operations for the Gaussian decomposition has been made.

One problem presents itself: the conditioning of the matrix $A - \sigma I$ can be estimated by the ratio of the largest eigenvalue of $A - \sigma I$ and the smallest eigenvalue of $A - \sigma I$. A priori, the matrix is badly conditioned and, therefore, the calculation of $x^{k+1/2}$ will be marked with a large error.

What should we do? Must we choose σ not too close to λ_j so that $A - \sigma I$ is better conditioned?

We need do nothing of the sort, as was shown by B. N. Parlett [65]. We will follow his analysis in the case when A is Hermitian. The real calculations can be modelled by

$$(A - \sigma I - E)\, \hat{x}^{k+1/2} = x^k + e^k,$$

where E models the rounding errors of the method, and e^k is the error in the vector x^k. We write

$$\hat{e}^k = e^k + E \hat{x}^{k+1/2}.$$

Consequently, instead of solving

$$(A - \sigma I)\, x^{k+1/2} = x^k,$$

we have in fact solved

$$(A - \sigma I)\, \hat{x}^{k+1/2} = x^k + \hat{e}^k,$$

and the error committed is given by

$$(A - \sigma I)^{-1} \hat{e}^k.$$

Using the decomposition of A over $\mathbb{K}v_j \oplus (\mathbb{K}v_j)^\perp = \mathbb{K}v_j \oplus \operatorname{Im} P_j$, we see that

$$(A - \sigma I)^{-1} \hat{e}^k = \frac{\left(v_j^* \hat{e}^k\right) v_j}{\lambda_j - \sigma} + (B_j - \sigma I)^{-1} P_j \hat{e}^k.$$

Suppose that there exists a constant M such that

$$\left|P_j \hat{e}^k\right| \leqslant M \left|v_j^* \hat{e}^k\right|,$$

which means that the components of $\hat{e}^k$ perpendicular to v_j are not too large compared to the components of $\hat{e}^k$ parallel to v_j. Let

$$m = \min \left\{ |\lambda_k - \sigma| : \lambda_k \neq \lambda_j \right\}.$$

We then have

$$\left|(B - \sigma I)^{-1} P \hat{e}^k\right| \leqslant \frac{1}{m} \left|P \hat{e}^k\right| \leqslant \frac{M}{m} \left|v_j^* \hat{e}^k\right| \leqslant \frac{M|\lambda_j - \sigma|}{m} \left|v_j^* (A - \sigma I)^{-1} \hat{e}^k\right|.$$

We see that, if $M|\lambda_j - \sigma| \ll m$, the error $\hat{e}^k$ is almost all in the direction of v^j, which is the eigenvector which we wanted to calculate.

Even if $(A - \sigma I)^{-1} \hat{e}^k$ is of the same order of magnitude as $x^{k+1/2}$ (since the matrix $A - \sigma I$ is badly conditioned), this is not serious for the calculation of v_j, it is even an advantage. We therefore obtain v_j in a few iterations.

13.4. QR method

13.4.1. The algorithm and its basic properties

Recall that every invertible matrix A admits a unique decomposition $A = QR$, where Q is unitary and R is upper triangular with a strictly positive diagonal (see Theorem 12.1.1).

The QR algorithm for finding the eigenvalues of some matrix A is defined as follows: we use a sequence of translation parameters σ_k, and we let

$$A_1 = A,$$
(13.4.1)
$$A_k - \sigma_k I = Q_k R_k,$$
$$A_{k+1} = \sigma_k I + R_k Q_k.$$

Thus, we effect the QR decomposition of matrix A_k, then we invert the order of the factors of the decomposition. The parameters σ_k will be chosen later, see below for the possible choices.

Note, first of all, that

(13.4.2) $$A_{k+1} = \sigma_k I + Q_k^* (A_k - \sigma_k I) Q_k = Q_k^* A_k Q_k.$$

Consequently, A_{k+1} is unitarily equivalent to A_k and, therefore, for all k, A_k is unitarily equivalent to A.

The QR method iteratively transforms A into one of its Schur forms. If the method converges nicely, the limit will be an upper triangular matrix on whose diagonal we will find the eigenvalues of A_∞, which are obviously the same as those of A. To find the eigenvectors of A, it necessary to do a bit more work, except if A is a normal matrix, that is, A commutes with its adjoint. In this case A is diagonalizable in an orthonormal basis and its Schur forms are also normal. If an upper triangular matrix commutes with its adjoint, it is diagonal. Then, it is clear that the eigenvectors can be obtained from the limit U of the products $Q_k \cdots Q_1$ as k tends to infinity; they are the column vectors of U.

However, the QR method does not always converge nicely, and its limit can be a block upper triangular matrix, with 2×2 blocks for the pairs of conjugate eigenvalues.

It is interesting to characterize the matrices $A = QR$ for which Q commutes with R. There is no simple general answer. If A is unitary, it then admits a QR decomposition with $Q = A$ and $R = I$, which obviously commute. In order to be able to conclude, we need an assumption on the spectrum of A:

Lemma 13.4.1. Let A be an invertible matrix whose eigenvalues are all simple and have distinct moduli. If the QR decomposition of A has the property

(13.4.3) $$A = QR = RQ,$$

then A is upper triangular and Q is diagonal.

Proof. We argue by induction on the spatial dimension. In dimension $d = 1$, there is nothing to prove. Assume the result to be true up to a certain dimension $d - 1$, and let A be a $d \times d$ matrix having all the properties stated in the lemma.

Let μ be an eigenvalue of R and let V be the corresponding eigenspace. We infer from the commutative property (13.4.3) that $QV = V$ and, therefore, there exists a basis of V consisting of eigenvectors of Q:

$$Qx_j = e^{i\phi_j} x_j.$$

This implies that the x_js are also eigenvectors for A. Consequently, if the eigenvalues of A are denoted by $\lambda_1, \ldots, \lambda_d$, the eigenvalues of R will be the absolute values of the λ_js and those of Q will be the phase factors $\lambda_j / |\lambda_j| = e^{i\psi_j}$.

The assumption on the eigenvalues of A implies that all the eigenvalues of R are distinct and, in particular, every eigenvector of R is also an eigenvector of Q. The first vector e_1 of the canonical basis is an eigenvector of R relative to the eigenvalue $R_{11} = |\lambda_j|$. Therefore, it is an eigenvector of Q:

$$Qe_1 = e^{i\psi_j} e_1.$$

Thus, the matrix Q is of the form

$$
Q = \begin{pmatrix} e^{i\psi_j} & Q_{12} & \cdots & Q_{1d} \\ 0 & Q_{22} & \cdots & Q_{2d} \\ \vdots & \vdots & & \vdots \\ 0 & Q_{d2} & \cdots & Q_{dd} \end{pmatrix},
$$

and the properties of unitary matrices imply that

$$
Q_{12} = \cdots = Q_{1d} = 0.
$$

We can write now Q and R in block form

$$
Q = \begin{pmatrix} e^{i\psi_j} & 0 \\ 0 & Q \end{pmatrix}, \quad R = \begin{pmatrix} |\lambda_j| & \rho^* \\ 0 & R_1 \end{pmatrix}.
$$

The commutation relation (13.4.3) now implies that

$$
\begin{pmatrix} \lambda_j & e^{i\psi_j}\rho^* \\ 0 & Q_1 R_1 \end{pmatrix} = \begin{pmatrix} \lambda_j & \rho^* Q_1 \\ 0 & R_1 Q_1 \end{pmatrix}
$$

and, in particular, $Q_1 R_1 = R_1 Q_1$. We apply the induction hypothesis and the lemma is proved. $\qquad\square$

Assume, indeed, that all the eigenvalues of A are distinct and that the sequences σ_k and A_k converge to the limits σ_∞ and A_∞, respectively. Moreover, assume that $A_\infty - \sigma_\infty I$ is invertible. Then, by continuity of the QR decomposition (Lemma 12.1.7), the matrices Q_k and R_k also converge to limits denoted by Q_∞ and R_∞, respectively. The set of unitary matrices is a compact group U_n; therefore, the set $U_n A$ is a compact set of matrices and thus, A_∞ is unitarily equivalent to A. Passing to the limit in eqns (13.4.1) and (13.4.2) gives the identities

(13.4.4) $A_\infty - \sigma_\infty I = Q_\infty^* (A_\infty - \sigma_\infty I) Q_\infty,$
(13.4.5) $A_\infty - \sigma_\infty I = Q_\infty R_\infty = R_\infty Q_\infty.$

Then, it is clear that, due to Lemma 13.4.1, the limit A_∞ is upper triangular.

13.4.2. Convergence in a special case

We are going to show the convergence of the QR method in a particular case:

Theorem 13.4.2. Let A be a positive definite Hermitian matrix. If the parameters of the translation are all zero, the sequence of matrices $\left(A_k\right)_{k\in\mathbb{N}}$ converges to a diagonal matrix A_∞.   ◇

Proof. The proof rests on a link between the Cholesky method and QR decomposition. Denote by $P(n, \mathbb{C})$ the set of $n \times n$ Hermitian positive definite matrices. We will need the square root mapping from $P(n, \mathbb{C})$ into itself, which associates each element B of $P(n, \mathbb{C})$ with a unique matrix $A \in P(n, \mathbb{C})$ such that $A^2 = B$. The square root is a differentiable function on $P(n, \mathbb{C})$, that we will denote by S. If necessary, we will also write $A = \sqrt{B}$. If B is diagonal then A is also. For the proof of the properties of the square root, see, for example, the exercises of Chapters 3 or 12.

Consider the following Cholesky iteration for $B \in P(n, \mathbb{C})$:

$$(13.4.6) \qquad\qquad\qquad B_0 = B,$$
$$(13.4.7) \qquad\qquad\qquad B_k = U_k^* U_k,$$
$$(13.4.8) \qquad\qquad\qquad B_{k+1} = U_k U_k^*.$$

Here, each matrix U_k belongs to $R^+(n, \mathbb{C})$, the group of upper triangular matrices with a strictly positive diagonal. We know that decomposition (13.4.7) is unique and that all matrices B_k are in $P(n, \mathbb{C})$. Furthermore, there is a simple relationship between this sequence of Cholesky iterations and the sequence of QR iterations defined as follows: let $A = \sqrt{B}$ and let

$$(13.4.9) \qquad\qquad\qquad A_0 = A,$$
$$(13.4.10) \qquad\qquad\qquad A_k = Q_k R_k,$$
$$(13.4.11) \qquad\qquad\qquad A_{k+1} = R_k Q_k.$$

The A_k are all unitarily equivalent to one another and belong to $P(n, \mathbb{C})$. We have the following identities:

$$A_k^2 = A_k^* A_k = R_k^* Q_k^* Q_k R_k = R_k^* R_k,$$
$$A_{k+1}^2 = A_{k+1} A_{k+1}^* = R_k Q_k Q_k^* R_k^* = R_k R_k^*.$$

Consequently, as the A_k are linked by these QR iterations, the B_k are linked by Cholesky iterations. By a uniqueness argument, we can identify B_k with A_k^2 (respectively, R_k and U_k), for all k, and we can deduce that the B_k are all unitarily equivalent to B. More precisely, since $R_{k+1} = Q_{k+1}^* R_k Q_k$,

$$(13.4.12) \qquad B_{k+1} = Q_k^* R_k^* Q_{k+1}^* Q_{k+1} R_k Q_k = Q_k^* B_k Q_k.$$

We will now establish the identities linking the coefficients of $B = R^* R$ and those of $B' = RR^*$, where R belongs to $R^+(n, \mathbb{C})$. We see that

$$(B)_{1j} = (R)_{11} (R)_{1j}, \quad (B)_{2j} = (\bar{R})_{12} (R)_{1j} + (R)_{22} (R)_{2j},$$

and, more generally, for $j \geqslant i$,

$$(13.4.13) \qquad\qquad (B)_{ij} = \sum_{\ell=1}^{i} (\bar{R})_{\ell i} (R)_{\ell j}.$$

Similarly,
$$(B')_{in} = (R)_{in} (R)_{nn},$$

and, more generally, for $i \leqslant j$,

(13.4.14)
$$(B')_{ij} = \sum_{\ell=j}^{n} (R)_{i\ell} (\bar{R})_{j\ell}.$$

Consequently, we have $(B)_{11} \leqslant (B')_{11}$ and, more precisely,

(13.4.15)
$$(B')_{11} - (B)_{11} = \sum_{i=2}^{n} |(R)_{1j}|^2.$$

We generalize eqn (13.4.15) by considering

$$\sum_{p=1}^{m} (B')_{pp} - \sum_{p=1}^{m} (B)_{pp}.$$

Note that

$$\sum_{p=1}^{m} (B)_{pp} = \sum_{i=1}^{m} \sum_{j=i}^{m} |(R)_{ij}|^2$$

and

$$\sum_{p=1}^{m} (B')_{pp} = \sum_{i=1}^{m} \sum_{j=i}^{n} |(R)_{ij}|^2.$$

We can therefore write

(13.4.16)
$$\sum_{p=1}^{m} (B')_{pp} - \sum_{p=1}^{m} (B)_{pp} = \sum_{i=1}^{m} \sum_{j=m+1}^{n} |(R)_{ij}|^2.$$

For $m = n$, the right-hand term of eqn (13.4.16) vanishes because the trace of B is equal to the trace of the equivalent matrix B'. Coming back to the sequence B_k, we deduce from eqn (13.4.16) that all of the sequences

(13.4.17)
$$\left\{ \sum_{p=1}^{m} (B_k)_{pp} \right\}_{k \in \mathbb{N}}$$

are increasing for $m = 1, \ldots, n$. By virtue of eqn (13.4.12), they are bounded and, consequently, they converge. Therefore, the sequences

$$\left\{ (B_k)_{pp} \right\}_{k \in \mathbb{N}}$$

converge for all $p = 1, \ldots, n$. Their respective limits are denoted by $(B_\infty)_{pp}$. Furthermore, the difference between two consecutive terms of the sequence tends to zero. For $p = 1$, eqn (13.4.15) implies that $(R_k)_{1j}$ tends to 0, for all $j \geqslant 2$, as k tends to infinity. Suppose that, for $i \leqslant p - 1$ and $j \geqslant i + 1$, the sequences $(R_k)_{ij}$ tend to zero. Then, to say that the difference between two consecutive terms of the sequence (13.4.17), indexed by p, tends to zero implies that $(R_k)_{pj}$ tends to zero for $j \geqslant p + 1$. Thus, all the non-diagonal terms of R_k tend to zero as k tends to infinity and, similarly, for the non-diagonal terms of B_k. We see that B_k converges to the diagonal matrix whose diagonal coefficients are those of $(B_\infty)_{pp}$ for $p = 1, \ldots, n$. $\qquad \square$

It is also possible to show the convergence of the product of the Q_k to a transformation matrix, allowing the diagonalization of A. Therefore, we also have the convergence of the eigenvectors. Many other cases can be treated, where the proofs are clearly more complicated. There does not currently exist a general proof of the convergence of the QR method, nor a counterexample either.

13.4.3. Effectiveness of QR

Under this naïve form, the algorithm, whose convergence we have just proved, is numerically very bad. Indeed, the QR decomposition of a matrix, viewed as a Gram–Schmidt orthogonalization, is slow and not very stable, as seen in Section 12.1.

On the other hand, if we linearize the QR algorithm in the neighbourhood of A_∞, we can see that, if all of the eigenvalues are pairwise distinct, the convergence is geometric, and its rate is given by

$$\max_{|\lambda_i| < |\lambda_j|} \frac{|\lambda_i|}{|\lambda_j|}.$$

If two eigenvalues are close to one another, the rate of convergence is very bad. We have therefore constructed a slow and unstable method, so there is no place for smugness.

To make the QR method effective, we transform the initial matrix into upper Hessenberg form (Section 12.1). We then saw that the upper Hessenberg form is invariant to the QR algorithm, which is realized by means of Householder transformations. This makes the problem more stable, and a little slower, but there is one fundamental improvement, which lies in the choice of translation parameters.

Once A is in Hessenberg form, the rate of convergence is given by

$$\max_i \left| \frac{\lambda_{i+1}}{\lambda_i} \right|.$$

If we choose a translation parameter close to λ_n, the rate of convergence of the coefficients in the last row is given by

$$\left| \frac{\lambda_n - \sigma}{\lambda_{n-1} - \sigma} \right|,$$

that is to say, very quick. When we consider that the last row has converged, we use deflation, that is, we strike out this last row and the last column and we work on an $(n-1) \times (n-1)$ matrix, to which we again apply a translation strategy. The precise definition of the translation strategies is the object of detailed studies in numerical linear algebra literature. It is implemented in the QR programs currently on the market.

When we look at the real life case, the QR method cannot converge to an upper triangular form if A has complex eigenvalues. In this case, the method converges to a block upper triangular form.

13.5. Exercises from Chapter 13

13.5.1. Spectral pathology

Exercise 13.5.1. Let λ and ϵ be real parameters, and let M be the matrix

$$M(\lambda, \epsilon) = \begin{pmatrix} \lambda & 1 \\ 0 & \lambda + \epsilon \end{pmatrix}.$$

What are the eigenvalues of $M(\lambda, \epsilon)$? When does $M(\lambda, \epsilon)$ possess a basis of eigenvectors?

Exercise 13.5.2. Calculate $\exp(tM(\lambda, \epsilon))$. The quickest way to do this is to note that the required matrix is the matrix of the linear mapping which assigns to the initial conditions x_1 and x_2 the solution, at time t, of the system of ordinary differential equations

$$\dot{X}_1 = \lambda X_1 + X_2, \qquad X_1(0) = x_1,$$
$$\dot{X}_2 = (\lambda + \epsilon) X_2, \qquad X_2(0) = x_2.$$

A first year course shows us how to explicitly calculate the solution to this system.

Exercise 13.5.3. Given the fixed parameter λ, compare $\exp(tM(\lambda, \epsilon))$ and $\exp(tM(\lambda, 0))$ as ϵ tends to zero. Is the behaviour of these expressions at $t = +\infty$ the same?

13.5.2. QR flow and Lax pairs

The Poisson bracket of two matrices A and B belonging to $\mathcal{M}_n(\mathbb{C})$ is defined by

$$[A, B] = AB - BA.$$

Exercise 13.5.4. Let M be a C^1 mapping from $\mathbb{R}$ to $\mathcal{M}_n(\mathbb{C})$. Show that the following two assertions are equivalent:

(i) There exists a C^1 mapping P from $\mathbb{R}$ to invertible matrices in $\mathcal{M}_n(\mathbb{C})$, such that
$$M(t) = P(t)^{-1} M_0 P(t), \quad \forall t \in \mathbb{R};$$

(ii) There exists a matrix $L(t)$, which should be determined, such that M satisfies the system of differential equations

$$\dot{M}(t) = [L(t), M(t)].$$

Exercise 13.5.5. Let U be a C^1 mapping from $\mathbb{R}$ to $\mathcal{M}_n(\mathbb{C})$. Show that the following two assertions are equivalent:

(i) The mapping U is to unitary matrices;

(ii) There exists a continuous mapping A from $\mathbb{R}$ to $\mathcal{M}_n(\mathbb{C})$, with values in the set of skew-Hermitian matrices, such that

$$\dot{U}(t) = A(t)U(t), \quad \forall t \in \mathbb{R}, \qquad U(0) \text{ is unitary.}$$

Exercise 13.5.6. Under the hypotheses of Exercise 13.5.4, show that $M(t)$ is unitarily equivalent to $M(0)$ if and only if, for all t, $L(t)$ is skew-Hermitian.

Exercise 13.5.7. Let R be a C^1 mapping from $\mathbb{R}$ to $\mathcal{M}_n(\mathbb{C})$. Show that the following two assertions are equivalent:

(i) R maps to invertible upper triangular matrices;

(ii) $R(0)$ is invertible and upper triangular, and there exists a continuous mapping S from $\mathbb{R}$ to the set of upper triangular matrices, such that

$$\dot{R}(t) = R(t)S(t), \quad \forall t \in \mathbb{R}.$$

Exercise 13.5.8. Show that the QR decomposition is a continuously differentiable mapping from the set of invertible matrices to the product of the sets of unitary matrices and of upper triangular matrices.

Exercise 13.5.9. Consider the following QR decomposition:

$$e^{tB} = U(t)S(t).$$

Show that U and S are continuously differentiable functions of t. What are the values of $U(0)$ and $S(0)$?

Show that there exists a continuously differentiable function L from $\mathbb{R}^+$ to the set of skew-symmetric matrices such that

$$\dot{U} = -UL.$$

Exercise 13.5.10. Define a sequence M_j by

$$M_0 = e^B, \quad M_j = Q_j R_j, \quad M_{j+1} = R_j Q_j.$$

Prove the identity

(13.5.1) $$M_k = Q_{k-1}^* \cdots Q_0^* M_0 Q_0 \cdots Q_{k-1}.$$

Exercise 13.5.11. Prove the identity

(13.5.2) $$e^{kB} = Q_0 \cdots Q_{k-1} R_{k-1} \cdots R_0,$$

and give the value of $U(k)$ in terms of the Q_js.

Exercise 13.5.12. Let M be the unique solution of the matrix-valued system

$$\dot{M} = [L, M], \quad M(0) = M_0.$$

Use identities (13.5.1) and (13.5.2) to prove that

$$M(k) = M_k.$$

Exercise 13.5.13. What is the limit of $M(t)$ as t tends to infinity?

Exercise 13.5.14. Let A be a symmetric positive definite matrix. Show that it can be written as the exponential of a symmetric matrix and conclude, from the previous study, that the unitary part U_k of the QR decomposition of the powers $A^k = U_k S_k$ gives the change of basis constructed during the QR algorithm starting from A.

14

Nonlinear equations and systems

14.1. From the existence of solutions to their construction

14.1.1. Existence and non-existence of solutions

Let f be a continuous mapping from an open subset of $\mathbb{R}^n$ to $\mathbb{R}^n$. We want to approximate numerically a solution of the system

$$(14.1.1) \qquad\qquad f(x) = 0,$$

on condition that such a solution exists.

Some very elementary examples can convince us that the system (14.1.1) does not always have a solution. Take, for instance,

$$f(x) = x^2 + 1,$$

with x a real variable. There is no real number x such that $f(x)$ vanishes. Take the following slightly more complicated example in two dimensions:

$$f(x) = f(x_1, x_2) = \begin{pmatrix} x_1^2 + x_2^2 - 1 \\ x_1 - 2 \end{pmatrix}.$$

A solution of this problem would have to belong to the circle $x_1^2 + x_2^2 = 1$ and to the straight line $x_1 = 2$, which is parallel to the x_2-axis. Figure 14.1 shows that the circle does not intersect the straight line. At this point, I would expect the reader to protest loudly and to say that I built naughty examples to prove my point. So, I will build even more elementary examples to make my point. Let us consider the linear equations

$$(14.1.2) \qquad\qquad Ax = b.$$

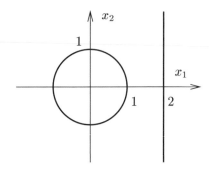

Figure 14.1: The straight line does not intersect the circle.

If A is not a square invertible matrix, we know a necessary and sufficient condition which enables us to solve the problem: there is a solution if and only if b belongs to the image of A. If we assume that the dimension of the target space is n and that the dimension of the image of A is strictly less than n, then, most of the time, we will be unable to solve the system (14.1.2). The reason for this is that the complement of the image of A in the target space is an open set with full Lebesgue measure.

Why do we care about the existence of solutions, anyway?

The next question comes immediately, this time, not on the mathematical side, but on the application side: 'My goodness, why should I care about the existence of solutions? I see them in my experiment/machinery/observation. I do not have time to waste with irrelevant abstract questions.' Usually, this kind of remark comes with a big laugh or a slightly commiserating look.

Dear physicist/engineer/performer of experiments: perish the thought that I should deny facts, but there is quite a distance between fact and theory, as you must well know. The existence of a solution is a property of equations, the observation of facts is material data. You claim that you can write a theory, with equations in it, which gives you a useful and, hopefully, faithful description of reality, and through which you are able to predict and understand nature. If I can prove mathematically that your equations do not have any solutions, shouldn't that say something to you about the value of your model? Maybe, you assumed that you had a stationary solution to your set of equations, and, if there is no solution, it might mean that, in fact, things are moving. Maybe you could improve your model, or improve your observations. On my side, I could quantify the amount of motion and we might conclude together that this motion is irrelevant, or that it is very slow and we are not interested in very large times. However, we would both be wiser with the extra knowledge.

14.1.2. Existence proofs translate into algorithms

Now that I have made a scientist's case for the existence theory of solutions, I'll defend the numerical analyst's case. There are basically two methods for obtaining the existence of solutions of a nonlinear equation or a system of nonlinear equations. One is based on topological methods and another one is based on differentiable methods.

The most elementary example of a topological method is the following: suppose that I have a continuous function f from a compact interval $[a, b]$ which is, say, positive at one end and negative at the other. Then, we know that there is a solution somewhere inside the interval. And how do we find it? We use the bisection method, which is cheap in one-dimensional space.

It is possible to obtain approximate numerical solutions by topological methods [2] in higher-dimensional space. However, this does not come cheap, since these methods require, for a given precision, a number of function evaluations which increases exponentially fast with the dimension.

The most elementary example of a method based on differentiability is the contracting fixed point method. It will work only if the function $x \mapsto x - f(x)$ is strictly contracting. However, then it will be quite efficient, and its performance does not depend on the spatial dimension, since the number of function evaluations is always one per step.

When only topological methods are available, numerical approximation of solutions is generally slow and painful. When differentiable methods are available, then the scientific computation is much more tractable. One of the most important differentiable methods comes up in the so-called perturbation situations. Here, the solution sought is close to a solution which is well known, or, more generally, it can be reached along a differentiable path. Then, a natural idea is to somewhere introduce an abstract version of time, and to use ordinary differential equations as a means to achieve the desired result. This is quite desirable, since there are many efficient methods for computing solutions of differential equations, as we will see in Chapters 16 and 17.

14.1.3. A long and exciting history

Historically, the Babylonians were the first to solve quadratic equations written with numbers. The ancient Greeks solved geometric problems with a ruler and compass, and these problems also reduce to quadratic equations. However, the solution of the cubic equation took much more time. The sixteenth century Italian mathematicians Scipione del Ferro, Niccolo Fontana, also known as Tartaglia, and Girolamo Cardano found the solution of this equation, and in the middle of the computation they introduced imaginary numbers. Subsection 14.3.1 enables the reader to find Cardano's formulae for themself.

The fourth degree equation was solved generally by Bombelli in 1572, and then the fairy tale starts, or maybe stops, depending on the point of view of the reader. In general, the polynomial equation of the fifth degree cannot be solved by radicals, i.e., by finite expressions involving arithmetic operations and taking roots of numbers. This impossibility was proved in the nineteenth century by Galois, who used a deep algebraic method, and who opened the way to an endless stream of beautiful mathematical thinking. However, from the practical point of view, the approximation of roots of polynomial equations remains a very important activity. Think only of the CAD applications: surfaces are represented by polynomials or, more generally, by rational parameterizations; a machine tool has to cut metal, just to the right shape, with not too many holes and each at the right place. How do we ascertain automatically that the machining will be correct? The computer program has to find intersections of surfaces, that is, to solve polynomial equations. But the circle closes itself: efficient computation relies more and more on computer algebra, in particular, in the area of fast multiplication. And computer algebra is... algebra. If every man is an island, mathematical humans tend to build bridges in places where they are not expected. The numerical analyst cares little whether the fifth degree equation can be solved by radicals, but she cares a lot about the speed and reliability of algorithms. If she had to use the most abstract mathematics to enhance the qualities of a numerical method, then she would, if time and space would allow it.

14.1.4. An overview of existence proofs

In the case of a single equation $(n = 1)$, if f is continuous on the compact interval $[a, b]$ of $\mathbb{R}$, a sufficient condition for f to vanish on at least one point of $[a, b]$ is that

$$(14.1.3) \qquad\qquad f(a) f(b) \leqslant 0.$$

The method employed to detect a zero of f under this condition is the classic bisection method. This method was recalled in Subsection 13.1.2 and it allows the localization of a zero of f in N steps with precision $2^{-N}(b - a)$.

The higher-dimensional generalizations rely on the so-called Brouwer's fixed point theorem, or other topological tools.

Theorem 14.1.1. Let g be a continuous mapping from the closed unit ball B_1 of $\mathbb{R}^n$ into itself. Then, there exists a point x of B_1 such that

$$g(x) = x. \qquad\qquad\diamond$$

In this theorem, the norm chosen matters little, since all the unit balls of $\mathbb{R}^n$ are homeomorphic. We deduce from this theorem the following corollary, where $(\cdot, \cdot)$ denotes the Euclidean scalar product:

Corollary 14.1.2. Let f be a mapping from the closed Euclidean unit ball B_1 of $\mathbb{R}^n$ into $\mathbb{R}^n$. Suppose that at the boundary the field f is reentrant, that is,

$$(14.1.4) \qquad\qquad (f(x), x) \leqslant 0, \quad \text{if } |x| = 1.$$

Then f has a zero in the ball B_1.

We recognize in eqn (14.1.4) a generalization of eqn (14.1.3).

To read an introduction to this type of question, the reader is referred to Subsection 14.3.2, where the degree is defined in dimension 2 and Brouwer's fixed point theorem is proved. We also refer the reader to the little book by A. Gramain [39]. The book [72] is a delightful introduction to fixed point theorems seen from a topological point of view. The book [63] is a more highbrow vision of topology; however, it is one of the classics of mathematics.

With another type of information, we could also prove the existence of a solution to eqn (14.1.1). The following classic case, frequently used in numerical analysis, is that when we can solve eqn (14.1.1) by minimization:

Lemma 14.1.3. Let F be a real C^1 function on $\mathbb{R}^n$ such that

$$(14.1.5) \qquad\qquad \lim_{|x|\to\infty} F(x) = +\infty.$$

Then, F attains its minimum at at least one point x of $\mathbb{R}^n$ and at this point $DF(x) = 0$.

Proof. There exists an $R > 0$ such that if $|x| \geqslant R$ then $F(x) \geqslant F(0) + 1$. The closed ball centred at 0 and of radius R is compact, and the function F reaches its minimum in it at some point x_0, where, in particular, $F(x_0) \leqslant F(0)$. The point x_0 is necessarily interior to the ball. If it was on the edge, we would have $F(x_0) \geqslant F(0) + 1$, which contradicts $F(x_0) \leqslant F(0)$. Consequently, $DF(x_0) = 0$. $\qquad\square$

This result is useful in the following case. If f is the gradient of a function F having the property (14.1.5), or if, more generally, f is a multiple of the gradient of such a functional F, we are tempted to use Lemma 14.1.3. We could also hope to use it by putting $F(x) = |f(x)|^2$ and seeking the absolute minimum of F. If this minimum is zero, we have succeeded. However, practical minimization calculations can be difficult for different reasons. On the one hand, F could have several minima, in which the iterative algorithm could get stuck. On the other hand, in n dimensions the calculation could have a dreadfully slow rate of convergence if it is badly done. The second difficulty forms the subject of an optimization course.

The first difficulty corresponds to a very open problem and is currently often treated by a probabilistic method known as simulated annealing. This method was invented to solve discrete problems, amongst others, the problem of the travelling salesman: find the shortest closed path passing through N towns without forgetting one or passing through the same one twice. This problem belongs to the category of problems called NP-complete, for which we do not know of an exact solution algorithm which requires fewer than $O(N^p)$ steps, with p an integer independent of N. In the simulated annealing method applied to the minimization of functionals, we modify the descent along the length of the gradient by a random walk, which grows smaller with time. It has been proved [33] that the limit of these iterations is, almost certainly, a point where the absolute minimum is attained. These methods allow the treatment of problems which are inaccessible without them. All of this is the subject of extremely active scientific work, which has applications in computing and in the modelling of the brain, amongst other fascinating subjects.

In some of the statements which have just been made, there is one underlying idea: the only systems that we know how to solve well are linear systems. We therefore want to get back to the linear case by different procedures. To arrive at the Brouwer theorem, we deform (in the topological sense of the term) the identity. The proof of Corollary 14.1.2 consists of making a differential deformation of a statement which is visibly true for a quadratic form.

The situation is much better when we have some local information. In the scalar case, suppose that we have a point x at which $f(x)$ is small, and that $|f'(y)| \geqslant M$ in the neighbourhood of x. If f is C^1 then f' has a constant sign in an interval $]x - \alpha, x + \alpha[$. To fix ideas, we suppose that

$$f'(y) \geqslant M > 0, \quad \forall y \in \,]x - \alpha, x + \alpha[\,.$$

If $x > y$, then

$$f(x) = f(y) + \int_y^x f'(b)\,\mathrm{d}b \geqslant f(y) + M(x - y)\,.$$

We see that, if $0 < f(x) < \alpha M$, then

$$f(x - \alpha) \leqslant f(x) - \alpha M < 0.$$

Consequently, f vanishes in the interval $]x - \alpha, x[$. The reader should treat the other cases, with $f(x) < 0$ or $f'(x) < 0$, by exploiting the symmetries of the problem. It is necessary to change f to $-f$ or to use $x + \alpha$ instead of $x - \alpha$.

To end this introduction, we can again note that a nonlinear problem has been already treated in this book. The search for eigenvalues and eigenvectors of linear operators is a nonlinear problem, since the eigenvalues of a sum of matrices are not generally the sum of the eigenvalues of the matrices.

14.2. Construction of several methods

14.2.1. The strictly contracting fixed point theorem

Let E be a metric space, equipped with a distance d. A strict contraction is a mapping g from E to itself for which there exists a constant $K < 1$ such that

$$(14.2.1) \qquad\qquad d(g(x), g(y)) \leqslant K d(x, y), \quad \forall x, y \in E.$$

We say that K is the ratio of the contraction g. The strictly contracting fixed point theorem is stated as follows:

Theorem 14.2.1. Let E be a complete metric space and g a strict contraction in E. Then:

(i) There exists a unique fixed point of g, that is, $x \in E$ such that $g(x) = x$.

(ii) For every initial data y^0, the sequence y^n defined by $y^{n+1} = g(y^n)$ converges to x.   ◇

This theorem is proved in all of the good books on analysis, and, in particular, in the book by J.-P. Ferrier [28, pp. 139–40].

We construct an algorithm to find the solutions of the equation $f(x) = 0$ using the fixed point theorem. First of all, we let

$$g(x) = x - f(x).$$

It is clear that it is equivalent to find a fixed point of g and to find a zero of f. A sufficient condition for g to have a fixed point in the interval $[a, b]$ is that firstly $g([a, b])$ is included in $[a, b]$ and secondly that g is a strict contraction with respect to K. Suppose that f is C^1 (and therefore g is also). Then, it is clear that the property (14.2.1) is equivalent to

$$|g'(x)| \leqslant K, \quad \forall x \in [a, b].$$

In other words, we have

$$|1 - f'(x)| \leqslant K, \quad \forall x \in [a, b].$$

This implies that, in particular, f' does not vanish and, therefore, that f is monotonic. We will give below a geometric interpretation of these iterations, including them in a more general case.

The above choice of g is restrictive. We can generalize it by introducing a constant λ and letting

(14.2.2) $$g(x) = x - \lambda f(x).$$

If λ is nonzero, we see that it is equivalent to seek either a zero of f or a fixed point of g. The sufficient conditions for g to have a fixed point in $[a, b]$ are that $g([a, b])$ is included in $[a, b]$ and that g is a strict contraction, giving

(14.2.3) $$|1 - \lambda f'(x)| \leqslant K < 1,$$

that is,

(14.2.4) $$1 - K \leqslant \lambda f'(x) \leqslant 1 + K.$$

This relation implies that f' does not change sign in $[a, b]$ and that λ is of the same sign as f'.

We give a geometric interpretation of the construction of the sequence of iterations

$$y^{n+1} = y^n - \lambda f(y^n).$$

We note that the line of slope μ passing through the point $(y^n, f(y^n))$ has the equation

$$\eta = f(y^n) + \mu(\xi - y^n).$$

It cuts the x-axis at ξ such that

$$\xi - y^n = -\frac{f(y^n)}{\mu},$$

that is,

$$\xi = y^n - \frac{f(y^n)}{\mu}.$$

If we let $\mu = 1/\lambda$, we see that the point y^{n+1} is obtained as the intersection of the x-axis with the line of slope $1/\lambda$ passing through the point $(y^n, f(y^n))$. We therefore start at a point $(y^0, f(y^0))$ and follow the line of slope $1/\lambda$ passing through this point until we reach the x-axis at the point y^1. From the point $(y^1, f(y^1))$, we follow the line of slope $1/\lambda$ until we reach the x-axis, which gives y^2, and so on. The method thus obtained is called the chord method. Figure 14.2 allows the visualization of these iterations.

14.2.2. Newton's method: geometric interpretation and examples

If we return to the relation (14.2.3), we see that the iterations will converge as quickly as the constant K that we take in eqn (14.2.3) is small. It is therefore natural to replace the constant multiplier λ appearing in eqn (14.2.2) by the

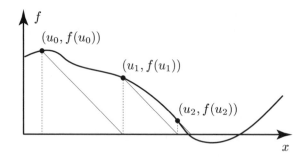

Figure 14.2: The chord method.

function $1/f'(x)$, and to let

$$(14.2.5) \qquad g(x) = x - \frac{f(x)}{f'(x)}.$$

If the function f' does not vanish at the zeros of f, it is equivalent to search for a zero of f and a fixed point of g. The choice (14.2.5) therefore leads us to consider Newton's method given by the iterations

$$(14.2.6) \qquad y^{n+1} = y^n - \frac{f(y^n)}{f'(y^n)}.$$

The geometric interpretation of these iterations is as follows. The line of slope $f'(y^n)$ passing through the point $(y^n, f(y^n))$ has the equation

$$\eta = f(y^n) + (\xi - y^n) f'(y^n)$$

and intersects the x-axis at

$$\xi = y^n - \frac{f(y^n)}{f'(y^n)} = y^{n+1}.$$

Graphically, we start at the point $(y^0, f(y^0))$ and follow the tangent to the graph of f, which intersects the x-axis at y^1. We restart from the point $(y^1, f(y^1))$ and follow the tangent up to the x-axis, where it intersects at y^2, and so on. Refer to Figure 14.3 for a clear picture of this process.

Before giving a theorem on the convergence of Newton's method, we treat explicitly a particular classic case of convergence, namely finding the square root of a positive number.

Given $a > 0$, we want to solve

$$f(x) = x^2 - a = 0.$$

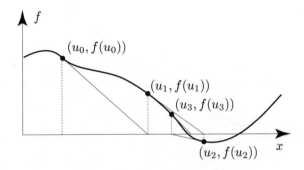

Figure 14.3: Newton's method.

We therefore have

$$g(x) = x - \frac{f(x)}{f'(x)} = x - \frac{x^2 - a}{2x} = \frac{1}{2}\left(x + \frac{a}{x}\right).$$

A drawing could suggest the behaviour of the iteration. We make two parallel drawings, one of the graph of f (Figure 14.4) and the other (Figure 14.5) of the graph of g. The drawings suggest that, if we start from $y^0 > 0$, then all the y^n are positive, $y^n \geqslant y^{n+1}$ for $n \geqslant 1$, and the sequence converges quickly to $\sqrt{a}$. This is exactly what we are going to prove.

We start at a point $y^0 > 0$. We are first of all going to see that all the y^n remain positive. This is a very elementary recurrence, since, if $y^n > 0$, then

$$y^{n+1} = \frac{1}{2}\left(y^n + \frac{a}{y^n}\right) > 0.$$

On the other hand, we calculate the difference $y^{n+1} - \sqrt{a}$ as follows:

$$y^{n+1} - \sqrt{a} = \frac{1}{2}\left(y^n + \frac{a}{y^n}\right) - \sqrt{a}$$

$$= \frac{(y^n)^2 - 2y^n\sqrt{a} + a}{2y^n} = \frac{(y^n - \sqrt{a})^2}{y^n} \geqslant 0.$$

We therefore see that, for $n \geqslant 1$, $y^n \geqslant \sqrt{a}$. Then, the decrease of the sequence,

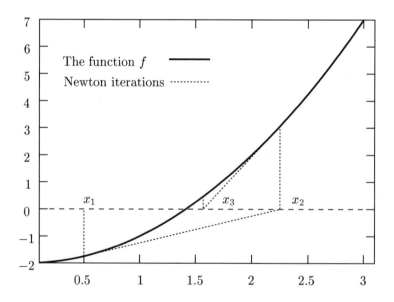

Figure 14.4: The function f and the Newton iterations.

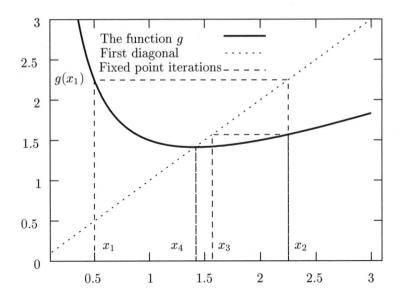

Figure 14.5: The function g and the fixed point iterations.

for $n \geq 1$, is proved by

$$y^n - y^{n+1} = y^n - \frac{1}{2}\left(y^n + \frac{a}{y^n}\right) = \frac{1}{2}\left(y^n - \frac{a}{y^n}\right) = \frac{(y^n)^2 - a}{2y^n} \geq 0.$$

We therefore have a decreasing sequence which is bounded by $\sqrt{a}$, for $n \geq 1$. It is therefore convergent and its limit is a positive or zero fixed point of g. This is therefore $\sqrt{a}$.

14.2.3. Convergence of Newton's method

Here we will prove the following convergence result, which can only be local:

Theorem 14.2.2. Let f be a C^2 function from the interval $[a, b]$ to $\mathbb{R}$, with $a < b$. Suppose that there exists $x \in [a, b]$ such that $f(x) = 0$ and $f'(x) \neq 0$. Then, there exists $\epsilon > 0$ such that, for every $y^0 \in [x - \epsilon, x + \epsilon]$, the sequence of Newton iterations defined, for $n \geq 0$, by

$$y^{n+1} = y^n - \frac{f(y^n)}{f'(y^n)}$$

is well defined, remains in the interval $[x - \epsilon, x + \epsilon]$, and converges to x as n tends to infinity. ◇

The proof of this theorem depends on the following lemma:

Lemma 14.2.3. Let $(r_n)_{n \geqslant 0}$ be a sequence of positive or zero numbers which satisfy

$$r_{n+1} \leqslant r_n^2.$$

If $r_0 < 1$ then this sequence converges to 0. Furthermore, we have

$$r_n \leqslant (r_0)^{2^n}.$$

Proof. Let $(s_n)_{n \geqslant 0}$ be the sequence defined by

$$s_0 = r_0, \quad s_{n+1} = s_n^2.$$

We check that the s_n bound from above the r_n. If $s_n \geqslant r_n$, then

$$s_{n+1} = s_n^2 \geqslant r_n^2 \geqslant r_{n+1}$$

and, therefore, by induction, $s_n \geqslant r_n$ for any n. We now show that the sequence s_n converges under the hypotheses of the lemma. We have

$$s_1 = s_0^2,$$
$$s_2 = s_1^2 = s_0^4$$

and, in general, as induction shows immediately,

$$s_n = (s_0)^{2^n}.$$

As $s_0 < 1$, it is obvious that the sequence s_n tends to 0 as n tends to infinity. The convergence is, moreover, a lot faster than geometric. $\qquad \square$

Proof of Theorem 14.2.2. We can compare y^n and x in the following way:

$$y^{n+1} - x = y^n - \frac{f(y^n)}{f'(y^n)} - x + \frac{f(x)}{f'(y^n)},$$

since $f(x) = 0$. Reducing to the same denominator, we have

(14.2.7) $$y^{n+1} - x = \frac{(y^n - x) f'(y^n) - f(y^n) + f(x)}{f'(y^n)}.$$

Since f' is continuous and does not vanish at x, there exist strictly positive numbers M and α such that

$$|f'(y)| \geqslant M, \quad \forall y \in [x - \alpha, x + \alpha].$$

On the other hand, the Taylor expansion gives

$$f(x) = f(y^n) + (x - y^n) f'(y^n) + \int_{y^n}^x f''(t)(x - t)\, dt.$$

As f is C^2,

$$\max_{[x-\alpha, x+\alpha]} |f''(y)| = L < \infty.$$

The absolute value of the numerator of the right-hand side of eqn (14.2.7) is equal to

$$\left| \int_{y^n}^{x} f''(t)(x-t)\,dt \right|$$

and is bounded by

$$\frac{L}{2} |y^n - x|^2.$$

If we let

$$r_n = \frac{L|y^n - x|}{2M},$$

we see that

$$r_{n+1} \leqslant r_n^2,$$

provided that y^n is in $[x-\alpha, x+\alpha]$. Let

$$\epsilon < \min\left(\alpha, \frac{2M}{L} \right).$$

Then, if $|y^n - x| \leqslant \epsilon$, Lemma 14.2.3 and its proof show us that $|y^{n+1} - x| \leqslant \epsilon$, and, therefore, by induction, the successive iterations are well defined. They converge to x much more quickly than geometrically. $\qquad\square$

In the case of Newton's method for the square root, we look at some iterations for $a = 2$ and starting at $x = 1$:

$$1.00000000000,$$
$$1.50000000000,$$
$$1.41666666666,$$
$$1.41421568628,$$
$$1.41421356238.$$

In the subsequent iterations the sequence is stationary. The value of $\sqrt{2}$ given by the machine used is 1.41421356237. The number of correct decimal places practically doubles at each Newton iteration. This remark will be made more precise later in the study of order.

Finally, we note that Newton's method is used to make proofs of existence for nonlinear problems in infinite dimensions by means of the Nash–Moser theorems. The global behaviour of the iterations of Newton's method for low degree polynomials in the complex plane is the object of a great deal of interest on the part of dynamical systems specialists. We are touching here on questions which are also of great current interest to 'pure' as much as to 'applied' mathematicians.

14.2.4. The secant method

The calculation of a first derivative can be very awkward, even if we have an analytic expression for the function to be differentiated. The interplay of the derivatives of compound functions can create very complicated objects. This is all the more true when f is implicit or it is obtained by an integration depending on a parameter. In this case, the calculation of derivatives can be formidable. We are therefore going to replace the derivative $f'(x_n)$ which appears in Newton's method by a finite difference, obtaining the secant method given by

$$(14.2.8) \qquad\qquad y^{n+1} = y^n - \frac{f(y^n)}{f[y^n, y^{n-1}]},$$

where $f[y^n, y^{n-1}]$ is the divided difference

$$f[y^n, y^{n-1}] = \frac{f(y^n) - f(y^{n-1})}{y^n - y^{n-1}}.$$

We now have a two-step method, which requires two starting values y^0 and y^1. Graphically, the two points $(y^0, f(y^0))$ and $(y^1, f(y^1))$ determine a line whose intersection with the x-axis gives the point y^2. We then take the line passing through $(y^1, f(y^1))$ and $(y^2, f(y^2))$, whose intersection with the x-axis gives y^3, and so on. Refer to Figure 14.6 to see the graphical behaviour of the iterations.

As for Newton's method, we have a local convergence theorem for the secant method:

Theorem 14.2.4. Let f be a C^2 function on the interval $[a, b]$, with $a < b$. Suppose that there exists a point x such that $f(x) = 0$ and $f'(x) \neq 0$. Then, there exists a number $\epsilon > 0$ such that, if y^0 and y^1 are in the interval $[x-\epsilon, x+\epsilon]$, the iterations of the secant method are all well defined, remain in the interval $[x - \epsilon, x + \epsilon]$, and converge to x.   ◇

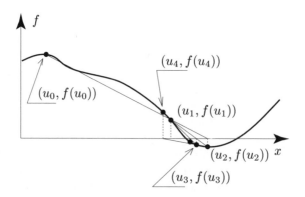

Figure 14.6: The secant method.

The proof of this result rests on a lemma of a similar nature to Lemma 14.2.3, but slightly more complicated.

Lemma 14.2.5. Let r_n be a sequence of positive reals such that

$$(14.2.9) \qquad r_{n+1} \leqslant r_n r_{n-1}.$$

Then, if $r_0 < 1$ and $r_1 < 1$, the sequence of the r_n is bounded by 1 and converges to 0. Furthermore, there exists a constant C, depending on the initial conditions, such that

$$(14.2.10) \qquad r_n \leqslant C r^{\rho^n},$$

where r is a number which is strictly less than 1 and

$$\rho = \frac{1 + \sqrt{5}}{2} \sim 1.618.$$

Proof. If r_n and r_{n-1} are bounded by 1, it is clear that r_{n+1} is also. The first assertion is therefore proved. Now define a sequence s_n by

$$(14.2.11) \qquad s_0 = r_0, \quad s_1 = r_1,$$
$$(14.2.12) \qquad s_{n+1} = s_n s_{n-1}.$$

Suppose that $r_n \leqslant s_n$ and that $r_{n-1} \leqslant s_{n-1}$. Then,

$$r_{n+1} \leqslant r_n r_{n-1} \leqslant s_n s_{n-1} = s_{n+1}.$$

By induction, the s_n form a sequence bounding r_n from above. If s_0 or s_1 is zero, then all the s_n are zero for $n \geqslant 1$ and the lemma is clear. Suppose, therefore, that neither s_0 or s_1 is zero and take the logarithm of the equality (14.2.12). We let

$$\ell_n = -\ln s_n$$

and we obtain

$$(14.2.13) \qquad \ell_{n+1} = \ell_n + \ell_{n-1}.$$

As in Section 11.2, we must write the characteristic equation of the relation (14.2.13), that is

$$(14.2.14) \qquad \rho^2 - \rho - 1 = 0.$$

The solutions of this equation are

$$\frac{1 + \sqrt{5}}{2} \quad \text{and} \quad -\frac{2}{1 + \sqrt{5}}.$$

Letting

$$\rho = \frac{1 + \sqrt{5}}{2},$$

the solutions of the recurrence relation (14.2.13) are all of the form

$$\ell_n = a\rho^n + b\left(-\rho\right)^{-n}.$$

The values of a and b are determined from the initial conditions (14.2.11). We must solve the system

$$\ell_0 = a + b,$$

$$\ell_1 = a\rho - \frac{b}{\rho}.$$

The solution of this system is given by

$$b = \frac{\rho\left(\rho\ell_0 - \ell_1\right)}{1 + \rho^2}, \quad a = \frac{\ell_0 + \rho\ell_1}{1 + \rho^2}.$$

The hypotheses on r_0 and r_1 imply that ℓ_0 and ℓ_1 are strictly positive, and, therefore, the coefficient a of ρ^n is strictly positive. Furthermore, the term in $(-\rho)^{-n}$ tends to 0 exponentially as n tends to infinity, and is bounded by $|b|$. We can therefore bound ℓ_n from below:

$$\ell_n \geqslant a\rho^n - |b|, \quad \forall n \in \mathbb{N}.$$

Returning to r_n, we see that

$$r_n \leqslant \exp\left(-a\rho^n\right)\exp\left(|b|\right).$$

If we let $r = \mathrm{e}^{-a}$ and $C = \mathrm{e}^{|b|}$, then we obtain the inequality (14.2.10).   □

Proof of Theorem 14.2.4.   As in the proof of the theorem on the convergence of Newton's method, we compare the iteration y^{n+1} with the solution x. Furthermore, we note that

$$f\left(y^n\right) = \left(y^n - x\right) f\left[y^n, x\right],$$

since $f(x) = 0$. We therefore have

$$\begin{aligned}
y^{n+1} - x &= y^n - \frac{f\left(y^n\right)}{f\left[y^n, y^{n-1}\right]} - x \\
&= y^n - \frac{\left(y^n - x\right) f\left[y^n, x\right]}{f\left[y^n, y^{n-1}\right]} - x \\
&= \left(y^n - x\right)\left(1 - \frac{f\left[y^n, x\right]}{f\left[y^n, y^{n-1}\right]}\right) \\
&= \left(y^n - x\right)\frac{f\left[y^n, y^{n-1}\right] - f\left[y^n, x\right]}{f\left[y^n, y^{n-1}\right]}.
\end{aligned}$$

Using the property (4.2.4) of divided differences, we see that

$$(14.2.15) \qquad y^{n+1} - x = \frac{(y^n - x)(x - y^{n-1}) f[y^n, y^{n-1}, x]}{f[y^n, y^{n-1}]}.$$

Let $[x - \alpha, x + \alpha]$ be an interval on which $|f'(x)| \geqslant M$. Since f' is continuous, it does not change sign on $[x - \alpha, x + \alpha]$, and, from the integral representation (4.2.5) of divided differences, we see that, if y^n and y^{n-1} are in $[x - \alpha, x + \alpha]$, then

$$\left| f[y^n, y^{n-1}] \right| \geqslant M.$$

On the other hand, let L be the upper bound of $|f''|$ on $[x - \alpha, x + \alpha]$. Again, from the integral representation of divided differences, we deduce that, if y^n and y^{n-1} are in $[x - \alpha, x + \alpha]$, then

$$\left| f[y^n, y^{n-1}, x] \right| \leqslant \frac{L}{2}.$$

Consequently, we see that

$$\left| y^{n+1} - x \right| \leqslant \frac{L}{2M} \left| y^n - x \right| \left| y^{n-1} - x \right|.$$

Let

$$r_n = \frac{L}{2M} \left| y^n - x \right|.$$

Let $\epsilon < \min(\alpha, 2M/L)$. Lemma 14.2.5 assures us that, if y^0 and y^1 are chosen in the interval $[x - \epsilon, x + \epsilon]$, then all the y^m remain in this interval and, furthermore, the sequence of errors $|y^n - x|$ tends to 0, following an estimate of the type (14.2.10). $\qquad \square$

We take the same numerical example for the secant method as we chose for Newton's method, and we seek a zero of $f(x) = x^2 - a$. The iterations are written

$$y^{n+1} = y^n - \frac{(y^n)^2 - a}{y^n + y^{n-1}}.$$

With $a = 2$ and the initial conditions $y^0 = 1.5$ and $y^1 = 1.4$, the successive iterations are given by

$$1.50000000000,$$
$$1.40000000000,$$
$$1.41379310345,$$
$$1.41421568628,$$
$$1.41421356206,$$
$$1.41421356237,$$

and the subsequent iterations are stationary. The calculation converges to machine precision in 6 iterations, instead of 4 in the case of Newton's method. On the other hand, it does not require the calculation of f'.

14.2.5. The golden ratio and Fibonacci's rabbits

Remark 14.2.6. The number $\rho \sim 1.618$, which appeared in the proof of Lemma 14.2.5, has been well known since antiquity, under the name of the golden ratio. It can be defined as the ratio between the length L and the width ℓ of a rectangle such that, if we remove from this rectangle a square of side ℓ, we find a rectangle similar to the first, see Figure 14.7.

We therefore have

$$\frac{L}{\ell} = \frac{\ell}{L - \ell}.$$

On letting $L = \rho\ell$, we find

$$\rho = \frac{1}{\rho - 1},$$

that is,

$$\rho^2 - \rho - 1 = 0,$$

which is exactly the characteristic eqn (14.2.14). Certain aesthetic theories consider the number ρ to be the most harmonious ratio between the sides of a rectangle, and painters and architects have composed their works with the aid of the golden ratio. It appeared in Euclid's elements (third century B.C.) and we believe that the pythagorians knew of it (500 B.C.). We also find it in the proportions of certain Egyptian pyramids. The pythagorian considerations on the mystique of numbers and the aesthetics of proportions have been largely

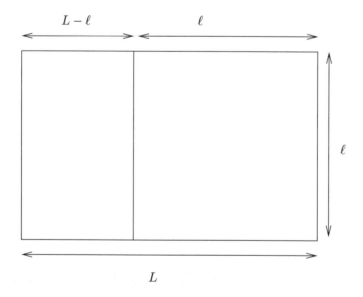

Figure 14.7: Geometrical interpretation of the golden section.

overtaken by the real beauty of mathematics. Besides, apparently constraining aesthetic theories leave much freedom to the designer. For example, the modulor of Le Corbusier was meant to be the heart of a system based upon the proportions of a human of size 1.83 m and the golden ratio. The opinion that we have of Le Corbusier should, quite obviously, be founded on the examination of his architectural work, rather than the underlying theories, in which no one is constrained to live.

The golden ratio is linked to the Fibonacci sequence of numbers proposed by Leonardo of Pisa (Fibonacci) in 1202 in the first quantitive model of biological population growth.

The following information on Fibonacci is found in [75], which contains some translations from Latin to English of choice morsels from classical mathematics texts.

Leonardo of Pisa was a merchant who had travelled widely in the Muslim world. In particular, he knew of the works of Al-Khowarizmi (whose name has been deformed into the word algorithm). His work is in the spirit of the Arab mathematicians of the time, but also shows an independent personal contribution. The *Liber Abaci* (1202, revised in 1228) was largely circulated in manuscript form, but was only published in 1857 in Rome, under the title *Scritti di Leonardo Pisan* [75].

One of the remarkable traits of this book is that in it Leonardo introduced and used the decimal system of positional numbering. The first chapter opens with the following sentence:

'Here are the nine figures of the Indians

$$9\,8\,7\,6\,5\,4\,3\,2\,1.$$

With these nine figures and with the sign 0 called zephirum, one can write any number, as we shall demonstrate later.'

In fact, *zephirum* transcribes the Arab word *as-sifr*, which is the literal translation of the Sanskrit word *sunya*, which signifies emptiness.

We often consider Fibonacci as the first notable western mathematician for having used decimal positional numbering in preference to roman numerals.

We now describe, in modern terms, the model of the growth of a population of rabbits that was proposed by Fibonacci. 'A man possesses a pair of (young) rabbits and a certain place entirely enclosed by walls ...'. These rabbits are going to reproduce. How many will there be at the end of a given number of months? The model of reproduction is the following. There are two types of rabbits: adult rabbits which can reproduce and bring into the world a pair of young rabbits every month, and young rabbits which cannot yet reproduce and become adults at the end of a month. Initially (at time 0), we have a pair of young rabbits and no adult rabbits, so that

$$j\,(0) = 1, \quad a\,(0) = 0.$$

At time 1, we find ourselves with $j(1) = 0$ pairs of young rabbits and $a(1) = 1$ pairs of adult rabbits. Inevitably, the pair of adult rabbits bring into the world a pair of young rabbits at time 2, and we will have

$$j(2) = 1, \quad a(2) = 1.$$

More generally, if at time n we have $j(n)$ pairs of young rabbits and $a(n)$ pairs of adult rabbits, we will have, at time $n + 1$,

$$a(n + 1) = a(n) + j(n) \qquad \text{adult rabbits,}$$
$$j(n + 1) = a(n) \qquad\qquad \text{young rabbits.}$$

Consequently, we have the following recurrence relation for $a(n)$:

$$a(n + 1) = a(n) + a(n - 1).$$

We recover the relation (14.2.13) with the initialization

$$a(0) = 0, \quad a(1) = 1.$$

The numbers which are solutions to this relation are the celebrated Fibonacci numbers. The first terms of the Fibonacci series are given by

$$1, 1, 2, 3, 5, 8, 13, 21, 34, 55, \ldots$$

Current work in biology on populations is a lot more sophisticated than this. It is largely based on probability and statistics and the theory of nonlinear differential systems. It forms the basis of all sorts of studies of great practical use: animal and plant ecology, the propagation of epidemics, choice of vaccination strategy, impact on mortality tables, and associated problems of insurance.

From the point of view of the forms of development in nature, the Fibonacci series has been proposed to explain the number of leaves per turn when the leaves are arranged in a spiral, or to describe the logarithmic spirals appearing in pine cones and sunflowers. This description is not very convincing for contemporary biologists. We can also refer to the magnificent little book by Hermann Weyl [77, pp. 77–8], which contains images of great quality and some developments on the Fibonacci series.

The limit of the ratio $a(n + 1)/a(n)$ is precisely the golden ratio. This irrational number is very badly approximated by rational numbers and, in some sense, is extremal in regard to this property. It is not, however, 'very irrational', since it is a quadratic number, that is, it is the root of a second-order equation with integer coefficients. The arithmetic properties of the Fibonacci numbers are the subject of contemporary studies. They occur in optimization, in the theory of dynamical systems, and in combinatorics. For this last subject, see [38].

14.2.6. Order of an iterative method

The order of an iterative method can be described in the following way. Let y^n be a sequence of approximations to a number (or, more generally, a vector) x. Consider the sequence of errors $e_n = |y^n - x|$. We have seen in the case of Newton's method that, when the starting point of the iteration is close enough to the zero, these errors satisfy the estimate

$$e_{n+1} \leqslant C e_n^2,$$

where C is a certain positive number. In the case of the secant method, we can prove that an equivalent error, for large n, is given by

$$e_n \sim C r^{\rho^n},$$

where $\rho \sim 1.618$ is the golden ratio. From this, we deduce that, asymptotically,

$$e_{n+1} \leqslant C^{1-\rho} \left(e_n\right)^\rho .$$

When a method follows from the fixed point algorithm, contracting with constant $K < 1$, the errors satisfy

$$e_{n+1} \leqslant K e_n,$$

as we can see by reading any proof of this classic result.

Generally, we will say that an iterative method is of order $\lambda > 1$ if λ is the supremum of real numbers for which there exists a constant C such that, asymptotically for large n,

$$e_{n+1} \leqslant C e_n^\lambda.$$

With this definition, Newton's method is of order 2 and the secant method is of order $\rho \sim 1.618$. As for the contracting method, it is of order 1. For this iterative method to converge, it is necessary that $C < 1$. For a method of order strictly greater than 1 to converge, we require only that the initial error is sufficiently small.

The order of an iterative method has a considerable effect on the speed of convergence. Let $p_n = -\log_{10} e_n$ be a measure of the number of correct decimal places of y^n. If we have, asymptotically,

$$e_{n+1} \sim C e_n^\lambda,$$

with $\lambda > 1$, then

$$-p_{n+1} \sim \log_{10} C - \lambda p_n,$$

and, therefore, asymptotically, y^{n+1} has λ times more correct decimal places than y^n. This explains the phenomenon demonstrated just after the end of the proof of Theorem 14.2.2.

14.2.7. Ideas on the solution of vector problems

Until now, we have only talked of the search for roots of scalar equations. We proceed, more or less, on the same lines when searching for the solution of vector problems. The function f is a C^1 function from an open subset of $\mathbb{R}^n$ to $\mathbb{R}^n$. The chord method can be generalized as follows. The sequence of iterations

$$y^{j+1} = y^j - M^{-1} f\left(y^j\right)$$

will be convergent if $y \mapsto y - M^{-1} f(y)$ is a strict contraction which maps a ball of $\mathbb{R}^n$ into itself.

Moreover, if f is C^2, we introduce Newton's method by letting

$$y^{j+1} = y^j - Df\left(y^j\right)^{-1} f\left(y^j\right),$$

where $Df(x)$ denotes the Jacobian of f with respect to x. This is a linear operator from $\mathbb{R}^n$ into itself. We can state the following convergence theorem for Newton's method in the vector case:

Theorem 14.2.7. Let f be a C^2 function from a closed ball B of $\mathbb{R}^n$ to $\mathbb{R}^n$. Suppose that f has a zero x in B and that $Df(x)$ is invertible. Then, there exists an $\epsilon > 0$ such that, for every initial condition y^0 satisfying $|y^0 - x| \leqslant \epsilon$, the sequence of Newton iterations is well defined and converges to 0 as j tends to infinity.   ◇

Proof. The proof is completely identical to that of the scalar case. We note that

$$y^{j+1} - x = y^j - x - Df\left(y^j\right)^{-1} f\left(y^j\right)$$
$$= Df\left(y^j\right)^{-1} \left(Df\left(y^j\right)\left(y^j - x\right) - f\left(y^j\right) + f\left(x\right)\right),$$

since f vanishes at x. There exist two strictly positive numbers M and α such that, for $|y - x| \leqslant \alpha$, we have

$$\left\|Df\left(y\right)^{-1}\right\| \leqslant M^{-1}.$$

The Taylor formula with integral remainder gives us, for every a and b in B,

$$f\left(b\right) - f\left(a\right) - Df\left(a\right)\left(b - a\right) = \int_0^1 \left(1 - s\right) D^2 f\left(a + s\left(b - a\right)\right)\left(b - a, b - a\right) ds,$$

and, consequently,

$$\left|f\left(x\right) - f\left(y^j\right) - Df\left(y^j\right)\left(x - y^j\right)\right| \leqslant \frac{1}{2} L\left|x - y^j\right|^2,$$

if L bounds $\|D^2 f(y)\|$ in the ball centred on x and of radius α. For $|y^j - x| \leqslant \alpha$, we then have the upper bound

$$\left|y^{j+1} - x\right| \leqslant \frac{L}{2M}\left|y^j - x\right|^2.$$

The rest of the proof is identical to the end of the proof of Theorem 14.2.2.   □

The generalization of the secant method to the vector case is much more tricky. Modern methods for the numerical solution of nonlinear equations call upon a lot of subtlety and shrewdness. The idea is either to make a Newton method or to make a chord type method, but to update the matrices M at the end of several steps. We do not necessarily choose to take M to be the inverse of $Df(y^j)$, but a matrix sufficiently close to this inverse to keep something of the good properties of Newton's method and have an order greater than 1. We thus obtain generalized Newton methods, which are tackled in [16], though always from an optimization perspective.

It is difficult to give accessible references on the solution of nonlinear systems of equations. It is a problem which is both very difficult and very open. Each time it is necessary to exploit the particular structure of the problem under consideration.

14.3. Exercises from Chapter 14

14.3.1. The Cardano formulae

Exercise 14.3.1. Show, by a change of variable of the form $y = x - x_0$, that every cubic equation

$$y^3 + ay^2 + by + c = 0$$

can be put under the form

(14.3.1) $$x^3 + px + q = 0.$$

Exercise 14.3.2. Show that, if

$$x_1 = u + v, \quad p = -3uv, \quad q = -\left(u^3 + v^3\right),$$

then x_1 is a solution of eqn (14.3.1).

Exercise 14.3.3. Write $U = u^3$ and $V = v^3$. Show that U and V are the roots of a quadratic equation and give these roots in terms of p and q.

Exercise 14.3.4. A priori, U and V each have three cubic roots. One could think that there are nine different combinations of the form $u + v$. How can a cubic equation have so many roots, if the polynomial does not vanish identically?

Show that the condition $p = -3uv$ implies that one can choose only three combinations among the nine, and give explicitly the three solutions of eqn (14.3.1). It may be convenient to use the complex cubic roots of 1, namely $j = (-1 + i\sqrt{3})/2$ and $j^2 = (-1 - i\sqrt{3})/2$, and to distinguish cases according to the sign of $4p^3 + 27q^2$.

Exercise 14.3.5. Devise a simple numerical example of eqn (14.3.1) with the following properties: all the roots of the equation are real and Cardano's formula uses complex numbers.

14.3.2. Brouwer's fixed point theorem in dimension 2

Exercise 14.3.6. Let f be a continuous and periodic function of period L from $\mathbb{R}$ to $\mathbb{C}$. Assume that f does not vanish. Show that there exists a continuous function ϕ from $\mathbb{R}$ to $\mathbb{R}$, defined up to an additive multiple of 2π, such that

$$\frac{f(x)}{|f(x)|} = \exp\left(i\phi(x)\right).$$

Show that $\phi(L) - \phi(0)$ is an integer multiple of 2π which does not depend on the choice of ϕ. We will say that ϕ is the phase of f and that the integer $(\phi(L) - \phi(0))/2\pi$ is the degree of f.

Exercise 14.3.7. Let Q_0 be the square $[0,1] \times [0,1]$ and let g be a continuous function from Q_0 to $\mathbb{R}^2$. Assume that g does not vanish on the boundary of Q_0 and that the boundary of Q_0 is parameterized in terms of arc length as follows:

$$(14.3.2) \qquad \psi_0(t) = \begin{cases} (t, 0)^\top & \text{if } 0 \leqslant t \leqslant 1; \\ (1, t-1)^\top & \text{if } 1 \leqslant t \leqslant 2; \\ (3-t, 1)^\top & \text{if } 2 \leqslant t \leqslant 3; \\ (0, 4-t)^\top & \text{if } 3 \leqslant t \leqslant 4. \end{cases}$$

This parameterization is extended to all of $\mathbb{R}$ by periodicity, of period 4. Denote by n_0 the degree of $g \circ \psi_0$. Subdivide Q_0 into four equal squares Q_1^α, with $\alpha \in \{0, 1\}^2$. Therefore,

$$Q_1^\alpha = \left[\frac{\alpha_1}{2}, \frac{\alpha_1 + 1}{2}\right] \times \left[\frac{\alpha_2}{2}, \frac{\alpha_2 + 1}{2}\right].$$

The boundary of Q_1^α is parameterized analogously to the boundary of Q_0 (see eqn (14.3.2)) by the functions ψ_1^α. If g does not vanish on any of the boundaries of the squares Q_1^α, show that the sum of the degrees n_1^α of the $g \circ \psi_1^\alpha$ is equal to n_0.

Exercise 14.3.8. For all $j \geqslant 1$ and all $\alpha \in \{0, \ldots, 2^j - 1\}^2$, let

$$Q_j^\alpha = \left[\frac{\alpha_1}{2^j}, \frac{\alpha_1 + 1}{2^j}\right] \times \left[\frac{\alpha_2}{2^j}, \frac{\alpha_2 + 1}{2^j}\right].$$

The boundary of Q_j^α is parameterized by ψ_j^α, as in eqn (14.3.2). Show that, if g does not vanish on the boundary of any of the Q_j^α, then the sum of the degrees of the $g \circ \psi_j^\alpha$ is equal to n_0.

Exercise 14.3.9. Assume that n_0 is not zero and that g does not vanish inside Q_0. Show that we have a contradiction by considering a sequence of nested squares $Q_j^{\alpha(j)}$, such that the degree of $g \circ \psi_j^{\alpha(j)}$ does not vanish for any j.

Exercise 14.3.10. Let f be a function from the unit ball $\{x_1^2 + x_2^2 \leqslant 1\}$ to itself. Consider the function

$$f_\varepsilon(x) = \frac{f(x)}{1 + \varepsilon}$$

and extend, as follows, this function to the whole square $[-1, 1]^2$:

$$f_\varepsilon(x) = f_\varepsilon\left(\frac{x}{|x|}\right) \quad \text{if} \quad x \in [-1, 1]^2 \setminus \{x_1^2 + x_2^2 \leqslant 1\}.$$

Show, with the help of the previous exercise, that f_ε has a fixed point in the unit ball. By taking the limit as ε tends to 0, derive Brouwer's theorem in dimension 2.

14.3.3. Comparison of two methods for calculating square roots

In this section, we study two methods for approximating the square root of a real positive number and we compare them for numerical efficiency.

Reducing the search to the interval $[1/4, 1]$

Exercise 14.3.11. Show that we can translate the problem of seeking the square root of a binary floating-point number $0.d_1 d_2 \cdots d_r \times 2^p$, with $d_1 = 1$ and $d_j = 0$ or 1, if $2 \leqslant j \leqslant r$, to the search for the square root of a binary floating-point number belonging to the interval $[1/4, 1[$.

Exercise 14.3.12. To get an initial estimate of $\sqrt{x}$, for $x \in [1/4, 1[$, we let

$$f_0(x) = \alpha x + \beta,$$

where f_0 is the best approximation to $\sqrt{x}$ in $[1/4, 1]$ in the maximum norm. There exist y_0, y_1, and y_2 in $[1/4, 1[$ such that $1/4 \leqslant y_0 < y_1 < y_2 \leqslant 1$ and

$$f_0(y_0) - \sqrt{y_0} = -f_0(y_1) + \sqrt{y_1} = f_0(y_2) - \sqrt{y_2},$$
$$\mu = \max_{x \in [1/4, 1]} |f_0(x) - \sqrt{x}| = |f_0(y_j) - \sqrt{y_j}|, \quad j = 0, 1, 2.$$

Calculate α and β, using the fact that $x \mapsto \sqrt{x}$ is concave on $[1/4, 1]$. Calculate μ.

Exercise 14.3.13. Let a be a strictly positive real number. Write down the equation for Newton's method which allows us to find the positive root of the equation

$$x^2 = a.$$

Exercise 14.3.14. We denote the iteration thus obtained by

$$x_{n+1} = g(x_n).$$

Carefully draw the graph of g and give the set of $x_0 \in \mathbb{R}_*^+$ for which this iterative method converges.

Exercise 14.3.15. We now intend to estimate the convergence rate of this Newton method for a in $[1/4, 1]$ and with $x_0 = f_0(a)$. Calculate the minimum of $f_0(a)$ in the interval $[1/4, 1]$. Carefully show that

$$\left| x_n - \sqrt{a} \right| \leqslant \mu^{2^n}.$$

Exercise 14.3.16. Single precision on a certain machine corresponds to a mantissa encoded with 21 bits, that is, $r = 21$ in the notation of Section 1.3. Double precision corresponds to $r = 52$. Calculate the value of n which allows us to achieve complete convergence in the case of single precision, then in double precision (error less than 2^{-r}). We give

$$\log_2 48 \simeq 5.585.$$

Calculate the number of multiplications and divisions necessary in each case, noting that division by 2 is particularly economic and should not be counted.

Acceleration of convergence

Consider the equation

$$f(x) = 0,$$

which we solve by the iterative method

$$x_{n+1} = g(x_n),$$

where $x \mapsto x - g(x)$ has the same roots as f. We intend to improve its order by the following argument. Let

$$e_n = g(y_n) - y_n, \quad y'_{n+1} = g(y_n), \quad e'_{n+1} = g(y'_{n+1}) - y'_{n+1}.$$

The straight line passing through the points (y_n, e_n) and (y'_{n+1}, e'_{n+1}) is well defined, if the method has not yet converged.

Exercise 14.3.17. Give the equation of the line passing through the points (y_n, e_n) and (y'_{n+1}, e'_{n+1}) in the coordinates (y, e). Calculate the x-coordinate y_{n+1} of its intersection with $y = 0$. For the next stage of the iteration, we choose this x-coordinate y_{n+1}. Show that y_{n+1} can be written in the form

$$y_{n+1} = y_n - \frac{e_n^2}{e'_{n+1} - e_n}$$

and also in the form

$$y_{n+1} = G(y_n),$$

with

$$G(y) = \frac{y g \circ g(y) - g(y)^2}{g \circ g(y) - 2g(y) + y}.$$

Exercise 14.3.18. We start with

$$g\left(x\right) = \frac{1}{2}\left(x + \frac{a}{x}\right).$$

Calculate G. What could be the limit of these iterations? What is the order of the method thus constructed?

Exercise 14.3.19. For which strictly positive initial conditions does the method converge? To make a complete study, it is convenient to change variables to $x = y/\sqrt{a}$, let $H(x) = G(y)/\sqrt{a}$, and to use the formula giving $\tanh 3\alpha$ in terms of $\tanh\alpha$.

Exercise 14.3.20. Show that each iteration of the method requires 4 multiplications and divisions.

Exercise 14.3.21. We want to estimate the error uniformly for $a \in [1/4, 1]$, beginning from $y_0 = f_0(a)$, the function constructed in Exercise 14.3.12. Let

$$\mu_j = \max_{a\in[1/4,1]}\left|y_j - \sqrt{a}\right|.$$

We recall that $\mu_0 = \mu$, which we calculated in Exercise 14.3.12. Show that

$$\mu_{j+1} \leqslant h\left(\mu_j\right),$$

where the function h is given by

$$h\left(t\right) = \frac{1}{1 - 2t}\frac{t^3}{1 - t + t^2}.$$

Exercise 14.3.22. Calculate the number of iterations necessary to reach $\sqrt{a}$ in single and in double precision. Calculate the number of multiplications and divisions necessary. Compare this with the result of Exercise 14.3.16 and deduce the method which should be chosen in practice. We give

$$46 \times 47 \times 48 = 103\,776, \qquad \frac{\ln 2}{\ln 10} \simeq 0.301.$$

14.3.4. Newton's method for finding the square roots of matrices

Newton's algorithm for the square root of a complex number

Exercise 14.3.23. Let a be a nonzero complex number and let f be a function from $\mathbb{C}$ to itself, defined by

$$f\left(z\right) = z^2 - a.$$

Write down Newton's method for finding the zeros of f.

Exercise 14.3.24. Let b be one of the square roots of a and let $(z_n)_{n \geqslant 0}$ be the sequence of iterations obtained by the Newton algorithm, starting from a given initial value z_0. What recurrence relation is satisfied by the sequence

$$w_n = \frac{z_n - b}{z_n + b},$$

when it is defined?

Exercise 14.3.25. From the preceding question, deduce that we can partition the complex plane $\mathbb{C}$ into three regions $\mathcal{R}_0$, $\mathcal{R}_+$, and $\mathcal{R}_-$, which are each invariant to iterations of the Newton algorithm. When z_0 is in $\mathcal{R}_+$ (respectively, $\mathcal{R}_-$, $\mathcal{R}_0$), the Newton algorithm converges to b (respectively, $-b$, does not converge). Describe these three regions both geometrically and analytically.

Square roots of matrices

We say that a square matrix A of order n has a square root if there exists a square matrix B of order n such that $B^2 = A$.

Exercise 14.3.26. Show that the matrix

$$J = \begin{pmatrix} 0 & 1 \\ 0 & 0 \end{pmatrix}$$

does not have a square root.

Exercise 14.3.27. Show that the matrix

$$K = \begin{pmatrix} 0 & 0 & 1 \\ 0 & 0 & 0 \\ 0 & 0 & 0 \end{pmatrix}$$

has an infinite number of square roots.

Exercise 14.3.28. Let A be an invertible upper triangular matrix. Show that there exists a choice of complex numbers $(B)_{jj}$ such that

$$(B)_{jj}^2 = (A)_{jj}$$

and

$$(B)_{jj} + (B)_{kk} \neq 0, \quad \forall j, \forall k \neq j.$$

Exercise 14.3.29. From the preceding question, deduce that every invertible matrix has a square root.

Newton's method for the square root of a matrix

Let A be an invertible matrix of order n and let F be the function defined on $\mathcal{M}_n(\mathbb{C})$ by

$$F(X) = X^2 - A.$$

Exercise 14.3.30. Calculate the differential $DF(A)H$ of F at the point A, with an increment of H.

Exercise 14.3.31. Write down Newton's method for F.

Exercise 14.3.32. Let T be an upper triangular matrix which satisfies the condition

$$(T)_{jj} + (T)_{kk} \neq 0, \quad \forall j, \ \forall k \neq j$$

and let C be a given matrix of order n. Show that the linear system

$$TH + HT = C$$

has a unique solution H, which is a square matrix of order n. Show that the system thus obtained is triangular, on the condition that the unknowns $(H)_{ij}$ are suitably numbered.

Exercise 14.3.33. Let B be a square root of A which satisfies

$$(B)_{jj} + (B)_{kk} \neq 0, \quad \forall j, \ \forall k \neq j.$$

Show that $DF(X)$ is invertible for X in the neighbourhood of B.

Exercise 14.3.34. Show that Newton's algorithm converges in the neighbourhood of B.

Exercise 14.3.35. How can we implement Newton's algorithm, as described in Exercise 14.3.31, in practice? Moreover, what do you think of the practical use of this algorithm?

A first alternative to the Newton algorithm

Exercise 14.3.36. Consider the following algorithms:

$$(14.3.3) \qquad Y_{k+1} = \frac{AY_k^{-1} + Y_k}{2},$$

$$Z_{k+1} = \frac{Z_k^{-1}A + Z_k}{2}.$$

Show that, if Y_0 (respectively, Z_0) commutes with A, then the same is true of all of the Y_k (respectively, Z_k), and that the sequence of Y_k (respectively, Z_k) is identical to the sequence of Newton iterations with initial condition Y_0 (respectively, Z_0).

Exercise 14.3.37. Suppose that A is diagonalizable and that all of its eigenvalues have a strictly positive real part. Show that, if we start from

(14.3.4) $$Y_0 = I_n,$$

the sequence of Y_k converges to a square root B of A whose eigenvalues all have strictly positive real parts.

Exercise 14.3.38. Let

$$G(Y) = \frac{AY^{-1} + Y}{2}.$$

Calculate $DG(B)H$.

Exercise 14.3.39. Working in a basis in which B is diagonal, calculate the spectral radius of the operator $Y \mapsto DG(B)Y$ and find a necessary and sufficient condition for which the spectral radius of $DG(B)$ is strictly less than 1.

Exercise 14.3.40. Deduce from the preceding results that the numerical method that we have described can only be stable if the matrix A has a condition number less that 9 in whichever norm is chosen. Here, the condition number of A is the product $\|A\| \, \|A^{-1}\|$ and $\| \cdot \|$ is the matrix norm.

A stable alternative to Newton's algorithm

Consider now the following algorithm:

$$P_0 = A,$$
$$Q_0 = I,$$
(14.3.5) $$P_{k+1} = \frac{P_k + Q_k^{-1}}{2},$$
$$Q_{k+1} = \frac{Q_k + P_k^{-1}}{2}.$$

Exercise 14.3.41. Show that P_k and Q_k commute with each other and with A, for every value of k for which they are defined.

Exercise 14.3.42. Suppose that A is diagonalizable and that all of its eigenvalues have a positive real part. Show that P_k tends to B and Q_k tends to B^{-1}, where B is the limit of the sequence of Y_k, defined by eqns (14.3.3) and (14.3.4), as k tends to infinity.

Exercise 14.3.43. We let

$$\mathcal{F}(P,Q) = \begin{pmatrix} \frac{1}{2}\left(P + Q^{-1}\right) \\ \frac{1}{2}\left(Q + P^{-1}\right) \end{pmatrix}.$$

Calculate

$$D\mathcal{F}\left(B, B^{-1}\right) \begin{pmatrix} H \\ K \end{pmatrix}.$$

What is the spectral radius of $D\mathcal{F}(B, B^{-1})$? What can we conclude about the stability of the method (14.3.5)?

15

Solving differential systems

The Cauchy problem for a system of differential equations consists of studying the solutions of the system

(15.0.1a) $$\dot{u}(t) = f(t, u(t)),$$
(15.0.1b) $$u(t_0) = u_0,$$

where the dot denotes differentiation with respect to time, the unknown is the *function* $t \mapsto u(t)$, and the data are the initial time t_0, the initial condition u_0, and the function $f : (s, v) \mapsto f(s, v)$. The time t belongs to an interval of $\mathbb{R}$ and the state u belongs to an open set of $\mathbb{R}^d$. In the context of the local theory which is seen in a course of differential calculus, we have a local theory in open sets. In our mathematically simpler context of global solutions, we consider the whole of $\mathbb{R}^d$.

15.1. Cauchy–Lipschitz theory

15.1.1. Idea of the proof of existence for ODEs

The essential point leading to the proof of the theorem which follows is that the operation of integration produces more regular functions than the operation of differentiation. More precisely, if u is a measurable and essentially bounded function on a compact interval of $\mathbb{R}$, we have no information on its derivative and do not even know if it is differentiable. On the other hand, all of its integrals are bounded, Lipschitz, and differentiable almost everywhere.

If we integrate eqn (15.0.1a) with respect to time, taking account of the initial condition (15.0.1b), we obtain

(15.1.1) $$u(t) = u_0 + \int_{t_0}^{t} f(s, u(s)) \, ds.$$

We are going to show that, under suitable conditions on f, we can solve eqn (15.1.1) by a strictly contracting fixed point theorem. Furthermore, in a function class that we will specify, eqn (15.1.1) is equivalent to eqns (15.0.1a) and (15.0.1b).

15.1.2. Cauchy–Lipschitz existence theorem

Theorem 15.1.1 (Cauchy–Lipschitz). Suppose that $[t_1, t_2]$ is a compact interval and that f is a continuous function from $[t_1, t_2] \times \mathbb{R}^d$ into $\mathbb{R}^d$ which satisfies the following property: there exists a constant L such that

$$(15.1.2) \qquad |f(t, v) - f(t, w)| \leqslant L |v - w|, \quad \forall t \in [t_1, t_2], \ \forall v, w \in \mathbb{R}^d.$$

Here, $|\cdot|$ denotes some norm on $\mathbb{R}^d$. Then, for any t_0 in $[t_1, t_2]$ and u_0 in $\mathbb{R}^d$, there exists a unique continuously differentiable function u from $[t_1, t_2]$ to $\mathbb{R}^d$ which satisfies eqns (15.0.1a) and (15.0.1b). ⋄

Proof. As we seek a u which satisfies eqn (15.1.1), we are going to consider the mapping $\mathcal{T}$ defined by

$$(15.1.3) \qquad (\mathcal{T}v)(t) = u_0 + \int_{t_0}^{t} f(s, v(s)) \, ds,$$

with the convention of an orientated integral. We will take account of the orientation by placing an absolute value on the outside of integrals when writing an upper bound for a norm.

Recall that $C^0([t_1, t_2])$ is the Banach space of real continuous functions on $[t_1, t_2]$. It is equipped with the norm $\max_{t \in [t_1, t_2]} |v(t)|$. We denote by $C^0([t_1, t_2]; \mathbb{R}^d)$ the space of continuous functions from the compact interval $[t_1, t_2]$ to $\mathbb{R}^d$. It is isomorphic to the product of d copies of the Banach space $C^0([t_1, t_2])$, and hence also a Banach space. It is equipped with the norm

$$(15.1.4) \qquad \|v\| = \max_{t \in [t_1, t_2]} |v(t)|.$$

For each t, $(\mathcal{T}v)(t)$ is a vector of $\mathbb{R}^d$ and, therefore, $\mathcal{T}$ is a mapping from a function space to a function space. If v is continuous on $[t_1, t_2]$, the mapping $s \mapsto f(s, v(s))$ is continuous from $[t_1, t_2]$ to $\mathbb{R}^d$. As the integral of a continuous function is a continuous function, $\mathcal{T}$ maps $C^0([t_1, t_2]; \mathbb{R}^d)$ to itself, and even to $C^1([t_1, t_2]; \mathbb{R}^d)$. We are going to show that $\mathcal{T}$ is a strict contraction in $C^0([t_1, t_2]; \mathbb{R}^d)$ if L is small enough. Now

$$\mathcal{T}v(t) - \mathcal{T}w(t) = \int_{t_0}^{t} (f(s, v(s)) - f(s, w(s))) \, ds.$$

Due to the hypothesis of the theorem, we have the upper bound

$$(15.1.5) \qquad |\mathcal{T}v(t) - \mathcal{T}w(t)| \leqslant \left| \int_{t_0}^{t} L \|v - w\| \, ds \right|,$$

from which we deduce

(15.1.6) $\|\mathcal{T}v - \mathcal{T}w\| \leqslant \|v - w\| L \max\left(|t_2 - t_0|, |t_1 - t_0|\right).$

Consequently, if $L\max(|t_2 - t_0|, |t_1 - t_0|)$ is strictly less that 1, $\mathcal{T}$ is a strict contraction in the complete metric space $C^0\left([t_1, t_2]; \mathbb{R}^d\right)$ and it possesses a unique fixed point.

This conclusion is not satisfactory, since Theorem 15.1.1 stated no condition on $L\max(|t_2 - t_0|, |t_1 - t_0|)$. We rid ourselves of this condition by using Picard's iterations, which we now define.

From eqn (15.1.5), we have that

$$|\mathcal{T}v\,(t) - \mathcal{T}w\,(t)| \leqslant L\,|t - t_0|\,\|v - w\|.$$

We show, by induction, that we have the general estimate

(15.1.7) $|\mathcal{T}^p v\,(t) - \mathcal{T}^p w\,(t)| \leqslant \dfrac{L^p\,|t - t_0|^p}{p!}\,\|v - w\|.$

Indeed, if eqn (15.1.7) holds, we see that

$$
\begin{aligned}
\left|\mathcal{T}^{p+1} v\,(t) - \mathcal{T}^{p+1} w\,(t)\right| &\leqslant \left|\int_{t_0}^{t} L\,|\mathcal{T}^p v\,(s) - \mathcal{T}^p w\,(s)|\,\mathrm{d}s\right| \\
&\leqslant \frac{L^{p+1}\,|t - t_0|^{p+1}}{(p + 1)!}\,\|v - w\|,
\end{aligned}
$$

which proves that the estimate (15.1.7) holds in general. Now, we have the identity

$$\sum_{p=0}^{\infty} \frac{L^p\,|t - t_0|^p}{p!} = \mathrm{e}^{L|t - t_0|},$$

which, in particular, shows us that

$$\lim_{p \to \infty} \frac{L^p\,|t - t_0|^p}{p!} = 0.$$

Therefore, for every L, we can find a p such that

$$\frac{L^p \max\left(|t_2 - t_0|^p, |t_1 - t_0|^p\right)}{p!} < 1.$$

This implies that $\mathcal{T}^p$ is a strict contraction in $C^0([t_1, t_2]; \mathbb{R}^d)$. Consequently, there exists a unique u such that

$$\mathcal{T}^p u = u.$$

We show that the set of fixed points of $\mathcal{T}^p$ is identical to the set of fixed points of $\mathcal{T}$. If u is a fixed point of $\mathcal{T}$, then it is also clearly a fixed point of $\mathcal{T}^p$. Conversely, if u is a fixed point of $\mathcal{T}^p$, we can apply $\mathcal{T}$ to the equation

$$\mathcal{T}^p u = u$$

to obtain

$$\mathcal{T}^{p+1} u = \mathcal{T} u,$$

which we can rewrite as

$$\mathcal{T}_{,}^{p} \left(\mathcal{T} u \right) = \mathcal{T} u.$$

Therefore, $\mathcal{T} u$ is a fixed point of $\mathcal{T}^p$. As the fixed point of $\mathcal{T}^p$ is unique, we have $\mathcal{T} u = u$, that is, eqn (15.1.1).

We have thus shown the existence of a unique fixed point of $\mathcal{T}$. As we remarked earlier, the image of $\mathcal{T}$ is included in the set $C^1([t_1, t_2]; \mathbb{R}^d)$. Therefore, u, the fixed point of $\mathcal{T}$, is continuously differentiable and we can differentiate the equality

$$u(t) = u_0 + \int_{t_0}^t f(s, u(s)) \, ds$$

with respect to time to give

$$\dot{u}(t) = f(t, u(t)),$$

and the initial condition

$$u(t_0) = u_0$$

is satisfied.

Conversely, if u is a C^1 solution of eqns (15.0.1a) and (15.0.1b), we integrate eqn (15.0.1a) with respect to time, taking account of the initial condition (15.0.1b), and obtain eqn (15.1.1), which completes the proof of our theorem. □

Note that Theorem 15.1.1 has an immediate generalization:

Corollary 15.1.2. Suppose that $]t_1, t_2[$ is a non-empty open interval of $\mathbb{R}$, finite or infinite, and that f is a continuous mapping from $]t_1, t_2[\times \mathbb{R}^d$ to $\mathbb{R}^d$ which satisfies the following property, for all compact intervals I in $]t_1, t_2[$:

$$(15.1.8) \qquad |f(t, v) - f(t, w)| \leqslant L(I) |v - w|, \quad \forall t \in I, \ \forall v, w \in \mathbb{R}^d.$$

Here, $|\cdot|$ denotes some norm on $\mathbb{R}^d$. Then, for any t_0 in $]t_1, t_2[$ and u_0 in $\mathbb{R}^d$, there exists a unique continuously differentiable function u from $]t_1, t_2[$ to $\mathbb{R}^d$ which satisfies eqns (15.0.1a) and (15.0.1b).

Proof. It is sufficient to note that, if $t_0 \in I \subset J$, with compact intervals I and J, we can define a solution u_I (respectively, u_J) of eqns (15.0.1a) and (15.0.1b) on the interval I (respectively, J). This solution is unique, and it is clear that the restriction of u_J to I is a solution of eqns (15.0.1a) and (15.0.1b). Since we have uniqueness, the restriction of u_J to I is equal to u_I. We can, therefore, continue a solution on a compact interval I into a solution on a growing union of compact intervals I_k, and we can choose the I_k such that their union is equal to the interval $]t_1, t_2[$. $\qquad\qquad\square$

Remark 15.1.3. We could have also taken u and f to have values in a complex vector space. The proof of existence under Cauchy–Lipschitz conditions is identical. To do this, it suffices to identify $\mathbb{C}^d$ with $\mathbb{R}^{2d}$. The complex theory is very useful, particularly in the case of linear differential equations. We can also construct a theory of systems of differential equations in the complex domain, that is, with a complex time. This theory is only interesting in the case where f is holomorphic and we study it by means of algebraic and topological tools. The theory of differential equations in the complex plane is entirely out of the scope of this course, although the point of departure is a theorem of existence and uniqueness completely analogous to that which has been proved here.

15.1.3. Systems of order 1 and of order p

Let g be a mapping from $[0,T] \times (\mathbb{R}^d)^p$ to $\mathbb{R}^d$. Consider the following differential system of order p:

$$(15.1.9) \qquad u^{(p)}(t) = g\left(t, u(t), \dot{u}(t), \ldots, u^{(p-1)}(t)\right).$$

This ordinary differential system can always be reduced to a system of the first order. Define, indeed,

$$Y = \begin{pmatrix} y_0 \\ y_1 \\ \vdots \\ y_{p-1} \end{pmatrix} \in \mathbb{R}^{dp}, \quad F(t,Y) = \begin{pmatrix} y_1 \\ y_2 \\ \vdots \\ g(t, y_0, y_1, \ldots, y_{p-1}) \end{pmatrix} \in \mathbb{R}^{dp}.$$

With this notation, the system (15.1.9) is equivalent to

$$\dot{U}(t) = F(t, U(t)),$$

provided that we perform the natural identification

$$U = \begin{pmatrix} u \\ \dot{u} \\ \vdots \\ u^{(p-1)} \end{pmatrix}.$$

15.1.4. Autonomous and non-autonomous systems, transformation of an autonomous system into a non-autonomous system

An ordinary differential system is said to be autonomous if the time variable does not appear explicitly in the left-hand side function f. The study of non-autonomous systems can be reduced to the study of autonomous systems, but there is a price to pay: this reduction adds one dimension to the system and possibly destroys its linear character. Consider the first-order system

$$\dot{u}(t) = f(t, u(t)).$$

It can be transformed into an autonomous system through the following transformation: define a function s by

$$s(t) = t.$$

If we let

$$Y = \begin{pmatrix} y \\ z \end{pmatrix} \in \mathbb{R}^d \times \mathbb{R}, \quad F(Y) = \begin{pmatrix} f(y, z) \\ 1 \end{pmatrix},$$

then

$$X(t) = \begin{pmatrix} x(t) \\ s(t) \end{pmatrix}$$

solves the differential system

$$\dot{X} = F(X),$$

which is, indeed, autonomous.

Even if we start from the simplest possible linear equation with non-constant coefficients

$$(15.1.10) \qquad \dot{x}(t) = a(t) x(t),$$

the system obtained by the previous transformation is not linear, since it can be written as

$$(15.1.11) \qquad \dot{X}_1 = a(X_2) X_1, \quad \dot{X}_2 = 1.$$

Moreover, the Cauchy–Lipschitz existence theorem has been proved under stronger assumptions on the system (15.1.11) than on the system (15.1.10).

15.2. Linear differential equations

15.2.1. Constant coefficient linear systems

The most simple example of a system of ordinary differential equations is the linear system with constant coefficients and no time-dependent forcing. We have a matrix $A \in \mathcal{M}_d(\mathbb{K})$, with constant coefficients in $\mathbb{K} = \mathbb{R}$ or $\mathbb{C}$, according to whether we are considering a real or complex problem, and we study

$$(15.2.1) \qquad\qquad \dot{u}(t) = Au(t).$$

We should check that we really have the conditions needed for the application of the Cauchy–Lipschitz theorem. We denote by $|\cdot|$ an arbitrary norm on $\mathbb{K}^d$. In this case, we have

$$f(t, u) = Au$$

and, as A is linear, it suffices to find L such that

$$|Au| \leqslant L\,|u|, \quad \forall u \in \mathbb{K}^d.$$

It is sufficient to take for L the norm of the matrix A which is subordinate to the norm $|\cdot|$, and we have existence and uniqueness of the solution to the problem (15.2.1), for every initial vaue u_0 and every initial time t_0. We consider now the following matrix-valued ordinary differential equation

$$(15.2.2a) \qquad\qquad \dot{M}(t) = AM(t),$$
$$(15.2.2b) \qquad\qquad M(0) = I.$$

If $\|\cdot\|$ denotes a matrix norm satisfying the algebraic property, the function $g(t, M) = AM$ satisfies the conditions of Corollary 15.1.2 (Cauchy–Lipschitz conditions). Indeed, we have

$$\|g(t, M)\| \leqslant \|A\|\,\|M\|$$

and, as g is linear with respect to M, this is enough for us. Consequently, the system (15.2.2) has a unique solution $M(t)$. If we consider now the vector function $u(t) = M(t - t_0)u_0$, we note that

$$\dot{u}(t) = \frac{\mathrm{d}}{\mathrm{d}t}\left(M(t - t_0)\,u_0\right) = \dot{M}(t - t_0)\,u_0$$
$$= (AM(t - t_0))\,u_0 = A\left(M(t - t_0)\,u_0\right) = Au(t).$$

Furthermore, $u(t_0) = u_0$. Since we have uniqueness for solutions to the system (15.2.1) with the initial condition u_0, its solution is equal to $M(t - t_0)u_0$.

The matrix function $M(t)$ has various interesting properties. First of all, it commutes with A. Indeed, we let

$$B(t) = AM(t) - M(t)A.$$

We then have

$$\dot{B}(t) = A^2 M(t) - AM(t)A = AB(t).$$

Since $B(0) = 0$, the uniqueness of the solution to eqn (15.2.2a), for all initial conditions, implies that $B(t) = 0$, for all t. The mapping $t \mapsto M(t)$ is a group homomorphism from $\mathbb{R}$ into the group of invertible matrices. Indeed, if s is some real number, consider

$$C(t) = M(t+s) - M(t)M(s).$$

We have

$$\dot{C}(t) = AM(t+s) - AM(t)M(s) = AC(t).$$

Since $C(0) = 0$, the uniqueness of solutions to eqn (15.2.2a) again implies that $C(t) = 0$.

15.2.2. Matrix exponentials

We can summarize the preceding results in the following lemma:

Lemma 15.2.1. The unique solution $M(t)$ of the matrix-valued differential eqns (15.2.2a) and (15.2.2b) is called the exponential of the matrix At, and it is denoted by e^{At}. It has the following properties:

$$(15.2.3) \qquad e^{At}A = Ae^{At}, \quad \forall t \in \mathbb{R}, \qquad e^{A(t+s)} = e^{At}e^{As}, \quad \forall s, t \in \mathbb{R}.$$

It is an analytic function of t which has the following series expansion:

$$(15.2.4) \qquad e^{At} = \sum_{j=0}^{\infty} \frac{A^j t^j}{j!},$$

with an infinite radius of convergence. We have the upper bound

$$(15.2.5) \qquad \left\| e^{At} \right\| \leqslant e^{\|A\| \, |t|}.$$

Furthermore, the unique solution of the system (15.2.1), which has the value u_0 at $t = t_0$, is given by $u(t) = e^{A(t-t_0)}u_0$.

Proof. It only remains to show the relations (15.2.4) and (15.2.5). We note that

$$\frac{d^p M}{dt^p}(0) = A^p.$$

Also, $\|A^p\| \leqslant \|A\|^p$ and the series with general term

$$\frac{\|A\|^p \, |t|^p}{p!}$$

converges and sums to $e^{\|A\|\|t\|}$, giving eqn (15.2.5). $\qquad \square$

We now study some particular properties of the function e^{At}, when A belongs to various sets of matrices:

Lemma 15.2.2. If A is Hermitian, e^{At} is Hermitian positive definite. If A is skew-Hermitian, e^{At} is unitary.

Proof. Assume that A is Hermitian. If we pass to the adjoint in eqn (15.2.2a), we get

$$\dot{M}\left(t\right)^* = M\left(t\right)^* A.$$

We note that, as $M(t)$ commutes with A, $M(t)^*$ also commutes with $A^* = A$. Also, from the relation $M(0)^* = I = M(0)$, we deduce that $M(t)^* = M(t)$, since we have uniqueness for the solutions to eqn (15.2.2a).

If x is an eigenvector of A corresponding to the eigenvalue $\lambda \in \mathbb{R}$, we see that

$$\frac{\mathrm{d}}{\mathrm{d}t} \mathrm{e}^{At} x = \mathrm{e}^{At} A x = \mathrm{e}^{At} \lambda x.$$

Consequently,

$$\mathrm{e}^{At} x = \mathrm{e}^{\lambda t} x.$$

Since we can decompose the space on a basis of eigenvectors of A, we see that all the eigenvalues of e^{At} are $\mathrm{e}^{\lambda_j t}$, where λ_j is an eigenvalue of A. This proves that e^{At} is positive definite.

If A is skew-Hermitian and if x is some vector, let

$$m\left(t\right) = x^* \left(\mathrm{e}^{At}\right)^* \mathrm{e}^{At} x.$$

We therefore have

$$\dot{m}\left(t\right) = x^* \left(\mathrm{e}^{At} A\right)^* \mathrm{e}^{At} x + x^* \left(\mathrm{e}^{At}\right)^* A \mathrm{e}^{At} x = x^* \left(\mathrm{e}^{At}\right)^* \left(A^* + A\right) \mathrm{e}^{At} x = 0.$$

Consequently, for the Hermitian norm $\sqrt{x^* x} = |x|$,

$$\left|\mathrm{e}^{At} x\right| = |x|,$$

for any t and x. This shows that e^{At} is an isometry and therefore unitary. $\square$

Lemma 15.2.3. We have the following relation:

$$(15.2.6) \qquad\qquad \det\left(\mathrm{e}^{At}\right) = \mathrm{e}^{t\,\mathrm{trace}(A)}.$$

Proof. We can construct this proof by using the explicit formula (15.2.4) and a triangulation of A. Instead, we prove it using the properties of differential equations, which gives us the opportunity to differentiate the determinant function, which we denote h. We have, for every matrix $B \in M_d(\mathbb{K})$,

$$Dh\left(I\right) \cdot B = \lim_{t \to 0} \frac{\det\left(I + tB\right) - 1}{t}.$$

We know that B is equivalent to an upper triangular matrix T, with P as the transformation matrix. Consequently,

$$Dh\left(I\right) \cdot B = \lim_{t \to 0} \frac{\det\left(I + tP^{-1}TP\right) - 1}{t}$$

$$= \lim_{t \to 0} \frac{\det\left(P^{-1}\left(I + tT\right)P\right) - 1}{t} = Dh\left(I\right) \cdot T.$$

Since T is upper triangular, we explicitly calculate $\det(I + tT)$, which is equal to $1 + t\operatorname{trace}(T) + O(t^2)$. Since the trace of a matrix is invariant to similarity,

(15.2.7) $$Dh\,(I) \cdot B = \operatorname{trace}(B).$$

From this, we deduce that

$$
\lim_{s \to 0} \frac{\det\left(e^{A(t+s)}\right) - \det\left(e^{At}\right)}{s} = \det\left(e^{At}\right) \lim_{s \to 0} \frac{\det\left(e^{As}\right) - I}{s}
$$

$$
= \det\left(e^{At}\right) Dh\,(I) \cdot \left.\frac{d}{ds}\left(e^{As}\right)\right|_{s=0}
$$

$$
= \det\left(e^{At}\right) \operatorname{trace}(A).
$$

Since $\det(e^{A0}) = \det(I) = 1$, we have only to integrate the differential equation

$$
\frac{d}{dt} \det\left(e^{At}\right) = \operatorname{trace}(A) \det\left(e^{At}\right), \quad \det\left(e^{A0}\right) = 0.
$$

This is done by inspection and gives the relation (15.2.6). □

From Lemma 15.2.3, we deduce that if A is a real skew-symmetric matrix then e^A is an orthogonal matrix with determinant 1. Therefore, e^A is a rotation matrix.

15.2.3. Duhamel's formula

If the linear constant coefficient system of differential equations that we are considering possesses a second term, which we suppose to be continuous, we can still solve it explicitly:

Lemma 15.2.4. Let g be a continuous function from $[0, T]$ to $\mathbb{R}^d$ and let $A \in \mathcal{M}_d(\mathbb{R})$. The system

$$
\dot{u}\,(t) = Au\,(t) + g\,(t), \quad u\,(0) = u_0
$$

possesses a unique solution, which is defined on $[0, T]$ and given by Duhamel's formula

(15.2.8) $$u\,(t) = e^{At}u_0 + \int_0^t e^{A(t-s)} g\,(s)\, ds.$$

Proof. Let $f(t, v) = Av + g(t)$. We see that

$$
|f\,(t, v) - f\,(t, w)| \leqslant \|A\|\,|v - w|.
$$

We certainly have the conditions to apply the Cauchy–Lipschitz theorem. Let $v(t) = e^{-At}u(t)$. We note that

$$
\dot{v}\,(t) = -e^{-At}Au\,(t) + e^{-At}\dot{u}\,(t)
$$

$$
= -e^{-At}Au\,(t) + e^{-At}\left(Au\,(t) + g\,(t)\right)
$$

$$
= e^{-At}g\,(t).
$$

We integrate the differential equation in v by inspection and obtain

$$v\left(t\right) = v_0 + \int_0^t e^{-As}g\left(s\right)\mathrm{d}s.$$

We immediately find formula (15.2.8). $\square$

15.2.4. Linear equations and systems with variable coefficients

We pass now to the case of linear systems with variable coefficients. Consider a continuous mapping A from $[0,T]$ to $\mathcal{M}_d(\mathbb{K})$, a continuous mapping g from $[0,T]$ to $\mathbb{K}^d$, and the system

$$(15.2.9) \qquad\qquad \dot{u}\left(t\right) = A\left(t\right)u\left(t\right) + g\left(t\right).$$

If we let $f(t,v) = A(t)v + g(t)$, the conditions of the Cauchy–Lipschitz theorem are fulfilled, with

$$L = \max_{t \in [0,T]} \left\|A\left(t\right)\right\|.$$

We therefore have existence and uniqueness of the solution to the system (15.2.9), for any initial condition at any initial time.

We now consider the scalar case, that is

$$(15.2.10) \qquad\qquad \dot{u}\left(t\right) = a\left(t\right)u\left(t\right) + g\left(t\right),$$

with a and g continuous on $[0,T]$. If $g = 0$, the equation is integrated by inspection and has as solution

$$(15.2.11) \qquad\qquad u\left(t\right) = u_0 \exp\left(a_1\left(t\right)\right),$$

where we let

$$(15.2.12) \qquad\qquad a_1\left(t\right) = \int_0^t a\left(s\right)\mathrm{d}s.$$

If g is not identically zero, we apply the *variation of parameters* method, which amounts to letting

$$(15.2.13) \qquad\qquad v\left(t\right) = u\left(t\right)\exp\left(-a_1\left(t\right)\right).$$

Consequently,

$$\begin{aligned}
\dot{v}\left(t\right) &= \dot{u}\left(t\right)\exp\left(-a_1\left(t\right)\right) - a\left(t\right)u\left(t\right)\exp\left(-a_1\left(t\right)\right) \\
&= \left(a\left(t\right)u\left(t\right) + g\left(t\right)\right)\exp\left(-a_1\left(t\right)\right) - a\left(t\right)u\left(t\right)\exp\left(-a_1\left(t\right)\right) \\
&= g\left(t\right)\exp\left(-a_1\left(t\right)\right).
\end{aligned}$$

We can therefore integrate by inspection and obtain

$$v(t) = v(0) + \int_0^t g(s) \exp(-a_1(s)) \, ds.$$

Noting that $v(0) = u_0$, and returning to the definition of u, we have

(15.2.14) $\qquad u(t) = u_0 \exp(a_1(t)) + \int_0^t \exp(a_1(t) - a_1(s)) g(s) \, ds.$

In higher dimensions the situation is less simple. Indeed, if $g = 0$, we do not have an expression for the solution of the system (15.2.9) making use of a matrix exponential. We are going to understand this by studying the differentiation of $e^{B(t)}$, where B is a C^1 mapping from $\mathbb{R}$ to $\mathcal{M}_d(\mathbb{K})$. We can find the derivative of this mapping as follows, denoting by $m(M)$ the exponential of M:

$$\frac{m(M + sN) - m(M)}{s}$$

$$= s^{-1} \left[I + M + sN + \frac{1}{2!} \left(M^2 + s(MN + NM) + s^2 N^2 \right) \right.$$

$$+ \frac{1}{3!} \left(M^3 + s(M^2 N + MNM + NM^2) \right.$$

$$\left. + s^2 \left(MN^2 + NMN + N^2 M \right) + s^3 N^3 \right) + \dots$$

$$\left. - \left(I + M + \frac{M^2}{2!} + \frac{M^3}{3!} + \dots \right) \right]$$

and, therefore,

$$Dm(M) \cdot N = \lim_{s \to 0} \frac{m(M + sN) - m(M)}{s}$$

$$= N + \frac{1}{2!}(MN + NM) + \frac{1}{3!}(M^2 N + MNM + NM^2) + \dots$$

We therefore have

$$\frac{d}{dt} e^{B(t)} = \dot{B}(t) + \frac{B(t)\dot{B}(t) + \dot{B}(t)B(t)}{2!}$$

$$+ \frac{B(t)^2 \dot{B}(t) + B(t)\dot{B}(t)B(t) + \dot{B}(t)B(t)^2}{3!} + \dots$$

Let

$$B(t) = \int_0^t A(s) \, ds.$$

Unless $A(t)$ and $B(t)$ commute for all t, we have little chance of being able to solve the system (15.2.9) by a formula of the type (15.2.11). This commutation condition is very restrictive and is generally not satisfied. For instance, if the eigenvalues of B are distinct on some interval, then this implies that its eigenspaces are constant on that interval.

Although the solutions are not explicit in dimension $n \geqslant 2$, we, nevertheless, have a very nice formalism which allows us to understand many things. Indeed, consider the matrix differential system

$$(15.2.15) \qquad \frac{\partial G(t,s)}{\partial t} = A(t)G(t,s), \quad G(s,s) = I.$$

It satisfies the Cauchy–Lipschitz criterion and, therefore, it possesses a unique solution. Note that $u(t) = G(t,t_0)u_0$ is the unique solution of the system (15.2.9) when g is identically zero. Indeed,

$$\dot{u}(t) = \frac{\partial G(t,t_0)}{\partial t} u_0 = A(t)G(t,t_0)u_0 = A(t)u(t).$$

Consequently, $G(t,s)$ is the solution operator. It associates the solution of the system (15.2.9), with $g = 0$, at time t to an initial condition u_0 at time s. In particular, the solution operator is a linear operator, which means that the j-th column vector of $G(t,s)$ is the value of the vector solution of the system (15.2.9), with $g = 0$, at time t, when the initial condition at time s is the j-th vector of the canonical basis. We call $G(t,s)$ the resolvent matrix of the differential system.

The family of matrices $G(t,s)$ has additive properties which generalize those of the exponential. Indeed, consider the function

$$B(\tau) = G(\tau,t)G(t,s) - G(\tau,s).$$

We have

$$\dot{B}(\tau) = \frac{\partial G(\tau,t)}{\partial \tau}G(t,s) - \frac{\partial G(\tau,s)}{\partial \tau}$$
$$= A(\tau)G(\tau,t)G(t,s) - A(\tau)G(\tau,s)$$
$$= A(\tau)B(\tau).$$

Since B is zero at $\tau = t$ and is the solution to a differential system satisfying the Cauchy–Lipschitz conditions, B must be identically zero. Therefore, we have the following relation, valid for all s, t, τ:

$$(15.2.16) \qquad G(\tau,t)G(t,s) = G(\tau,s).$$

Physically, this expresses a causality relation: the state of the system at the instant τ is entirely determined by the state of the system at some other instant.

With the aid of $G(t,s)$, we will be able to solve the problem with a time-dependent forcing term, and the formula is analogous to the expression (15.2.8):

Lemma 15.2.5. Let g be a continuous function from $[0,T]$ to $\mathbb{R}^d$. The system

$$\dot{u}(t) = A(t)u(t) + g(t), \quad u(t_0) = u_0,$$

with data $t_0 \in [0, T]$ and $u_0 \in \mathbb{R}^d$, possesses a unique solution, defined on $[0, T]$ and given by Duhamel's formula

$$(15.2.17) \qquad u(t) = G(t, t_0) u_0 + \int_{t_0}^t G(t, s) g(s) \, ds.$$

Proof. Since $G(\cdot, s)$ is the solution of a differential system which satisfies the conditions of the Cauchy–Lipschitz theorem, it is a C^1 function. If we differentiate the formula (15.2.17) with respect to time, taking account of the relation (15.2.15), we obtain

$$\dot{u}(t) = A(t) G(t, t_0) u_0 + G(t, t) g(t) + \int_{t_0}^t A(t) G(t, s) g(s) \, ds$$
$$= A(t) u(t) + g(t).$$

As u also satisfies the initial condition $u(t_0) = u_0$ at $t = t_0$, the lemma is proved. Observe that eqn (15.2.17) also makes sense for $t \in [0, t_0]$ if the integral is oriented. $\qquad \square$

Just as we calculated the determinant of e^{At}, we are going to determine a differential equation satisfied by $\det(G(t, s))$, and then deduce its value. Indeed,

$$\frac{\det(G(t + h, s)) - \det(G(t, s))}{h} = \frac{\det(G(t + h, t) G(t, s)) - \det(G(t, s))}{h}$$
$$= \frac{\det(G(t + h, t)) - 1}{h} \det(G(t, s)).$$

We therefore need to calculate the derivative

$$\frac{\partial}{\partial t} \det(G(t, s)) \bigg|_{t=s}.$$

Using eqn (15.2.7) and the theorem on the derivative of composite functions,

$$\frac{\partial}{\partial t} \det(G(t, s)) \bigg|_{t=s} = \text{trace}(A(s)).$$

From this, we obtain the result

$$\frac{\partial}{\partial t} \det(G(t, s)) = \text{trace}(A(t)) \det(G(t, s)),$$

which we integrate by inspection to obtain

$$(15.2.18) \qquad \det(G(t, s)) = \exp\left(\int_s^t \text{trace}(A(\sigma)) \, d\sigma\right).$$

We now give an example of the information that we can obtain by these techniques. Suppose that $A(t)$ is a continuous function from $[0, T]$ to $\mathcal{M}_d(\mathbb{R})$

and that it is skew-symmetric. Then, all the $G(t, s)$ are isometries. To see this, it is sufficient to differentiate the function $t \mapsto x^* G(t, s)^* G(t, s) x$ and find

$$\frac{\partial}{\partial t} x^* G(t, s)^* G(t, s) x$$
$$= x^* G(t, s)^* A(t)^* G(t, s) x + x^* G(t, s)^* A(t) G(t, s) x = 0.$$

We are going to use this information to bound the solutions of the system (15.2.9). We deduce from formula (15.2.17) that

$$|u(t)| \leqslant |u_0| + \int_{t_0}^t |g(s)| \, ds.$$

This relation is a lot finer than the estimate that we get from Gronwall's lemma, which we will prove below. In particular, it does not use the norm of $A(t)$, but only some of its qualitative properties.

The result (15.2.18), which follows from the relation (15.2.7), has a geometric interpretation. We examine the two-dimensional case. If A is some 2×2 matrix, there exists an orthogonal matrix P such that $T = P^{-1}AP$ is upper triangular. We have

$$T = \begin{pmatrix} \alpha & \beta \\ 0 & \gamma \end{pmatrix}.$$

Now, the area of the parallelogram constructed from the vectors

$$\begin{pmatrix} 1 + \alpha t \\ 0 \end{pmatrix} \quad \text{and} \quad \begin{pmatrix} \beta t \\ 1 + \gamma t \end{pmatrix}$$

is equal to the area of the rectangle constructed from the vectors

$$\begin{pmatrix} 1 + \alpha t \\ 0 \end{pmatrix} \quad \text{and} \quad \begin{pmatrix} 0 \\ 1 + \gamma t \end{pmatrix},$$

which is itself equal to the determinant of the matrix

$$I + t \begin{pmatrix} \alpha & 0 \\ 0 & \gamma \end{pmatrix}.$$

Refer to Figure 15.1 to visualize the rectangle and the parallelogram.

We see that the off-diagonal terms of the matrix A do not contribute to the determinant of the matrix $I + tA$. They have a purely shearing effect, which does not modify volumes. We come across shears as geometric basis transformations in incompressible fluid mechanics, which include most liquid flows. Contemplating the turbulence which appears in your favourite river, outside periods of drought, shows that very complicated things can happen, even with transformations which conserve volumes.

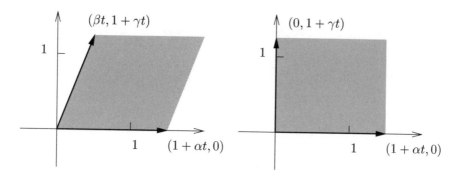

Figure 15.1: Shearing does not modify volumes.

15.2.5. Gronwall's lemma

Gronwall's lemma is a result which allows us to deduce an estimate from a differential inequality. There are many forms of Gronwall's lemma and it is difficult to give a form which has the maximum generality. The general study of differential inequalities is, moreover, an active area of research. We will content ourselves, therefore, with a form which will suffice in the area of Cauchy–Lipschitz theory.

Lemma 15.2.6 (Gronwall's lemma). Let u be a continuously differentiable function from $[0, T]$ to $\mathbb{R}^d$ and let ϕ and ψ be integrable functions on $[0, T]$ which are positive or zero almost everywhere. Suppose that u satisfies the differential inequality

$$(15.2.19) \qquad |\dot{u}(t)| \leqslant \phi(t) + \psi(t) |u(t)|$$

almost everywhere on $[0, T]$. Then, if we let

$$(15.2.20) \qquad \Psi(t) = \int_0^t \psi(s) \, \mathrm{d}s,$$

u satisfies the estimate

$$(15.2.21) \qquad |u(t)| \leqslant |u_0| \, \mathrm{e}^{\Psi(t)} + \int_0^t \phi(s) \, \mathrm{e}^{\Psi(t) - \Psi(s)} \, \mathrm{d}s, \quad \forall t \in [0, T].$$

Proof. We begin with a 'formal proof'. Let

$$|u(t)| = g(t).$$

We then have

$$(15.2.22) \qquad \dot{g}(t) \leqslant |\dot{u}(t)| \leqslant \phi(t) + \psi(t) g(t).$$

Consequently,

$$\dot{g}(t) - \psi(t)\, g(t) \leqslant \phi(t),$$

which we multiply by $e^{-\Psi(t)}$ to give

$$\frac{d}{dt}\left(g(t)\, e^{-\Psi(t)}\right) \leqslant \phi(t)\, e^{-\Psi(t)}.$$

This inequality can be integrated by inspection, leading to

$$g(t)\, e^{-\Psi(t)} - g(0) \leqslant \int_0^t \phi(s)\, e^{-\Psi(s)}\, ds.$$

Multiplying this last relation by $e^{\Psi(t)}$, we obtain the desired result.

Although this calculation contains all the essential ideas of the proof, it is not correct in the preceding form. Indeed, the first line (15.2.22) rests on the inequality

$$\frac{d}{dt}\,|u(t)| \leqslant |\dot{u}(t)|,$$

which is easy to justify in one dimension or if the norm $|\cdot|$ is differentiable away from 0, which is the case for the Euclidean norm. For this, we must use the left and right derivatives where u vanishes. On the other hand, if the norm is not everywhere differentiable away from 0, this inequality is much more difficult to justify. In particular, $t \mapsto |u(t)|$ is not differentiable in the usual sense of the term. We can, nevertheless, completely justify this type of inequality by calling on ideas from convex analysis. This type of proof is outside the scope of a degree-level course, and also of this book, and is not necessary anyway, since there is a way to get around this difficulty. Indeed, the formal proof allows us to find the bound (15.2.21). We are going to show, by a connectedness argument, that we have a bound of the type (15.2.21), but with a parameter $\epsilon > 0$ that we are then going to make tend to zero. Let

$$(15.2.23) \qquad h(t,\epsilon) = e^{\Psi(t)}\left(|u_0| + \epsilon\right) + \int_0^t e^{\Psi(t)-\Psi(s)}\phi(s)\, ds.$$

We note that $t \mapsto h(t,\epsilon)$ is a continuous function and that

$$|u(0)| < h(0,\epsilon).$$

Furthermore, h satisfies the linear differential equation

$$(15.2.24) \qquad \dot{h}(t,\epsilon) = \phi(t) + h(t,\epsilon)\,\psi(t),$$

since we recognize a formula of the type (15.2.14) in eqn (15.2.23). By continuity, there exists a maximal interval $[0,\tau]$ in which

$$|u(t)| \leqslant h(t,\epsilon), \quad \forall t \in [0,\tau].$$

We are going to show that $\tau = T$. Indeed, since the interval is maximal, if $\tau < T$ then we see that

$$(15.2.25) \qquad |u(\tau)| = h(\tau, \epsilon).$$

Furthermore, if $t \leqslant \tau$,

$$|u(t)| \leqslant |u_0| + \int_0^t |\dot{u}(s)|\, ds$$

$$\leqslant |u_0| + \int_0^t (\phi(s) + \psi(s)\,|u(s)|)\, ds$$

$$\leqslant |u_0| + \int_0^t (\phi(s) + \psi(s)\,h(s, \epsilon))\, ds.$$

From eqn (15.2.24), we obtain

$$\int_0^t \psi(s)\,h(s, \epsilon)\, ds = \int_0^t \left(\dot{h}(s, \epsilon) - \phi(s)\right) ds$$

$$= h(t, \epsilon) - |u_0| - \epsilon - \int_0^t \phi(s)\, ds.$$

Finally,

$$|u(t)| \leqslant h(t, \epsilon) - \epsilon.$$

For $t = \tau$, this relation contradicts eqn (15.2.25) and we see that, for any $t \in [0, T]$,

$$|u(t)| \leqslant h(t, \epsilon).$$

We conclude the proof by passing to the limit as ϵ tends to 0. $\qquad\square$

Note the analogy between the inequality (15.2.19) and the variation of parameter formula (15.2.14). We agreed to only make legal operations with the inequalities and it is because of this that we made the positivity hypotheses in Lemma 15.2.6.

15.2.6. Applications of Gronwall's lemma

We are, first of all, going to show that the solution of eqns (15.0.1a) and (15.0.1b) depends continuously on the set of data, that is on u_0, t_0, and f.

Lemma 15.2.7. We denote by C_L the set of continuous functions f on $[t_1, t_2] \times \mathbb{R}^d$ which satisfy

$$|f(t, u) - f(t, v)| \leqslant L\,|u - v|, \quad \forall t \in [t_1, t_2],\ \forall u, v \in \mathbb{R}^d.$$

Then, the mapping which takes $(f, t_0, u_0) \in C_L \times [t_1, t_2] \times \mathbb{R}^d$ to the solution of eqns (15.0.1a) and (15.0.1b) is continuous. Furthermore, we have the estimate

$$|u(t) - v(t)| \leqslant e^{L(t-t_0)} \left(|u_0 - v_0| + |t_0 - s_0| \max_{s \in [t_1, t_2]} |\dot{v}(s)|\right)$$

$$(15.2.26) \qquad\qquad\qquad + \int_{t_0}^t |g(s, v(s)) - f(s, v(s))|\, ds,$$

for $t \geqslant t_0$.

Proof. We write

$$\dot{u}(t) = f(t, u(t)),$$
$$\dot{v}(t) = g(t, v(t))$$

and subtract the second equation from the first to find, on letting $w(t) = u(t) - v(t)$,

$$|\dot{w}(t)| \leqslant |f(t, u(t)) - f(t, v(t))| + |f(t, v(t)) - g(t, v(t))|$$
$$\leqslant L|w(t)| + |f(t, v(t)) - g(t, v(t))|.$$

Consequently, if we let

$$\phi(t) = |f(t, v(t)) - g(t, v(t))|,$$

we can apply Gronwall's lemma and find

$$|w(t)| \leqslant e^{L(t-t_0)}|u_0 - v(t_0)| + \int_{t_0}^{t} e^{L(t-s)}\phi(s)\,\mathrm{d}s.$$

It remains for us to estimate $|u_0 - v(t_0)|$, which we bound by $|u_0 - v_0| + |v_0 - v(t_0)|$. Now, v is differentiable with respect to time and its derivative is bounded. We can therefore conclude the result of the lemma. $\qquad\square$

15.2.7. Smoother solutions

The preceding results suggest that much better properties hold if we suppose that f is a more regular function. Indeed, suppose that f is C^p, with $p \geqslant 1$. The relation (15.0.1a) shows that u is the composite of a C^p function with a C^1 function. It is, therefore, a C^1 function. By an immediate recurrence, u will be a C^{p+1} function. In the following, we will need a result which is a little more precise.

Lemma 15.2.8. Let f be a C^p function and let $(f_k)_{0 \leqslant k \leqslant p}$ be the sequence of functions defined by

(15.2.27)
$$f_0(t, u) = f(t, u),$$
$$f_{k+1}(t, u) = \frac{\partial f_k}{\partial t}(t, u) + D_2 f_k(t, u) f(t, u).$$

Then, if u is C^1 and satisfies the differential eqn (15.0.1a), u is C^{p+1} and satisfies

(15.2.28)
$$\frac{\mathrm{d}^{k+1}u}{\mathrm{d}t^{k+1}}(t) = f_k(t, u(t)), \quad \forall k \in \{0, \ldots, p\}.$$

Proof. We claim that each of the functions f_k is C^{p-k}. For $k = 0$, eqn (15.2.28) is clear. Suppose that eqn (15.2.28) is true for a certain $k < p$. Then, u is C^{k+1}, $t \mapsto f_k(t, u(t))$ is at least C^1, and we can differentiate eqn (15.2.28) to obtain

$$\frac{d}{dt} \frac{d^{k+1} u}{dt^{k+1}}(t) = \frac{d}{dt} f_k(t, u(t)) = \frac{\partial f_k}{\partial t}(t, u(t)) + D_2 f_k(t, u(t)) \dot{u}(t)$$
$$= f_{k+1}(t, u(t)).$$

This concludes the proof. □

We are also going to apply Gronwall's lemma to the differentiable dependence, with respect to the initial conditions, of the solution of eqns (15.0.1a) and (15.0.1b).

Lemma 15.2.9. Let f be C^p and satisfy the Cauchy–Lipschitz conditions. Then, the solution of eqns (15.0.1a) and (15.0.1b) is C^p with respect to the initial conditions.

Proof. Denote by $u(t) = S(t, u_0)$ the unique solution of eqns (15.0.1a) and (15.0.1b) and write
$$u(t; h) = S(t, u_0 + h v_0),$$
with some v_0 in $\mathbb{R}^d$. We deduce from the estimate (15.2.26) that
$$|u(t; h) - u(t)| \leqslant |v_0| h e^{L(t-t_0)}.$$
If we formally differentiate the relations
$$\dot{u}(t; h) = f(t, u(t; h)), \quad u(t_0; h) = u_0 + h v_0$$
with respect to h, we find that the derivative w must satisfy
$$\dot{w}(t) = D_2 f(t, u(t)) w(t), \quad w(t_0) = v_0.$$
We are going to show that w really is the derivative that we are looking for. Let
$$z(t; h) = u(t; h) - u(t) - h w(t).$$
Then, $z(t; h)$ satisfies the differential equation
$$\dot{z}(t; h) = D_2 f(t, u(t)) (u(t; h) - u(t)) - h D_2 f(t, u(t)) w(t)$$
$$+ o(|u(t; h) - u(t)|)$$
$$= D_2 f(t, u(t)) z(t; h) + o(h).$$
By applying Gronwall's lemma, we see that
$$|z(t; h)| = o(h),$$
which is precisely the definition of differentiability. The derivative $D_2 S(t; u_0)$ is the mapping $G(t, t_0)$ defined as in eqn (15.2.15), with $A(t) = D_2 f(t, u(t))$. We find recursively the successive derivatives of $u_0 \mapsto S(t; u_0)$. The details are left to the reader. □

15.3. Exercises from Chapter 15

15.3.1. Lyapunov function for a 2×2 linear system

Let M be a 2×2 real matrix and let A be a 2×2 real positive definite symmetric matrix. We suppose that, for every $x \neq 0$, we have

$$(15.3.1) \qquad\qquad (AMx, x) < 0.$$

Let $(\cdot, \cdot)$ denote the canonical scalar product on $\mathbb{R}^2$.

Exercise 15.3.1. Show that all of the eigenvalues of M have strictly negative real part.

Exercise 15.3.2. Conversely, if all of the eigenvalues of M have negative real part, we wish to construct a 2×2 real matrix, symmetric and positive definite, for which eqn (15.3.1) holds. Show that there exists a constant $k > 0$ and a constant $K > 0$ such that

$$\left\| e^{Mt} \right\| \leqslant k e^{-Kt},$$

where $\| \cdot \|$ is some operator norm.

Exercise 15.3.3. Let $\| \cdot \|$ be the Euclidean norm on $\mathbb{R}^2$. We let

$$\|x\|_e^2 = \int_0^\infty \left| e^{Mt} x \right|^2 \, \mathrm{dt}.$$

Show that $\| \cdot \|_e$ defines a Euclidean norm and give the corresponding scalar product $(\cdot, \cdot)_e$.

Exercise 15.3.4. Calculate the derivative with respect to time s of

$$\left\| e^{Ms} x \right\|_e^2 .$$

Exercise 15.3.5. Deduce, from the preceding question, that the positive definite symmetric matrix A such that

$$(x, y)_e = (Ax, y)$$

has the property (15.3.1).

15.3.2. A delay differential equation

In this subsection, we are going to study the following differential system with delay:

$$(15.3.2) \qquad\qquad \dot{u}(t) = f(t, u(t), u(t - \tau)).$$

The data is as follows. The space $\mathbb{R}^d$ is equipped with a norm, denoted by $| \cdot |$. The function f is defined and continuous on $\mathbb{R} \times \mathbb{R}^d \times \mathbb{R}^d$. Furthermore, it satisfies the following Lipschitz condition: there exists a positive or zero constant L such that, for every (u_1, v_1) and (u_2, v_2) in $\mathbb{R}^d \times \mathbb{R}^d$ and for every $t \in \mathbb{R}$,

$$(15.3.3) \qquad |f(t, u_1, v_1) - f(t, u_2, v_2)| \leqslant L\left(|u_1 - u_2| + |v_1 - v_2|\right).$$

The number τ satisfies

$$(15.3.4) \qquad \tau > 0.$$

This is referred to as the delay. The initial condition is a *function* ϕ, defined on the time interval $[t_0 - \tau, t_0]$. We will suppose that

$$(15.3.5) \qquad \phi \text{ is continuous from } [t_0 - \tau, t_0] \text{ to } \mathbb{R}^d.$$

Existence and uniqueness

Exercise 15.3.6. Let $T > t_0$ and let E be the set of continuous functions u from $[t_0 - \tau, T]$ to $\mathbb{R}^d$ such that

$$u(t) = \phi(t), \quad \forall t \in [t_0 - \tau, t_0].$$

Verify that E, equipped with the distance

$$d(u_1, u_2) = \max\left\{|u_1(t) - u_2(t)| : t_0 \leqslant t \leqslant T\right\},$$

is a complete metric space.

Exercise 15.3.7. We define an integral operator $\mathcal{T}$ by

$$(15.3.6) \qquad (\mathcal{T}u)(t) = \begin{cases} \phi(t) & \text{if } t \leqslant t_0 \ ; \\ \phi(t_0) + \int_{t_0}^{t} f(s, u(s), u(s - \tau))\, ds & \text{otherwise.} \end{cases}$$

Show that $\mathcal{T}$ is well defined on the whole of E and that the image of E by $\mathcal{T}$ is contained in E.

Exercise 15.3.8. Let $u \in E$ and let $\mathcal{T}u = w$. Show that the restriction of w to $]t_0, T[$ is continuously differentiable and that the derivative dw/dt has a right limit at t_0 and a left limit at T. We then say that w is continuously differentiable on $[t_0, T]$.

Exercise 15.3.9. Show that the following two assertions are equivalent for a function $u \in E$ which is continuously differentiable on $[t_0, T]$:

 (i) u satisfies the system with delay (15.3.2);

 (ii) u is a fixed point of the equation

$$(15.3.7) \qquad \mathcal{T}u = u.$$

Exercise 15.3.10. We want to solve the system (15.3.2) by a fixed point method. Estimate $|\mathcal{T}u_1(t) - \mathcal{T}u_2(t)|$ as a function of t and of

$$\max_{t_0 \leqslant s \leqslant t} |u_1(s) - u_2(s)|.$$

Exercise 15.3.11. Deduce, from the preceding question, that, for each integer p, we have the estimate

$$d(\mathcal{T}^p u_1, \mathcal{T}^p u_2) \leqslant \frac{(2LT)^p}{p!} d(u_1, u_2).$$

Exercise 15.3.12. State and prove the existence and uniqueness theorem relative to the system (15.3.2), under the hypotheses (15.3.3)–(15.3.5). Take care not to forget the initial condition ϕ in the statement.

15.3.3. A second-order ordinary differential equation

The space $\mathbb{R}^d$ is equipped with an arbitrary norm, denoted by $|\cdot|$.

The aim of this subsection is to study the system of ordinary differential equations

$$(15.3.8) \qquad\qquad \ddot{u}(t) = f(t, u(t)).$$

We will suppose, in everything that follows, that f is a continuous function from $[0, T] \times \mathbb{R}^d$ to $\mathbb{R}^d$. Furthermore, there exists a constant L, positive or zero, such that, for every u_1 and u_2 in $\mathbb{R}^d$,

$$(15.3.9) \qquad\qquad |f(t, u_1) - f(t, u_2)| \leqslant L|u_1 - u_2|.$$

Exercise 15.3.13. Show that, for all initial data u_0 and v_0 in $\mathbb{R}^d$ and for every initial time t_0 in $[0, t]$, there exists a unique C^2 function u from $[0, T]$ to $\mathbb{R}^d$ such that eqn (15.3.8) holds with the initial conditions

$$(15.3.10) \qquad\qquad u(t_0) = u_0 \quad \text{and} \quad \dot{u}(t_0) = v_0.$$

The problem is reduced to the first-order case by writing $v(t) = \dot{u}(t)$ and

$$(15.3.11) \qquad\qquad z(t) = \begin{pmatrix} u(t) \\ v(t) \end{pmatrix},$$

and then writing a first-order system satisfied by z in the form

$$(15.3.12) \qquad\qquad \dot{z}(t) = \phi(t, z(t)).$$

Show that, if f is C^m, the solutions of eqns (15.3.8) and (15.3.10) are C^{m+2}.

The numerical part of this problem is continued in Subsection 16.5.5.

16

Single-step schemes

Not all ordinary differential equations have explicit solutions, even when calling on very complicated special functions and allowing for some finite number of quadratures (integrations). This is even more true of systems of differential equations. There are far more equations that we cannot integrate explicitly than those that we can.

To gain some information on the behaviour of a differential system we can attempt to find some qualitative information for large time. This is the objective of the theory of dynamical systems. In addition, we can use techniques of numerical approximation. These two areas of study are not generally tackled by the same mathematicians, even though they are related.

Dynamical systems specialists never fail to admire the complicated images delivered to them by physicists, for example, in turbulence, or the chaotic reactions studied in chemistry. The origin of the qualitative theory of differential systems was initially motivated by a question from astronomy: is the solar system stable? Could it be that one day, which can be proved to be far in the future, our beautiful planet will go plunging into the sun, or, on the contrary, escape from it?

We note that this question is completely open. Amateur astronomers and lovers of the paradoxical should get hold of the delectable work [7], illustrated with the author's drawings. It is not exactly an easy text, but in small doses a degree-level mathematics student can tackle it. For a more physics-related approach, consult the book by Bergé, Pomeau, and Vidal [8], which is clearly more mathematically elementary than the preceding work. It is so well written that the ideas appear easy, which they are not.

We should not forget the 'Bible' of differential equations seen from the qualitative point of view: *Ordinary differential equations* by V. I. Arnol'd [3]. This book by Arnol'd is extraordinarily enlightening, and, if possible, add to your reading list *Mathematical methods of classical mechanics* by the same author [4]. We do not pretend that these books are elementary, but it is not in the nature of mathematics for all beautiful things to come easily.

It is not unusual to present results on the behaviour of a differential system, and, in particular, on its qualitative behaviour for large time, based on numerical simulations. In general, we cannot prove the convergence, on an infinite time interval, of the numerical approximations which we will study below. In estimates of convergence a quantity generally appears which grows exponentially as a function of the length of the time interval, over which we integrate, as we will see in Theorem 16.1.6 below. Proving that the qualitative behaviour (for large time) of discrete approximations of a system gives information pertinent to the behaviour of the system is, in itself, a problem of dynamical systems.

Conversely, the choice of approximation scheme of a differential system depends on the qualitative analysis that we make of the system, however rudimentary. If we expect, for example, to have solutions which are uniformly bounded in time, we will try to find a method which conserves this property well enough— and this is not always easy.

Finally, and this is not tackled at all in this book, the truncation errors are not negligible when we have many iterations. A way of modelling them consists of assuming them to be independent random perturbations, which must be justified since we are *a priori* in a situation which is perfectly deterministic.

It is advisable to be aware that a simulation conducted without precautions over long time intervals is in more danger of reflecting the (bad) properties of the approximation, the arithmetic vices of the machine, and the odd habits of the programmer than the behaviour of the system which we want to understand.

We must, therefore, pose many questions about a numerical result, especially if it is pretty and in nice colours.

16.1. Single-step schemes: the basics

Since this book does not aim to be an encyclopaedia, we start with the theory of single-step schemes, with uniform time steps. These are recurrence relations of the form

$$(16.1.1) \qquad U_{k+1} = U_k + hF(t_k, U_k, h).$$

We will attempt, as much as possible, to denote discretized quantities by capital letters. The time step h is strictly positive and belongs to the interval $]0, h^*]$. The time t_k is defined by

$$(16.1.2) \qquad t_0 \text{ given}, \quad t_{k+1} = t_k + h.$$

When we consider a given interval of time, we will have more points of discretization for smaller h. We denote by $J(h)$ the maximum index of discretization. This is the largest integer k less than or equal to $(T - t_0)/h$. The vector U_k of $\mathbb{R}^d$ is an approximation of $u(t_k)$, if u is the solution of eqn (15.0.1), provided that the initial condition U_0 of the scheme approximates the initial condition u_0 of the differential system. The function F is defined on $[t_0, T] \times \mathbb{R}^n \times [0, h^*]$.

16.1.1. Convergence, stability, consistency

We now give some definitions. We are interested in the approximation of a system

$$(16.1.3) \qquad \dot{u}(t) = f(t, u(t)), \quad \forall t \in [t_0, T],$$

with the initial condition given on $\mathbb{R}^d$

$$(16.1.4) \qquad u(t_0) = u_0.$$

We will always suppose that f satisfies the Cauchy–Lipschitz conditions.

Definition 16.1.1. The approximation of eqns (16.1.3) and (16.1.4) defined by the one-step scheme (16.1.1) is said to be convergent if, for any initial u_0,

$$(16.1.5) \qquad \lim_{\substack{h \to 0 \\ U_0 \to u_0}} \max_{0 \leqslant k \leqslant J(h)} |u(t_k) - U_k| = 0.$$

We do not suppose that the scheme has the exact initial condition of the ODE. Indeed, on one hand, there could be a truncation error in the initial condition. On the other hand, the initial condition may not be known exactly, being itself the result of a calculation, or it could be obtained by a sampling process. This is necessarily the case when we discretize partial differential equations.

Generally, just as the solution of a differential system is continuous with respect to the initial data, the solution approximated by a one-step scheme must be continuous with respect to a perturbation of the initial conditions.

Convergence, as we will see, results from two properties. The first, stability, is a property of the scheme. This ensures that the scheme does not amplify too much the errors created at each step. The other, consistency, describes a relation between the scheme and the differential system. It implies that the scheme does not differ much from the solution locally.

Definition 16.1.2. Scheme (16.1.1) is said to be stable if there exists a constant M such that, for all $U_0 \in \mathbb{R}^d$, for all $V_0 \in \mathbb{R}^d$, for all $h \leqslant h^*$, and for every sequence of vectors ϵ_j, the sequences U_j and V_j defined by the relations

$$(16.1.6) \qquad \begin{aligned} U_{j+1} &= U_j + hF(t_j, U_j, h), \\ V_{j+1} &= V_j + hF(t_j, V_j, h) + \epsilon_j \end{aligned}$$

satisfy the estimate

$$(16.1.7) \qquad |U_j - V_j| \leqslant M \left(|U_0 - V_0| + \sum_{k=0}^{j-1} |\epsilon_k| \right), \quad \forall j \leqslant J(h).$$

Definition 16.1.3. A scheme (16.1.1) is said to be consistent with the system (16.1.3) if, for every solution to system (16.1.3), we have

$$(16.1.8) \qquad \lim_{h \to 0} \sum_{0 \leqslant j \leqslant J(h)-1} |u(t_{j+1}) - u(t_j) - hF(t_j, u(t_j), h)| = 0.$$

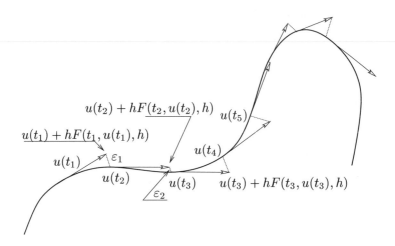

Figure 16.1: Graphical representation of the local error.

Refer to Figure 16.1 for a graphical interpretation of the second of these properties, the vector $u(t_{j+1}) - u(t_j) - hF(t_j, u(t_j), h)$ represents the error that we make by replacing $u(t_{j+1})$ by the quantity calculated with the aid of the scheme. This is what we call the local error.

The following theorem is simple and essential. It is generally known by the name of 'consistency plus stability implies convergence'.

Theorem 16.1.4. Let f be a function satisfying the Cauchy–Lipschitz conditions and let F be a continuous function of $t \in [t_0, T]$, $u \in \mathbb{R}^d$, and $h \in [0, h^*]$, which defines a one-step scheme (16.1.1). If this one-step scheme is consistent with system (16.1.3) and it is stable, then it is convergent.   ◇

Proof. We let
$$V_j = u(t_j).$$
Then,
$$V_{j+1} - V_j - hF(t_j, V_j, h) = \epsilon_j$$
is the local error. We can apply the inequality (16.1.7) and we have
$$|U_j - V_j| \leqslant M \left(|U_0 - u(t_0)| + \sum_{k=0}^{j-1} |\epsilon_k| \right).$$

From the consistency hypothesis, we see that $U_j - V_j$ tends uniformly to zero with respect to j as h tends to 0.   □

Reducing the proof of convergence to the verification of consistency and stability has a double advantage. On one hand, experimentally this corresponds to different behaviours. A stable scheme which is not consistent certainly calculates something, but not what we are looking for. On the other hand, an unstable but consistent scheme calculates a solution which could be initially close to the one we seek, but which separates from it quickly. Often this happens in an oscillatory manner, more rapidly as the step size is reduced, quickly ending up in an overflow.

From the theoretical point of view, this approach allows us to divide the difficulties and makes the proofs more clear.

16.1.2. Necessary and sufficient condition of consistency

We now give conditions which assure stability and convergence.

Theorem 16.1.5. Let F be a continuous function of $t \in [t_0, T]$, $u \in \mathbb{R}^d$, and $h \in [0, h^*]$, defined by a one-step scheme (16.1.1). A necessary and sufficient condition so that the scheme is consistent with system (16.1.3) is that

$$(16.1.9) \qquad F(t, u, 0) = f(t, u), \quad \forall t \in [t_0, T], \; \forall u \in \mathbb{R}^d. \qquad \diamond$$

Proof. The local error ϵ_j is given by

$$\epsilon_j = u(t_{j+1}) - u(t_j) - hF(t_j, u(t_j), h).$$

We can rewrite this in the form

$$\epsilon_j = \int_{t_j}^{t_{j+1}} [f(s, u(s)) - f(t_j, u(t_j))]\, ds + h[f(t_j, u(t_j)) - F(t_j, u(t_j), 0)]$$

$$+ h[F(t_j, u(t_j), 0) - F(t_j, u(t_j), h)].$$

Let

$$\alpha_j = \int_{t_j}^{t_{j+1}} [f(s, u(s)) - f(t_j, u(t_j))]\, ds,$$

$$\beta_j = h[f(t_j, u(t_j)) - F(t_j, u(t_j), 0)],$$

$$\gamma_j = h[F(t_j, u(t_j), 0) - F(t_j, u(t_j), h)].$$

Let ω be the modulus of continuity of $t \mapsto f(t, u(t))$ and let ω_1 be the modulus of continuity of $(t, h) \mapsto F(t, u(t), h)$. We can estimate α_j and γ_j by means of ω and ω_1:

$$|\alpha_j| \leqslant h\omega(h) \quad \text{and} \quad |\gamma_j| \leqslant \omega_1(h).$$

Consequently,

$$|\epsilon_j| \leqslant |\beta_j| + h(\omega(h) + \omega_1(h)) \quad \text{and} \quad |\beta_j| \leqslant |\epsilon_j| + h(\omega(h) + \omega_1(h)).$$

Suppose that the scheme is consistent. Then,

$$\sum_{0 \leqslant j \leqslant J(h)-1} |\beta_j| \leqslant hJ(h)(\omega(h) + \omega_1(h)) + \sum_{j \leqslant J(h)-1} |\epsilon_j|.$$

Consequently,

$$\lim_{h \to 0} \sum_{j \leqslant J(h)-1} |\beta_j| = 0.$$

However, the sums

$$\sum_{j \leqslant J(h)-1} h \left| f\left(t_j, u\left(t_j\right)\right) - F\left(t_j, u\left(t_j\right), 0\right) \right|$$

define the rectangle rule for the continuous function

$$t \mapsto \left| f\left(t, u\left(t\right)\right) - F\left(t, u\left(t\right), 0\right) \right|$$

on the interval $[t_0, t_0 + hJ(h)]$. Consequently, they converge to

$$\int_{t_0}^{T} \left| f\left(s, u\left(s\right)\right) - F\left(s, u\left(s\right), 0\right) \right| \mathrm{d}s.$$

We see that

$$\int_{t_0}^{T} \left| f\left(s, u\left(s\right)\right) - F\left(s, u\left(s\right), 0\right) \right| \mathrm{d}s = 0,$$

and, consequently, for every solution u of eqn (15.0.1a),

$$f\left(t, u\left(t\right)\right) = F\left(t, u\left(t\right), 0\right).$$

Since there is a solution of eqn (15.0.1b) passing through each pair $(t', u') \in [t_0, T] \times \mathbb{R}^d$, we see that

$$F\left(t', u', 0\right) = f\left(t', u'\right), \quad \forall \left(t', u'\right) \in [t_0, T] \times \mathbb{R}^d.$$

We have therefore shown that consistency implies that eqn (16.1.9) holds.

Conversely, if eqn (16.1.9) holds, β_j vanishes and we have the upper bound

$$|\epsilon_j| \leqslant h \left(\omega\left(h\right) + \omega_1\left(h\right) \right).$$

From this we have that

$$\sum_{j \leqslant J(h)-1} |\epsilon_j| \leqslant (T - t_0) \left(\omega\left(h\right) + \omega_1\left(h\right) \right).$$

We therefore have consistency.   □

16.1.3. Sufficient condition for stability

We now give a sufficient condition for stability:

Theorem 16.1.6. For a scheme to be stable, it is sufficient that there exists a constant Λ such that

$$|F(t, u, h) - F(t, v, h)| \leqslant \Lambda |u - v|, \quad \forall t \in [t_0, T], \ \forall u, v \in \mathbb{R}^d, \ \forall h \in [0, h^*].$$
(16.1.10)

Furthermore, the constant M which appears in inequality (16.1.7) can be taken to be equal to $e^{\Lambda(T - t_0)}$. ◇

The proof of this result will need a discrete form of Gronwall's lemma, which we will now prove:

Lemma 16.1.7 (Discrete Gronwall's lemma). Let Λ and h be two given positive numbers and let $(a_j)_{j \geqslant 0}$ and $(b_j)_{j \geqslant 0}$ be two sequences of positive numbers or zero which satisfy the inequality

(16.1.11) $$a_{j+1} \leqslant (1 + \Lambda h) a_j + b_j.$$

Then,

(16.1.12) $$a_j \leqslant e^{\Lambda jh} a_0 + \sum_{k=0}^{j-1} b_k e^{\Lambda(j-k-1)h}.$$

Proof. We are going to make an exponential appear in inequality (16.1.11) by noticing that

(16.1.13) $$1 + x \leqslant e^x, \quad \forall x \in \mathbb{R}^+.$$

Consequently, inequality (16.1.11) implies

(16.1.14) $$a_{j+1} \leqslant e^{\Lambda h} a_j + b_j.$$

As in the proof of Gronwall's lemma (Lemma 15.2.6), we make a change of unknown function:

$$a_j = \alpha_j e^{\Lambda jh}.$$

Substituting into inequality (16.1.11) we have

$$\alpha_{j+1} e^{\Lambda(j+1)h} \leqslant \alpha_j e^{\Lambda jh} e^{\Lambda h} + b_j.$$

The sequence of α_j therefore satisfies the inequalities

$$\alpha_{j+1} \leqslant \alpha_j + b_j e^{-\Lambda(j+1)h},$$

which gives us

$$\alpha_j \leqslant \alpha_0 + \sum_{k=0}^{j-1} b_k e^{-\Lambda(k+1)h}.$$

Changing back to the a_j we obtain inequality (16.1.12). ☐

Proof of Theorem 16.1.6. From eqn (16.1.6), if we subtract U_{j+1} from V_{j+1} and apply the triangle inequality, we get

$$|V_{j+1} - U_{j+1}| \leqslant |V_j - U_j| + h\,|F(t_j, V_j, h) - F(t_j, U_j, h)| + |\epsilon_j|\,.$$

From hypothesis (16.1.10), we therefore have the inequality

$$|V_{j+1} - U_{j+1}| \leqslant (1 + \Lambda h)\,|V_j - U_j| + |\epsilon_j|\,.$$

We can therefore apply the discrete form of Gronwall's lemma to obtain

$$|U_j - V_j| \leqslant e^{\Lambda j h}\,|U_0 - V_0| + \sum_{k=0}^{j-1} e^{\Lambda(j-k-1)h}\,|\epsilon_k|\,.$$

Since $jh \leqslant hJ(h) \leqslant T - t_0$, we can conclude the result of the lemma.   □

Note that the constant $M = e^{\Lambda(T-t_0)}$ could be absolutely enormous. Indeed, if we take $\Lambda = 10$ (which is not enormous) and $T - t_0 = 10$, then $M = e^{100}$, so that

$$M \sim 2.7 \times 10^{43}.$$

Consequently, to have a relative error of $O(1)$ in the solution, we require a relative error of $O(10^{-43})$ in the initial conditions. This also implies that the sum of the truncation errors must never exceed 10^{-43} times the absolute value of the initial data. Therefore, we must typically work in quadruple precision. This is enormous, and we will see later that this choice of the constant M is the best possible in many cases. It can still be possible, however, that when f has solutions which remain bounded for all time, a judicious choice of scheme leads to better estimates. They depend on the particular properties of f and of the scheme, and cannot be simply deduced from this general theorem.

16.2. Order of a one-step scheme

It is not sufficient for schemes to converge, they must also converge sufficiently quickly to be of practical interest. We are therefore going to define a notion of order for one-step schemes.

Definition 16.2.1. Let p be an integer greater or equal to 1. A scheme (16.1.1) is said to be of order p if, for every solution u of eqn (16.1.3), there exists a positive C such that

$$(16.2.1) \qquad \sum_{0 \leqslant j \leqslant J(h)-1} |u(t_{j+1}) - u(t_j) - hF(t_j, u(t_j), h)| \leqslant Ch^p.$$

We immediately get a finer convergence result:

Theorem 16.2.2. Let f be a function satisfying the Cauchy–Lipschitz conditions, and let F be a continuous function of $t \in [t_0, T]$, $u \in \mathbb{R}^d$, and $h \in [0, h^*]$, defining a one-step scheme (16.1.1). If this one-step scheme is of order p, if it is stable, and if $|u_0 - U_0| \leqslant C' h^p$, then

$$(16.2.2) \qquad \max_{k \leqslant J(h)} |u(t_k) - U_k| \leqslant M (C + C') h^p. \qquad \diamond$$

The proof of this result follows from that of Theorem 16.1.4 and is left to the reader.

We now give a necessary and sufficient condition for a scheme to be of order p. For this, we recall the notation of Lemma 15.2.8: if f is a C^p function of t and u, we write

$$f_0(t, u) = f(t, u), \qquad f_{k+1}(t, u) = \frac{\partial f_k(t, u)}{\partial t} + D_2 f_k(t, u) f(t, u) \quad \text{if } k \leqslant p.$$
$$(16.2.3)$$

We recall that, if u is a solution of scheme (16.1.1) we have

$$(16.2.4) \qquad \frac{d^{k+1} u}{dt^{k+1}}(t) = f_k(t, u(t)).$$

We then have the following result:

Theorem 16.2.3. Let f be a C^p function with respect to the set of its variables and let F be a continuous mapping from $[t_0, T] \times \mathbb{R}^d \times [0, h^*]$ to $\mathbb{R}^d$, which is p-times differentiable with respect to h and whose p derivatives with respect to h are continuous functions of all the variables. Then the scheme (16.1.1) is of order p with respect to the system (16.1.3) if and only if, for every k between 0 and $p - 1$, we have

$$(16.2.5) \qquad \frac{\partial^k F}{\partial h^k}(t, u, 0) = \frac{f_k(t, u)}{k + 1}, \quad \forall t \in [t_0, T], \ \forall u \in \mathbb{R}^d. \qquad \diamond$$

Proof. We write the local error ϵ_j with the aid of Taylor's formula with integral remainder, in the form

$$\begin{aligned}
\epsilon_j &= u(t_{j+1}) - u(t_j) - hF(t_j, u(t_j), h) \\
&= u(t_j) + h\frac{du}{dt}(t_j) + \ldots + \frac{h^p}{p!}\frac{d^p u}{dt^p}(t_j) \\
&\quad + \int_{t_j}^{t_{j+1}} \frac{(t_{j+1} - s)^p}{p!}\frac{d^{p+1} u}{dt^{p+1}}(s)\, ds - u(t_j) \\
&\quad - h\Bigg[F(t_j, u(t_j), 0) + \ldots + \frac{h^{p-1}}{(p-1)!}\frac{\partial^{p-1} F}{\partial h^{p-1}}(t_j, u(t_j), 0) \\
&\qquad\qquad + \int_0^h \frac{(h-s)^{p-1}}{(p-1)!}\frac{\partial^p F}{\partial h^p}(t_j, u(t_j), s)\, ds \Bigg].
\end{aligned}$$

We let, for $0 \leqslant k \leqslant p-1$,

$$\beta_j^k = \frac{f_k\left(t_j, u\left(t_j\right)\right)}{k+1} - \frac{\partial^k F}{\partial h^k}\left(t_j, u\left(t_j\right), 0\right).$$

We then have

$$\epsilon_j = h\left(\sum_{k=1}^{p} \frac{h^k}{k!} \beta_j^{k-1}\right) + \int_{t_j}^{t_{j+1}} \frac{(t_{j+1}-s)^p}{p!} f_p\left(s, u\left(s\right)\right) \mathrm{d}s$$

$$- h \int_0^h \frac{(h-s)^{p-1}}{(p-1)!} \frac{\partial^p F}{\partial h^p}\left(t_j, u\left(t_j\right), s\right) \mathrm{d}s.$$

If condition (16.2.5) holds, as $f_p(t, u(t))$ and $(\partial^p F/\partial h^p)(t, u(t), h)$ are uniformly bounded on $[t_0, T]$ and $[t_0, T] \times [0, h^*]$, respectively, then

$$|\epsilon_j| \leqslant Ch^{p+1},$$

where C is a constant which depends only on f, F, and u. Consequently,

$$\sum_{j \leqslant J(h)-1} |\epsilon_j| \leqslant Ch^{p+1} J\left(h\right) \leqslant C'h^p.$$

Conversely, if the scheme is of order $p \geqslant 1$, it is, in particular, consistent and, therefore, $F(t, u, 0) = f(t, u)$, which implies that β_j^0 vanishes for all j. We have, on using the triangle inequality and the bounds on β_j^k, for $k \geqslant 2$,

$$\sum_{j \leqslant J(h)-1} h^2 \left|\beta_j^1\right| \leqslant C_1 \left(h^p + h^2\right).$$

Dividing by h, we immediately see that

$$\lim_{h \to 0} \sum_{j \leqslant J(h)-1} h \left|\beta_j^1\right| = 0.$$

Reasoning as in Theorem 16.1.5, we see that

$$\lim_{h \to 0} \sum_{j \leqslant J(h)-1} h \left|\beta_j^1\right| = \int_{t_0}^{T} \left|\frac{f_1\left(t, u\left(t\right)\right)}{2} - \frac{\partial F}{\partial h}\left(t, u\left(t\right), 0\right)\right| \mathrm{d}s,$$

from which we deduce that

$$\frac{\partial F}{\partial h}\left(t, u\left(t\right), 0\right) = \frac{f_1\left(t, u\left(t\right)\right)}{2},$$

for all t in $[t_0, T]$ and u in $\mathbb{R}^d$.

More generally, we argue by induction. If condition (16.2.5) holds up to index $k-1$ then

$$h^{k+1} \sum_{j \leqslant J(h)-1} \left| \beta_j^k \right| \leqslant C_k \left(h^p + h^{k+1} \right).$$

Therefore,

$$\lim_{h \to 0} \sum_{j \leqslant J(h)-1} h \left| \beta_j^k \right| = 0 = \int_{t_0}^{T} \left| \frac{f_k \left(t, u \left(t \right) \right)}{k+1} - \frac{\partial^k F}{\partial h^k} \left(t, u \left(t \right), 0 \right) \right| \mathrm{d}t.$$

Consequently, condition (16.2.5) holds up to index k. $\square$

16.3. Explicit and implicit Euler schemes

16.3.1. The forward Euler scheme

The simplest scheme is the explicit Euler scheme:

$$(16.3.1) \qquad U_{n+1} = U_n + hf \left(t_n, U_n \right).$$

We say that it is explicit since, contrary to certain schemes which we will see later, the calculation of U_{n+1} does not depend on the solution of a nonlinear system, but only on an evaluation of f. In this case, the function F is defined by

$$F \left(t, u, h \right) = f \left(t, u \right).$$

It is immediate that the scheme is stable and consistent, and therefore convergent. It is of order 1 if f is C^1, but is not generally of order 2. Indeed,

$$\frac{f_1 \left(t, u \right)}{2} - \frac{\partial F}{\partial h} \left(t, u, 0 \right) = \frac{1}{2} \left(\frac{\partial f}{\partial t} + D_2 f \left(t, u \right) f \left(t, u \right) \right),$$

which does not generally vanish.

We apply the Euler scheme to the differential equation

$$\dot{u} = \lambda u,$$

where λ is a given real number. If we choose $t_0 = 0$, $u_0 = 1$, and $h = T/k$, we see that

$$U_{j+1} = \left(1 + \lambda h \right) U_j,$$

which gives us

$$U_j = \left(1 + \frac{\lambda T}{k} \right)^j,$$

and Theorem 16.1.4 allows us to recover the well-known Eulerian formula [25]

$$\lim_{k \to \infty} \left(1 + \frac{\lambda T}{k} \right)^k = \mathrm{e}^{\lambda T}.$$

If we choose $\lambda = 10$ and $T = 10$ we see that the constant M, which appears in Theorem 16.1.6, cannot be less than e^{100}. Indeed, by a continuity argument, the perturbation on the initial conditions of the scheme cannot have an effect which is less than the perturbation on the initial conditions of the equation. Since the solution operator of $\dot{u} = 10u$ is multiplication by e^{10t}, we see that, at time $T = 10$, the perturbation a on the initial conditions is multiplied by e^{10}. It is in this sense that the estimate of Theorem 16.1.6 is optimal.

Stiffness

On the other hand, if we choose $\lambda = -10$, we have an unsatisfactory situation since the exact solution tends very quickly to zero, although the approximate solution satisfies only a very coarse estimate.

Numerically, furthermore, since all the solutions of $\dot{u} = -10u$ are bounded for $t \geqslant 0$, we would like the same to be true of the numerical solutions. For this, it is sufficient that the factor $1 - 10h$ be bounded by 1 in absolute value, that is,

$$h \leqslant 0.2.$$

This condition could be considered too restrictive on long intervals of time since we know that the solution must tend very quickly to zero, and thus we use up a large amount of computer time to little effect.

We say that the ordinary differential equation under consideration is stiff.

Now, there is another way to approximate an exponential by an Eulerian formula. Indeed,

$$\lim_{k \to \infty} \left(1 - \frac{\lambda T}{k} \right)^{-k} = e^{\lambda T}.$$

This corresponds to the scheme

$$U_{j+1} \left(1 - \lambda h \right) = U_j,$$

or, again,

$$U_{j+1} = U_j + \lambda h U_{j+1}.$$

In this case, if $\lambda < 0$, we will obtain bounded solutions without restriction on the time step.

16.3.2. Backwards Euler scheme

It is because of this advantage that we introduce the backwards Euler scheme or implicit Euler scheme:

$$(16.3.2) \qquad\qquad U_{j+1} = U_j + hf \left(t_{j+1}, U_{j+1} \right).$$

This is an implicit scheme: to find U_{j+1} it is necessary to solve a system which is generally nonlinear. However, this is not difficult since we begin with a good approximation of the solution, namely the value of u at the preceding time step.

We show that this can be put in the form (16.1.1) of one-step schemes. Write

$$(16.3.3) \qquad v = u + hf(s,v).$$

Let

$$g(s,u,h,v) = u + hf(s,v).$$

As f satisfies the Cauchy–Lipschitz conditions, g is a strict contraction with respect to its argument u, provided that $hL < 1$. We therefore choose an h^* such that $h^*L < 1$, and we work from now on with $h \leqslant h^*$. Therefore, there exists a unique solution of eqn (16.3.3), which we denote by $G(s,u,h)$. This function G is continuous with respect to all of its arguments and Lipschitz with respect to u. Indeed, if

$$v_1 = u_1 + h_1 f(s_1, v_1) \quad \text{and} \quad v_2 = u_2 + h_2 f(s_2, v_2),$$

we subtract the second equation from the first and we apply the triangle inequality, to give

$$|v_1 - v_2| \leqslant |u_1 - u_2| + h_1 |f(s_1, v_1) - f(s_1, v_2)|$$
$$+ h_1 |f(s_1, v_2) - f(s_2, v_2)| + |h_1 - h_2| |f(s_2, v_2)|.$$

We fix v_2, s_2, and h_2, to obtain

$$(1 - h_1 L)|v_1 - v_2| \leqslant |u_1 - u_2| + h_1 |f(s_1, v_2) - f(s_2, v_2)| + |h_1 - h_2| |f(s_2, v_2)|.$$

Consequently,

$$|v_1 - v_2| \leqslant (1 - h^*L)^{-1} \left(|u_1 - u_2| + h_1 |f(s_1, v_2) - f(s_2, v_2)| \right.$$
$$\left. + |h_1 - h_2| |f(s_2, v_2)| \right).$$

We see that when (s_1, v_1, h_1) tends to (s_2, v_2, h_2), u_1 tends to u_2. Furthermore, if $s_1 = s_2$ and $h_1 = h_2 = h$, we have the inequality

$$(16.3.4) \qquad |v_1 - v_2| = |G(s, u_1, h) - G(s, u_2, h)| \leqslant (1 - hL)^{-1} |u_1 - u_2|.$$

Relation (16.3.2) can now be rewritten as

$$U_{j+1} = G(t_j + h, U_j, h).$$

Consequently,

$$U_{j+1} = U_j + hf(t_j + h, G(t_j + h, U_j, h)).$$

Therefore, we have defined F by

$$(16.3.5) \qquad F(t, u, h) = f(t + h, G(t + h, u, h)).$$

It now remains to study the properties of F. This is a function which is continuous in all of its arguments. $G(t, u, 0)$ is the solution of

$$v = u + 0f(t, v),$$

therefore, $G(t, u, 0) = u$ and $F(t, u, 0) = f(t, G(t, u, 0)) = f(t, u)$. We therefore have consistency. Furthermore,

$$|F(t, u_1, h) - F(t, u_2, h)| \leqslant L |G(t + h, u_1, h) - G(t + h, u_2, h)|$$
$$\leqslant L(1 - hL)^{-1} |u_1 - u_2| \leqslant L(1 - h^*L)^{-1} |u_1 - u_2|.$$

The function F is Lipschitz. Consequently, the backwards Euler scheme is stable.

16.3.3. θ-method

In the same way, we can combine the explicit Euler scheme (16.3.1) and the implicit Euler scheme: thus, we have the θ-method

$$U_{j+1} = U_j + h[\theta f(t_{j+1}, U_{j+1}) + (1 - \theta) f(t_j, U_j)].$$

For $\theta = 0$ we recover the explicit Euler scheme and for $\theta = 1$ we obtain the implicit Euler scheme. With the same techniques we can show that the scheme is well defined for $h \leqslant h^*$, that it is stable, and that it is consistent. If f is C^1 the scheme is of order 1. If f is C^2 and $\theta = 1/2$ it is of order 2.

We have already seen that the explicit Euler scheme is of order 1.

We now move on to the case of the implicit Euler scheme. With the notation used previously, $G(s, u, h)$ is the unique solution v of eqn (16.3.3). If f is C^1 the implicit function theorem implies that G is C^1 with respect to its arguments for $(u, s, h) \in \mathbb{R}^d \times [t_0, T] \times [0, h^*]$. We have seen previously that $F(t, u, h) = f(t + h, G(t + h, u, h))$, and that $F(t, u, 0) = f(t, G(t, u, 0)) = f(t, u)$. Therefore, the scheme is of order 1. Suppose now that f is C^2. We calculate the partial derivative of F with respect to h:

$$\frac{\partial F(t, u, h)}{\partial h} = \frac{\partial}{\partial t} f(t + h, G(t + h, u, h))$$
$$+ D_2 f(t, G(t + h, u, h)) \frac{\partial G}{\partial h}(t + h, u, h),$$

and, therefore,

$$\frac{\partial F}{\partial h}(t, u, 0) = \frac{\partial}{\partial t} f(t, u) + D_2 f(t, u) \frac{\partial G}{\partial h}(t, u, 0).$$

It remains for us to calculate the partial derivative of G with respect to h. For this, we differentiate the relation

$$G(t, u, h) = u + hf(t, G(t, u, h))$$

with respect to h and we find

$$\frac{\partial G}{\partial h}(t, u, h) = f(t, G(t, u, h)) + h \frac{\partial}{\partial h}[f(t, G(t, u, h))],$$

and it is pointless to calculate the expression between square brackets since we are only interested in what happens at $h = 0$. Finally, we find

$$\frac{\partial G}{\partial h}(t, u, 0) = f(t, u).$$

We then note that

$$\frac{\partial F}{\partial h}(t, u, 0) - \frac{f_1(t, u)}{2} = \frac{f_1(t, u)}{2}.$$

The implicit Euler scheme can, therefore, only be of order 2 for very particular values of the function f, namely those for which f_1 is identically zero.

We refer to Subsection 16.5.1 on the θ-method to see the details of a slightly more complicated order calculation.

16.4. Relation with quadrature formulae: Runge–Kutta formulae

We are going to look for ways to interpret the formulae already studied. We can rewrite the differential system (16.1.3) and (16.1.4) in the integral form between t_j and t_{j+1} as follows:

$$u(t_{j+1}) = u(t_j) + \int_{t_j}^{t_{j+1}} \dot{u}(s)\,\mathrm{d}s.$$

We can therefore obtain an approximation of $u(t_{j+1}) - u(t_j)$ by using a quadrature formula. If we use the left rectangle rule we will have

$$u(t_{j+1}) - u(t_j) \simeq h\dot{u}(t_j),$$

which leads us to the explicit Euler scheme. By using the right rectangle rule we have

$$u(t_{j+1}) - u(t_j) \simeq h\dot{u}(t_{j+1}),$$

which leads us to the implicit Euler scheme. Finally, by using a rule with knots at the extremities of the interval, we write

$$u(t_{j+1}) - u(t_j) \simeq \tau\dot{u}(t_j) + \theta\dot{u}(t_{j+1}).$$

For such a rule to be of order 0 as an integration formula, it is necessary that $\tau + \theta = h$. We therefore obtain the θ-method. The case $\theta = h/2$, which corresponds to the trapezium rule, leads to the Crank–Nicolson method, which is more accurate.

To construct schemes based on integration formulae we argue as follows: We consider q knots c_j belonging to $[0,1]$, distinct or not, arranged in increasing order, and the quadrature formulae

$$(16.4.1) \qquad \int_0^1 f(t)\,\mathrm{d}t \simeq \sum_{j=1}^{q} b_j f(c_j)$$

and

$$(16.4.2) \qquad \int_0^{c_i} f(t)\,\mathrm{d}t \simeq \sum_{j=1}^{q} a_{ij} f(c_j)\,.$$

The quadrature formula (16.4.2) differs a little from the formulae which we studied previously in that we consider points situated at the exterior of the interval of integration. There is no additional difficulty when we make this hypothesis.

A Runge–Kutta formula consists of constructing an approximation based on these quadrature formulae, that is, if we let $t_{k,i} = t_k + c_i h$,

$$(16.4.3) \qquad U_{k,i} = U_k + h \sum_{j=1}^{q} a_{ij} f(t_{k,j}, U_{k,j})$$

and

$$(16.4.4) \qquad U_{k+1} = U_k + h \sum_{j=1}^{q} b_j f(t_{k,j}, U_{k,j})\,.$$

We note that relations (16.4.3) allow us to determine the $U_{k,j}$ explicitly if the a_{ij} are zero for $j \geqslant i$. We then say that the Runge–Kutta method is explicit. If the a_{ij} are zero for $j > i$, but certain a_{ii} are not, the method is called semi-implicit, as we can solve each of the eqns (16.4.3) in turn by a nonlinear equation solver. Finally, if there are pairs (i,j), with $j > i$, for which a_{ij} is not zero, eqns (16.4.3) form a system of nq coupled nonlinear equations, n being the dimension of the space.

We generally put the coefficients a_{ij}, b_j, and c_j in a table of the following form:

$$
\begin{array}{c|cccc}
c_1 & a_{11} & a_{12} & \cdots & a_{1q} \\
c_2 & a_{21} & a_{22} & \cdots & a_{2q} \\
\vdots & \vdots & \vdots & & \vdots \\
c_q & a_{q1} & a_{q2} & \cdots & a_{qq} \\
\hline
 & b_1 & b_2 & \cdots & b_q
\end{array}
$$

16.4.1. Examples of Runge–Kutta schemes

With this presentation, the explicit Euler scheme has the following table:

$$\begin{array}{c|c} 0 & 0 \\ \hline & 1 \end{array}$$

The implicit Euler scheme has the following table:

$$\begin{array}{c|c} 1 & 1 \\ \hline & 1 \end{array}$$

The θ-method has the following table:

$$\begin{array}{c|cc} 0 & 0 & 0 \\ 1 & 1-\theta & \theta \\ \hline & 1-\theta & \theta \end{array}$$

The modified Euler method, also known as the Runge–Kutta method of order 2, is constructed by means of a geometric argument. The Euler method consists of making a step in the direction of the tangent. We note that, as the trajectory turns, the speed which allows it to reach the state $u(t_{j+1})$ is closer to the speed at $u(t_j + h/2)$ than to the speed at $u(t_j)$. Consequently, we let

(16.4.5)
$$U_{j+1/2} = U_j + \frac{h}{2} f(t_j, U_j),$$
$$U_{j+1} = U_j + hf(t_{j+1/2}, U_{j+1/2}).$$

Refer to Figure 16.2 to see this construction.

We show that eqn (16.4.5) defines a scheme of order 2. We have

$$F(t, u, h) = f(t + h/2, u + (h/2) f(t, u)).$$

We suppose that f is C^2. Then,

$$F(t, u, 0) = f(t, u)$$

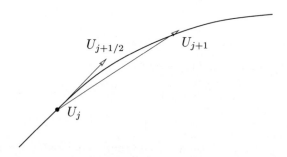

$U_{j+1/2}$ U_{j+1}

U_j

Figure 16.2: How to understand the modified Euler method geometrically.

and

$$\frac{\partial F}{\partial h}(t,u,0) = \frac{1}{2}\frac{\partial}{\partial t}f(t,u) + \frac{D_2 f(t,u)\, f(t,u)}{2} = \frac{f_1(t,u)}{2}.$$

The table corresponding to this method is given by

$$
\begin{array}{c|cc}
0 & 0 & 0 \\
1/2 & 1/2 & 0 \\
\hline
 & 0 & 1
\end{array}
$$

An analogous method is the Heun method, given by

$$U_{j,1} = U_j + hf(t_j, U_j),$$

$$U_{j+1} = U_j + \frac{h}{2}f(t_j, U_j) + \frac{h}{2}f(t_{j+1}, U_{j,1}).$$

The corresponding Runge–Kutta table is

$$
\begin{array}{c|cc}
0 & 0 & 0 \\
1 & 1 & 0 \\
\hline
 & 1/2 & 1/2
\end{array}
$$

This method is of order 2, as the reader may verify.

Finally, the classic Runge–Kutta method is an explicit method of order 4 given by the table

$$
\begin{array}{c|cccc}
0 & 0 & 0 & 0 & 0 \\
1/2 & 1/2 & 0 & 0 & 0 \\
1/2 & 0 & 1/2 & 0 & 0 \\
1 & 0 & 0 & 1 & 0 \\
\hline
 & 1/6 & 1/3 & 1/3 & 1/6
\end{array}
$$

or also by

$$U_{n,1} = U_n,$$

$$U_{n,2} = U_{n,1} + \frac{h}{2}f(t_n + h/2, U_{n,1}),$$

(16.4.6)
$$U_{n,3} = U_{n,1} + \frac{h}{2}f(t_n + h/2, U_{n,2}),$$

$$U_{n,4} = U_n + hf(t_n + h/2, U_{n,3}),$$

$$U_{n+1} = U_n + \frac{h}{6}[f(t_n, U_{n,1}) + 2f(t_n + h/2, U_{n,2})$$
$$+ 2f(t_n + h/2, U_{n,3}) + f(t_n + h, U_{n,4})].$$

We find Runge–Kutta methods of all orders tabulated in the literature. For q from 1 to 4, the maximum order of an explicit Runge–Kutta method is q. For q from 5 to 7, it is $q - 1$. For q greater than or equal to 8, it is $q - 2$. On the other hand, for every q, the maximal order of an implicit Runge–Kutta method is $2q$. This phenomenon, as well as the better behaviour of implicit Runge–Kutta methods on stiff problems, that is, those of type $\dot{u} = -\lambda u$, with λ large and positive, is sufficient to justify the study of implicit Runge–Kutta methods.

16.5. Exercises from Chapter 16

16.5.1. Detailed study of the θ-scheme

We denote by $\|\cdot\|$ some norm on $\mathbb{R}^d$ which we fix throughout this subsection. Let f be a C^2 function from $\mathbb{R} \times \mathbb{R}^d$ to $\mathbb{R}^d$. We suppose that there exists a positive constant L such that

$$(16.5.1) \qquad \|f(t, u) - f(t, v)\| \leqslant L \|u - v\|, \quad \forall t \in \mathbb{R}, \ \forall u \in \mathbb{R}^d, \ \forall v \in \mathbb{R}^d.$$

Exercise 16.5.1. Show that there exists a strictly positive constant k_0 such that, for every k satisfying $|k| < k_0$, for every y in $\mathbb{R}^d$, and for every t in $\mathbb{R}$, the system of nonlinear equations in x

$$(16.5.2) \qquad\qquad\qquad x = y + kf(t, x)$$

has a unique solution.

Exercise 16.5.2. We will denote the unique solution of the system (16.5.2) by

$$(16.5.3) \qquad\qquad\qquad x = G(t, y, k).$$

What is the value of $G(t, y, 0)$?

Exercise 16.5.3. Show that, if $|k| < k_0$, G is C^2 with respect to t, y, and k. Apply the implicit function theorem to $K(t, x, y, k) = x - y - kf(t, x)$ and show that the solution obtained is C^2 in $\mathbb{R} \times \mathbb{R}^d \times]-k_0, k_0[$.

Exercise 16.5.4. Calculate the partial derivative of G with respect to t, y, and k, respectively, at $k = 0$. To do this calculation, substitute $G(t, y, k)$ for x in the system (16.5.2) and note that the derivatives with respect to t and y can be deduced from the expression for $G(t, y, 0)$.

Exercise 16.5.5. We wish to study the following θ-scheme where θ is an arbitrary real:

$$(16.5.4) \qquad U_{n+1} = U_n + h \left[\theta f(t_n, U_n) + (1 - \theta) f(t_{n+1}, U_{n+1}) \right].$$

Here, we have let

$$t_n = t_0 + nh.$$

What scheme do we obtain if $\theta = 1$? And if $\theta = 0$?

Exercise 16.5.6. Show that there exists, for every real θ, a value $h_0(\theta)$ such that, for every $h \in \,]0, h_0(\theta)[$ and for every U_n in $\mathbb{R}^d$, U_{n+1} is uniquely defined by eqn (16.5.4). *We will suppose that this condition is satisfied in the rest of this subsection.* Express U_{n+1} as a function of t_n, U_n, and h with the aid of G. We let

$$(16.5.5) \qquad\qquad U_{n+1} = H\left(t_n, U_n, h\right).$$

Exercise 16.5.7. Calculate $H(t, u, 0)$. Show that

$$\frac{\partial H}{\partial t}(t, u, 0) = 0, \quad D_u H\left(t, u, 0\right) = I_d, \quad \frac{\partial H}{\partial h}(t, u, 0) = f\left(t, u\right).$$

Exercise 16.5.8. Define a function F such that

$$(16.5.6) \qquad\qquad U_{n+1} = U_n + hF\left(t_n, U_n, h\right),$$

by using f, θ, and H. What is the value of $F(t, u, 0)$ and $\partial F(t, u, 0)/\partial h$? Show that F is the solution of the implicit equation

$$(16.5.7) \qquad F\left(t, u, h\right) = \theta f\left(t, u\right) + (1 - \theta)\, f\left(t + h, u + hF\left(t, u, h\right)\right).$$

Exercise 16.5.9. For which values of θ is the scheme (16.5.4) of first order? And of second order?

Exercise 16.5.10. Prove that, for sufficiently small h, scheme (16.5.4) is stable. To this end, we can use relation (16.5.7) to prove that the function F is Lipschitz for sufficiently small h.

16.5.2. Euler scheme with variable step size and asymptotic error estimates

In this subsection we define a variable step size version of the Euler scheme for systems of ordinary differential equations. Then, we asymptotically estimate the error when the time step tends to zero.

The space $\mathbb{R}^d$ is equipped with some arbitrary norm denoted by $\|\cdot\|$. We denote by $\mathcal{L}_d$ the space of continuous linear mappings from $\mathbb{R}^d$ into itself.

In all that follows we consider a time interval $[T_0, T_1]$. We will consider functions f from $[T_0, T_1] \times \mathbb{R}^d$ to $\mathbb{R}^d$ which satisfy the hypotheses of Theorem 15.1.1, that is, that f is continuous with respect to the set of its arguments and it satisfies eqn (15.1.2) for $t \in [T_0, T_1]$ and $u \in \mathbb{R}^d$.

Variable-step size Euler scheme

Let $\left(t_n\right)_{0 \leqslant n \leqslant J}$ be a sequence which increases in time in the interval T_0 to T_1:

$$T_0 = t_0 < t_1 < t_2 < \cdots < t_J = T_1.$$

We let
$$h_n = t_{n+1} - t_n, \quad h = \max_{0 \leqslant n \leqslant J-1} h_n.$$

We define a variable-step Euler scheme by letting
$$U_{n+1} = U_n + h_n f\left(t_n, U_n\right).$$

Exercise 16.5.11. Let u_0 be a given initial condition and $u(t)$ be the solution of the problem

(16.5.8) $$\dot{u}\left(t\right) = f\left(t, u\left(t\right)\right),$$

(16.5.9) $$u\left(T_0\right) = u_0.$$

We let ω be the modulus of continuity of $t \mapsto f(t, u(t))$; ω is a continuous function from $\mathbb{R}^+$ into itself, increasing, vanishing at zero, and such that
$$\left|f\left(t, u\left(t\right)\right) - f\left(s, u\left(s\right)\right)\right| \leqslant \omega\left(|t - s|\right), \quad \forall s, t \in [T_0, T_1].$$

Estimate the vector

(16.5.10) $$\epsilon_n = u\left(t_{n+1}\right) - u\left(t_n\right) - h_n f\left(t_n, u\left(t_n\right)\right)$$

as a function of ω and h_n.

Exercise 16.5.12. Show that
$$\lim_{h \to 0} \sum_{n=0}^{J-1} |\epsilon_n| = 0.$$

Exercise 16.5.13. We suppose that U_n and V_n are defined by the data U_0 and V_0, respectively, together with the following recurrences:

(16.5.11) $$U_{n+1} = U_n + h_n f\left(t_n, U_n\right),$$

(16.5.12) $$V_{n+1} = V_n + h_n f\left(t_n, V_n\right) + \alpha_n,$$

where α_n is an arbitrary sequence of vectors in $\mathbb{R}^d$. Show that we have the relation
$$\left|U_{n+1} - V_{n+1}\right| \leqslant e^{Lh_n} \left|U_n - V_n\right| + |\alpha_n|.$$

Deduce that there exists a constant M such that
$$\left|U_n - V_n\right| \leqslant M\left(\left|U_0 - V_0\right| + \sum_{j=0}^{n-1} |\alpha_n|\right).$$

To do this, let $\gamma_n = |U_n - V_n| \exp\left(-L(t_n - T_0)\right)$ and write a recurrence inequality in terms of the γ_n.

Exercise 16.5.14. We suppose U_n to be defined by the recurrence (16.5.11). Show that we have convergence, that is
$$\lim_{\substack{h \to 0 \\ U_0 \to u_0}} \max_{1 \leqslant n \leqslant J} \left|U_n - u\left(t_n\right)\right| = 0.$$

Asymptotic error estimates

We suppose here that η is a continuous function of $t \in [T_0, T_1]$ which is strictly positive and that f is a C^1 function which satisfies eqn (15.1.2). We will suppose, from now on, that the time steps are given by

$$(16.5.13) \qquad\qquad h_n = \eta(t_n)(h + o(h)).$$

We will also suppose that the initial condition satisfies

$$(16.5.14) \qquad\qquad U_0 - u(0) = h(v + o(1)).$$

Exercise 16.5.15. Show that every solution of eqns (16.5.8) and (16.5.9) is C^2 on $[T_0, T_1]$.

Exercise 16.5.16. Under hypothesis (16.5.13), show that there exists an expression E such that the consistency error defined by eqn (16.5.10) can be written in the form

$$\epsilon_n = h_n h E(t_n, u(t_n), \eta(t_n)) + h_n o(h).$$

Exercise 16.5.17. We define the error at the time t_n by

$$\beta_n = U_n - u(t_n).$$

Verify that there exists a constant C such that

$$|\beta_n| \leqslant Ch,$$

for sufficiently small h.

Exercise 16.5.18. We let

$$\delta_n = \frac{\beta_n}{h}.$$

Show that δ_n satisfies a recurrence relation of the form

$$\delta_{n+1} = \delta_n + h_n B(t_n)\delta_n + h_n C(t_n) + h_n o(1),$$

where $B(t)$ is a continuous mapping from $[T_0, T_1]$ to $\mathcal{L}_d$, which should be determined, and $C(t)$ is a continuous mapping from $[T_0, T_1]$ to $\mathbb{R}^d$ which should also be determined. Write with care the expression for $h_n o(1)$.

Exercise 16.5.19. Let g be a mapping from $[T_0, T_1] \times \mathbb{R}^d$ to $\mathbb{R}^d$ of the following form

$$g(t, u) = B(t)u + C(t),$$

where B is continuous from $[T_0, T_1]$ to $\mathcal{L}_d$ and C is continuous from $[T_0, T_1]$ to $\mathbb{R}^d$. Show that this function satisfies the hypotheses of the Cauchy–Lipschitz theorem.

Exercise 16.5.20. We suppose that eqns (16.5.13) and (16.5.14) are satisfied. Show that, if w is the solution of

(16.5.15)
$$\dot{w}(t) = B(t)w(t) + C(t), \quad t \in [T_0, T_1],$$
$$w(T_0) = v,$$

then

$$\lim_{h \to 0} \frac{\beta_n - hw(t_n)}{h} = 0.$$

16.5.3. Numerical schemes for a delay differential equation

We define a numerical scheme to solve the system (15.3.2) in the following way: We fix the time step h and we let $n_1 = \lfloor \tau/h \rfloor$ (the integer part of τ/h). Given a function F from $\mathbb{R} \times \mathbb{R}^d \times \mathbb{R}^d \times [0, h^*]$ to $\mathbb{R}^d$, we let $t_m = t_0 + mh$ and we define a one-step scheme for the delay eqn (15.3.2) by

(16.5.16) $\quad U_{-m} = \phi(t_0 - mh) \quad$ if $0 \leqslant m \leqslant n_1$,

(16.5.17) $\quad U_{m+1} = U_m + hF(t_m, U_m, U_{m-n_1}, h) \quad$ if $0 \leqslant m \leqslant T/h - 1$.

Exercise 16.5.21. Let u be a solution of the system (15.3.2) and suppose that F is continuous on its domain of definition. Furthermore, suppose that

(16.5.18) $\quad F(t, u, v, 0) = f(t, u, v), \quad \forall t \in \mathbb{R}, \ \forall u, v \in \mathbb{R}^d.$

Show that the sum of the local errors

$$C_h = \sum_{0 \leqslant m \leqslant [T/h]-1} |u(t_{m+1}) - u(t_m) - hF(t_m, u(t_m), u(t_{m-n_1}), h)|$$

tends to 0 as h tends to 0.

Exercise 16.5.22. Given the number Λ, which is positive or zero, consider

$$\alpha_{m+1} = \alpha_m + h\Lambda\alpha_m + h\Lambda\alpha_{m-n_1} + \epsilon_m,$$

where we suppose

$$\alpha_j = 0 \quad \text{and} \quad -n_1 \leqslant m \leqslant 0$$

and the ϵ_m are positive numbers or zero for $0 \leqslant m \leqslant M - 1$. Show that the α_m are positive or zero for every m from 1 to M.

Exercise 16.5.23. Let

$$\alpha_m = (1 + h\Lambda)^m \beta_m.$$

Write down the difference equation satisfied by β_m. Show that the sequence of the β_m is increasing.

Exercise 16.5.24. Using the fact that the sequence of the β_m is increasing, show that we have the inequality

$$\beta_m \leqslant \left(\beta_0 + \sum_{j=0}^{m-1} \epsilon_m \right) \exp\left(mhK_h \right),$$

where K_h is a number tending to $\Lambda \exp(-\Lambda\tau)$ as h tends to 0.

Exercise 16.5.25. We now suppose that α_m satisfies the difference *inequality*

$$\alpha_{m+1} \leqslant \alpha_m + h\Lambda\alpha_m + h\Lambda\alpha_{m-n_1} + \epsilon_m.$$

Show that it satisfies an estimate of the type

$$\alpha_m \leqslant \exp\left(mh\left(\Lambda + K_h \right) \right) \left(\sum_{j=0}^{m-1} \epsilon_m \right).$$

Exercise 16.5.26. Suppose that the function F defining the one-step scheme (16.5.17) satisfies the estimate

$$(16.5.19) \qquad \left| F\left(t, u, v, h \right) - F\left(t, u_1, v_1, h \right) \right| \leqslant \Lambda\left(\left| u - u_1 \right| + \left| v - v_1 \right| \right),$$

for any $t \in \mathbb{R}$, u, v, u_1, $v_1 \in \mathbb{R}^d$, and any $h \in [0, h^*]$. Then, consider the perturbed finite difference scheme

$$(16.5.20) \qquad V_{-m} = \phi\left(t_0 - mh \right) \quad \text{if } 0 \leqslant m \leqslant n_1,$$
$$(16.5.21) \qquad V_{m+1} = V_m + hF\left(t_m, V_m, V_{m-n_1}, h \right) + \zeta_m \quad \text{if } 0 \leqslant m \leqslant T/h - 1,$$

where the ζ_m are arbitrary vectors belonging to $\mathbb{R}^d$. Show that there exists a constant C depending only on $T - t_0$, τ, and Λ such that

$$\left| U_m - V_m \right| \leqslant C \sum_{j=0}^{m-1} \left| \zeta_j \right|.$$

Exercise 16.5.27. Show that, if F is a continuous function of its arguments which satisfies eqns (16.5.18) and (16.5.19), then the scheme defined by eqns (16.5.16) and (16.5.17) converges to the solution of the system (15.3.2) with initial condition ϕ.

Exercise 16.5.28. Consider the scheme

$$(16.5.22) \qquad U_{m+1} = U_m + hf\left(t_m, U_{m+1}, U_{m-n_1} \right).$$

Show that this scheme, which is implicit in U_{m+1}, can be put into the form of eqn (16.5.17). Give the conditions to be able to solve the problem

$$u = v + hf\left(t, u, w \right),$$

and denote its solution by $G(t, v, w, h)$. Express F using G.

Exercise 16.5.29. Show that scheme (16.5.22) is convergent.

16.5.4. Alternate directions

Alternate direction methods

Let T be a strictly positive number and f and g be functions from $[0, T] \times \mathbb{R}^d$ to $\mathbb{R}^d$, in the first part equipped with an arbitrary norm denoted by $|\cdot|$. We suppose that these two functions satisfy the hypotheses of the Cauchy–Lipschitz theorem, that is,

(16.5.23) $\qquad$ f and g are continuous on their domain of definition

and there exists a constant $L > 0$ such that

(16.5.24)
$$\max \left(|f(t, u) - f(t, v)|, |g(t, u) - g(t, v)| \right) \leqslant L |u - v|,$$
$$\forall u \in \mathbb{R}^d, \ \forall v \in \mathbb{R}^d, \ \forall t \in [0, T].$$

We let
$$e = f + g,$$

and in this problem we will look for numerical methods which allow us to integrate the differential system

(16.5.25)
$$\frac{du}{dt}(t) = e(t, u(t)),$$

with the initial condition

(16.5.26)
$$u(0) = u_0,$$

taking account of the particular properties of f and g.

We suppose that F (respectively, G) is a continuous function from $[0, T] \times \mathbb{R}^d \times [0, h^*]$ to $\mathbb{R}^d$, defining a one-step method which is consistent with f (respectively, g). We recall that the necessary and sufficient condition (other than regularity conditions which we will not worry about here) for a scheme to be of order p is that we have

$$\frac{\partial^m F}{\partial h^m}(t, u, 0) = \frac{1}{m+1} f_m(t, u), \quad \forall t \in [0, T], \ \forall u \in \mathbb{R}^d, \ \forall m \leqslant p - 1.$$

Here the f_m have been defined by the recurrence

$$f_0(t, u) = f(t, u), \quad f_{m+1}(t, u) = D_u f_m(t, u) f(t, u) + \frac{\partial f_m}{\partial t}(t, u).$$

Furthermore, we will suppose that F and G are Lipschitz with respect to u, uniformly in t and h, with a Lipschitz constant Λ.

Exercise 16.5.30. We define a numerical scheme by

(16.5.27)
$$U^{n+1/2} = U^n + hF(t_n, U^n, h),$$
$$U^{n+1} = U^{n+1/2} + hG\left(t_n, U^{n+1/2}, h\right).$$

Show that the scheme (16.5.27) defines a one-step scheme. Give the function E which defines it explicitly.

Exercise 16.5.31. Show that E defines a scheme which is stable and consistent.

Exercise 16.5.32. We suppose that F and G are sufficiently regular and that they define schemes of order 2. Is the scheme defined by E, in general, of order 2?

Exercise 16.5.33. We define another numerical scheme by

(16.5.28)
$$U^{n+1/4} = U^n + \frac{h}{2}F\left(t_n, U^n, \frac{h}{2}\right),$$
$$U^{n+1/2} = U^{n+1/4} + \frac{h}{2}G\left(t_n, U^{n+1/4}, \frac{h}{2}\right),$$
$$U^{n+3/4} = U^{n+1/2} + \frac{h}{2}G\left(t_n + \frac{h}{2}, U^{n+1/2}, \frac{h}{2}\right),$$
$$U^{n+1} = U^{n+3/4} + \frac{h}{2}F\left(t_n + \frac{h}{2}, U^{n+3/4}, \frac{h}{2}\right).$$

Show that the scheme (16.5.28) defines a one-step scheme. We denote by H the function with defines it, and we can use the notation

$$\tilde{E}(t, u, h) = G(t, u, h) + F(t, u + hG(t, u, h), h).$$

Exercise 16.5.34. Show that H defines a stable scheme.

Exercise 16.5.35. Suppose that F and G define schemes of order 2. Show that this is also true of H.

Applications

Exercise 16.5.36. We equip $\mathbb{R}^d$ with the Euclidean norm. Let A be a $d \times d$ positive definite symmetric matrix. Show that, for every $\lambda > 0$, the linear system

(16.5.29) $(I + \lambda A)x = (I - \lambda A)b$

possesses a unique solution and that it satisfies

$$|x| \leqslant |b|.$$

Argue using a basis of eigenvectors of A.

Exercise 16.5.37. We solve the differential system

(16.5.30) $\frac{du}{dt} = -Au,$

by means of the scheme

(16.5.31) $U^{n+1} = U^n - \frac{h}{2}A\left(U^{n+1} + U^n\right).$

Show that this scheme is of order 2. In this exercise, we restrict ourselves to the particular form $f(t, u) = Au$ given.

Exercise 16.5.38. Show that the scheme (16.5.31) is stable and that the constant M which appears in the definition of stability can be taken to be equal to 1. Use Exercise 16.5.36.

Exercise 16.5.39. Suppose that A and B are $d \times d$ positive definite symmetic matrices. Furthermore, suppose that A is tridiagonal and there exists a permutation matrix P_σ such that $P_\sigma^* B P_\sigma$ is also tridiagonal. We let $C = A + B$ and we wish to choose a scheme which allows us to solve the linear differential system

$$\frac{du}{dt} = -Cu,$$

with maximum efficiency, knowing that C is band-p. Compare the following two schemes, which are both of order 2:

(16.5.32)
$$\left(I + \frac{h}{4}A\right) U^{n+1/4} = \left(I - \frac{h}{4}A\right) U^n,$$

$$\left(I + \frac{h}{4}B\right) U^{n+1/2} = \left(I - \frac{h}{4}B\right) U^{n+1/4},$$

$$\left(I + \frac{h}{4}A\right) U^{n+3/4} = \left(I - \frac{h}{4}A\right) U^{n+1/2},$$

$$\left(I + \frac{h}{4}B\right) U^{n+1} = \left(I - \frac{h}{4}B\right) U^{n+3/4}$$

and

(16.5.33)
$$\left(I + \frac{h}{2}(A + B)\right) V^{n+1} = \left(I - \frac{h}{2}(A + B) V^n\right).$$

Which has the least cost in terms of arithmetic operations? Give an estimate of the number of operations in each case. Assume that $d \gg p \gg 1$. This condition is met by many of the classical cases of discretization of partial differential equations.

Exercise 16.5.40. We augment the system (16.5.30) by a nonlinear term g satisfying the hypotheses of the Cauchy–Lipschitz theorem with a Lipschitz constant $L = O(1)$. Therefore, we now have the new system

(16.5.34)
$$\frac{du}{dt}(t) = -Au(t) + g(t, u(t)).$$

Show that we can choose a scheme of order 1 of the type given in eqn (16.5.27), without solving nonlinear equations, for which the stability constant M is $O(e^{LT})$ and is independent of the norm of matrix A, which is always symmetric and positive definite.

Exercise 16.5.41. Under the conditions of Exercise 16.5.40 show that we can choose a scheme of order 2 of the type given in eqn (16.5.28) for the system (16.5.34) without solving nonlinear equations, for which the stability constant M is $O(e^{LT})$ and is independent of the norm of matrix A, which is still symmetric and positive definite.

16.5.5. Numerical analysis of a second-order differential equation

This problem continues from Subsection 15.3.3, and uses the same notation.

Exercise 16.5.42. Let w be a C^6 function on $[0, T]$. Show that there exists a choice of reals $\alpha_0, \alpha_1, \alpha_{-1}$ and $\beta_0, \beta_1, \beta_{-1}$ such that

$$
\begin{aligned}
\epsilon\,(t, h) = {} & \alpha_1 w\,(t + h) + \alpha_0 w\,(t) + \alpha_{-1} w\,(t - h) \\
& - h^2\,[\beta_1 \ddot{w}\,(t + h) + \beta_0 \ddot{w}\,(t) + \beta_{-1} \ddot{w}\,(t - h)]
\end{aligned}
$$

(16.5.35)

is uniformly $O(h^6)$ on $[0, T]$. Normalize α_1 by letting $\alpha_1 = 1$.

Exercise 16.5.43. Let $t_j = jh$ and consider the numerical scheme

$$
U_{j+1} - 2U_j + U_{j-1} = \frac{h^2}{12}\,(f\,(t_{j+1}, U_{j+1}) + 10 f\,(t_j, U_j) + f\,(t_{j-1}, U_{j-1}))\,.
$$

(16.5.36)

Show that, for every U_j and U_{j-1}, U_{j+1} is well defined, provided that h is at most equal to a certain h^*, which you should determine.

Exercise 16.5.44. Let $\mathcal{O}$ be an open subset of $\mathbb{R}^m$ and let N be a continuous function defined on $\mathcal{O} \times \mathbb{R}^d$ with values in $\mathbb{R}^d$. The variable ξ ranges over $\mathcal{O}$ and the variable v ranges over $\mathbb{R}^d$. Suppose that there exists a constant $K < 1$ such that, for all ξ in $\mathcal{O}$, v and $\hat{v}$ in $\mathbb{R}^d$,

$$
|N\,(\xi, v) - N\,(\xi, \hat{v})| \leqslant K\,|v - \hat{v}|\,.
$$

Show that the fixed point v of $v \mapsto N(\xi, v)$ is a continuous function G of ξ. Suppose that $\mathbb{R}^m$ has a decomposition as a direct sum of the two subspaces $X_1 \oplus X_2$ and that ξ decomposes as $\xi_1 + \xi_2$, with $\xi_1 \in X_1$ and $\xi_2 \in X_2$. Show that, if there exists a constant K' such that for all $\xi = \xi_1 + \xi_2$, $\hat{\xi} = \hat{\xi}_1 + \xi_2$,

$$
\left|N\,(\xi_1 + \xi_2, v) - N\left(\hat{\xi}_1 + \xi_2, v\right)\right| \leqslant K'\left|\xi_1 - \hat{\xi}_1\right|\,,
$$

then G is Lipschitz with respect to ξ_1 and calculate its Lipschitz constant.

Exercise 16.5.45. Denote the fixed point of $v \mapsto x + hf(t, u + hv)/12$ by $G(t, x, u, h)$, for $h \leqslant h^*$. Calculate the constants L_1 and L_2 such that, for all $u, \hat{u}, x, \hat{x}$ in $\mathbb{R}^d$, all t in $[0, T]$, and all h in $[0, h^*]$,

(16.5.37) $|G\,(t, x, u, h) - G\,(t, \hat{x}, \hat{u}, h)| \leqslant L_1\,|x - \hat{x}| + L_2\,|u - \hat{u}|\,.$

Exercise 16.5.46. Let

$$
\frac{U_{j+1} - U_j}{h} = V_{j+1}\,.
$$

Show that we can rewrite the scheme (16.5.36) in the form of a system of equations, the unknowns being V_{j+1} and the data being U_j, V_j, t_j, and h.

Exercise 16.5.47. We introduce the function

$$H\left(t,u,v,h\right) = \frac{1}{12}\left(10f\left(t,u\right) + f\left(t-h,u-hv\right)\right).$$

Determine, with the aid of G and H, the function $C(t,u,v,h)$ such that

$$V_{j+1} = C\left(t_j, U_j, V_j, h\right).$$

Give the function Φ which allows the scheme (16.5.36) to be put in the form of a one-step scheme:

$$Z_j = \begin{pmatrix} U_j \\ V_j \end{pmatrix}, \quad Z_{j+1} = Z_j + h\Phi\left(t_j, Z_j, h\right).$$

Exercise 16.5.48. Show that this scheme is stable and consistent.

Exercise 16.5.49. Consider the particular case where $n = 1$, $f(t,u) = u$. Determine Φ explicitly and show that the scheme is not of order 2.

Exercise 16.5.50. We return to the general case and consider the solution $\hat{U}_{j+1}$ of

$$(16.5.38) \qquad \begin{aligned} \hat{U}_{j+1} - 2\hat{U}_j + \hat{U}_{j-1} = \frac{h^2}{12}\Big(f\left(t_{j+1}, \hat{U}_{j+1}\right) + 10f\left(t_j, \hat{U}_j\right) \\ + f\left(t_{j-1}, \hat{U}_{j-1}\right)\Big) + h\epsilon_j, \end{aligned}$$

where $(\epsilon_j)_j$ is a sequence of vectors in $\mathbb{R}^d$. Express $\hat{V}_{j+1} = (\hat{U}_{j+1} - \hat{U}_j)/h$ as a function of $\hat{U}_j$, $\hat{V}_j$, t_j, h, and ϵ_j with the aid of the functions G and H.

Exercise 16.5.51. Let

$$\hat{Z}_j = \begin{pmatrix} \hat{U}_j \\ \hat{V}_j \end{pmatrix}.$$

Estimate $|\hat{Z}_{j+1} - Z_{j+1}|$ as a function of $|\hat{Z}_j - Z_j|$.

Exercise 16.5.52. Suppose that f is C^4 and let $\hat{U}_j = u(t_j)$. Using Gronwall's lemma and the estimate of $\epsilon(t,h)$ defined by eqn (16.5.35), prove that there exists, for every solution of eqns (15.3.8) and (15.3.10), a constant C such that

$$|U_j - u\left(t_j\right)| \leqslant Ch^4, \quad \forall j \leqslant \frac{T}{h},$$

provided that U_0 and U_1 are suitably chosen.

17

Linear multistep schemes

17.1. Constructing multistep methods

When calculating an approximation at time t_{n+1}, the Runge–Kutta methods do not use information on the results obtained at times prior to t_n; on the contrary, the multistep methods will use this information systematically. Therefore, a multistep method will be described by the data of $2q+2$ numbers α_j and β_j, $0 \leqslant j \leqslant q$, and will be written as

$$(17.1.1) \qquad \sum_{j=0}^{q} \alpha_j U_{n+j} = h \sum_{j=0}^{q} \beta_j f\left(t_{n+j}, U_{n+j}\right).$$

In what follows, we will systematically use the notation

$$(17.1.2) \qquad F_j = f\left(t_j, U_j\right).$$

We consider only the constant time step case and, therefore,

$$t_j = t_0 + jh,$$

as for the study performed previously for the Runge–Kutta schemes. The theory of variable step for the one-step schemes is easy; it has been treated for the Euler scheme in Subsection 16.5.2. The theory of variable-step multistep schemes goes beyond the level of this book, and the reader is invited to read [19] or other books on the numerical analysis of ordinary differential equations, such as [43, 52].

In order to fix the effective number of time steps used in eqn (17.1.1), we shall assume that

$$(17.1.3) \qquad \alpha_q \neq 0, \quad |\alpha_0| + |\beta_0| \neq 0.$$

We also have to initialize the values $U_0, \ldots, U_{q-1}$. This initialization will be performed in such a way that for u to be a solution of eqn (15.0.1) we have

$$U_j - u\left(t_j\right) = O\left(h^{p+1}\right), \quad 0 \leqslant j \leqslant q - 1,$$

with p being the order of the method under consideration, as defined below in Section 17.2.

The mathematical theory of the stability of multistep methods has some subtle aspects, whilst the theory of order is completely straightforward. This is in complete contrast with the mathematical theory of Runge–Kutta schemes.

We will start by giving some of the best known classes of multistep methods, then we will successively study the order, the stability, and the convergence of multistep methods.

17.1.1. Adams methods

Adams, astronomy, and computation

John Couch Adams was an astronomer; he is the co-discoverer of the planet Neptune, together with Urbain Le Verrier; the existence of an unknown planet beyond Uranus had been proposed to explain the irregularities in the orbit of Uranus. Adams was twenty-four years old at the time he finished his calculation. Initially, he had been encouraged by Airy, the Astronomer Royal, and by Challis who was the director of the Cambridge Observatory. But Airy adopted a discouraging attitude, and Adams did not publish before Le Verrier's announcement. Indeed, Le Verrier had embarked on the same task as Adams, both being unaware of the other's work, and he announced the position of the new planet in November 1845. The Paris Observatory started a search but did not persevere. After Airy had received the announcement of Le Verrier, which gave about the same result as the calculation by Adams, he convinced Challis to search for the planet. In July and August 1846, Challis saw the planet, but did not recognize it.

In despair about the situation, Le Verrier wrote to the young Johann Galle, who was an astronomer in the Berlin Royal Observatory. Galle received permission from the director of his observatory to look for the planet and, indeed, during the night of 23rd September 1846, together with his assistant Heinrich d'Arrest, Galle found a planet at less than one degree of arc from the position predicted by Le Verrier and less than three degrees of arc from the position predicted by Adams.

This discovery made a lasting impression, since it evidenced the power of computations in the discovery of physical phenomena. It is said about Le Verrier that 'he discovered a star with the tip of his pen, without any instruments other than the strength of his calculations alone'.

However, in this particular case, there is more legend than fact. Indeed, Adams and Le Verrier had assumed that the distance from Neptune to the Sun was double the distance from Uranus to the Sun, whilst this ratio is but 1.57. The revolution period determined by Adams was 227 years instead of the 165 observed; there were a number of other false hypotheses.

It is only thanks to a remarkable series of coincidences that these many errors compensated one another. They were pointed out by several astronomers during the following decades.

A more detailed story of the discovery of Neptune and, more generally, a description of the evolution of ideas in celestial mechanics can be found in the excellent and popular book by Ivars Peterson [66].

Adams conceived numerical methods because he needed them in order to integrate numerically what cannot be expressed explicitly by algebraic and analytic means. At the end of his life, he came up with the method to be described below.

The astronomers introduced the notion of slow and fast variables: a fast variable would be, for instance, the position of the Earth as a function of time, and a slow variable would be the eccentricity of its orbit, or the length of its major semi-axis, or the obliquity of the Earth, i.e., the angle of the Earth's axis of rotation with respect to the ecliptic plane, i.e., the plane of its orbit. If a planet has a fixed obliquity, its seasons and its climate will be relatively stable. Due to perturbations from other planets, the parameters of the orbit of the earth change, but on a time scale which is large with respect to the period, i.e., one year; the slow variables can also be integrated numerically. The slow variables have been called secular variables by astronomers, since their effect can be observed only on very long time scales: for a human being, a century (*seculum* in Latin) is a very long scale; for the universe, the matter is quite different—Jacques Laskar calculated numerically the evolution of slow variables, with time steps of 500 years, and he found that the orbits of the planets are basically unpredictable after 100 million years. The following is a summary of his findings:

'Large-scale chaos is present everywhere in the solar system. It plays a major role in the sculpting of the asteroid belt and in the diffusion of comets from the outer region of the solar system. All the inner planets probably experienced large-scale chaotic behaviour for their obliquities during their history. The Earth's obliquity is presently stable only because of the presence of the Moon, and the tilt of Mars undergoes large chaotic variations from 0° to about 60°. On a billion-year time scale, the orbits of the planets themselves present strong chaotic variations which can lead to the escape of Mercury or collision with Venus in less than 3.5 Gyr. The organization of the planets in the solar system thus seems to be strongly related to this chaotic evolution, reaching at all times a state of marginal stability, that is, practical stability on a time scale comparable to its age.' (Jacques Laskar, [57].)

17.1.2. The multistep methods of Adams

The idea of the explicit Adams methods, also called Adams–Bashforth methods, is very simple: if u solves eqn (15.0.1) we have

$$u(t_{n+1}) - u(t_n) = \int_{t_n}^{t_{n+1}} f(t, u(t)) \, dt.$$

Then replace $f(t, u(t))$ by the interpolation polynomial $P \in \mathbb{P}_{q-1}$ which takes the value $F_j = f(t_j, U_j)$ at t_j, for $n - q + 1 \leqslant j \leqslant n$. Employing eqn (4.5.11), P is given by

$$(17.1.4) \qquad P\left(t_n + hs\right) = \sum_{i=0}^{q-1} (-1)^i \binom{-s}{i} \left(\nabla^i F\right)_n.$$

Hence, we obtain

$$\int_{t_n}^{t_{n+1}} P(t)\, \mathrm{d}t = h \int_0^1 \sum_{i=0}^{q-1} (-1)^i \binom{-s}{i} \left(\nabla^i F\right)_n \mathrm{d}s.$$

Define

$$\gamma_i = \int_0^1 (-1)^i \binom{-s}{i}\, \mathrm{d}s.$$

Thus, the class of explicit Adams methods with q steps or $q + 1$ levels is given by

$$(17.1.5) \qquad U_{n+1} - U_n = h \sum_{i=0}^{q-1} \gamma_i \left(\nabla^i F\right)_n.$$

The coefficients γ_i may be calculated recursively; the easy and interesting way to get them is to use a generating function:

$$(17.1.6) \qquad \gamma(x) = \sum_{i \geqslant 0} \gamma_i x^i.$$

A priori, the above series is a formal series, and we know nothing about its convergence. Quoting Herbert Wilf [78], 'a generating function is a clothes-line on which we hang a sequence of numbers for display'. However, there exists perfectly rigorous mathematical theory which gives sense to an expression of the form (17.1.6). It suffices to know that the only permissible operations are those where any arithmetic operations involve only a finite number of terms of the formal series which we are computing. In particular, it is possible to multiply two formal series by generalizing the multiplication rule for polynomials:

$$(17.1.7) \qquad \left(\sum_{i \geqslant 0} \gamma_i x^i\right)\left(\sum_{j \geqslant 0} \delta_j x^j\right) = \sum_{k \geqslant 0} \left(\sum_{n=0}^{k} \gamma_n \delta_{k-n}\right) x^k.$$

Then define

$$\phi(x, s) = \sum_{i \geqslant 0} x^i \binom{-s}{i} (-1)^i.$$

This expression is the series expansion with respect to x of $(1-x)^{-s}$, with s being an arbitrary real number. This is a convergent expansion for $|x| < 1$ and for all complex s. Therefore, if $|s| < 1$, we have

$$\gamma(x) = \int_0^1 (1-x)^{-s}\, ds = \frac{-x}{(1-x)\ln(1-x)},$$

and hence,

$$-\frac{\ln(1-x)}{x}\gamma(x) = \frac{1}{1-x}.$$

The multiplication rule for series now gives

$$\gamma_0 = 1,$$

$$\frac{\gamma_0}{2} + \gamma_1 = 1,$$

(17.1.8) $$\vdots$$

$$\frac{\gamma_0}{n+1} + \ldots + \frac{\gamma_{n-1}}{2} + \gamma_n = 1.$$

The first explicit Adams methods are thus given by the following formulae:

$$U_{n+1} = U_n + hF_n,$$

$$U_{n+1} = U_n + h\left(\frac{3}{2}F_n - \frac{1}{2}F_{n-1}\right),$$

$$U_{n+1} = U_n + h\left(\frac{23}{12}F_n - \frac{16}{12}F_{n-1} + \frac{5}{12}F_{n-2}\right),$$

$$U_{n+1} = U_n + h\left(\frac{55}{24}F_n - \frac{59}{24}F_{n-1} + \frac{37}{24}F_{n-2} - \frac{9}{24}F_{n-3}\right).$$

The particular case $q = 1$ is simply the explicit Euler method.

It is a known fact that the values of an interpolation polynomial outside of the interval enclosed between the extreme interpolation knots are not a very good approximation of the interpolated functions. The Adams–Moulton methods, or implicit Adams methods, consist of approximating the function $f(t, u(t))$ by a polynomial interpolating the $f(t_j, U_j)$ for $n - q + 1 \leqslant j \leqslant n + 1$. Once again, using eqn (4.5.11), we now obtain

$$P(t_n + hs) = \sum_{i=0}^q (-1)^i \binom{-s+1}{i} (\nabla^i F)_{n+1}.$$

Let us define

$$\gamma_i^* = (-1)^i \int_0^1 \binom{-s+1}{i}\, ds.$$

The family of implicit Adams methods is therefore given by

$$U_{n+1} = U_n + h \sum_{i=0}^{q} \gamma_i^* \left(\nabla^i F\right)_{n+1}.$$

A recurrence relation for the γ_i^* is proved in Exercise 17.5.1 along the lines of the derivation of eqns (17.1.8).

The first implicit Adams methods are given by

$$U_{n+1} = U_n + hF_{n+1},$$

$$U_{n+1} = U_n + h \left(\frac{1}{2}F_{n+1} + \frac{1}{2}F_n\right),$$

$$U_{n+1} = U_n + h \left(\frac{5}{12}F_{n+1} + \frac{8}{12}F_n - \frac{1}{12}F_{n-1}\right),$$

$$U_{n+1} = U_n + h \left(\frac{9}{24}F_{n+1} + \frac{19}{24}F_n - \frac{5}{24}F_{n-1} - \frac{1}{24}F_{n-2}\right).$$

The case $q = 1$ corresponds to the implicit Euler method, and the case $q = 2$ corresponds to the Crank–Nicolson method.

17.1.3. Backward differentiation

In the backward differentiation method, we interpolate u, and not f, at the points t_i, for $n - q + 1 \leqslant i \leqslant n + 1$. The corresponding interpolation polynomial is

$$Q\left(t_n + sh\right) = \sum_{i=0}^{q} (-1)^i \binom{-s+1}{i} \left(\nabla^i U\right)_{n+1}$$

and we impose a collocation relation:

$$Q'\left(t_{n+1}\right) = f\left(t_{n+1}, U_{n+1}\right).$$

Therefore, we will have

$$\sum_{j=0}^{q} \delta_j^* \left(\nabla^j U\right)_{n+1} = hF_{n+1},$$

and the coefficients δ_j^* are given by

$$\delta_j^* = (-1)^j \frac{\mathrm{d}}{\mathrm{d}s} \binom{-s+1}{j}\Big|_{s=1}.$$

A direct calculation gives

$$\delta_0^* = 0, \qquad \delta_j^* = \frac{1}{j}, \quad \forall j \geqslant 1.$$

The first backward differentiation methods are:

$$U_{n+1} - U_n = hF_{n+1},$$

$$\frac{3}{2}U_{n+1} - 2U_n + \frac{1}{2}U_{n-1} = hF_{n+1},$$

$$\frac{11}{6}U_{n+1} - 3U_n + \frac{3}{2}U_{n-1} - \frac{1}{3}U_{n-2} = hF_{n+1},$$

$$\frac{25}{12}U_{n+1} - 4U_n + 3U_{n-1} - \frac{4}{3}U_{n-2} + \frac{1}{4}U_{n-3} = hF_{n+1},$$

$$\frac{137}{60}U_{n+1} - 5U_n + 5U_{n-1} - \frac{10}{3}U_{n-2} + \frac{5}{4}U_{n-3} - \frac{1}{5}U_{n-4} = hF_{n+1},$$

$$\frac{147}{60}U_{n+1} - 6U_n + \frac{15}{2}U_{n-1} - \frac{20}{3}U_{n-2} + \frac{15}{4}U_{n-3} - \frac{6}{5}U_{n-4} + \frac{1}{6}U_{n-5} = hF_{n+1}.$$

17.1.4. Other multistep methods

The Nyström extrapolation formulae are constructed in the same fashion as the Adams–Bashforth formulae, except for the change in the integration interval: the starting point is

$$u(t_{n+1}) = u(t_{n-1}) + \int_{t_{n-1}}^{t_{n+1}} f(s, u(s))\, ds,$$

and $f(s, u(s))$ is replaced by the interpolation polynomial $P \in \mathbb{P}_{q-1}$ which takes the values F_j at t_j, $n - q + 1 \leqslant j \leqslant n$. The Nyström extrapolation methods are of the form

(17.1.9) $$U_{n+1} = U_{n-1} + h \sum_{j=0}^{q-1} \kappa_j \left(\nabla^j F\right)_n,$$

and the first of these methods are as follows:

(17.1.10)
$$U_{n+1} = U_{n-1} + 2hF_n,$$
$$U_{n+1} = U_{n-1} + \frac{h}{3}\left(7F_n - 2F_{n-1} + F_{n-2}\right),$$
$$U_{n+1} = U_{n-1} + \frac{h}{3}\left(8F_n - 5F_{n-1} + 4F_{n-2} - F_{n-3}\right).$$

The first of these methods is called the midpoint method, and it is much used.

The corresponding implicit construction gives the so-called Milne–Simpson formulae; they are of the form

(17.1.11) $$U_{n+1} - U_{n-1} = h \sum_{j=0}^{q} \kappa_j^* \left(\nabla^j F\right)_{n+1},$$

and the first of these formulae are

$$(17.1.12) \qquad U_{n+1} - U_{n-1} = 2hF_{n+1},$$

$$(17.1.13) \qquad U_{n+1} - U_{n-1} = 2hF_n,$$

$$(17.1.14) \qquad U_{n+1} - U_{n-1} = \frac{h}{3}\left(F_{n+1} + 4F_n + F_{n-1}\right).$$

Formula (17.1.12) is of little interest because it is an implicit Euler method, with double step, and two staggered grids, namely the even numbered times and the odd numbered times. Formula (17.1.13) is, again, the midpoint formula, obtained as a Nyström formula. Formula (17.1.14) reduces to Simpson's integration formula, should f depend only on time. This is the reason for the name of this class of multistep formulae. Milne also provided the following rule:

$$(17.1.15) \qquad U_{n+1} - U_{n-3} = \frac{h}{3}\left(8F_n - 4F_{n-1} + 8F_{n-2}\right),$$

to be used as a predictor formula in conjunction with eqn (17.1.14). For the use of predictor formulae, see Subsection 17.5.4.

17.2. Order of multistep methods

17.2.1. The order is nice and easy for multistep methods

Let u be a real function of class C^1. The consistency error in the multistep scheme (17.1.1) is the quantity

$$\varepsilon(t, u, h) = \sum_{j=0}^{q} \alpha_j u(t + jh) - h \sum_{j=0}^{q} \beta_j \dot{u}(t + jh).$$

Definition 17.2.1. A multistep method is said to be of order p if the consistency error vanishes uniformly for all polynomials of degree at most p. A method of order 1 is said to be consistent.

There are equivalent ways of formulating the definition of the order:

Theorem 17.2.2. The following assertions are equivalent:

(i) The multistep scheme (17.1.1) is of order p;

(ii) The following algebraic relations hold:

$$(17.2.1) \qquad \sum_{j=1}^{q} \alpha_j = 0,$$

$$(17.2.2) \qquad \sum_{j=0}^{q} j^l \alpha_j - l \sum_{j=0}^{q} j^{l-1} \beta_j = 0, \quad \forall l = 1, \dots, p.$$

In the relation (17.2.2), for $l = 1$, we use the convention $0^0 = 1$.

(iii) For all $t > 0$ and for all $h^* > 0$, there exists a number C such that, for all functions u of class C^{p+1}, the consistency error is estimated as follows:

$$|\varepsilon(t, u, h)| \leqslant Ch^{p+1} \max\left\{\left|u^{(p+1)}(s)\right|, \ t \leqslant s \leqslant t + qh\right\}, \quad \forall h \leqslant h^*.$$
(17.2.3)

(iv) Let ρ and σ be the polynomials

(17.2.4) $$\rho(x) = \sum_{j=0}^{q} \alpha_j x^j, \quad \sigma(x) = \sum_{j=0}^{q} \beta_j x^j.$$

Then, in a neighbourhood of $x = 0$,

(17.2.5) $$\rho(e^x) - x\sigma(e^x) = O\left(x^{p+1}\right). \qquad\qquad\qquad \diamond$$

Proof. (i) $\Longleftrightarrow$ (ii). If a multistep scheme is of order p, we choose $u(x) = x^l$, and we find that, for $l = 0$, the consistency error is given by

$$\sum_{j=0}^{q} \alpha_j,$$

which must therefore vanish. For $l \in \{1, \ldots, p\}$, the consistency error is

$$h^l \left(\sum_{j=0}^{q} j^l \alpha_j - \sum_{j=0}^{q} l j^{l-1} \beta_j\right).$$

Conversely, if the algebraic relations (17.2.1) and (17.2.2) hold, then the consistency error for the monomials x^l, where $0 \leqslant l \leqslant p$, vanishes.

(i) $\Longrightarrow$ (iii). Let u be a function of class C^{p+1}. Then the Taylor expansion of u at t is given by

$$u(t + hs) = P(hs) + h^{p+1} I(t, s, h),$$

where P is the truncated Taylor expansion at t and I is the integral term given by

$$I(t, s, h) = \frac{1}{p!} \int_0^s (s - s')^p u^{(p+1)}(t + hs') \, ds'.$$

Similarly, we have

$$\dot{u}(t + hs) = \dot{P}(hs) + h^p I_1(t, s, h),$$

where I_1 is given by

$$I_1(t, s, h) = \frac{1}{(p-1)!} \int_0^s (s - s')^{p-1} u^{(p+1)}(t + hs') \, ds'.$$

Then, due to the order assumption, we see that the local truncation error is given by

$$\varepsilon\left(t, u, h\right) = h^{p+1}\left(\sum_{j=0}^{q} \alpha_j I\left(t, jh, h\right) - \sum_{j=0}^{q} \beta_j I_1\left(t, jh, h\right)\right).$$

If we choose the number C to be equal to

$$C = \sum_{j=1}^{q}\left(|\alpha_j|\frac{j^{p+1}}{(p+1)!} + |\beta_j|\frac{j^p}{p!}\right),$$

then expression (17.2.3) is proved.

(iii) $\Longrightarrow$ (iv). Take $u(t) = e^t$, then

$$\varepsilon\left(0, u, h\right) = \rho\left(e^h\right) - h\sigma\left(e^h\right),$$

and the conclusion is immediate.

(iv) $\Longrightarrow$ (ii). The Taylor expansion of $\rho(e^x) - x\sigma(e^x)$ at $x = 0$ is given by

$$\rho\left(e^x\right) - x\sigma\left(e^x\right) = \sum_{j=0}^{q}\alpha_j\sum_{l=0}^{p}\frac{(jx)^l}{l!} - \sum_{j=0}^{q}\sum_{l=0}^{p-1}\frac{(jx)^l x}{l!} + O\left(x^{p+1}\right),$$

and the conclusion is immediate. $\qquad\qquad\square$

17.2.2. Order of some multistep methods

The order of the Adams methods is very easy to find. As a consequence of Theorem 17.2.2, it suffices to calculate $\rho(e^h) - h\sigma(h)$, i.e., the consistency error in the case of the exponential, to understand the error. The value of $\rho(e^h)$ is $e^{qh} - e^{(q-1)h}$, and the value of $\sigma(e^h)$ is the primitive of the interpolation polynomial $P(\cdot, h)$ of $t \mapsto e^t$ at the points $0, \ldots, (q-1)h$. We know, from Theorem 4.3.1, that the error committed here is

(17.2.6) $$e^t - P\left(t, h\right) = O\left(h^q\right),$$

when t belongs to the interval $[(q-1)h, qh]$. When we integrate eqn (17.2.6) on the interval $[(q-1)h, qh]$, we find that

$$e^{qh} - e^{(q-1)h} - \int_{(q-1)h}^{qh} P\left(s, h\right) \mathrm{d}s = O\left(h^{q+1}\right).$$

Therefore, the Adams–Bashforth methods on $q+1$ levels are of order q.

The same argument also proves that the Adams–Moulton methods on $q+1$ levels are of order $q+1$.

A very similar argument shows that the backward differentiation method on $q+1$ levels is of order q.

17.3. Stability of multistep methods

The stability of multistep methods needs a linear algebra preparation, like an infantry battle needs an artillery preparation—at least at the time of WWI.

17.3.1. Multistep methods can be very unstable

Consider the multistep scheme given by the coefficients

$$(17.3.1) \qquad \alpha_2 = 1, \quad \alpha_1 = 1, \quad \alpha_0 = 2, \quad \beta_2 = \frac{1}{4}, \quad \beta_1 = 2, \quad \beta_0 = \frac{3}{4}.$$

The reader may check that this scheme is of order 3. We apply this scheme to the differential equation

$$\dot{u} = -u,$$

with the initial data

$$U_0 = 1, \quad U_1 = \exp{(-h)}.$$

The results of the numerical simulation for $h = 1/50$ and $h = 1/100$ are shown in Figure 17.1. The coordinates are clipped so as to make visible the onset of instability.

Not only does instability start earlier with a smaller time step but it is also much larger, as can be seen from a plot of the logarithm of the absolute value of the two numerical solutions for the above two time steps, see Figure 17.2.

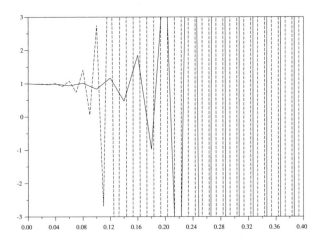

Figure 17.1: Numerical solution for the scheme (17.3.1), with a time step of $1/50$ (solid line) and a time step of $1/100$ (dashed line).

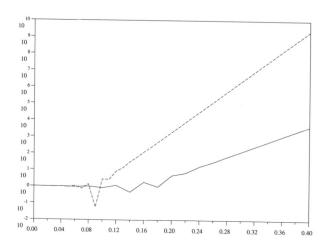

Figure 17.2: The logarithms of the absolute value of the numerical solutions of the multistep scheme defined by the coefficients (17.3.1), for $h = 1/50$ (solid line) and for $h = 1/100$ (dashed line).

This behaviour is the trademark of instability.

What is the culprit? If we define

$$a = \frac{2 - 3h/4}{1 + h/4}, \quad b = -\frac{1 + 2h}{1 + h/4},$$

we may rewrite the numerical scheme in the form

$$\begin{pmatrix} U_{n+1} \\ U_{n+2} \end{pmatrix} = A(h) \begin{pmatrix} U_n \\ U_{n+1} \end{pmatrix},$$

with $A(h)$ being the 2×2 matrix given by

$$A(h) = \begin{pmatrix} 0 & 1 \\ a & b \end{pmatrix}.$$

The matrix $A(h)$ is a continuous function of h; for $h = 0$, it is equal to

$$A(0) = \begin{pmatrix} 0 & 1 \\ 2 & -1 \end{pmatrix},$$

and its spectrum is $\{1, -2\}$. Therefore, by continuity of the spectrum with respect to the matrix, for small values of h, the eigenvalues of $A(h)$ are close to those of $A(0)$. Practically, this means that the component along the eigenvector, relative to the eigenvalue -2, is multiplied by -2 at each time step, while the other component keeps the same magnitude. Therefore, any initial inaccuracy is generally multiplied by 2 at each time step and, thus, we should expect that for a halved time step, the instability increases twice as fast. This is exactly what Figure 17.2 tells us.

17.3.2. The stability theory for multistep methods

Stable matrices

We will first study the so-called stable matrices, whose properties are subsumed in the next lemma:

Lemma 17.3.1. The following three assertions are equivalent for a square complex matrix $A \in \mathcal{M}_d(\mathbb{C})$:

(i) There exists a vector norm N on $\mathbb{C}^d$ such that, for the corresponding matrix norm $\|\cdot\|_N$ on $\mathcal{M}_d(\mathbb{C})$, A satisfies the estimate

$$\|A\|_N \leqslant 1;$$

(ii) In $\mathcal{M}_d(\mathbb{C})$, the non-negative powers of A are bounded uniformly;

(iii) The eigenvalues of A are of modulus 1 and the algebraic multiplicity of the eigenvalues of modulus 1 is equal to their geometric multiplicity. In other words, the corresponding Jordan blocks are of dimension 1.

Proof. The implication (i) $\implies$ (ii) is immediate.

Assume now (ii) holds. The Jordan decomposition of A is of the form

$$A = P^{-1}JP,$$

with P being a regular matrix, and J is the Jordan form of A:

$$J = \begin{pmatrix} J(\lambda_1, m_1) & 0 & \cdots & 0 \\ 0 & J(\lambda_2, m_2) & \cdots & 0 \\ \vdots & & \ddots & \vdots \\ 0 & 0 & \cdots & J(\lambda_k, m_k) \end{pmatrix}.$$

The Jordan blocks are given by

$$J(\lambda, m) = \lambda I_m + N_m, \quad N_m = \begin{pmatrix} 0 & 1 & 0 & \cdots & 0 & 0 \\ 0 & 0 & 1 & \cdots & 0 & 0 \\ \vdots & & & \ddots & \vdots & \vdots \\ 0 & 0 & 0 & \cdots & 1 & 0 \\ 0 & 0 & 0 & \cdots & 0 & 1 \\ 0 & 0 & 0 & \cdots & 0 & 0 \end{pmatrix}.$$

Since the matrices A^n are uniformly bounded with respect to n, the matrices $J^n = (PAP^{-1})^n = PA^nP^{-1}$ are also uniformly bounded with respect to n. But J^n is equal to

$$
J^n = \begin{pmatrix} J(\lambda_1, m_1)^n, & 0 & \cdots & 0 \\ 0 & J(\lambda_2, m_2)^n & \cdots & 0 \\ \vdots & & \ddots & \vdots \\ 0 & 0 & \cdots & J(\lambda_k, m_k)^n \end{pmatrix}.
$$

A classical calculation gives

$$
J(\lambda, m)^n = \lambda^n I_m + \sum_{i=1}^{m-1} C_n^i \lambda^{n-i} N_m^i.
$$

As the matrices $I, N, \ldots, N^{m-1}$ are linearly independent, the boundedness immediately implies that all of the λ_j are of modulus at most 1. If there is an eigenvalue of modulus 1 for which there is a corresponding Jordan block $J(\lambda_j, m_j)$ of dimension at least 2, then the coefficient of N_{m_j} in the n-th power of $J(\lambda_j, m_j)$ is equal to $n\lambda_j^{n-1}$, which cannot be bounded. Hence, the geometric multiplicity of the eigenvalues of modulus 1 is equal to their algebraic multiplicity.

Assume now that (iii) holds. In order to construct the norm N, we split the vector space $\mathbb{C}^d$ into two complementary subspaces: V_1 is the direct sum of the eigenspaces relative to the eigenvalues of A of modulus 1; $V_{<1}$ is the sum of the generalized eigenspaces relative to the eigenvalues of A of modulus strictly inferior to 1. Thus, we may write $\mathbb{C}^d$ as a direct sum as follows:

$$
\mathbb{C}^d = V_1 \oplus V_{<1}.
$$

The projections on the factors V_1 and $V_{<1}$ of this direct sum are denoted by P_1 and $P_{<1}$, respectively.

Let $v_1, \ldots, v_\ell$ be a basis of eigenvectors of A in V_1 and define a norm on V_1 by

$$
x = \sum_{i=1}^{\ell} \xi_i v_i, \quad N_1(x) = \sum_{i=1}^{\ell} |\xi_i|.
$$

The spectral radius of the restriction of A to $V_{<1}$ is strictly inferior to 1, and we know from Lemma 11.1.5 that there exists a vector norm $N_{<1}$ on $V_{<1}$ such that the restriction of A to this space is of norm strictly inferior to 1. We now define

$$
N(x) = N_1(P_1 x) + N_{<1}(P_{<1} x),
$$

and the result is proved. $\square$

We will need the following characterization of a stable block diagonal matrix:

Corollary 17.3.2. Let A be a block diagonal matrix:

$$A = \begin{pmatrix} A_1 & 0 & \cdots & 0 \\ 0 & A_2 & \cdots & 0 \\ \vdots & & \ddots & \vdots \\ 0 & 0 & \cdots & A_k \end{pmatrix}.$$

The matrix A is stable if and only if the blocks A_k are stable.

Proof. In view of assertion (iii) of Lemma 17.3.1, the statement is immediate.
$\qquad\qquad\qquad\qquad\qquad\qquad\qquad\qquad\qquad\qquad\qquad\qquad\qquad\qquad\qquad$ □

The case of the companion matrix A given by

$$(17.3.2) \qquad A = \begin{pmatrix} 0 & 1 & 0 & \cdots & 0 \\ 0 & 0 & 1 & \cdots & 0 \\ \vdots & & & \ddots & \vdots \\ 0 & 0 & 0 & \cdots & 1 \\ -a_0 & -a_1 & -a_2 & \cdots & -a_{q-1} \end{pmatrix}$$

is of particular importance.

Lemma 17.3.3. Let A be the companion matrix given by eqn (17.3.2), and let P be its characteristic polynomial given by

$$P(x) = x^q + \sum_{j=0}^{q-1} a_j x^j.$$

Then A is a stable matrix if and only if the following two conditions are satisfied:

(i) All the roots of P are of modulus at most 1;

(ii) The roots of modulus 1 are simple.

Proof. The proof of this result has been the object of Subsection 3.3.6. We give here a direct proof, whereby we exhibit a Jordan basis for the companion matrix.

Let the extended binomial coefficients be as in eqn (4.5.9). Define the polynomial-valued vector $V(x, m)$ by the list of its components:

$$V_l(x, m) = \binom{l-1}{m} x^{l-1-m}, \quad \forall l = 1, \dots, q.$$

Then, an elementary calculation gives

$$(AV(x, m))_l = \binom{l}{m} x^{l-m}, \quad \forall l = 1, \dots, q-1,$$

$$(AV(x, m))_q = -\frac{P^{(m)}(x)}{m!} + \binom{q}{m} x^{q-m}.$$

If λ is a root of P of multiplicity at least $m + 1$, then $P^{(m)}(\lambda)$ vanishes, and we get the relation

$$AV(\lambda, m) - \lambda V(\lambda, m) = V(\lambda, m - 1),$$

due to the binomial identity. Now let $\{\lambda_1, \ldots, \lambda_k\}$ be the list of roots of P, without repetition; the multiplicity of λ_i is called m_i. The vectors $V(\lambda_i, m)$, $0 \leqslant m \leqslant m_i - 1$, $1 \leqslant i \leqslant k$ are independent. Suppose that there exists a linear combination

$$(17.3.3) \qquad \sum_{i=1}^{k} \sum_{m=0}^{m_i - 1} \xi_{i,m} V(\lambda_i, m) = 0.$$

When the product

$$(A - \lambda_1 I)^{m_1 - 1} (A - \lambda_2 I)^{m_2} \cdots (A - \lambda_k I)^{m_k}$$

is applied to this relation, the only remaining term is

$$\xi_{1, m_1 - 1} V(\lambda_1, m_1 - 1),$$

which must vanish and, hence, the scalar $\xi_{1, m_1 - 1}$ must vanish. All the coefficients $\xi_{1,i}$, $0 \leqslant i \leqslant m_1 - 2$ will also vanish, as can be proved by successively applying the operators

$$(A - \lambda_1 I)^{i} (A - \lambda_2 I)^{m_2} \cdots (A - \lambda_k I)^{m_k}, \quad i = m_1 - 2, \ldots, 0,$$

and an obvious induction on the index of the eigenvalues enables us to see that the coefficients of the linear combination (17.3.3) must all vanish. Since we have the right number of vectors, we have produced an explicit Jordan basis for A.

Now that we know that the dimension of the Jordan blocks of A is exactly the multiplicity of the roots of its characteristic polynomial, the proof is achieved.

□

Stability theorem for multistep schemes

We can now prove a stability result for multistep schemes, which is completely analogous in its method to that of Theorem 16.1.6.

Theorem 17.3.4. Let f map continuously $\mathbb{R} \times \mathbb{R}^d$ to $\mathbb{R}^d$ and assume that it is uniformly Lipschitz continuous with respect to its second argument. Its Lipschitz constant will be denoted by L. Let eqn (17.1.1) define a multistep method and assume that the polynomial ρ, defined by eqn (17.2.4), has all of its roots in the unit disk, while its roots of modulus 1 are simple. Then, there exists a number h^* and, for all $T > 0$, there exists a number C such that, if $0 < h \leqslant h^*$ and U_n

and $\tilde{U}_n$ are sequences satisfying

(17.3.4)
$$\sum_{j=0}^{q} \alpha_j U_{n+j} = h \sum_{j=0}^{q} \beta_j F_{n+j},$$
$$\sum_{j=0}^{q} \alpha_j \tilde{U}_{n+j} = h \sum_{j=0}^{q} \beta_j \tilde{F}_{n+j} + \varepsilon_n,$$

then the following estimate holds for $q \leqslant n \leqslant T/h$:

(17.3.5)
$$|U_n - \tilde{U}_n| \leqslant C \left(\sum_{j=0}^{q-1} |U_j - \tilde{U}_j| + \sum_{j=0}^{n-q} |\varepsilon_j| \right). \qquad \diamond$$

Proof. Due to hypothesis (17.1.3), we may define

$$\alpha'_j = \frac{\alpha_j}{\alpha_q}, \quad \beta'_j = \frac{\beta_j}{\alpha_q}.$$

Let A be the companion matrix

(17.3.6)
$$A = \begin{pmatrix} 0 & 1 & 0 & \cdots & 0 \\ 0 & 0 & 1 & \cdots & 0 \\ \vdots & & & \ddots & \vdots \\ 0 & 0 & 0 & \cdots & 1 \\ -\alpha'_0 & -\alpha'_1 & -\alpha'_2 & \cdots & -\alpha'_{q-1} \end{pmatrix}.$$

Due to the assumptions of our theorem, it is a stable matrix. Now let B be the $qd \times qd$ matrix

$$B = \begin{pmatrix} 0 & I_d & 0 & \cdots & 0 \\ 0 & 0 & I_d & \cdots & 0 \\ \vdots & & & \ddots & \vdots \\ 0 & 0 & 0 & \cdots & I_d \\ -\alpha'_0 I_d & -\alpha'_1 I_d & -\alpha'_2 I_d & \cdots & -\alpha'_{q-1} I_d \end{pmatrix}.$$

A reordering of the canonical basis of $\mathbb{R}^{qd}$ shows that B is similar to a $d \times d$ block diagonal matrix with constant diagonal block equal to the matrix A. Therefore, according to Corollary 17.3.2, B is also a stable matrix. We let $|\cdot|$ be a vector norm on $\mathbb{R}^{qd}$ for which the corresponding subordinate matrix norm of B is at most equal to 1.

Define vectors in $\mathbb{R}^{qd}$ as follows:

(17.3.7)
$$V_n = \begin{pmatrix} U_n \\ \vdots \\ U_{n+q-1} \end{pmatrix}, \quad \tilde{V}_n = \begin{pmatrix} \tilde{U}_n \\ \vdots \\ \tilde{U}_{n+q-1} \end{pmatrix}.$$

Define functions ϕ and ψ from $\mathbb{R} \times \mathbb{R}^{qd} \times [0, h^*]$ to $\mathbb{R}^{qd}$ by

$$x = \begin{pmatrix} x_1 \\ \vdots \\ x_{q-1} \\ x_q \end{pmatrix}, \quad \phi(t, x, h) = \begin{pmatrix} 0 \\ \vdots \\ 0 \\ \beta'_q f(t + qh, x_q) \end{pmatrix},$$

$$\psi(t, x, h) = \begin{pmatrix} 0 \\ \vdots \\ 0 \\ \sum_{j=0}^{q-1} \beta'_j f(t + jh, x_{j+1}) \end{pmatrix}.$$

The functions ϕ and ψ are clearly Lipschitz continuous with respect to their second argument; their respective Lipschitz constants will be denoted L_ϕ and L_ψ. Finally, we let η_n be the vector of $\mathbb{R}^{qd}$ whose first $q-1$ d-dimensional blocks vanish, the last block being equal to ε_n. With these notations, relations (17.3.4) can be rewritten as

$$V_{n+1} = BV_n + h\phi(t_n, V_{n+1}, h) + h\psi(t_n, V_n, h),$$
$$\tilde{V}_{n+1} = B\tilde{V}_n + h\phi(t_n, \tilde{V}_{n+1}, h) + h\psi(t_n, \tilde{V}_n, h) + \eta_n.$$

If we subtract the second of these equalities from the first, if we apply the triangle inequality, and if we recall that B is of norm at most 1, we find

$$(17.3.8) \quad \left|V_{n+1} - \tilde{V}_{n+1}\right| \leqslant \left|V_n - \tilde{V}_n\right| + hL_\phi\left|V_{n+1} - \tilde{V}_{n+1}\right| + hL_\psi\left|V_n - \tilde{V}_n\right| + \left|\varepsilon_n\right|.$$

This is an example of the application of the discrete form of Gronwall's lemma, and the conclusion follows. $\square$

There is a converse to Theorem 17.3.4:

Theorem 17.3.5. If a multistep scheme is stable, then the polynomial ρ satisfies the conditions of Theorem 17.3.4. $\diamond$

Proof. Consider the two scalar sequences U_n and $\tilde{U}_n$, defined by

$$U_n = 0, \quad \forall n \geqslant 0, \qquad \sum_{j=0}^{q} \alpha_j \tilde{U}^i_{n+j} = 0, \quad \forall i \geqslant 0,$$

where f is chosen to be equal to 0. We take $\tilde{U}^i_j = \delta_{ij}$ for all $i, j = 1, \ldots, q-1$. Then, the hypothesis of stability implies that there exists, for all $h \in \,]0, h^*]$ and for all $i = 0, \ldots, q-1$, a number C_i such that the following estimate holds:

$$\left|\tilde{U}^i_n\right| \leqslant C_i, \quad \forall n \in \{0, \ldots, T/h\}.$$

This means that, if A is the companion matrix (17.3.2), then

$$\left|A^n \left(\tilde{U}^i_j\right)_{0 \leqslant j \leqslant q-1}\right|$$

is bounded independently of i and n and, consequently, A^n is bounded independently of n. Then Lemmas 17.3.1 and 17.3.3 give the conclusion. $\square$

17.3.3. Stability of some multistep schemes

The stability of the Adams methods immediately results from the fact that the polynomial ρ for an Adams method with $q + 1$ levels is

$$\rho\left(x\right) = x^q - x^{q-1},$$

then it is clear that the criteria of Lemma 17.3.3 are satisfied.

It can be proved that the backward differentiation methods are stable for $q = 1, \ldots, 6$, but unstable for $q = 7$.

17.4. Convergence of multistep schemes

The convergence theory comes very easily from the stability and order theories.

Theorem 17.4.1. Let eqn (17.1.1) define a consistent and stable multistep scheme of order p. Let f be a function of class C^p from $\mathbb{R} \times \mathbb{R}^d$ to $\mathbb{R}$, which is Lipschitz continuous with respect to its second argument. Let u_0 be given in $\mathbb{R}^d$, let T be a strictly positive number, and let u be the solution of the system (15.0.1). Then, if the initialization of the numerical scheme satisfies, for sufficiently small h, the estimate

$$\sum_{j=0}^{q-1} |U_j - u\left(jh\right)| \leqslant Ch^p,$$

the scheme is convergent and of order p in the following sense: there exists a number C such that

$$(17.4.1) \qquad \max_{0 \leqslant nh \leqslant T} |U_n - u\left(nh\right)| \leqslant Ch^p. \qquad \diamond$$

Proof. Due to the regularity result in Lemma 15.2.8, the solution u of the system (15.0.1) is of class C^{p+1}. We let

$$\tilde{U}_n = u\left(t_n\right) \quad \text{and} \quad \varepsilon_n = \varepsilon\left(t_n, u, h\right).$$

Then, the hypothesis on the order tells us that

$$|\varepsilon_n| \leqslant Ch^{p+1}.$$

Then, estimate (17.4.1) is an immediate consequence of the stability estimate (17.3.5). $\qquad \square$

17.4.1. Initializing multistep methods

The above analysis stressed the importance of the quality of the initialization. The initial error is carried throughout the calculation and, therefore, we need to get very good approximations of the first q data. One approach is to use a Taylor formula expansion to obtain them, together with the calculation of the derivatives of the solution according to eqn (15.2.27). Another option is to use a Runge–Kutta method of high order.

17.4.2. Solving in the implicit case

When β_q does not vanish, each step of a multistep scheme requires the resolution of

$$(17.4.2) \qquad \alpha_q U_{n+q} - h\beta_q f\left(t_{n+q}, U_{n+q}\right) = \sum_{j=0}^{q-1} \left(-\alpha_j U_{n+j} + h\beta_j F_{n+j}\right).$$

The right-hand side of eqn (17.4.2) is known from the previous steps; let us call it z_{n+q}. The left-hand side contains the (generally) nonlinear function $u \mapsto \alpha_q u - h\beta_q f(t, u)$. However, if f is Lipschitz continuous, then, for small h, $u \mapsto (h\beta_q f(t_{n+q}, u) + z_{n+q})/\alpha_q$ is a strict contraction, and it will suffice to make a few iterations of the following form:

$$(17.4.3) \qquad U_{n+q}^{r+1} = \frac{1}{\alpha_q}\left(h\beta_q f\left(t_{n+q}, U_{n+q}^r\right) + z_{n+q}\right),$$

to obtain a reasonable approximation to the solution of eqn (17.4.2) which is sought here.

In fact, if we fix *a priori* the number of iterations of the form (17.4.3), and the process is used to obtain the first approximation, we get the so-called predictor–corrector methods which are studied in more detail in Subsection 17.5.4.

17.5. Exercises from Chapter 17

17.5.1. Short exercises

Exercise 17.5.1. Show that the coefficients γ_i^* of the Adams–Moulton methods satisfy the following relations:

$$\gamma_0^* = 1,$$

$$\frac{\gamma_0^*}{2} + \gamma_1^* = 0,$$

$$\vdots$$

$$\frac{\gamma_0^*}{n+1} + \ldots + \frac{\gamma_{n-1}^*}{2} + \gamma_n^* = 0.$$

Hint: reproduce the formal series derivation of eqns (17.1.8), with appropriate changes.

Exercise 17.5.2. Give the recurrence satisfied by the coefficients κ_i in the Nyström methods (17.1.9) and verify the coefficients appearing in eqn (17.1.10). What is the order of a Nyström method? Is it stable?

Exercise 17.5.3. Give the recurrence satisfied by the coefficients κ_j^* appearing in eqn (17.1.11); calculate the coefficients up to $j = 3$. What does this result mean for the order of the formula (17.1.14)?

Exercise 17.5.4. Study the order and the stability of the methods (17.1.14) and (17.1.15).

17.5.2. An alternative formulation of the order condition

Exercise 17.5.5. Show that, for any consistent multistep method, the polynomial ρ has 1 as a root.

Exercise 17.5.6. Let x_0 be a simple root of ρ. Show that, as a map from $\mathbb{R}^2$ to itself, the map $z \mapsto \rho(z) - h\sigma(z)$ is of class C^2, and calculate its first derivative at $z = x_0$. Show that, for sufficiently small h, this derivative is invertible. Show, with the help of the implicit function theorem, that, for sufficiently small h, the polynomial $\rho - h\sigma$ has one simple root in a neighbourhood of x_0, and that this simple root, denoted by $x(h)$ is a function of h, of class C^2. Calculate $\dot{x}(0)$.

Exercise 17.5.7. If you know about analytic functions, show that $x(h)$ is an analytic function of h in a neighbourhood of $h = 0$. This fact is not needed to answer the next questions.

Exercise 17.5.8. Assume that $x = 1$ is a simple root of ρ. Show that, for real h, $x(h)$ is real.

Exercise 17.5.9. Let $r(x, h) = \rho(e^x) - h\sigma(e^x)$. Show that, in a neighbourhood of $(1, 0)$, $\partial_1 r(x, h)$ is bounded away from 0 and that $\partial_1^2 r(x, h)$ is bounded.

Exercise 17.5.10. Show that the method is of order p if and only if

$$x(h) = e^h + O\left(h^{p+1}\right).$$

Hint: write a Taylor formula with integral remainder to estimate $r(x(h), h) - r(e^h, h)$, and use characterization (17.2.5) of order and Exercise 17.5.9.

17.5.3. Weak instability

Exercise 17.5.11. Simulate numerically the solution of $\dot{x} = -x$, $x(0) = 1$ with

  (i) The Adams–Moulton method with three levels;

  (ii) The Milne–Simpson method (17.1.14).

Initialize using the exact solution, and test several time steps and several time intervals. Plot the difference between the computed solution and the exact solution. What do you observe?

Exercise 17.5.12. Consider a stable consistent multistep method such that the polynomial ρ has a simple root x_0 of absolute value 1 which is not equal to 1. Let $x(h, \lambda)$ be the simple root of $\rho - \lambda h\sigma$ which is in the neighbourhood of x_0. Calculate the derivative of x with respect to h at $h = 0$. Such a method is called weakly unstable.

Exercise 17.5.13. In the case of the Milne–Simpson method (17.1.14), calculate this derivative for $x_0 = -1$.

Exercise 17.5.14. Consider the differential equation $\dot{x} = -\lambda x$. Show that, if $\Re\lambda > 0$ and $\Re(\lambda\sigma(x_0)/x_0\rho'(x_0))$ is strictly larger than $\Re\lambda$, then the error $u(t_n) - U_n$ contains a term of magnitude $O(1)\exp(nh\lambda\sigma(x_0)/x_0\rho'(x_0))$, which dominates the solution for sufficiently large nh.

Exercise 17.5.15. Explain the numerical observation of Exercise 17.5.11, in view of Exercises 17.5.13 and 17.5.14.

17.5.4. Predictor–corrector methods

A predictor–corrector method is defined by the data of two multistep schemes: an explicit one, with data $\bar{\alpha}_j$, $0 \leqslant j \leqslant q$ and $\bar{\beta}_j$, $0 \leqslant j \leqslant q - 1$, and an implicit one with data α_j, $0 \leqslant j \leqslant q$ and β_j, $0 \leqslant j \leqslant q$. For simplicity, and without loss of generality, we shall assume that $\bar{\alpha}_q = \alpha_q = 1$. We assume the first of these multistep methods to be of order $\bar{p}$ and the second to be of order p.

Given $U_n, \ldots, U_{n+q-1}$, the prediction step gives

$$U^0_{n+q} = -\sum_{j=0}^{q-1} \bar{\alpha}_j U_{n+j} + h \sum_{j=0}^{q-1} \bar{\beta}_j F_{n+j}.$$

There can be $N \geqslant 1$ evaluation and correction steps, given by

$$R_{n+q} = -\sum_{j=0}^{q-1} \alpha_j U_{n+j} + h \sum_{j=0}^{q-1} \beta_j F_{n+j},$$

$$F^r_{n+q} = f\left(t_{n+q}, U^r_{n+q}\right), \quad U^{r+1}_{n+q} = R_{n+q} + h\beta_q F^r_{n+q}, \quad \forall r = 0, \ldots, N - 1.$$

The final step can be either

$$U_{n+q} = U^N_{n+q}, \quad F_{n+q} = F^{N-1}_{n+q},$$

in which case we have a $P(EC)^N$ scheme, or

$$U_{n+q} = U^N_{n+q}, \quad F_{n+q} = F\left(t_{n+q}, U_{n+q}\right),$$

in which case we have a $P(EC)^N E$ scheme.

Exercise 17.5.16. Let u be a solution of the differential equation $\dot{u} = f(t, u)$, with f satisfying the assumption of the Cauchy–Lipschitz theorem and is of class $C^{\bar{p}+1}$. Let $\tilde{U}_n = u(t_n)$ and $\tilde{F}_n = f(t_n, \tilde{U}_n)$. Define

(17.5.1) $$\tilde{U}^0_{n+q} = -\sum_{j=0}^{q-1} \bar{\alpha}_j \tilde{U}_{n+j} + h \sum_{j=0}^{q-1} \bar{\beta}_j \tilde{F}_{n+j},$$

and for all $r \geqslant 0$,

$$\tilde{F}^r_{n+q} = f\left(t_{n+q}, \tilde{U}^r_{n+q}\right),$$

$$\tilde{U}^{r+1}_{n+q} = -\sum_{j=0}^{q-1} \alpha_j \tilde{U}_{n+j} + h\sum_{j=0}^{q-1} \beta_j \tilde{F}_{n+j} + h\beta_q \tilde{F}^r_{n+q}.$$

Derive the following estimate, for all $r \geqslant 0$:

$$\left|\tilde{U}^r_{n+q} - \tilde{U}_{n+q}\right| = O\left(h^{\bar{p}+r+1}\right).$$

Exercise 17.5.17. We start with the study of the *PECE* method. Define a block vector W_{n+1} by

$$W_{n+1} = \begin{pmatrix} U^0_{n+q} \\ U_n \\ \vdots \\ U_{n+q} \end{pmatrix}.$$

Show that the *PECE* method can be written as

(17.5.2) $$W_{n+1} = LW_n + hG\left(W_{n+1}, t_{n+1}, h\right),$$

where L and G have the following properties: L is a block matrix of the form

$$L = (a_{ij}I_d)_{1\leqslant i,j\leqslant q+2};$$

G is a function given in block form as

$$\left(G\left(W,t,h\right)\right)_i = \sum_{j=1}^{q+2} b_{ij} f\left(t + v_j h, (W)_j\right),$$

where the notation $(W)_j$ stands for the j-th block of W, and v is a vector of integers given by

$$v_1 = q - 1, \qquad v_j = j - 3, \quad \forall j = 2,\ldots,q+2;$$

finally, G is Lipschitz continuous with respect to its second argument.

Exercise 17.5.18. Extend Corollary 17.3.2 to the case of a block triangular matrix. Let $L_{3:q+2}$ be the matrix constructed from L by chopping off its first two rows of blocks and its first two columns of blocks. Show that L is a stable matrix if and only if $L_{3:q+2}$ is a stable matrix. Infer that the PECE method is stable if and only if the corrector method is stable.

Exercise 17.5.19. Assume that the corrector method is stable. Let $\tilde{U}_n$, $\tilde{U}_n^0$, $\tilde{F}_n$, and $\tilde{F}_n^0$ be defined as in Exercise 17.5.16. Define

$$
\tilde{W}_{n+q} = \begin{pmatrix} \tilde{U}_{n+q}^0 \\ \tilde{U}_n \\ \vdots \\ \tilde{U}_{n+q} \end{pmatrix}.
$$

Derive the estimate

$$
\left| \tilde{W}_{n+1} - L\tilde{W}_n - hG\big(t_{n+1}, \tilde{W}_{n+1}, h\big) \right| = O\left(h^{\bar{p}+1} \right).
$$

Exercise 17.5.20. Assume that f is sufficiently differentiable and uniformly Lipschitz continuous. Also assume that initially

$$
\sum_{j=0}^{q-1} \left| u\left(t_j\right) - U_j \right| = O\left(h^p\right).
$$

Use the principle of the proof of convergence of multistep methods (Theorem 17.4.1) to show that the $PECE$ method is of order $\min(p, \bar{p}+1)$.

Exercise 17.5.21. Using the techniques of Exercise 17.5.20, show that the $P(EC)^N E$ method is of order $\min(\bar{p}+N, p)$.

Exercise 17.5.22. For the PEC method, the W_n will be

$$
W_{n+1} = \begin{pmatrix} U_n^0 \\ \vdots \\ U_{n+q}^0 \\ U_n \\ \vdots \\ U_{n+q} \end{pmatrix}.
$$

Show that all the above arguments can be translated to this case. Show that this method is of order $\min(p, \bar{p})$.

Exercise 17.5.23. Define the vector W_{n+1} for a $P(EC)^N$ method. Give the stability condition and the order of such a method.

17.5.5. One-leg methods

Exercise 17.5.24. Let the data $\alpha_0, \ldots, \alpha_q$ and $\beta_0, \ldots, \beta_q$ define a consistent and stable method. Show that

$$
q\alpha_q + (q-1)\alpha_{q-1} + \ldots + \alpha_1 \neq 0,
$$

and infer that

$$
\beta_q + \ldots + \beta_0 \neq 0.
$$

Exercise 17.5.25. Assume, without loss of generality, that $\alpha_q = 1$. Let

$$\beta = \sum_{j=0}^{q} \beta_j$$

and define a one-leg method by

(17.5.3) $$\frac{1}{\beta} \sum_{j=0}^{q} \alpha_j U_{n+j} = hf\left(t_{n+q}, \frac{\sum_{j=0}^{q} \beta_j U_{n+j}}{\beta}\right).$$

Define

$$U_{n+q}^{\dagger} = \frac{1}{\beta} \sum_{j=0}^{q} \beta_j U_{n+j}.$$

Prove the identity

$$U_{n+q}^{\dagger} = \sum_{j=0}^{q-1} \frac{(\beta_j - \alpha_j \beta_q) U_{n+j}}{\beta} + \beta_q hf\left(t_{n+q}, U_{n+q}^{\dagger}\right).$$

Exercise 17.5.26. Define

$$W_{n+1} = \begin{pmatrix} U_n \\ \vdots \\ U_{n+q-1} \\ U_{n+q}^{\dagger} \end{pmatrix}.$$

Show that the one-leg method (17.5.3) can be written in the form

$$W_{n+1} = LV_n + hG\left(t_{n+1}, W_{n+1}, h\right),$$

and give precisely the matrix L and the function G.

Exercise 17.5.27. Show that L is a stable matrix if and only if the original multistep method is stable.

Exercise 17.5.28. What can you say about the order of the one-leg method (17.5.3)?

18

Towards partial differential equations

In this chapter, we present the elementary theory and the numerical analysis of some partial differential equations. In general, the numerical analysis of partial differential equations requires much more functional analysis than this book aims to present. However, for the simplest partial differential equations, we can work without adding any analytical tools to the ones already used.

We will consider two different kinds of partial differential equations: the advection and wave equations on one hand and the heat equation on the other hand. The analysis of the numerical methods and most of the theory will be performed in one-dimensional space. This is certainly not general; however, enough significant numerical phenomena can be analysed in these cases to make the study relevant.

An important notation must be introduced here. As an alternative to the fractional notation for the partial derivatives

$$\frac{\partial u}{\partial t}, \quad \frac{\partial u}{\partial x}, \quad \text{and} \quad \frac{\partial u}{\partial x_j}$$

we will often use the subscripted notations u_t, u_x, and u_{x_j} or $u_{,j}$, respectively.

18.1. The advection equation

18.1.1. The advection equation and its physical origin

The advection equation is the partial differential equation which resembles most an ordinary differential equation, and its solution requires only ordinary differential equations. This is the reason why we start with it, with a strong emphasis on its physical origin.

The simplest setting for the advection equation is in the full d-dimensional space $\mathbb{R}^d$, with time running from 0 to infinity or from 0 to T. As is classical, we let ∇u denote the spatial gradient of a function u, which is defined as

$$(18.1.1) \qquad \nabla u = \left(\frac{\partial u}{\partial x_1} \quad \cdots \quad \frac{\partial u}{\partial x_d} \right).$$

Given a vector field $a(x,t)$ on $\mathbb{R}^d \times \mathbb{R}^+$ or on $\mathbb{R}^d \times [0,T]$, the advection equation is

$$(18.1.2) \qquad \frac{\partial u}{\partial t} + a \cdot \nabla u = f.$$

The physical origin of this equation is the key for understanding its properties. Equation (18.1.2) describes the transport of matter by a fluid, or analogous physical phenomena. Here, a is the velocity, at time t, of a fluid particle situated at the point x; this is called the Eulerian description of a fluid motion. It is very important to realize that a can be independent of space and time, and yet the fluid will move. Indeed, if, at the time t, a fluid particle is located at the point x and has the velocity a, then, at the time $t' \neq t$, it must be located at the point $x + a(t' - t)$ and, unless a vanishes, this means that the particle has actually moved.

If the velocity depends on time and space, the trajectory of the particle which was at x_0 at the time t_0 can be described in terms of differential equations. If $X(t)$ is its position at time t, we must have

$$(18.1.3) \qquad \dot{X}(t) = a\left(X(t), t \right).$$

If a is smooth enough, for t close enough to t_0, there is a unique solution of eqn (18.1.3) satisfying the initial condition

$$(18.1.4) \qquad X(t_0) = x_0.$$

If we want to emphasize the fact that X depends also on t_0 and x_0, we shall write it as $X(t; t_0, x_0)$. The description of a fluid motion by the motion of the individual particles of fluid is also called its Lagrangian description.

The mapping X is called the flow of the vector field a.

When a does not depend on the time t and is locally Lipschitz continuous, the dependency of X on t and t_0 takes the simpler form

$$(18.1.5) \qquad X(t; t_0, x_0) = Y(t - t_0, x_0),$$

where Y is the solution of

$$\frac{\partial}{\partial t} Y(t; x_0) = a\left(Y(t; x_0) \right),$$

with the initial condition

$$Y(t, x_0) = x_0.$$

We check this fact by observing that

$$Y\left(t - t_0; x_0\right)|_{t=t_0} = x_0$$

and that $t \mapsto Y(t - t_0, x_0)$ satisfies the differential equation

$$\frac{\partial}{\partial t} Y\left(t - t_0; x_0\right) = a\left(Y\left(t - t_0; x_0\right)\right).$$

Then, by the uniqueness of the solution of differential equations, we obtain the relation (18.1.5).

If a depends on time and space and is Lipschitz continuous, then we have the following relation:

$$(18.1.6) \qquad X\left(t_2; t_1, X\left(t_1; t_0, x_0\right)\right) = X\left(t_2; t_0, x_0\right), \quad \forall t_0, t_1, t_2, x_0.$$

This is proved by observing that the function $\xi : t \mapsto X(t; t_1, X(t_1; t_0, x_0))$ satisfies the differential equation

$$\dot{\xi}\left(t\right) = a\left(\xi\left(t\right)\right),$$

together with the following initial condition at t_1:

$$\xi\left(t_1\right) = X\left(t_1; t_0, x_0\right).$$

This means that ξ and $t \mapsto X(t; t_0, x_0)$ satisfy the same system of ordinary differential equations and coincide at time t_1. Hence, due to the uniqueness of solutions to systems of ordinary differential equations, they coincide at all times.

An obvious consequence of relation (18.1.6) is that the mapping $x_0 \mapsto X(t; t_0, x_0)$ has an inverse, which is given by $x \mapsto X(t_0; t, x)$:

$$(18.1.7) \qquad X\left(t; t_0, X\left(t_0; t, x\right)\right) = x, \quad X\left(t_0; t, X\left(t; t_0, x_0\right)\right) = x_0.$$

If, instead of assuming that a is Lipschitz continuous, we had assumed that it is only locally Lipschitz continuous, we would still have relation (18.1.6), but only for the space coordinates x_0 and the times t_0, t_1, and t_2 for which the different expressions in eqn (18.1.6) are defined.

In fluid mechanics, the trajectory of a fluid particle is known as a streamline and a region limited by streamlines, and possibly by planes of constant time, is known as a stream tube.

The translation to mathematical language thus says that a streamline is the image of the mapping $t \mapsto X(t; t_0, x_0)$ and that a stream tube is the region $\{X(t; t_0, x_0) : t \in [t_1, t_2], x_0 \in U\}$, with U a region of space. It is limited by the planes $t = t_1$ and $t = t_2$, and by the streamlines through (t_0, x_0), where x_0 runs through the boundary of the region U. All of these objects are represented in Figure 18.1.

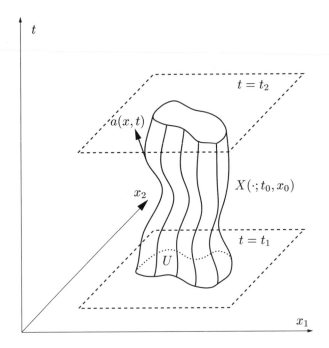

Figure 18.1: Streamlines and a stream tube.

18.1.2. Solving the advection equation

Suppose that the fluid transports a colorant, which has a certain initial density $u_0(x)$ at time t_0. For instance, we could stain the fluid with fluoresceine, in order to track streamlines, and then add a very small quantity of fluid in the neighbourhood of some point, so that the density of colorant vanishes away from that point.

As a first approximation, we may assume that there is no diffusion. This assumption means that the density of colorant is constant along the streamlines, provided that there are no sources of colorant in the domain under consideration. Assuming that u is a function of class C^1 of $x \in \mathbb{R}^d$ and $t \in \mathbb{R}$, we differentiate the function $t \mapsto u(X(t; t_0, x_0), t)$ as follows:

$$\frac{\partial}{\partial t}\left(u\left(X\left(t; t_0, x_0\right), t\right)\right)$$

$$= \sum_{j=1}^{d} \frac{\partial u}{\partial x_j}\left(X\left(t; t_0, x_0\right), t\right) \frac{\partial X_j}{\partial t}\left(t; t_0, x_0\right) + \frac{\partial u}{\partial t}\left(X\left(t; t_0, x_0\right), t\right)$$

$$= \left(u_t + a \cdot \nabla u\right)\left(X\left(t; t_0, x_0\right), t\right).$$

Therefore, it is equivalent for a function of class C^1 to be constant along streamlines and to satisfy the relation

$$u_t + a \cdot \nabla u = 0.$$

In particular, if the initial condition at time t_0 is given by a function u_0 over $\mathbb{R}^d$, we must have, for all t, t_0, and x_0, the following relation:

(18.1.8) $$u\left(X\left(t; t_0, x_0\right), t\right) = u_0\left(x_0\right).$$

Since we know how to invert $x_0 \mapsto X(t; t_0, x_0)$, due to the relations (18.1.7), we transform eqn (18.1.8) into

(18.1.9) $$u\left(x, t\right) = u_0\left(X\left(t_0; t, x\right)\right).$$

It is important to observe that the relation (18.1.9) makes sense even if u_0 is not of class C^1; then, we say that we have a weak or generalized solution of eqn (18.1.2), with $f = 0$. This is, indeed, a solution in the sense of distributions or a generalized solution. The meaning of these words is analysed in Subsection 18.5.8.

When a is a constant vector, the solution (18.1.9) has the following very simple form:

(18.1.10) $$u\left(x, t\right) = u_0\left(x - at\right).$$

Let us solve now eqn (18.1.2) for any continuous function f, when the initial data are of class C^1.

Theorem 18.1.1. Let a be a vector field over $\mathbb{R}^d \times [0, T]$ which is uniformly Lipschitz continuous with respect to the first variable. Assume that the function f is continuous over $\mathbb{R}^d \times [0, T]$ and that the initial data u_0 is continuously differentiable over $\mathbb{R}^d$. Then, there exists a unique solution to eqn (18.1.2) satisfying the initial condition

$$u\left(x, 0\right) = u_0\left(x\right), \quad x \in \mathbb{R}^d,$$

and it is given by

(18.1.11) $$u\left(x, t\right) = u_0\left(X\left(t_0; t, x\right)\right) + \int_{t_0}^{t} f\left(X\left(s; t, x\right), s\right) \mathrm{d}s. \qquad \diamond$$

Proof. It is immediate that, if eqn (18.1.2) holds, then

(18.1.12) $$\frac{\partial}{\partial t}\left(u\left(X\left(t; t_0, x_0\right), t\right)\right) = f\left(X\left(t; t_0, x_0\right), t\right)$$

and, therefore, by a direct integration,

$$u\left(X\left(t; t_0, x_0\right), t\right) = u_0\left(x_0\right) + \int_{t_0}^{t} f\left(X\left(s; t_0, x_0\right), s\right) \mathrm{d}s.$$

Due to the change of variable $x = X(t; t_0, x_0)$, whose inverse is given by $x_0 = X(t_0; t, x)$ (see the relation (18.1.7)), and to the identity (18.1.6) applied at the times t_0, s, and t, we finally obtain expression (18.1.11). The uniqueness is immediate: if u and v are solutions of eqn (18.1.2) with the same initial condition u_0, then $u - v$ is a solution of eqn (18.1.2) with vanishing right-hand side and initial condition; it satisfies the relation

$$\frac{\partial}{\partial t} \left((u - v) \left(X\left(t; t_0, x_0\right), t\right)\right) = 0,$$

with vanishing initial condition, which immediately implies that $u - v$ vanishes on every streamline.   □

Observe that the expression (18.1.11) makes sense even if f and u_0 satisfy weaker regularity hypotheses.

18.1.3. More general advection equations and systems

The formula (18.1.11) enables us also to solve semilinear advection equations in d variables and hyperbolic systems in 1 variable.

Replace, indeed, $f(x, t)$ in the right-hand side of eqn (18.1.2) by a function $f(x, t, u)$. Then, eqn (18.1.12) becomes

$$\frac{\partial}{\partial t} \left(u\left(X\left(t; t_0, x_0\right), t\right)\right) = f\left(X\left(t; t_0, x_0\right), t, u\left(X\left(t; t_0, x_0\right), t\right)\right),$$

which is simply an ordinary differential equation whose unknown function $v(t) = u(X(t; t_0, x_0), t)$ satisfies

$$\dot{v}\left(t\right) = f\left(v\left(t\right), t\right).$$

Consider now the following hyperbolic system of n linear equations in one dimension:

(18.1.13) $$\frac{\partial u}{\partial t} + M \frac{\partial u}{\partial x} = f,$$

where M is an $n \times n$ matrix, which is continuously differentiable with respect to x, t and has the following strict hyperbolicity property:

M has n distinct real eigenvalues $\lambda_j\left(x, t\right)$, $1 \leqslant j \leqslant n$, $\forall x, t$.

Then, it is possible to find n eigenvectors $r_j(x, t)$, for $1 \leqslant j \leqslant n$, of $M(x, t)$ and n eigenvectors $l_j(x, t)$, for $1 \leqslant j \leqslant n$, of $M^\top(x, t)$ such that

$$M r_j = \lambda_j r_j, \quad l_j^\top M = \lambda_j l_j^\top, \quad l_j^\top r_k = \delta_{jk},$$

and the vectors r_j and l_j are continuously differentiable with respect to x and t.

This fact is proved in Subsection 18.5.1.

Now decompose the unknown vector u as

$$u(x,t) = \sum_{j=1}^{n} v_j(x,t) r_j(x,t).$$

With this notation, eqn (18.1.13) now becomes, after substitution of the new expression for u and multiplication on the left by $l_j^\top$,

(18.1.14) $$v_{j,t} + \lambda_j v_{j,x} + l_j^\top \sum_{k=1}^{n} (r_{k,t} + M r_{k,x}) v_k = l_j^\top f.$$

Thus, we have obtained a system which is essentially composed of equations of the type (18.1.2), coupled via terms of order 0.

Since the terms of order 0 already contain a dependency with respect to u, it makes sense to assume also that f depends on u. Therefore, it will be a function $f(x,t,u)$.

In order to look for a solution, we reduce eqn (18.1.14) to an integral equation, with the help of expression (18.1.11). Denoting by $S_j(t;t_0)$ the transformation given by

$$(S_j(t;t_0)u)(x) = u(X_j(t;t_0,x_0)),$$

with X_j the flow associated to r_j, we may rewrite eqn (18.1.14) as

$$v_j(\cdot,t) = S_j(t;t_0) v_j(\cdot,t_0) + \int_{t_0}^{t} S_j(t;s) g_j(\cdot,s,v)\,\mathrm{d}s,$$

where the functions g_j are defined as

$$g_j(x,s,v) = l_j(x,t)^\top f(x,s,v) - l_j(x,t)^\top \sum_{k=1}^{n} (r_{k,t} + M r_{k,x})(x,t) v_k(x,t).$$

If the mappings g_j are Lipschitz continuous in v, uniformly with respect to $x \in \mathbb{R}$ and to $t \in [0,T]$, then the method of proof of existence for the Cauchy–Lipschitz theorem by Picard iterations works. For instance, in the functional space $C_b^0(\mathbb{R})^d$ of bounded continuous functions on $\mathbb{R}^d$, the mapping

$$v \mapsto (S_j(t;t_0) g_j(\cdot,\cdot,v))_{1 \leqslant j \leqslant n}$$

is Lipschitz continuous from $C^0(\mathbb{R})^d$ to itself.

Of course, the solution obtained by this process is a generalized solution; the exact significance of this term is explained for the advection equation in Subsection 18.5.8. I hope that, at this stage, the reader will agree to believe me, or else, refer to more advanced work, such as [26].

18.2. Numerics for the advection equation

18.2.1. Definition of some good and some bad schemes

Since we have an explicit expression (18.1.11) for the solution of eqn (18.1.2), the reader may well wonder at this point why it is necessary to find a numerical method for solving eqn (18.1.2). After all, we just have to find the streamlines, which are also called characteristics, and apply on these characteristics our favourite quadrature formula to find an approximation of expression (18.1.11).

This is, indeed, the essence of the so-called method of characteristics, which remains a method of choice for solving the advection equation.

But, if we wish to solve a slightly more complicated problem, such as the system (18.1.13), other options, rather than the method of characteristics, are reasonable, and even more so in higher spatial dimensions, where they become necessary. The wave equation or the elasticity system do not reduce to coupled advection equations, though they share propagation properties with the advection equation. For them, there is no simple equivalent of the method of characteristics.

The advection equation should be considered as a toy system on which it is useful to test ideas before applying them to more complicated situations. This is the reason why we study the finite difference methods used for the approximation of the very simple eqn (18.1.2). Even in dimension 1 and with a constant velocity a, we shall see that there is food for thought.

The simplest ideas can be used to construct a discrete approximation of

$$(18.2.1) \qquad\qquad u_t + au_x = 0.$$

We replace u_t and u_x by finite differences and the variables x and t by discrete variables. According to a traditional notation, we let U_j^n be an approximation of $u(j\delta x, n\delta t)$.

To fix ideas, we let a be a strictly positive number. The results for a negative are subsequently deduced by transforming x into $-x$, as can be immediately seen.

However, we have multiple choices; to keep the computational effort at a minimum, we settle for an explicit scheme in time. This means that we replace u_t by

$$\frac{U_j^{n+1} - U_j^n}{\delta t}.$$

But, what about the space difference? We may take

$$(18.2.2) \qquad\qquad \frac{U_j^n - U_{j-1}^n}{\delta x},$$

$$(18.2.3) \qquad\qquad \frac{U_{j+1}^n - U_j^n}{\delta x},$$

or

$$(18.2.4) \qquad \frac{U^n_{j+1} - U^n_{j-1}}{2\delta x},$$

and each of these choices will lead to different numerical schemes with widely differing properties.

In order to have a first approach to the properties of these schemes, we introduce the notion of a stencil. It is the set of points $((j' - j)\delta x, (n' - n)\delta t)$ such that (j', n') are used in the computation of U^{n+1}_j. Thus, for instance, the scheme corresponding to the choice (18.2.2) is written as

$$(18.2.5) \qquad \frac{U^{n+1}_j - U^n_j}{\delta t} + a\frac{U^n_j - U^n_{j-1}}{\delta x} = F^n_j,$$

with stencil made out of the points $(0,0)$, $(-\delta x, 0)$, and $(0, \delta t)$. This scheme is called the upwind scheme, since the position of the discretization point, used for differentiating spatially, is translated against the wind: a is the velocity of the flow, understood as the flow of a fluid, and we go up on the streamlines to construct the finite difference.

The scheme corresponding to the choice (18.2.3) is the downwind scheme. It is written as

$$(18.2.6) \qquad \frac{U^{n+1}_j - U^n_j}{\delta t} + a\frac{U^n_{j+1} - U^n_j}{\delta x} = F^n_j$$

and its stencil is made out of the points $(0,0)$, $(\delta x, 0)$, and $(0, \delta t)$. Finally, the centered scheme, corresponding to the choice (18.2.4), is written as

$$(18.2.7) \qquad \frac{U^{n+1}_j - U^n_j}{\delta t} + a\frac{U^n_{j+1} - U^n_{j-1}}{2\delta x} = F^n_j$$

and its stencil is made out of the points $(-\delta x, 0)$, $(\delta x, 0)$, and $(0, \delta t)$.

We already know that the solution of eqn (18.2.1) is given by the relation (18.1.10), with u taken to be equal to $u_0(x)$ at the initial time $t = 0$.

Therefore, if our initial data vanish outside the interval $[-1, 1]$, at time t, and the right-hand side f vanishes, the solution must vanish outside the interval $[-1 + at, 1 + at]$. If we approximate our initial data by $U(j, 0)$ which vanishes for $|j|\delta x \leqslant 1$, then we can use the stencil to understand where the numerical approximation will necessarily vanish and where it might be different from 0. It will be reasonable to take a vanishing right-hand side in either scheme (18.2.5), (18.2.6) or (18.2.7).

Graphical arguments will impose necessary conditions on the type of scheme and on the numerical parameters. If these conditions are not satisfied, there is no hope whatsoever of convergence.

An important number is the CFL (Courant–Friedrichs–Lewy) number, defined as

$$\lambda = \frac{a\delta t}{\delta x}.$$

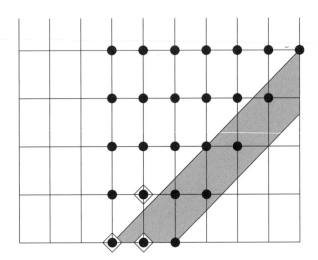

Figure 18.2: Upwind numerical scheme (18.2.5), with $a\delta t/\delta x > 1$. The black circles denote the points where the numerical approximation can be nonzero and the shaded region is that where the exact solution can be nonzero. The stencil is indicated by white squares.

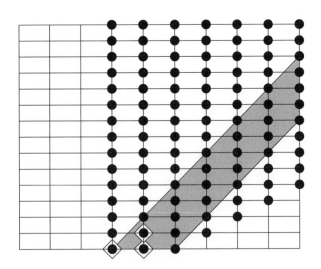

Figure 18.3: Upwind numerical scheme (18.2.5), with $a\delta t/\delta x \leqslant 1$, using the same graphical conventions as in Figure 18.2.

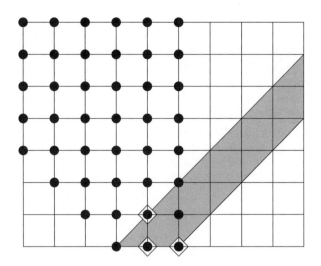

Figure 18.4: Downwind numerical scheme (18.2.6), with $a\delta t/\delta x = 1$, using the same graphical conventions as in Figure 18.2.

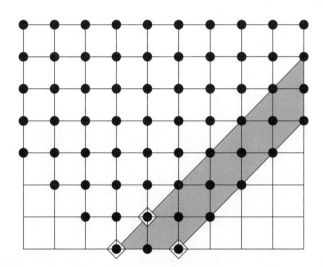

Figure 18.5: Central numerical scheme (18.2.7), with $a\delta t/\delta x = 1$, using the same graphical conventions as in Figure 18.2.

For the upwind numerical scheme (18.2.5), if $\lambda > 1$, then the numerical method cannot converge. We see on Figure 18.2 that the region where the solution can be different from zero is not completely included in the region where the numerical solution can be different from 0. On the other hand, if $\lambda \leqslant 1$, as in Figure 18.3, we see that we do not run into the same difficulty, and we will show below that this condition is indeed sufficient for convergence.

If we now take the numerical scheme (18.2.6), the same graphical considerations as above show that there is no situation in which this numerical scheme can ever be convergent. Indeed, the numerical solution propagates in the opposite direction to what we expected, see Figure 18.4. Therefore, downwind discretization leads to disastrous results, but they are not difficult to recognize numerically, as is shown in Exercise 18.5.9.

In the last case, i.e. scheme (18.2.7), the graphical representation of Figure 18.5 shows that it is necessary that λ be at most equal to 1. We will see below that this is a very mediocre numerical scheme, since this condition is not sufficient. However, we need more analysis to understand this phenomenon.

18.2.2. Convergence of the scheme (18.2.6)

The proof of the convergence of the scheme (18.2.6) goes through the same logical steps as the proof of the convergence of the one-step and multistep schemes for ordinary differential equations, i.e., consistency plus stability imply convergence. The new feature here is that we need a functional space, and this is the reason why the numerical analysis of partial differential equations is more complicated than the numerical analysis of ordinary differential equations. Moreover, the convergence is usually proved upon assuming that the initial data are smoother than what is needed for the existence of solutions.

We choose to take an initial condition u_0 of class C^1, with Lipschitz-continuous first derivative. We also choose a Lipschitz-continuous and C^1 right-hand side f. Then, the function $u(x, t)$ given by the relation (18.1.11) becomes, in this particular case,

$$(18.2.8) \qquad u(x, t) = u_0(x - at) + \int_0^t f(x - at + as, s)\, \mathrm{d}s.$$

Under our smoothness assumption on u_0 and f, it is clear that u is of class C^1 and satisfies the partial differential eqn (18.2.1) and the initial condition

$$u(\cdot, 0) = u_0.$$

Let us first state and prove consistency:

Lemma 18.2.1. Let u_0 be of class C^1, with Lipschitz-continuous first derivative, and let f be Lipschitz continuous over $\mathbb{R} \times [0, T]$. The local consistency error defined by

$$\varepsilon_j^n = \frac{u(j\delta x, (n+1)\delta t) - u(j\delta x, n\delta t)}{\delta t} + a\frac{u(j\delta x, n\delta t) - u((j-1)\delta x, n\delta t)}{\delta x}$$

satisfies the estimate

$$(18.2.9) \qquad \left|\varepsilon_j^n\right| \leqslant C\left(\delta x + \delta t\right),$$

where C depends only on the Lipschitz constants of u_0' and f.

Proof. Let L_0 be the Lipschitz constant of u_0' and let L_1 be the Lipschitz constant of f. By Taylor expansions, we have the following estimates:

$$\left|u_0\left(x - at\right) - u_0\left(x - \delta x - at\right) - \delta x u_0'\left(x - at\right)\right| \leqslant L_0 \frac{\delta x^2}{2},$$

$$\left|u_0\left(x - at - a\delta t\right) - u_0\left(x - at\right) + a\delta t u_0'\left(x - at\right)\right| \leqslant L_0 \frac{a^2 \delta t^2}{2},$$

and similarly

$$\left|\int_t^{t+\delta t} f\left(x - at - a\delta t + as, s\right) ds - \delta t f\left(x, t\right)\right| \leqslant a L_1 \frac{\delta t^2}{2},$$

$$\left|\int_0^t \left(f\left(x - at - a\delta t + as, s\right) - f\left(x - at + s, s\right)\right) ds\right| \leqslant a L_1 \delta t,$$

$$\left|\int_0^t \left(f\left(x - at + as, s\right) - f\left(x - \delta x - at + as, s\right)\right) ds\right| \leqslant L_1 T \delta x.$$

If we summarize all of these estimates, we immediately obtain the relation (18.2.9). □

We now turn to the stability statement:

Lemma 18.2.2. Assume that

$$(18.2.10) \qquad 0 \leqslant \frac{a\delta t}{\delta x} = \lambda \leqslant 1.$$

Let $G(n, j)$ be a bounded sequence indexed by $j \in \mathbb{Z}$ and $n \in [0, T/\delta t]$, and let W be the sequence defined recursively from $j \mapsto W_j^0$ by

$$(18.2.11) \qquad \frac{W_j^{n+1} - W_j^n}{\delta t} + a \frac{W_j^n - W_j^{n-1}}{\delta x} = G_j^n.$$

There exists a constant C such that, for all δt and δx satisfying the inequality (18.2.10), the following estimate holds:

$$(18.2.12) \qquad \sup_j \left|W_j^m\right| \leqslant C\left(\sup_j \left|W_j^0\right| + \sum_{n=0}^{m-1} \delta t \sup_j \left|F_j^m\right|\right), \qquad \forall m \in [1, T/\delta t].$$

Proof. We write eqn (18.2.11) as

$$W_j^{n+1} = (1 - \lambda) W_j^n + \lambda W_{j-1}^n + \delta t G_j^n.$$

The condition (18.2.10) is equivalent to

$$|1 - \lambda| + |\lambda| \leqslant 1,$$

and hence, by the triangle inequality,

$$\sup_j \left| W_j^{n+1} \right| \leqslant \sup_j \left| W_j^n \right| + \delta t \sup_j \left| G_j^n \right|,$$

which immediately implies the estimate (18.2.12), with $C = 1$.   □

Remark 18.2.3. Let us give an articulate definition of the concept of stability. A scheme, depending on a time step δt and a space step δx, associates to discrete data at time t, depending on δx, a set of new data at the later time $t + \delta t$. These data belong to a normed space $B(\delta x)$ and, if we consider a linear partial differential equation with time-independent coefficients and vanishing right-hand side, we will usually describe the transition from data at time t to data at time $t + \delta t$ by a linear operator $P(\delta x, \delta t)$, from $B(\delta x)$ to itself, which is called a propagator. The requirement of stability can now be phrased as the following condition: there exist constants C_1 and C_2 such that the following relation holds:

$$(18.2.13) \qquad \left\| P\left(\delta x, \delta t\right)^n \right\|_{\mathcal{L}(B(\delta x))} \leqslant C_1 e^{C_2 n \delta t}$$

for all δx, for all δt belonging to an interval starting at 0 and ending possibly at a value dependent on δx, and for all integers n. The norm in the relation (18.2.13) is the operator norm.

I would like to emphasize a very important fact: the constants C_1 and C_2 must not depend on δt, δx, and n for stability to be true. We say that a numerical scheme is conditionally stable if we need to limit δt as a function of δx for the relation (18.2.13) to hold. We say that we have unconditional stability if the upper limit of the interval where we take our δt does not depend on δx.

Of course, the condition $\|P(\delta x, \delta t)\| \leqslant 1 + C_2 \delta t$ implies the relation (18.2.13). However, condition (18.2.13) is stated in order to treat the situation where the linear operator $U(\delta x, \delta t)$ does not satisfy nice conditions, but, nevertheless, the product of n copies of the propagator can still be controlled.

How would we modify condition (18.2.13) to treat the time-dependent situation? Then, instead of having a time-independent transition from data at time t to data at time $t + \delta t$, we have a time-dependent transition $P(t, \delta x, \delta t)$, and we state an analogous property, this time for a time-ordered product of transition operators. We must also allow for variable time steps.

However, such a situation cannot be treated conveniently in a general setting. To obtain any substantial results, we need to specify which partial differential equation or system we are interested in, since the results depend on the details of the data and coefficients and, in particular, on their regularity.

The last observation is relative to the space $X(\delta x)$. When we consider an advection equation in $\mathbb{R}$, a natural space for finite difference approximation is $\ell^\infty(\mathbb{Z})$, the space of bounded sequences indexed by $\mathbb{Z}$. However, there are other options: we could also use the space $\ell^2(\mathbb{Z})$ of square-integrable sequences; we shall see below that the centered scheme (18.2.7) is unstable in $\ell^\infty(\mathbb{Z})$ but stable in $\ell^2(\mathbb{Z})$. Therefore, one should keep in mind that functional spaces are tools which enable us to measure and understand the behaviour of mathematical objects. Using different functional spaces means asking different questions and, quite often, also getting different answers.

We now give a proof of convergence, which includes an order for the approximation error:

Theorem 18.2.4. Assume that u_0 and f have the regularity described in Lemma 18.2.1 and that the initial data and the right-hand side satisfy

$$\sup_j \left| U_j^0 - u_0\left(j\delta x\right) \right| \leqslant C\delta x, \qquad \sup_{0 \leqslant n\delta t \leqslant T} \sup_j \left| F_j^n - f\left(j\delta x, n\delta t\right) \right| \leqslant C\left(\delta t + \delta x\right).$$

Assume, moreover, that the space step δx and the time step δt satisfy the CFL condition:

$$a\delta t \leqslant \delta x.$$

Then, there exists a number C' for which the following estimate holds:

$$(18.2.14) \quad \sup_j \left| U_j^n - u\left(j\delta x, n\delta t\right) \right| \leqslant C'\left(\delta t + \delta x\right), \quad \forall j \in \mathbb{Z}, \; \forall n \in [0, T/\delta t].$$

In particular, the numerical scheme (18.2.5) is convergent. $\qquad \diamond$

Proof. It suffices to define

$$W_j^n = U_j^n - u\left(j\delta x, n\delta t\right)$$

and

$$G_j^n = F_j^n + \varepsilon_j^n.$$

Then, the conclusion (18.2.14) is an immediate consequence of Lemmas 18.2.1 and 18.2.2. $\qquad \square$

How do we see that the scheme (18.2.7) is mediocre? First, a simple argument shows that stability in the supremum norm, i.e., in $\ell^\infty(\mathbb{Z})$, is not very likely: if we define U_j^0 by

$$U_j^0 = \begin{cases} -1 & \text{if } j = -1; \\ 1 & \text{if } j = 0 \text{ or } j = 1; \\ 0 & \text{otherwise,} \end{cases}$$

then

$$\sup_{j \in \mathbb{Z}} |U_j^1| = 1 + \lambda.$$

This means that for the scheme to be stable it would be sufficient to have

$$\frac{a\delta t}{\delta x} = \lambda \leqslant C_1 \delta t.$$

However, this condition implies that δx is bounded away from 0. Of course, this is not sufficient to conclude instability, and so we seek sequences U_j^0 which are eigenfunctions of the propagator $P(\delta x, \delta t)$. Trying trigonometric functions, we see that, if ξ is an arbitrary real number and

$$U_j^0 = \sin(j\xi\delta x),$$

then a simple computation gives

$$U_j^1 = (1 + \lambda \cos(\xi\delta x)) \sin(j\xi\delta x),$$

and, therefore,

$$U_j^n = (1 + \lambda \cos(\xi\delta x))^n \sin(j\xi\delta x).$$

Therefore,

$$\|P(\delta x, \delta t)^n\|_{\mathcal{L}(\ell^\infty(\mathbb{Z}))} \geqslant \sup_{\xi \in \mathbb{R}} |1 + \lambda \cos(\xi\delta x)|^n = (1 + \lambda)^n.$$

This proves that the scheme (18.2.7) is unstable in $\ell^\infty(\mathbb{Z})$.

However, it is stable in $\ell^2(\mathbb{Z})$ under the CFL condition

$$\delta t \leqslant C\delta x^2,$$

as proved in Exercise 18.5.12.

Several cures are possible for the defects of the scheme (18.2.7). First, we may replace it by the Lax–Friedrichs scheme

$$(18.2.15) \qquad U_j^{n+1} = \frac{U_{j-1}^n + U_{j+1}^n}{2} + \frac{a\delta t \left(U_{j+1}^n - U_{j-1}^n\right)}{2\delta x}$$

or by the Lax–Wendroff scheme

$$(18.2.16) \qquad \frac{U_j^{n+1} - U_j^n}{\delta t} = a \frac{U_{j+1}^n - U_{j-1}^n}{2\delta x} - \frac{a^2\delta t}{2\delta x^2} \frac{U_{j+1}^n - 2U_j^n + U_{j-1}^n}{2\delta x^2}.$$

Both of the schemes (18.2.15) and (18.2.16) are stable in $\ell^\infty(\mathbb{Z})$ and in $\ell^2(\mathbb{Z})$, provided that the CFL number is at most equal to 1, and their convergence can be proved along the lines of the proof of Theorem 18.2.4. These questions are taken up in Subsections 18.5.4 and 18.5.5, which use a very straightforward analysis.

Another possibility is to use ideas in multistep schemes for ordinary differential equations; thus we get the leap-frog scheme given by

$$(18.2.17) \qquad \frac{U_j^{n+1} - U_j^{n-1}}{2\delta t} + a\frac{U_{j+1}^n - U_{j-1}^n}{2\delta x} = 0.$$

This time it is necessary to use the ℓ^2 theory to prove the convergence of this scheme. Stability is studied in Subsection 18.5.6 and the convergence is left for a more advanced course, where Sobolev spaces and distributions can be freely used.

18.3. The wave equation in one dimension

The wave equation is defined as

$$(18.3.1) \qquad \frac{1}{c^2}\frac{\partial^2 u}{\partial t^2} - \frac{\partial^2 u}{\partial x^2} = f.$$

It is quite interesting to derive it from a model of springs and masses; historically, this derivation goes back to the eighteenth century.

In one-dimensional space, the wave equation can be understood as a system of advection equations, about which we know everything. However, here we shall also be interested in boundary conditions. The numerical analysis would be quite straightforward if we had the appropriate functional tools. We will satisfy ourselves with proving the ℓ^2 stability of one standard numerical scheme. The convergence and consistency results should be left for more advanced books, since they use more functional analysis. However, the reader is strongly advised to simulate numerically the solution of the wave equation so as to get a feeling for the phenomena that take place.

18.3.1. Masses and springs

We shall approximate a strongly stretched string by a discrete mechanical system. We consider N material points, each of mass m/N, which are separated by identical springs of length at rest $L_0/(N+1)$. The end springs are fixed at the points of abscissa 0 and $L > L_0$.

It is assumed that the small springs are made of the same linear homogeneous material. In other words, we could take an homogeneous spring of length L_0 at rest and cut it into $N+1$ identical pieces. The assumption of linearity means that, when subjected to a force f, a spring of length ℓ_0 at rest stretches by an amount $\delta\ell$ proportional to f. On the other hand, the extension is also inversely proportional to the length at rest. Suppose that we apply a force f to a spring of length $2\ell_0$ at rest. The tension is constant along the spring, but the first section of length ℓ_0 at rest is stretched by $\delta\ell/2$, and so is the second section. Here it is the homogeneity assumption which imposes that the stretching is uniformly distributed along the length of the spring.

Therefore, the stiffness of a spring is inversely proportional to its length, which is a very intuitive statement. It is clearly more difficult to extend by a given length a very short spring than a very long spring of the same material. Thus, our small springs have stiffness

$$\kappa\,(N+1)\,/L_0,$$

with the number κ describing the physical properties of the spring.

Assume that the mass indexed by j has coordinates $([jL/(N+1)]+x_j, y_j, z_j)$, and that the extremities of the system are tied at the points with coordinates

$$(0,0,0) = (x_0, y_0, z_0) \quad \text{and} \quad (L,0,0) = (L + x_{N+1}, y_{N+1}, z_{N+1})\,.$$

It will be convenient to use the notation

$$\ell = \frac{L}{N+1}, \quad \ell_0 = \frac{L_0}{N+1}, \quad \delta\ell = \ell - \ell_0.$$

The elastic potential energy of the deformed system of springs is

$$V\,(x,y,z) = \frac{\kappa\,(N+1)}{2L_0} \sum_{j=0}^{N} \Big[\big((x_{j+1} - x_j + \ell)^2$$
$$+ (y_{j+1} - y_j)^2 + (z_{j+1} - z_j)^2 \big)^{1/2} - \ell_0 \Big]^2,$$

since the length of the deformed spring between mass j and mass $j+1$ is

$$\Big((x_{j+1} - x_j + \ell)^2 + (y_{j+1} - y_j)^2 + (z_{j+1} - z_j)^2 \Big)^{1/2},$$

with appropriate modifications for the end springs. The kinetic energy of the system is given by

$$T\,(\dot{x}, \dot{y}, \dot{z}) = \frac{m}{2N} \sum_{j=1}^{N} \big(\dot{x}_i^2 + \dot{y}_i^2 + \dot{z}_i^2 \big)\,.$$

The equations of motion are

$$\frac{m}{N} \ddot{x}_j + \frac{\partial}{\partial x_j} V\,(x,y,z) = 0,$$
$$\frac{m}{N} \ddot{y}_j + \frac{\partial}{\partial y_j} V\,(x,y,z) = 0,$$
$$\frac{m}{N} \ddot{z}_j + \frac{\partial}{\partial z_j} V\,(x,y,z) = 0.$$

In this generality, we obtain a highly nonlinear problem about which we are cannot say much. However, if we are interested in the small vibrations close to equilibrium, we shall be content with an approximation to the potential energy of order at most 2. Thus, we have the following expansion:

$$
\left[\left((x_{j+1} - x_j + \ell)^2 + (y_{j+1} - y_j)^2 + (z_{j+1} - z_j)^2 \right)^{1/2} - \ell_0 \right]^2
$$

$$
= \delta\ell^2 + 2\delta\ell \, (x_{j+1} - x_j) + (x_{j+1} - x_j)^2 + \frac{\delta\ell}{\ell} \left((y_{j+1} - y_j)^2 + (z_{j+1} - z_j)^2 \right)
$$

$$
+ \text{ higher-order terms.}
$$

Therefore, the potential energy, under the hypothesis of small deformations, can be written as

$$
\frac{\kappa \, (N+1)}{2L_0} \sum_{j=0}^{N} \left((x_{j+1} - x_j)^2 + \frac{\delta\ell}{\ell} (y_{j+1} - y_j)^2 + \frac{\delta\ell}{\ell} (z_{j+1} - z_j)^2 \right).
$$

In order to make our equations more palatable, we introduce other physical quantities. The linear density of mass is denoted by ρ, so that the total mass of the spring is ρL. Observe that we have taken the stretched length as the reference. The tension per unit length of the spring at equilibrium is f_0 given by

$$
f_0 = \frac{\kappa \, (L - L_0)}{L_0}.
$$

With these notations, the equations of motion for small vibrations can now be written as follows:

$$
(18.3.2a) \qquad \frac{L\rho}{N} \ddot{x}_j - \frac{\kappa \, (N+1)}{L_0} (x_{j+1} - 2x_j + x_{j-1}) = 0,
$$

$$
(18.3.2b) \qquad \frac{L\rho}{N} \ddot{y}_j - \frac{(N+1) \, f_0}{L} (y_{j+1} - 2y_j + y_{j-1}) = 0,
$$

$$
(18.3.2c) \qquad \frac{L\rho}{N} \ddot{z}_j - \frac{(N+1) \, f_0}{L} (z_{j+1} - 2z_j + z_{j-1}) = 0.
$$

What is interesting is that the above three equations are decoupled; eqns (18.3.2b) and (18.3.2c) are identical, whilst eqn (18.3.2a) is different. Of course, since we dropped all of the annoying nonlinear coupling terms, we may have lost the most interesting features of the problem; but one has to start somewhere...

If we multiply eqns (18.3.2) by N/L, we recognize that the expression

$$
\frac{(x_{j+1} - 2x_j + x_{j-1}) \, N \, (N+1)}{L^2}
$$

and its analogues are very close to a central finite difference of the second order. Therefore, as N tends to infinity, the formal limit of eqns (18.3.2) is

$$
\rho \frac{\partial^2 x}{\partial t^2} - \frac{\kappa L}{L_0} \frac{\partial^2 x}{\partial s^2} = 0, \quad \rho \frac{\partial^2 y}{\partial t^2} - f_0 \frac{\partial^2 y}{\partial s^2} = 0,
$$

the equation for z being identical to the equation for y. Here s is the spatial coordinate. Moreover, we expect x, y, and z to vanish for $s = 0$ and $s = L$.

18.3.2. Elementary facts about the wave equation

The Cauchy problem for the wave equation in $\mathbb{R}$ consists of solving eqn (18.3.1) together with the initial conditions

$$(18.3.3) \qquad u(x,0) = u_0(x), \quad u_t(x,0) = u_1(x), \quad x \in \mathbb{R}.$$

It turns out that there is a completely explicit solution, due to D'Alembert.
    Define a new variable v by

$$(18.3.4) \qquad\qquad\qquad v = \frac{1}{c} u_t + u_x.$$

We see immediately that v satisfies the equation

$$(18.3.5) \qquad\qquad\qquad \frac{1}{c} v_t - v_x = f.$$

This means that the wave equation in one dimension reduces to two successive advection equations. We integrate eqn (18.3.5), using the relation (18.1.11), and we find

$$(18.3.6) \quad v(x,t) = \frac{u_1(x+ct)}{c} + u_{0,x}(x+ct) = c \int_0^t f(x + c(t-s), s)\, ds.$$

Then, we integrate eqn (18.3.4), which gives

$$(18.3.7) \qquad u(x,t) = u_0(x-ct) + c \int_0^t v(x + c(s-t), s)\, ds.$$

We substitute into the relation (18.3.7) the value of v given by eqn (18.3.6). We observe that

$$u_0(x-ct) + c \int_0^t \left(\frac{u_1(x+2cs-ct)}{c} + u_{0,x}(x+2cs-ct) \right) ds$$

$$= \frac{u_0(x-ct) + u_0(x+ct)}{2} + \frac{1}{2c} \int_{x-ct}^{x+ct} u_1(y)\, dy.$$

Similarly, the expression involving f is given by

$$c^2 \int_0^t \int_0^s f(x + c(2s-t-r), r)\, dr\, ds$$

and, after the change of variable $y = x + c(2s-t-r)$, this expression becomes

$$\frac{c}{2} \int_0^t \int_{x-cs}^{x+cs} f(y,s)\, dy\, ds.$$

Finally, the solution of eqn (18.3.1), together with the Cauchy data (18.3.3), is given by

$$
\begin{aligned}
u\left(x,t\right) = {} & \frac{u_0\left(x-ct\right)+u_0\left(x+ct\right)}{2} + \frac{1}{2c}\int_{x-ct}^{x+ct} u_1\left(y\right)\mathrm{d}y \\
& + \frac{c}{2}\int_0^t \int_{x-cs}^{x+cs} f\left(y,s\right)\mathrm{d}y\,\mathrm{d}s.
\end{aligned}
$$

(18.3.8)

Formula (18.3.8) has been obtained under the assumption that f is continuous, u_1 is of class C^1, and u_0 is of class C^2, and it produces a classical solution, i.e., a solution of class C^2 of the wave equation. However, formula (18.3.8) makes sense under much milder assumptions. But, in that case, it does not give a classical solution, but a generalized solution.

If the right-hand side f of eqn (18.3.1) vanishes, the solution is the superposition of two functions $g(x - ct)$ and $h(x + ct)$, whose values can be found from the initial conditions, see Exercise 18.5.25.

When the data have symmetries, these symmetries are transmitted to the solution. Thus, if the data are even, odd or periodic with respect to space, then the solution has the same properties. These facts are proved in Exercises 18.5.26 and 18.5.27. These symmetries enable us to solve the wave equation on an interval, with homogeneous Dirichlet boundary conditions (Exercise 18.5.28) or Neumann boundary conditions (Exercise 18.5.29).

The notion of domain of influence or of dependence is deeper and more important. The solution at the point (x,t) depends only on the data in the set $\{(y,s) : s \leqslant t - |x - y|/c\}$, called the cone of dependence of the point (x,t). Conversely, the data at (x,t) can influence only the solution at points in the set $\{(y,s) : s \geqslant t + |x - y|/c\}$ and this set is called the cone of influence.

These two properties make very precise the fact of propagation in the wave equation: the effect of a disturbance at (x,t) cannot be felt at (y,s) unless $s \geqslant t + |x - y|/c$, i.e., such a disturbance does not propagate faster than c. However, in dimension 1, this effect may linger for all time. If we lived in spatial dimension 1, it would be quite inefficient to transmit information by sound, since it would not be very well localized. It is easy to experience this effect in our three-dimensional world, for example, talk in a long corridor with hard walls, such as a mine gallery, or a large metal pipe, and listen to the sound!

18.3.3. A numerical scheme for the wave equation

We consider the following elementary numerical scheme for the wave equation:

$$
\frac{1}{c^2}\frac{U_j^{n+1} - 2U_j^n + U_j^{n-1}}{\delta t^2} - \frac{U_{j+1}^n - 2U_j^n + U_{j-1}^n}{\delta x^2} = F_j^n.
$$

(18.3.9)

We will prove the following stability result:

Theorem 18.3.1. If $c\delta t/\delta x \leqslant 1$, the scheme (18.3.9) is stable in the ℓ^2 norm. More precisely, if the right-hand side F_j^n vanishes and, for all $\alpha \in\ ]0,1[$, there exists a constant C such that, for all δt and δx satisfying the relation $c\delta t/\delta x \leqslant 1 - \alpha$, then

$$\sup_{n \geqslant 2} \sum_{j \in \mathbb{Z}} |U_j^n|^2\, \delta x \leqslant C \sum_{j \in \mathbb{Z}} \left(|U_1^n|^2 + |U_0^n|^2 \right) \delta x. \qquad \diamond$$

Proof.   The proof of this result uses the following fact: if $a = (a_j)_{j \in \mathbb{Z}}$ belongs to $\ell^2(\mathbb{Z})$, i.e.,

$$\|a\| = \left(\sum_{j \in \mathbb{Z}} |a_j|^2\, \delta x \right)^{1/2} < \infty,$$

then we can define the Fourier transform of the sequence a by

$$(18.3.10) \qquad \hat{a}\,(\xi) = \sum_{j \in \mathbb{Z}} a_j e^{-2i\pi j \xi \delta x}\, \delta x.$$

Definition (18.3.10) is obviously a discretization of the Fourier transform of a function. However, we need only the theory developed in Chapter 7 and a change of scale to obtain the inversion formula

$$a_j = \int_0^{\delta x} \hat{a}\,(\xi)\, e^{2i\pi j \xi \delta x}\, \mathrm{d}\xi.$$

Moreover, after a change of scale, the Parseval identity (7.1.16) can be rewritten as

$$(18.3.11) \qquad \sum_{j \in \mathbb{Z}} |a_j|^2\, \delta x = \int_0^{1/\delta x} |\hat{a}\,(\xi)|^2\, \mathrm{d}\xi.$$

Another property will be important: the Fourier transform of the sequence b, defined by $b_j = a_{j+1}$, is readily computed as

$$\sum_{j \in \mathbb{Z}} a_{j+1} e^{-2i\pi j \xi \delta x}\, \delta x = \sum_{j \in \mathbb{Z}} a_{j+1} e^{-2i\pi (j+1)\xi \delta x}\, \delta x\, e^{2i\pi \xi \delta x}$$

$$= \hat{a}\,(\xi)\, e^{2i\pi \xi \delta x}.$$

We apply a Fourier transform in j to the scheme (18.3.9), defining $\hat{U}^n(\xi)$ to be the Fourier transform of $(U_j^n)_j$:

$$(18.3.12) \qquad \frac{1}{c^2} \frac{\hat{U}^{n+1} - 2\hat{U}^n + \hat{U}^{n-1}}{\delta t^2} = \frac{\left(e^{2i\pi \xi \delta x} + e^{-2i\pi \xi \delta x} - 2 \right) \hat{U}^n\,(\xi)}{\delta x^2}.$$

We let

$$\omega\,(\xi) = \frac{2c^2 \delta t^2\,(1 - \cos\,(2\pi \xi \delta x))}{\delta x^2}.$$

With this notation, we can rewrite relation (18.3.12) in the matrix form

$$\begin{pmatrix} \hat{U}^{n+1} \\ \hat{U}^n \end{pmatrix} = \begin{pmatrix} 2 - \omega & -1 \\ 1 & 0 \end{pmatrix} \begin{pmatrix} \hat{U}^n \\ \hat{U}^{n-1} \end{pmatrix}.$$

Then, if we can prove that all the positive powers of the matrix

$$M(\xi) = \begin{pmatrix} 2 - \omega(\xi) & -1 \\ 1 & 0 \end{pmatrix}$$

are bounded, our result is a consequence of the identity (18.3.11). The characteristic polynomial of this matrix is

$$X^2 - X(2 - \omega) + 1$$

and its roots, the eigenvalues of M, are of modulus at most 1 for all ξ, if and only if $\omega \leqslant 4$, as the reader can check. Then, they are the conjugate complex numbers λ_+ and λ_- given by

$$\lambda_\pm = \frac{1}{2}\left(2 - \omega \pm i\sqrt{(4 - \omega)\omega}\right).$$

However, we cannot conclude from the bound on the powers of $\lambda_\pm$ that the powers of M are bounded, since M is not a normal matrix.

The powers of M are given by

$$M^n = \frac{1}{\lambda_+ - \lambda_-} \begin{pmatrix} 1 & -\lambda_- \\ -1 & \lambda_+ \end{pmatrix} \begin{pmatrix} \lambda_+^n & 0 \\ 0 & \lambda_-^n \end{pmatrix} \begin{pmatrix} \lambda_+ & \lambda_- \\ 1 & 1 \end{pmatrix}.$$

It is clear now that, if $1 - c\delta t/\delta x$ is bounded away from 0, then M^n is bounded independently of $n \geqslant 1$ and of $\xi \in \mathbb{R}$. This concludes the proof of the theorem.

□

18.4. The heat equation and separation of variables

In this last section, we are going to apply a number of the techniques already described to the solution of another partial differential equation, the heat equation. Furthermore, we are going to show how to approximate it numerically.

18.4.1. Derivation of the heat equation

We begin with a little physics to understand how the heat equation is derived. The *Théorie Analytique de la Chaleur* by J. Fourier (1822) [32] is the classic work in this area. The so-called Fourier series was known of by Daniel Bernoulli and by Euler, but the Fourier integral is really due to Fourier, contrary to the usual rule that new results and concepts rarely carry the name of their author. Nevertheless, Cauchy made considerable contributions to Fourier theory, to the point that, if Fourier was the first to announce the inversion formula which bears his name, then Cauchy published its proof before Fourier.

We begin by the modelling phase which is inherent in the understanding of every physical problem. To give the explanations which follow, I am aided on the one hand by Fourier's book and on the other hand by *Thermodynamique* by G. Bruhat [11].

When we put two solid bodies, at temperatures Θ_1 and Θ_2, in contact for a long time within an isolated enclosure their temperatures tend to equilibrate. If the two bodies are of the same mass and the same composition, the final temperature is half the sum of the initial temperatures. If the compositions are identical but the masses m_1 and m_2 are different, the final temperature is

$$\Theta_{\text{final}} = \frac{m_1\Theta_1 + m_2\Theta_2}{m_1 + m_2}.$$

If the compositions are different, the final temperature is

$$\Theta_{\text{final}} = \frac{m_1 C_1 \Theta_1 + m_2 C_2 \Theta_2}{m_1 C_1 + m_2 C_2}.$$

The numbers C_1 and C_2 are the specific heats of the two bodies. They are physical characteristics of the bodies and they describe the capacity of the bodies to store energy in the form of heat. The measurements for solids are made at constant pressure so that these are specific heats at constant pressure. By definition, the quantity of heat stored in a body, of specific heat C and of mass m, which changes temperature from Θ to the temperature $\Theta + \Delta\Theta$ is $mC\Delta\Theta$. This is a positive or negative quantity which has the dimensions of energy.

We move on now to the notion of heat flux. Imagine a homogeneous body which fills the interval between two infinite parallel planes P_1 and P_2 separated by unit distance. These planes are maintained at the temperatures Θ_1 and Θ_2, respectively. If the body is in a steady state, its temperature is constant in each of the planes parallel to the boundary planes and it is an affine function of the distance x from the boundary plane P_1, as shown in Figure 18.6

The quantity of heat which crosses any plane parallel to P_1 and of unit area during one second is independent of the distance of this surface from P_1. If $\Theta_1 - \Theta_2$ is one degree and the planes are separated by unit distance then this heat flux is equal to a certain constant K which depends on the chosen units. Consequently, if S denotes the area of the surface across which the heat passes and the two planes are separated by distance L, the quantity of heat passing through the surface in time Δt is

$$\frac{K\Delta t\,(\Theta_1 - \Theta_2)\,S}{L}.$$

Suppose now that the distribution of temperature in the slice situated between the two planes is not steady, but is, nevertheless, constant in each plane

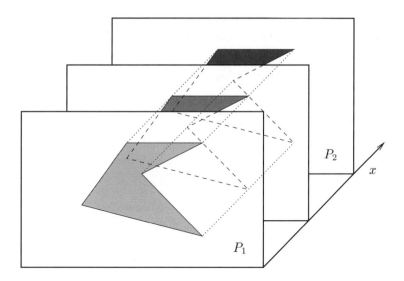

Figure 18.6: Parallel planes having distinct temperatures and a prism of unit cross-sectional area.

parallel to P_1. Also, suppose that the plane is cut into J slices parallel to P_1, each of thickness $h = \Delta x$. As the temperature is constant in the planes parallel to P_1, there is no heat flux, except in the x-direction. We can, therefore, consider a prismatic domain D of unit cross-sectional area and with sides perpendicular to P_1 and P_2. There is no heat flux across the sides of D. In each slice numbered j, and for each $jh \leqslant x \leqslant (j+1)h$, we are going to consider the temperature to be the constant value Θ_j. The quantity of heat coming from slice $j+1$ and entering slice j is therefore

$$\frac{K\Delta t\,(\Theta_{j+1} - \Theta_j)}{\Delta x}$$

in the interval of time Δt. In the same way, the quantity of heat entering slice j and coming from slice $j-1$ is

$$\frac{K\Delta t\,(\Theta_{j-1} - \Theta_j)}{\Delta x}$$

during the interval of time Δt. During the time interval Δt, the quantity of heat

(18.4.1) $$\Delta Q_j = \frac{K\Delta t\,(\Theta_{j+1} - 2\Theta_j + \Theta_{j-1})}{\Delta x}$$

enters slice j and so the temperature of this slice will increase by $\Delta\Theta_j$, given by

(18.4.2) $$\Delta\Theta_j = \frac{\Delta Q_j}{Cm}.$$

Here m is the mass of the slice of thickness Δx. This has the value $d\Delta x$, since the volume of this slice is Δx, its base having unit area, and d is the density of the body considered. We therefore have

$$(18.4.3) \qquad \frac{\Delta \Theta_j}{\Delta t} = \frac{K}{Cd} \frac{\Theta_{j+1} - 2\Theta_j + \Theta_{j-1}}{\Delta x^2}.$$

If we pass to the formal limit in the relation (18.4.3), that is, if we make Δt and Δx tend to zero, we have

$$(18.4.4) \qquad \frac{\partial \theta}{\partial t} = \frac{K}{Cd} \frac{\partial^2 \theta}{\partial x^2},$$

which is the heat equation. This equation is satisfied for x in $[0, d]$ and for $t \in [0, T]$. We are also going to take account of the conditions at the boundary: on the planes P_1 and P_2 we fix the temperature at the values θ_1 and θ_2 respectively, that is

$$(18.4.5) \qquad \theta(0, t) = \theta_1, \quad \theta(L, t) = \theta_2.$$

We move on to a model of a cylindrical bar of length L, immersed in an infinite medium of fixed temperature $\bar{\Theta}$. The cross-section of the bar is not necessarily circular. We can simplify the modelling by supposing that the temperature is constant in each cross-section of the bar. This is a reasonable approximation if the bar is not very thick. For the transfers between the elements of the bar we can reuse the preceding model, but we must additionally take account of the transfer with the exterior medium, which is given by

$$\Delta Q'_j = -K'\ell\Delta x\Delta t \left(\Theta_j - \bar{\Theta}\right)$$

during the time Δt. Here ℓ is the perimeter of the cross-section of the bar and K' is a constant which describes the efficiency of the transfer with the exterior medium. It goes without saying that this loss of heat is proportional to the area $\ell\Delta x$ of the element of the bar that we are considering. We will then have

$$Cd\Delta x\Delta \Theta_j = K\Delta t \frac{\Theta_{j+1} - 2\Theta_j + \Theta_{j-1}}{\Delta x} - K'\ell\Delta x\Delta t(\Theta_j - \bar{\Theta}),$$

and, therefore,

$$\frac{\Delta \Theta_j}{\Delta t} = \frac{K}{Cd} \frac{\Theta_{j+1} - 2\Theta_j + \Theta_{j-1}}{\Delta x^2} - \frac{K'\ell}{Cd} \left(\Theta_j - \bar{\Theta}\right).$$

It will be useful to rewrite this relation highlighting the discrete time $n\Delta t$. It then becomes

$$(18.4.6) \qquad \frac{\Theta_j^{n+1} - \Theta_j^n}{\Delta t} = \frac{K}{Cd} \frac{\Theta_{j+1}^n - 2\Theta_j^n + \Theta_{j-1}^n}{\Delta x^2} - \frac{K'\ell}{Cd} \left(\Theta_j^n - \bar{\Theta}\right).$$

Passing to the formal limit, as we did to obtain eqn (18.4.4), we have

$$(18.4.7) \qquad \frac{\partial \theta}{\partial t} = \frac{K}{Cd} \frac{\partial^2 \theta}{\partial x^2} - \frac{K'\ell}{Cd} (\theta - \bar{\Theta}).$$

We can pose the same boundary conditions as previously.

To simplify the solution of our problem, we are going to suppose that $\theta_1 = \theta_2$ is a temperature independent of time and, by means of a translation of the temperature scale, suppose that this temperature is zero. Our conditions at the ends of the bar thus become

$$(18.4.8) \qquad \theta(0, t) = \theta(L, t) = 0.$$

We will make a second simplification which consists of supposing that the exterior temperature is equal to the temperature at the extremities and, therefore, $\bar{\Theta} = 0$.

We are also going to rewrite our constants so that eqn (18.4.7) becomes

$$(18.4.9) \qquad \frac{\partial \theta}{\partial t} = a \frac{\partial^2 \theta}{\partial x^2} - b\theta,$$

with a strictly positive and b positive or zero. This amounts to choosing

$$a = \frac{K}{Cd} \quad \text{and} \quad b = \frac{K'\ell}{Cd},$$

with the convention that $b = 0$ when we are in the situation of an infinite medium.

We are first of all going to show that eqn (18.4.9) possesses a solution if we know the initial temperature distribution and if we fix the conditions (18.4.8) at the ends. Then, we are going to justify passing to the limit in the relation (18.4.6) and we will show that eqn (18.4.6) is a numerical scheme which actually allows us to approximate the solutions of eqn (18.4.9).

18.4.2. Seeking a particular solution by separation of variables

We seek a solution of eqn (18.4.9) in the form

$$(18.4.10) \qquad \theta(x, t) = X(x) T(t).$$

We say that such a solution is in separated variables. In this case

$$\frac{\partial \theta}{\partial t} - a \frac{\partial^2 \theta}{\partial x^2} + b\theta = X(x) T'(t) - aX''(x) T(t) + bX(x) T(t) = 0.$$

If we divide the last equality by XT, which we suppose to be nonzero, we have

$$\frac{T'(t)}{T(t)} + b = a \frac{X''(x)}{X(x)}.$$

To the left of the equals sign, we find a function which depends only on t and, to the right of this sign, we find a function which depends only on x. For them to be equal, it is necessary and sufficient that there exists a constant λ such that

(18.4.11)
$$\frac{T'(t)}{T(t)} + b = -a\lambda,$$

(18.4.12)
$$\frac{X''(x)}{X(x)} = -\lambda.$$

Furthermore, we are going to impose that our particular solution satisfies the boundary conditions (18.4.8), which gives us here

(18.4.13)
$$X(0) = X(L) = 0.$$

It is clear that eqn (18.4.12) has solutions of the form

$$X(x) = \alpha e^{zx} + \beta e^{-zx},$$

provided that $\lambda = -z^2$. The boundary condition $X(0) = 0$ requires that

$$\alpha + \beta = 0$$

and the boundary condition $X(L) = 0$ implies that

$$e^{zL} - e^{-zL} = 0,$$

if we always exclude the uninteresting case $\lambda = 0$. We therefore have

$$zL = i\, m\pi,$$

with m in $\mathbb{Z}$. It follows that X is necessarily of the form

$$X(x) = \alpha \sin\left(\frac{m\pi x}{L}\right).$$

From this we have the following value of λ:

$$\lambda = \frac{m^2\pi^2}{L^2}$$

and, therefore, the following expression for T:

$$T(t) = \beta \exp\left(-\frac{am^2\pi^2 t}{L^2} - bt\right).$$

We have thus found a particular solution of the heat equation of the form

(18.4.14)
$$\theta_m(x, t) = \sin\left(\frac{m\pi x}{L}\right) \exp\left(-\frac{am^2\pi^2 t}{L^2} - bt\right),$$

and this particular solution satisfies the boundary conditions (18.4.8). Note the parallels between the calculation which has just been done with that of Subsection 11.2.1 for a finite difference matrix.

18.4.3. Solution by Fourier series

As θ_m is a solution of eqn (18.4.9), every linear combination of functions θ_m is a solution of eqn (18.4.9). Taking an infinite linear combination of such functions amounts to writing the solutions in the form of a series and studying their convergence.

We therefore seek $\theta(x,t)$ of the form

$$(18.4.15) \qquad \theta(x,t) = \sum_{m=1}^{\infty} c_m \theta_m(x,t).$$

For the initial conditions to be satisfied we must have

$$(18.4.16) \qquad \theta(x,0) = \sum_{m=1}^{\infty} c_m \sin\left(\frac{m\pi x}{L}\right).$$

In other words, it suffices that $\theta(\cdot,0)$ can be expanded as a sine series.

By a theory analogous to the theory of complex exponential Fourier series, we see that, if $\theta(\cdot,0)$ is in $L^2(0,L)$, then the coefficients c_m are defined by

$$c_m = \frac{2}{L}\int_0^L \theta(y,0)\sin\left(\frac{m\pi y}{L}\right)dy$$

and the series defined by the right-hand side of eqn (18.4.16) converges in $L^2(0,L)$ to its sum, which is $\theta(\cdot,0)$. Under this hypothesis, we have a very strong regularity result:

Lemma 18.4.1. Suppose that

$$(18.4.17) \qquad \sum_{m=0}^{\infty} c_m^2 < +\infty.$$

The relation (18.4.15) defines a function θ which is infinitely differentiable on the set $[0,L] \times]0,\infty[$ and which satisfies

$$(18.4.18) \qquad \int_0^L \theta(x,t)^2 \, dx \leqslant \int_0^L \theta(x,0)^2 \, dx$$

and

$$(18.4.19) \qquad \lim_{t \to 0}\int_0^L (\theta(x,t) - \theta(x,0))^2 \, dx = 0.$$

Proof. If $u(\cdot,0)$ is in $L^2(0,L)$ then, for every $t > 0$ and for every $p \in \mathbb{N}$,

$$\sum_{m=1}^{\infty} |c_m| \exp\left(-\frac{am^2\pi^2 t}{L^2} - bt\right)$$

$$\leqslant \left(\sum_{m=1}^{\infty} c_m^2\right)^{1/2}\left(\sum_{m=1}^{\infty}\exp\left(-\frac{2am^2\pi^2 t}{L^2} - 2bt\right)\right)^{1/2} < +\infty,$$

since the sum of the exponential terms clearly converges more rapidly than geometrically, if $t > 0$. Therefore, for each $t > 0$, the series (18.4.15) converges uniformly towards its limit, which is therefore continuous. In fact, we have more, since, for every $p > 0$,

$$\sum_{m=1}^{\infty} |c_m| \, m^p \exp\left(-\frac{am^2\pi^2 t}{L^2} - bt\right)$$

$$\leqslant \left(\sum_{m=1}^{\infty} c_m^2\right)^{1/2} \left(\sum_{m=1}^{\infty} m^{2p} \exp\left(-\frac{2am^2\pi^2 t}{L^2} - 2bt\right)\right)^{1/2} < +\infty$$

and, consequently, the series

$$\sum_{m=1}^{\infty} c_m \left(\frac{m\pi}{L}\right)^p \sin^{(p)}\left(\frac{m\pi x}{L}\right) \exp\left(-\frac{am^2\pi^2 t}{L^2} - bt\right)$$

converges uniformly and defines the derivative $\partial^p\theta/\partial x^p$, for every $t > 0$. This shows us that, for every fixed $t > 0$, the function θ given by eqn (18.4.15) is infinitely differentiable with respect to x, for every $t > 0$. We note that the following series is uniformly convergent:

$$(-1)^p \sum_{m=1}^{\infty} c_m \left(\frac{am^2\pi^2}{L^2} + b\right)^p \sin\left(\frac{m\pi x}{L}\right) \exp\left(-\frac{am^2\pi^2 t}{L^2} - bt\right).$$

By the application of Lebesgue's theorem on the differentiation of integrals depending on a parameter to the case of series, we see that the above expression defines $\partial^p\theta/\partial t^p$. This shows us that, for each $t > 0$, $\partial^p\theta/\partial t^p$ is a continuous function. The reader can easily convince herself that it is also a C^∞ function, noting the convergence of the expressions defining $\partial^{p+q}\theta/\partial x^p \partial t^q$. Finally, another application of Lebesgue's theorem for series shows us that all of the expressions

$$(-1)^q \sum_{m=1}^{\infty} c_m \left(\frac{m\pi}{L}\right)^p \left(\frac{am^2\pi^2}{L^2} + b\right)^q \sin^{(p)}\left(\frac{m\pi x}{L}\right) \exp\left(-\frac{am^2\pi^2 t}{L^2} - bt\right)$$

define continuous functions on $[0, L] \times \,]0, \infty[$.

A theorem from differential calculus allows us to confirm that, if a function of n variables is separately differentiable with respect to each of its n arguments and if its derivatives are continuous functions, then this function is continuously differentiable with respect to the set of its variables. We have thus proved that θ is infinitely differentiable in $[0, L] \times \,]0, \infty[$.

The estimate (18.4.18) comes from Plancherel's formula:

$$\int_0^L \theta\,(x,t)^2 \, \mathrm{d}x = \frac{L}{2} \sum_{m=1}^{\infty} c_m^2 \exp\left(-\frac{2am^2\pi^2 t}{L^2} - 2bt\right)$$

$$\leqslant \frac{L}{2} \sum_{m=1}^{\infty} c_m^2 = \int_0^L \theta\,(x,0)^2 \, \mathrm{d}x.$$

We obtain the continuity relation (18.4.19) by rewriting Plancherel's formula as follows:

$$\int_0^L \left(\theta\left(x,t\right) - \theta\left(x,0\right)\right)^2 \, dx = \frac{L}{2} \sum_{m=1}^{\infty} c_m^2 \left(1 - \exp\left(-\frac{am^2\pi^2t}{L^2} - bt\right)\right)^2.$$

We then note that, by the application of Lebesgue's convergence theorem for series, we obtain the desired result. □

18.4.4. Relation between the heat equation and the discrete model

The relation (18.4.6) can be seen as the time discretization by an Euler scheme of the matrix differential equation

$$\frac{d\Theta}{dt} = A\Theta,$$

where the $J \times J$ matrix A is given by

$$(18.4.20) \qquad A = \frac{a}{\Delta x^2} \begin{pmatrix} -2 & 1 & 0 & \cdots & 0 \\ 1 & -2 & 1 & & 0 \\ \vdots & \ddots & \ddots & \ddots & \vdots \\ 0 & & 1 & -2 & 1 \\ 0 & & 0 & 1 & -2 \end{pmatrix} - bI.$$

The negative of the matrix with the factor $a\Delta x^{-2}$ has already been studied in Subsection 11.2.1. The matrix A is obviously symmetric and its eigenvalues are

$$-4\frac{a}{\Delta x^2} \sin^2\left(\frac{m\pi}{2\left(J+1\right)}\right) - b.$$

These eigenvalues are strictly negative and we therefore have

$$\Theta\left(t\right) = e^{tA}\Theta\left(0\right).$$

We are going to show that the discrete solution (18.4.6) is a good approximation to the continuous solution.

Theorem 18.4.2. Let θ be the solution of eqn (18.4.9) with an initial condition $\theta(x,0)$ which satisfies

$$\sum_{m=1}^{\infty} m^{10} c_m^2 < +\infty,$$

and let Θ_j^n be defined by the relation (18.4.6) and the initial condition

$$\Theta_j^0 = \theta\left(j\Delta x,0\right).$$

Then, there exists a constant C'', which depends only on θ and the data of the problem $\theta(\cdot, 0)$, a, and b, such that

$$\left| \Theta_j^n - \theta\left(j\Delta x, n\Delta t\right) \right| \leqslant C'' \Delta x^2 e^{xb\Delta t},$$

provided that we have the inequality

$$\Delta t \leqslant \frac{\Delta x^2}{2a}. \qquad \diamond$$

Proof.   We rewrite the relation (18.4.6) using the constants a and b as follows:

$$\frac{\Theta_j^{n+1} - \Theta_j^n}{\Delta t} = a\frac{\Theta_{j+1}^n - 2\Theta_j^n + \Theta_{j-1}^n}{\Delta x^2} - b\Theta_j^n.$$

Let

$$\Psi_j^n = \theta\left(j\Delta x, n\Delta t\right),$$

if θ is the exact solution of eqn (18.4.9).

   We calculate

$$\varepsilon_j^n = \frac{\Psi_j^{n+1} - \Psi_j^n}{\Delta t} - a\frac{\Psi_{j+1}^n - 2\Psi_j^n + \Psi_{j-1}^n}{\Delta x^2} + b\Psi_j^n.$$

We are going to suppose for this that $\theta(x, 0)$ is sufficiently regular, so that $\theta(x, t)$ has the following properties:

$$\max_{(x,t)\in[0,L]\times\mathbb{R}^+}\left\{\left|\frac{\partial^4\theta}{\partial x^4}(x,t)\right| + \left|\frac{\partial^2\theta}{\partial t^2}(x,t)\right|\right\} \leqslant C < \infty.$$

We can fulfil these conditions by demanding that

$$\sum_{m=1}^{\infty} m^{10}c_m^2 < +\infty,$$

as the reader can verify. Under these conditions, a Taylor expansion shows that

$$\theta\left(j\Delta x, (n+1)\Delta t\right) - \theta\left(j\Delta x, n\Delta t\right) = \Delta t\frac{\partial\theta}{\partial t}\left(j\Delta x, n\Delta t\right) + O\left(\Delta t^2\right)$$

and

$$\theta\left((j+1)\Delta x, n\Delta t\right) - 2\theta\left(j\Delta x, n\Delta t\right) + \theta\left((j-1)\Delta x, n\Delta t\right)$$
$$= \frac{\partial^2\theta}{\partial x^2}\left(j\Delta x, n\Delta t\right) + O\left(\Delta x^4\right).$$

In the above two expressions, the terms $O(\Delta t^2)$ and $O(\Delta x^4)$ are bounded by $C\Delta t^2$ and $C\Delta x^4$, respectively, where C is a constant independent of j and n. We can therefore write, with the notation (18.4.20),

$$\Theta^{n+1} - \Psi^{n+1} = (I + A\Delta t)(\Theta^n - \Psi^n) - \Delta t\,\varepsilon^n$$

We note that

$$
I + \frac{a\Delta t}{\Delta x^2}
\begin{pmatrix}
-2 & 1 & 0 & \cdots & 0 \\
1 & -2 & 1 & & 0 \\
\vdots & \ddots & \ddots & \ddots & \vdots \\
0 & & 1 & -2 & 1 \\
0 & & 0 & 1 & -2
\end{pmatrix}
$$

has positive elements provided that

(18.4.21) $$\lambda = \frac{a\Delta t}{\Delta x^2} \leqslant \frac{1}{2},$$

a condition that we will suppose to be satisfied from now on. Under this conditions, we have the following estimate for ε_j^n:

(18.4.22) $$\left| \varepsilon_j^n \right| \leqslant C' \Delta x^2.$$

We explicitly calculate $\Theta_j^{n+1} - \Psi_j^{n+1}$ as follows:

$$
\Theta_j^{n+1} - \Psi_j^{n+1} = \left(\Theta_{j-1}^n - \Psi_{j-1}^n \right)(1 + \lambda) + \left(\Theta_j^n - \Psi_j^n \right)(1 - 2\lambda) \\
+ \left(\Theta_{j+1}^n - \Psi_{j+1}^n \right)(1 + \lambda) - b\Delta t \left(\Theta_j^n - \Psi_j^n \right) - \Delta t\, \varepsilon_j^n.
$$

We then see that, using the condition (18.4.21),

$$
\max_j \left| \Theta_j^{n+1} - \Psi_j^{n+1} \right| \leqslant (1 + b\Delta t) \max_j \left| \Theta_j^n - \Psi_j^n \right| + \Delta t \max_j \left| \varepsilon_j^n \right|.
$$

By applying the discrete form of Gronwall's lemma (Lemma 16.1.7) and the estimate (18.4.22), we obtain

$$
\max_j \left| \Theta_j^n - \Psi_j^n \right| \leqslant e^{nb\Delta t} \left[\max_j \left| \Theta_j^0 - \Psi_j^0 \right| + C'' \Delta x^2 \right].
$$

Here C'' is a constant which depends only on a, b, and θ. This allows us to conclude the proof of convergence. $\qquad\square$

There are many more exciting and important results in the field of the numerical analysis of partial differential equations; not one, but several more books are needed. If this chapter has led the reader to ask for more, my aim will have been fulfilled.

18.5. Exercises from Chapter 18

18.5.1. The eigenvectors of a strictly hyperbolic matrix

Exercise 18.5.1. Let $\mathcal{R}$ be the set of $n \times n$ real matrices with n distinct real eigenvalues. For $M \in \mathcal{R}$, let $\lambda_j(M)$, $1 \leqslant j \leqslant n$, be the eigenvalues of M, arranged in increasing order. Show that the projection $P_j(M)$ onto the eigenspace relative to the eigenvalue λ_j is a continuous function of M. Conclude, therefore, that, for all matrices $M_0 \in \mathcal{R}$, it is possible to find a neighbourhood of M_0 and a choice of eigenvectors $r_j(M)$ and $l_j(M)$ depending continuously on M and having the properties

$$\left. \begin{aligned} M r_j(M) &= \lambda_j(M) r_j(M) \\ l_j(M)^\top M &= \lambda_j(M) l_j(M)^\top \end{aligned} \right\} \qquad \forall j = 1, \ldots, n,$$

$$l_j(M)^\top r_k(M) = \delta_{jk}, \qquad\qquad \forall j, k = 1, \ldots, n.$$

Hint: start with $r_j(M_0)$ and $l_j(M_0)$ having the required properties. Define $r_j(M) = P_j(M) r_j(M_0)$ for M close enough to M_0 and construct the corresponding vectors l_j.

Exercise 18.5.2. Show that, in fact, the dependence of P_j on M is of class C^∞.

Exercise 18.5.3. Assume that M is a mapping from an open subset $\mathcal{O}$ of $\mathbb{R}^d$ to $\mathcal{M}_n(\mathbb{R})$ which is of class C^1 and takes its values in $\mathcal{R}$. Show, with the help of a partition of unity, that it is possible to find, globally in $\mathcal{O}$, eigenvectors r_j and l_j which are functions of class C^1 over $\mathcal{O}$ and which satisfy the conditions

$$\left. \begin{aligned} M(x) r_j(x) &= \lambda_j(M(x)) r_j(x) \\ l_j(x)^\top M(x) &= \lambda_j(M(x)) l_j(x)^\top \end{aligned} \right\} \qquad \forall j = 1, \ldots, n,$$

$$l_j(x)^\top r_k(x) = \delta_{jk}, \qquad\qquad \forall j, k = 1, \ldots, n.$$

18.5.2. More on the upwind scheme

In this section, we consider various supplementary properties of the upwind scheme (18.2.5).

Exercise 18.5.4. Run some numerical simulations on the upwind scheme, with several different initial data and several different choices of the CFL number. Recommended initial data are:

- smooth functions, for instance piecewise polynomial, with high enough overall differentiability;

- square functions such as

$$u_0(x) = \begin{cases} 1 & \text{if } 0 \leqslant x \leqslant 1; \\ 0 & \text{otherwise.} \end{cases}$$

We did not work on boundary conditions for the advection equation. Due to the finite velocity of propagation, it is enough to simulate until the numerical wave hits the boundary of the integration domain, provided that the initial data have compact support. The integration time depends, therefore, on the distance from the support of the initial data to the ends of the spatial interval of integration.

Exercise 18.5.5. Assume that f vanishes and that u_0 is bounded and continuously differentiable with uniformly continuous derivative. Show that the upwind scheme converges.

Hint: use the modulus of continuity of u_0 to get estimates on the local consistency error.

Exercise 18.5.6. Assume that a depends on $x \in \mathbb{R}$ and $t \in [0, T]$. If a is strictly positive, define an upwind scheme by

$$\frac{U_j^{n+1} - U_j^n}{\delta t} + a\left(j\delta x, n\delta t\right) \frac{U_j^n - U_{j-1}^n}{\delta x} = 0.$$

If a is bounded, show that there is a constant $C > 0$ such that, for $\delta t / \delta x \leqslant C$, this scheme is stable. Find sufficient conditions on the regularity of a, u_0, f, and

$$\sup_j \left| U_j^n - u\left(j\delta x, n\delta t\right) \right|$$

which ensure the convergence of this scheme in $C^0(\mathbb{R} \times [0, T])$, i.e.,

$$\lim_{\substack{\delta x \to 0 \\ \delta t / \delta x \leqslant C}} \sum_{j \in \mathbb{Z}, n\delta t \leqslant T} \left| U_j^n - u\left(j\delta x, n\delta t\right) \right| = 0.$$

Hint: this is really the proof of Theorem 18.2.4.

Exercise 18.5.7. Generalize the study of Exercise 18.5.6 by introducing a right-hand side f, which will be assumed to be smooth enough to perform a convergence proof.

Exercise 18.5.8. In this exercise, a is a bounded function on $\mathbb{R} \times [0, T]$. We define an upwind scheme which changes according to the direction of the wind:

$$\frac{U_j^{n+1} - U_j^n}{\delta t} + \max\left(a\left(j\delta x, n\delta t\right), 0\right) \frac{U_j^n - U_{j-1}^n}{\delta x}$$
$$+ \min\left(a\left(j\delta x, n\delta t\right), 0\right) \frac{U_{j+1}^n - U_j^n}{\delta x} = 0.$$

Show that this scheme is stable and prove its convergence under sufficient conditions of regularity.

Exercise 18.5.9. Assume that a is strictly positive and that the following initial data is given for the downwind scheme (18.2.6):

$$U_j^0 = \begin{cases} 1 & \text{if } j = 0; \\ 0 & \text{otherwise.} \end{cases}$$

Let λ be the CFL number. Calculate explicitly the solution of the downwind scheme, show that it oscillates strongly, and show that it satisfies the following equivalence:

$$\sup_j \left| U_j^n \right| \sim \frac{C\left(2\lambda + 1\right)^{n+1}}{\sqrt{n}},$$

for n large.

Hint: apply the binomial formula and Stirling's asymptotic formula for the factorial.

18.5.3. Fourier analysis of difference schemes for the advection equation

Exercise 18.5.10. Let U_j be a square-integrable sequence indexed by $\mathbb{Z}$. Define its Fourier transform as

$$\hat{U}(\xi) = \sum_{j \in \mathbb{Z}} e^{-2i\pi j \xi \delta x} U_j \delta x.$$

Show that the mapping $U \mapsto \hat{U}$ is an isometry from $\ell^2(\mathbb{Z})$, equipped with the norm

$$|U| = \left(\sum_{j \in \mathbb{Z}} |U_j|^2 \, \delta x \right)^{1/2},$$

to $L^2(0, 1/\delta x)$, equipped with the standard norm

$$|\hat{U}| = \left(\int_0^{1/\delta x} \left| \hat{U}(\xi) \right|^2 \, d\xi \right)^{1/2}.$$

Hint: this is a Fourier series statement, with scale parameters differing from the standard ones used in Chapter 7.

Exercise 18.5.11. Let τ be the operator in $\ell^2(\mathbb{Z})$ defined by

$$(\tau U)_j = U_{j+1}.$$

Calculate the Fourier transform of τU.

Exercise 18.5.12. Consider the schemes (18.2.5)–(18.2.7). Denoting by $\hat{U}^n(\xi)$ the Fourier transform of $j \mapsto U_j^n$, give the transformation $\hat{U}^n(\xi) \mapsto \hat{U}^{n+1}(\xi)$ for each of these schemes. Show that it is described by a multiplication by a function depending on ξ. Show that the scheme (18.2.5) is stable in $\ell^2(\mathbb{Z})$ under the CFL condition, that the scheme (18.2.6) is never stable, and that the scheme (18.2.7) is stable under the condition

$$\delta t \leqslant C \delta x^2.$$

18.5.4. The Lax–Friedrichs scheme

The Lax–Friedrichs scheme for the advection eqn (18.2.1) is defined by the relation (18.2.15).

Exercise 18.5.13. Run some numerical simulations to understand the behaviour of the Lax–Friedrichs scheme. What happens if you take as initial data a square function? What happens if you take a smooth function? How is this different from the behaviour of the numerical approximation obtained by the upwind scheme?

Exercise 18.5.14. Show that the Lax–Friedrichs scheme is stable in $\ell^\infty(\mathbb{Z})$ and in $\ell^2(\mathbb{Z})$, if $\lambda = |a|\delta t/\delta x$ is at most equal to 1.

Exercise 18.5.15. Calculate the consistency error for the Lax–Friedrichs scheme and show the convergence of this scheme in $\ell^\infty(\mathbb{Z})$.

18.5.5. The Lax–Wendroff scheme

The Lax–Wendroff scheme for the advection eqn (18.2.1) is defined by the relation (18.2.16).

Exercise 18.5.16. Run some numerical simulations to get a feeling for what kind of approximation the Lax–Wendroff scheme gives, preferably on the same initial data and with the same initial data as for the Lax–Friedrichs scheme. What differences do you observe?

Exercise 18.5.17. Show that the Lax–Wendroff scheme is stable in $\ell^\infty(\mathbb{Z})$ and in $\ell^2(\mathbb{Z})$, if $\lambda = |a|\delta t/\delta x$ is at most equal to 1.

Exercise 18.5.18. Calculate the consistency error for the Lax–Wendroff scheme and show that it is of higher order in δx than the consistency error for the Lax–Friedrichs scheme or the upwind scheme. Show the convergence of the Lax–Wendroff scheme in $\ell^\infty(\mathbb{Z})$.

18.5.6. Stability of the leap-frog scheme

Consider the leap-frog scheme (18.2.17). Denote by $\hat{U}^n(\xi)$ the Fourier transform of $j \mapsto U_j^n$.

Exercise 18.5.19. Run the leap-frog scheme on the same type of initial data as for the Lax–Friedrichs or the Lax–Wendroff scheme. You have to initialize two vectors of data, U_j^0 and U_j^1; it is convenient to take simply $U_j^0 = u_0(j\delta x)$ and $U_j^1 = u_0(j\delta x - a\delta t)$.

Exercise 18.5.20. Show that there is a matrix $M(\xi)$ such that

$$\begin{pmatrix} \hat{U}^{n+1}(\xi) \\ \hat{U}^n(\xi) \end{pmatrix} = M(\xi) \begin{pmatrix} \hat{U}^n(\xi) \\ \hat{U}^{n-1}(\xi) \end{pmatrix}$$

and give the explicit expression for this matrix.

Exercise 18.5.21. Calculate the eigenvalues of $M(\xi)$. Deduce from this computation that a necessary condition for stability of the leap-frog scheme is that $\lambda = |a|\delta t/\delta x$ is at most equal to 1.

Exercise 18.5.22. Calculate a matrix $S(\xi)$ which diagonalizes $M(\xi)$. Show that $S(\xi)$ and $S(\xi)^{-1}$ are bounded uniformly in ξ if and only if λ is strictly less than 1.

Exercise 18.5.23. Show that a sufficient condition for stability in $\ell^2(\mathbb{Z})$ of the leap-frog scheme is that λ is strictly less than 1.

Exercise 18.5.24. What happens when λ is exactly equal to 1?

18.5.7. Elementary questions on the wave equation

Exercise 18.5.25. Knowing that the solution of eqn (18.3.1) is of the form

$$w\,(x - ct) + z\,(x + ct)$$

when f vanishes, find the values of w and z so as to satisfy the initial conditions (18.3.3).

Exercise 18.5.26. Assume that u_0 and u_1 are periodic, with period L, on $\mathbb{R}$ and that, for all t, $x \mapsto f(x,t)$ is periodic, with period L, on $\mathbb{R}$. Show that the solution of the wave eqn (18.3.1) with the initial data (18.3.3) is periodic, with period L, with respect to x.

Exercise 18.5.27. Assume that u_0, u_1, and f are even (respectively, odd) with respect to x. Show that u, the solution of eqn (18.3.1) with the initial data (18.3.3), is even (respectively, odd).

Exercise 18.5.28. Given v_0 and v_1 on $[0, L]$ and g on $[0, L] \times [0, \infty)$, define functions u_0, u_1, and f by the conditions

$$u_0\big|_{[0,L]} = v_0, \quad u_1\big|_{[0,L]} = v_1, \quad f\big|_{[0,L) \times [0,\infty)} = g,$$

and the requirement that u_0, u_1, and $x \mapsto f(x,t)$ be odd and periodic, with period $2L$. Show that the expression (18.3.8) provides a function u which is spatially odd and of period $2L$. What conditions of regularity must be imposed on v_0, v_1, and f so that u is of class C^2? Then, show that the restriction of u to $[0, L] \times [0, \infty)$ solves the wave equation on $[0, L] \times (0, \infty)$, with Dirichlet boundary conditions, i.e., $u(0,t) = u(L,t) = 0$, for all $t \geqslant 0$.

Exercise 18.5.29. Under the conditions of Exercise 18.5.28, find the symmetry necessary for solving the wave equation with Neumann boundary conditions, i.e., $u_x(0,t) = u_x(L,t) = 0$.
Hint: you need a very small modification of the symmetries used in Exercise 18.5.28.

18.5.8. Generalized solutions for the advection equation

Exercise 18.5.30. Let C_0^k be the set of functions of class C^k on $\mathbb{R}^2$ which vanish outside of a compact set. Give examples of nonzero elements of this set for any order k.
Hint: reread Section 6.1.

Exercise 18.5.31. Let u_0 belong to $L^1_{\text{loc}}(\mathbb{R})$, i.e., the restriction of u_0 to each bounded set of $\mathbb{R}$ is integrable, and define a function u by $u(x,t) = u_0(x - at)$. Show that, for all functions ϕ in $C^1_0(\mathbb{R}^2)$, the following identity holds:

$$\int_{\mathbb{R}^2} u\,(\phi_t + a\phi_x)\,\mathrm{d}x\,\mathrm{d}t = 0.$$

Exercise 18.5.32. Let f belong to $L^1_{\text{loc}}(\mathbb{R}^2)$, i.e., assume that f is measurable and that the integral of its absolute value on any compact set of $\mathbb{R}^2$ is finite. Show that, if, for all ϕ in C^k_0, the expression

$$\int_{\mathbb{R}^2} f\phi\,\mathrm{d}x_1\,\mathrm{d}x_2$$

vanishes, then f also vanishes.
Hint: the proof relies on the same idea as the proof of Lemma 6.1.2. It is useful to show, for instance, that, for all functions χ in $C^1_0(\mathbb{R}^2)$ and all functions ω in $C^1_0(\mathbb{R})$ whose integral is equal to 1, the function

$$(x,t) \mapsto \int_{-\infty}^{t} \chi\,(x,s)\,\mathrm{d}s - \omega\,(t) \int_{\mathbb{R}} \chi\,(x,s)\,\mathrm{d}s$$

belongs to $C^1_0(\mathbb{R}^2)$.

Exercise 18.5.33. Let f be as in Exercise 18.5.32. Assume that, for all functions ϕ in $C^1_0(\mathbb{R}^2)$, the following relation holds:

$$(18.5.1) \qquad\qquad \int_{\mathbb{R}^2} f\phi_t\,\mathrm{d}x\,\mathrm{d}t = 0.$$

Then, show that $f(x,t)$ is, almost everywhere on $\mathbb{R}^2$, equal to a function $g(x)$.

Exercise 18.5.34. Assume that the relation (18.5.1) holds only for functions with support in $\mathbb{R} \times \,]0,\infty[$. Then, show that the analogous conclusion holds: almost everywhere on $\mathbb{R} \times \,]0,\infty[$, f is equal to a function depending only on x.

Exercise 18.5.35. Let u belong to $L^1_{\text{loc}}(\mathbb{R} \times [0,\infty))$ and assume that, for all functions $\phi \in C^1_0(\mathbb{R}^2)$ whose support is included in $\mathbb{R} \times \,]0,\infty[$, the following relation holds:

$$\int_{\mathbb{R}\times]0,\infty[} u\,(\phi_t + a\phi_x)\,\mathrm{d}x\,\mathrm{d}t = 0.$$

Define the following new variables and function:

$$y = x - at, \quad s = t, \quad v\,(y,s) = u\,(x,t).$$

Show that v is, almost everywhere on $\mathbb{R} \times \,]0,\infty[$, equal to a function of space only and deduce that $u(x,t)$ is of the form $u_0(x - at)$. Conclude that the data given by the value of u on the line $t = 0$ determines uniquely the generalized solution of the advection equation with vanishing right-hand side.

18.5.9. Advection–diffusion equation

Exercise 18.5.36. Consider the equation

$$u_t + au_x - \varepsilon u_{xx} = 0, \quad x \in \mathbb{R}, \ t \in \,]0,T[$$

with initial data

$$u(x,0) = u_0(x).$$

Write an explicit numerical scheme which uses a centred scheme for the second-order differentiation in space and a centred or upwind difference for the advection term. What must the Courant–Friedrichs–Lewy condition be in these two cases? Use Fourier analysis to study the ℓ^2 stability.

Exercise 18.5.37. Assume that the space variable remains in the interval $]0,1[$ and that Dirichlet boundary conditions are given at the boundary:

$$u(0,t) = u(L,t) = 0.$$

Run numerical simulations for both schemes, choosing successively $\varepsilon = 1$, $\varepsilon = 10^{-2}$, and $\varepsilon = 10^{-4}$. What happens and how do you explain it?

References

[1] Abhyankar, S. S. (1976). Historical ramblings in algebraic geometry and related algebra. *Amer. Math. Monthly*, **83**(6), 409–48.

[2] Allgower, E. L. and Georg, K. (1990). *Numerical Continuation Methods. Springer Series in Computational Mathematics*, Vol. 13. Springer–Verlag, Berlin. An introduction.

[3] Arnol'd, V. I. (1992). *Ordinary Differential Equations*. Springer–Verlag, Berlin. Translated from the third Russian edition by R. Cooke.

[4] Arnol'd, V. I. (1997). *Mathematical Methods of Classical Mechanics.* Springer–Verlag, New York. Translated from the 1974 Russian original by K. Vogtmann and A. Weinstein. Corrected reprint of the second (1989) edition.

[5] Atkinson, K. E. (1989). *An Introduction to Numerical Analysis* (2nd edn). John Wiley & Sons, New York.

[6] Baker, Jr, G. A. (1975). *Essentials of Padé Approximants.* Academic Press, New York.

[7] Béletski, V. (1986). *Essai sur le Mouvement des Corps Cosmiques.* Mir, Moscow.

[8] Bergé, P., Pomeau, Y., and Vidal, C. (1986). *Order within Chaos.* John Wiley & Sons, New York. Towards a deterministic approach to turbulence. With a preface by D. Ruelle. Translated from the French by L. Tuckerman.

[9] Bézier, P. (1986). *Courbes et Surfaces. Mathématiques et CAO.* Hermès, Paris.

[10] Brezinski, C. (1980). *Padé-Type Approximation and General Orthogonal Polynomials.* Birkhäuser Verlag, Basel.

[11] Bruhat, G. (1939). *Cours de Physique Générale. Thermodynamique.* Masson, Paris.

[12] Businger, P. and Golub, G. H. (1965). Handbook series linear algebra. Linear least squares solutions by Householder transformations. *Numer. Math.*, **7**, 269–76.

[13] Butcher, J. C. (1987). *The Numerical Analysis of Ordinary Differential Equations*. John Wiley & Sons, Chichester. Runge–Kutta and general linear methods.

[14] Canuto, C., Hussaini, M. Y., Quarteroni, A., and Zang, T. A. (1988). *Spectral Methods in Fluid Dynamics*. Springer–Verlag, New York.

[15] Chatelin, F. (1993). *Eigenvalues of Matrices*. John Wiley & Sons, Chichester. With exercises by M. Ahués and the author. Translated from the French and with additional material by W. Ledermann.

[16] Ciarlet, P. G. (1989). *Introduction to Numerical Linear Algebra and Optimisation*. Cambridge University Press. With the assistance of B. Miara and J.-M. Thomas. Translated from the French by A. Buttigieg.

[17] Cooley, J. W. and Tukey, J. W. (1965). An algorithm for the machine calculation of complex Fourier series. *Math. Comp.*, **19**, 297–301.

[18] Cox, M. G. (1972). The numerical evaluation of *b*-splines. *J. Inst. Math. Appl.*, **10**, 134–49.

[19] Crouzeix, M. and Mignot, A. L. (1984). *Analyse Numérique des Équations Différentielles*. Masson, Paris.

[20] de Boor, C. (1972). On calculating with *b*-splines. *J. Approx. Theory*, **6**, 50–62.

[21] de Boor, C. (1978). *A Practical Guide to Splines*. Springer, New York.

[22] de Casteljau, P. (1985). *Formes à Pôles. Mathématiques et CAO*. Hermès, Paris.

[23] Dierckx, P. (1995). *Curve and Surface Fitting with Splines*. Clarendon Press, Oxford.

[24] Dieudonné, J. (1970). *Treatise on Analysis*, Vol. II. Academic Press, New York. Translated from the French by I. G. Macdonald. Pure and Applied Mathematics, Vol. 10-II.

[25] Euler, L. (1983). *Einleitung in die Analysis des Unendlichen. Teil 1*. Springer–Verlag, Berlin. Reprint of the 1885 edition. With an introduction by W. Walter.

[26] Evans, L. C. (1998). *Partial Differential Equations*. American Mathematical Society, Providence, RI.

[27] Feller, W. (1968). *An Introduction to Probability Theory and its Applications*, Vol. I (3rd edn). John Wiley & Sons, New York.

[28] Ferrier, J.-P. (1994). *Mathématiques pour la Licence* (2nd edn). Masson, Paris. Complex variables, differential and tensor calculus, integral calculus, Fourier analysis.

[29] Fine, N. J. (1988). *Basic Hypergeometric Series and Applications*. American Mathematical Society, Providence, RI. With a foreword by G. E. Andrews.

[30] Forsythe, G. E. (1970). Pitfalls in computations, or why a math book is not enough. *Amer. Math. Monthly*, **77**, 931–56.

[31] Fourier, J. (1955). *Analytical Theory of Heat*. Dover Publications, New York. Translated, with notes, by A. Freeman.

[32] Fourier, J. (1988). *Théorie Analytique de la Chaleur*. Éditions Jacques Gabay, Paris. Reprint of the 1822 original.

[33] Gidas, B. (1987). Simulations and global optimization. In *Random Media (Minneapolis, Minn., 1985)*, pp. 129–45. Springer, New York.

[34] Golub, G. (1965). Numerical methods for solving linear least-squares problems. *Numer. Math.*, **7**, 206–16.

[35] Golub, G. H. and van Loan, C. F. (1996). *Matrix Computations* (3rd edn). Johns Hopkins University Press, Baltimore, MD.

[36] Gottlieb, D. and Orszag, S. A. (1977). *Numerical analysis of spectral methods: theory and applications*. SIAM, Philadelphia, PA. CBMS–NSF Regional Conference Series in Applied Mathematics, No. 26.

[37] Gottlieb, D., Hussaini, M. Y., and Orszag, S. A. (1984). Theory and applications of spectral methods. In *Spectral Methods for Partial Differential Equations (Hampton, Va., 1982)*, pp. 1–54. SIAM, Philadelphia, PA.

[38] Graham, R. L., Knuth, D. E., and Patashnik, O. (1994). *Concrete Mathematics* (2nd edn). Addison–Wesley, Reading, MA. A foundation for computer science.

[39] Gramain, A. (1984). *Topology of Surfaces*. BCS Associates, Moscow, Idaho. Translated from the French by L. F. Boron, C. O. Christenson, and B. A. Smith.

[40] Gregory, R. T. and Karney, D. L. (1978). *A Collection of Matrices for Testing Computational Algorithms*. Robert E. Krieger Publishing Co., Huntington, NY. Corrected reprint of the 1969 edition.

[41] Hackbusch, W. (1985). *Multigrid Methods and Applications*. Springer–Verlag, Berlin.

[42] Hairer, E. and Wanner, G. (1996). *Solving Ordinary Differential Equations*, Vol. II (2nd edn). Springer–Verlag, Berlin. Stiff and differential-algebraic problems.

[43] Hairer, E., Nørsett, S. P., and Wanner, G. (1993). *Solving Ordinary Differential Equations*, Vol. I (2nd edn). Springer–Verlag, Berlin. Non-stiff problems.

[44] Hardy, G. H., Littlewood, J. E., and Pólya, G. (1988). *Inequalities*. Cambridge University Press. Reprint of the 1952 edition.

[45] Henrici, P. (1962). *Discrete Variable Methods in Ordinary Differential Equations*. John Wiley & Sons, New York.

[46] Hildebrandt, S. and Tromba, A. (1985). *Mathematics and Optimal Form*. Scientific American Library, New York.

[47] Householder, A. S. (1958). Unitary triangularization of a non-symmetric matrix. *J. Assoc. Comput. Mach.*, **5**, 339–42.

[48] Ifrah, G. (1994). *Histoire Universelle des Chiffres. L'Histoire des Hommes Racontée par le Chiffre et le Calcul. Bouquins*, Vol. 1. Robert Laffont, Paris.

[49] Ifrah, G. (1994). *Histoire Universelle des Chiffres. L'Histoire des Hommes Racontée par le Chiffre et le Calcul. Bouquins*, Vol. 2. Robert Laffont, Paris.

[50] Ifrah, G. (2000). *The Universal History of Numbers*. John Wiley & Sons, New York. From pre-history to the invention of the computer. Translated from the 1994 French original by D. Bellos, E. F. Harding, S. Wood, and I. Monk.

[51] Isaacson, E. and Keller, H. B. (1966). *Analysis of Numerical Methods*. John Wiley & Sons, New York.

[52] Iserles, A. (1996). *A First Course in the Numerical Analysis of Differential Equations*. Cambridge University Press.

[53] Knuth, D. E. (1975). *The Art of Computer Programming*, Vol. 1 (2nd edn). Addison–Wesley, Reading, MA. Fundamental algorithms, Addison–Wesley Series in Computer Science and Information Processing.

[54] Knuth, D. E. (1981). *The Art of Computer Programming*, Vol. 2 (2nd edn). Addison–Wesley, Reading, MA. Semi-numerical algorithms, Addison–Wesley Series in Computer Science and Information Processing.

[55] Korovkin, P. P. (1986). *Inequalities*. Mir, Moscow. Translated from the Russian by S. Vrubel. Reprint of the 1975 edition.

[56] Kreiss, H.-O. (1978). *Numerical Methods for Solving Time-Dependent Problems for Partial Differential Equations*. Presses de l'Université de Montréal.

[57] Laskar, J. (1995). Large scale chaos and marginal stability in the solar system. In *XIth International Congress of Mathematical Physics (Paris, 1994)*, pp. 75–120. Internat. Press, Cambridge, MA.

[58] Lawson, C. L. and Hanson, R. J. (1995). *Solving Least-Squares Problems*. SIAM, Philadelphia, PA. Revised reprint of the 1974 original.

[59] Lelong-Ferrand, J. and Arnaudiès, J.-M. (1977). *Cours de Mathématiques. Tome 1*. Dunod, Paris.

[60] Malliavin, P. and Airault, H. (1994). *Intégration et Analyse de Fourier. Probabilités et Analyse Gaussienne* (2nd edn). Masson, Paris.

[61] Meyer, Y. (1992). *Wavelets and Operators*. Cambridge University Press. Translated from the 1990 French original by D. H. Salinger.

[62] Miller, Jr, W. (1977). *Symmetry and Separation of Variables*. Addison–Wesley, Reading, MA. With a foreword by R. Askey, Encyclopedia of Mathematics and its Applications, Vol. 4.

[63] Milnor, J. W. (1965). *Topology from the Differentiable Viewpoint*. The University Press of Virginia.

[64] Nikiforov, A. F. and Uvarov, V. B. (1988). *Special Functions of Mathematical Physics*. Birkhäuser Verlag, Basel. A unified introduction with applications. Translated from the Russian, with a preface by R. P. Boas, and a foreword by A. A. Samarskiĭ.

[65] Parlett, B. N. (1998). *The Symmetric Eigenvalue Problem*. SIAM, Philadelphia, PA. Corrected reprint of the 1980 original.

[66] Peterson, I. (1993). *Newton's Clock: Chaos in the Solar System*. W. H. Freeman, New York.

[67] Piegl, L. and Wayne, T. (1997). *The NURBS Book. Monographs in visual communication*. Springer, Berlin.

[68] Ralston, A. and Rabinowitz, P. (1978). *A First Course in Numerical Analysis* (2nd edn). McGraw–Hill, New York. International Series in Pure and Applied Mathematics.

[69] Rice, J. R. (1966). Experiments on Gram–Schmidt orthogonalization. *Math. Comp.*, **20**, 325–8.

[70] Richtmyer, R. D. and Morton, K. W. (1994). *Difference Methods for Initial Value Problems* (2nd edn). Robert E. Krieger Publishing Co., Malabar, FL.

[71] Schumaker, L. L. (1981). *Spline Functions: Basic Theory*. John Wiley & Sons, New York. Pure and Applied Mathematics, a Wiley–Interscience Publication.

[72] Shashkin, Yu. A. (1991). *Fixed Points. Mathematical World*, Vol. 2. American Mathematical Society, Providence, RI. Translated from the Russian by V. Minachin (V. V. Minakhin).

[73] Stoer, J. and Bulirsch, R. (1993). *Introduction to Numerical Analysis* (2nd edn). Springer–Verlag, New York. Translated from the German by R. Bartels, W. Gautschi, and C. Witzgall.

[74] Strang, G. (1980). *Linear Algebra and its Applications* (2nd edn). Academic Press, New York.

[75] Struik, D. J. (ed.) (1986). *A Source Book in Mathematics, 1200–1800*. Princeton University Press, NJ. Reprint of the 1969 edition.

[76] von Kármán, T. (1963). *The Wind and Beyond*. Little, Brown and Co., Boston.

[77] Weyl, H. (1989). *Symmetry*. Princeton University Press, Princeton, NJ. Reprint of the 1952 original.

[78] Wilf, H. S. (1994). *Generating Functionology*. Academic Press.

[79] Wilkinson, J. H. (1959). The evaluation of the zeros of ill-conditioned polynomials. I, II. *Numer. Math.*, **1**, 150–80.

Index